Rasch Analysis in the Human Sciences

William J. Boone • John R. Staver
Melissa S. Yale

Rasch Analysis in the Human Sciences

Springer

William J. Boone
Miami University
Oxford, OH, USA

John R. Staver
Purdue University
West Lafayette, IN, USA

Melissa S. Yale
Irving, TX, USA

Additional material to this book can be downloaded from http://extras.springer.com

ISBN 978-94-007-6856-7 ISBN 978-94-007-6857-4 (eBook)
DOI 10.1007/978-94-007-6857-4
Springer Dordrecht Heidelberg New York London

Library of Congress Control Number: 2013953542

Printed on acid-free paper

Springer is part of Springer Science+Business Media (www.springer.com)

To William Fitzpatrick Boone

Preface

Where did this book all begin? How did it come to be? As a former geologist, geophysicist, and high school science teacher, I always appreciated how important it was to confidently measure. It was critical to know exactly when and exactly where a P-wave arrived at a seismograph. And if I wanted to understand what my students had mastered (and not mastered), I had to present the right mix of test items. Of course these issues would be the ones that I would learn how to deeply consider as a PhD student of Benjamin Wright at the University of Chicago.

At the University of Chicago, with the aid of the likes of Ben Wright, Mike Linacre, and frequent visitors such as David Andrich, Carl Granger, Richard Smith, and of course the spirit of George Rasch, I learned that it was possible to bring the rigor of scientific measurement to the measurement needed in the social sciences. It took me a while, and I am ever learning, but now I think I grasp the types of issues Ben and Mike had mastered when I first arrived at 59th and Kimbark in the fall of 1989.

This book is one I had thought about off and on for over 10 years now. It is a book meant to explain a selected set of topics in Rasch measurement which seem to be issues that come up over and over in the undergraduate evaluation courses I lead, the Rasch graduate classes I teach, and the Rasch workshops I conduct. Of course not all Rasch topics are presented, and those topics selected are described in a general way with the hope that the material can be grasped by mostly all undergraduates, graduate students, researchers, and practitioners. There are numerous more technical books (e.g., from MESA Press and JAM Press) which readers completing *Rasch Measurement in the Human Sciences* can later read to further expand their Rasch understanding.

The work presented in these chapters would have not made it to paper without the influence of many individuals, first and foremost Mike Linacre the author of Winsteps. Mike has *always* been willing to patiently answer questions and provides amazingly timely (less than 24 h) support for his Rasch Winsteps program. Through the years as I have broadened and deepened my understanding of Rasch and Winsteps, Mike has been the one I have turned to for help and insight. Mike, without you, I would not have progressed as I have. Thank you Mike!

The other key influence for this book is Ben Wright. Ben, always an advocate of measurement, was of course brilliant (his Person-Maps [aka Wright Maps] are impacting research in so many fields). Ben was unwavering in his measurement clarity, with or without the meter stick in hand. Ben was very generous in his time he spent with me (and all of the MESA Program's students and visitors).

My coauthors John Staver and Melissa Yale, naturally, have been a critical component of this book. Through 5+ rounds of edits and ideas, John and Melissa made what one reads in this book possible. Thank you John and Melissa for joining me in this endeavor! Without your help, this book would not have been completed.

The text, figures, and tables are ones which have gone through a number of iterations. I want to personally thank Everett Smith, Greg Stone, Donna Sturges Tatum, Tobias Viering, and Mike Linacre for reviewing chapters of this book. The comments each provided were of great help. And of course one must have text on paper, but without a good publisher all can be lost. I want to thank Springer and in particular Springer's Bernadette Olmer for her help and encouragement through the writing process. I also want to thank Springer's Marianna Pascale. I am also in debt to the many individuals who have provided data sets for this book.

A broad range of people have influenced this book in many ways. In particular I wish to mention my colleague Ross Nehm and my German colleagues Hans Fischer, Knut Neumann, Birgit Neuhaus, and Andrea Moeller. Ross introduced me to Hans and Knut, and with that link I have had countless trips to Germany to work with these and other researchers who wish to measure "mit Rasch." I also want to thank Xiufeng Liu with whom I have had many Rasch conversations. Xiufeng asked me to coedit a book with him, and that collaboration was one that I greatly appreciated. Thank you so much Xiufeng!!

Finally, I wish to express my gratitude to a number of individuals who have also influenced the completion of this book in many ways. Those listed give of themselves, are listeners, enjoy learning, and add joy to the journey of life. They are interesting and they are interested: Valerie Chase, Dan Shepardson, Jerry Krockover, Kim Metcalf, Roy Forbes, Melanie Jüttner, Annika Ohle, Suzi Seale, Alton McWorter, Sandra Abell, William Werner Boone, Eileen Boone, Jane Herweh, Bob Inkrot, Joe Finke, Mike Wilger, Dennis Koenig, Richard Reed, Alan Bell, Charles Johnson, Kim Fisher, John Holmes, John Jordan, Louis Morrison, Greg Bilbrey, Dave Fopay, Carl Bauer, Sue Dix, Mike Dix, Mike Roth, Jale Çakıroğlu, Özgül Yılmaz-Tüzün, and Rose Wetterau. Also I would like to pay tribute to the faculty and staff of Cincinnati's St. Xavier High School for their unwavering commitment to education and student growth.

I hope readers will learn and marvel as they read and learn about Rasch measurement. Basic application of Rasch measurement techniques will allow you to develop rigorous measurement devices, monitor data quality, compute measures for statistical tests, and communicate findings in a manner which brings meaning to measures.

Cincinnati, OH, USA William J. Boone (boonewjd@gmail.com)

Contents

Chapter 1
What Is Rasch Measurement and How Can Rasch Measurement Help Me?

Isabelle and Ted: Two Colleagues Conversing

Ted: Isabelle, I am just starting this huge research project. I will need to design some pure multiple-choice tests, some partial credit tests, and some rating scale surveys. After I design those instruments, I will need to "show" that the instruments are valid and reliable. Then after all of that I need to compute scale scores that will be used for my parametric statistical tests. Where in the world do I start, I am at a loss.

Isabelle: Ted, don't worry! These issues can all be tackled by applying Rasch measurement theory. The theory was developed by George Rasch (a Danish mathematician) and then applied by other individuals such a Ben Wright, Mike Linacre, Richard Smith, and David Andrich. Rasch measurement helps you learn how to think as you develop and revise measurement instruments. Also, the theory allows you to learn how to evaluate the reliability and validity of instruments and to compute so-called scale scores, which are the "measures" that Ben Wright would often refer to.

Isabelle: Here is a book authored by Boone, Staver, and Yale that I suggest you just page through. Then we can work our way through it; how does that sound?

A Week Later

Ted: Isabelle, this book that you gave me, it looks a little different than other books on Rasch in the library. What's going on?

Isabelle: Calm down Ted. I know you are a physics geek, but have you paged through the book?

Ted: I have. It seems to be all about application, more of a "how-to" book. It seems to be a book that helps someone who doesn't know anything about Rasch. Also, I don't think that you need to know very much math or statistics.

Isabelle: That's the beauty of it. Anything else?

Ted: One thing that I really like is how most chapters end. The authors present exercises at the end of most chapters where I get to practice what they presented in the text. Also, I have a lot of colleagues in fields such as medicine, business, psychology, and education who will be able to use the book. Really, it looks as if most of the exercises immediately relate to these other fields.

W.J. Boone et al., *Rasch Analysis in the Human Sciences*,
DOI 10.1007/978-94-007-6857-4_1, © Springer Science+Business Media Dordrecht 2014

Why Did We Write This Book?

We wrote this book to provide immediate guidance to researchers within and beyond science education (e.g., researchers in psychology, education, medicine, market research) who wish to use Rasch measurement techniques to (1) design instruments (e.g., surveys, tests), (2) analyze data sets, (3) better understand published work (more advanced books, articles) on Rasch measurement, (4) increase the quality of their own research presentations and publications, and (5) enhance grants they might submit to funding agencies. Chapters begin with a "mind capture" – the dialog between our friends Isabelle and Ted. Chapters are guided with core science lesson planning tips often used by Hans Andersen, professor emeritus at Indiana University and past president of the National Science Teachers Association: "Where am I going, how will I get there, how will I know when I have arrived?" (H. O. Andersen, personal communication, fall semester, 1967). Additionally, each chapter includes a sample data set of the type of data often collected, analyzed, and reported in the human sciences. Winsteps (Linacre, 2012a) allows readers to conduct a Rasch analysis (provided data sets (extras.springer.com) can be evaluated with a free scaled down version of Winsteps, named Ministeps) (Linacre, 2012b), but also readers can edit program code for a reader's own particular situation (e.g., sample code might be provided for 15 survey items, but a reader might have an 11-item survey). In each chapter we devote, when appropriate, step-by-step guidance for running Winsteps. Finally, in each chapter we share common questions that colleagues have posed to us in workshops, and we provide tips on how to communicate Rasch findings in a research article. Our goal is to provide enough guidance so that readers are able to write text explaining their own Rasch analysis.

Finally, Rasch measurement is certainly quantitative, but it is also very qualitative. We believe that when researchers read this book, they will immediately see not only the qualitative nature of Rasch measurement, but they will also realize that rigorous requirements of Rasch measurement directly address many of the weaknesses of some quantitative work in the social sciences. Rasch measurement requires qualitative reflection. Data are not just run through a program and numbers computed to the thousandth place; rather, theory is used to guide an analysis. If theory is not confirmed, then reflection takes place. In some instances, odd responses by respondents are investigated in detail (perhaps one particular type of high-performing respondent consistently misses an easy test item, and as a result a researcher decides it is important to conduct an open-ended interview). In some cases the reflection that Rasch theory demands might result in the realization that only part of a data set should be evaluated.

Rasch Measurement?

Throughout this book readers will learn about this technique called Rasch measurement. In a nutshell, what is Rasch measurement? When we conduct workshops, we might sketch out the tasks that Ted mentions to Isabelle in which (for a research project) he has to design a number of instruments (rating scale surveys, multiple-choice tests, partial credit tests). On top of the design issue, Ted knows he will need to revise instruments and be able to confidently defend the reliability and validity of the instruments. After all of that work, he knows he will need to be careful how he utilizes each individual's responses to the set of items presented in each instrument. As readers will see (and practice) in this book, we will talk you through all of these issues in the text. We will explain how Rasch measurement helps you carefully design and revise a measurement instrument and carefully compute "measures" that can be confidently used with parametric statistical tests.

An entire book can be written solely about the theory of Rasch measurement and the mathematical equations that express Rasch measurement. Rasch measurement, in part, can be thought of as a technique by which measurement scales (e.g., surveys, tests) can be built and used in research. The technique is based upon thinking about what makes common measurement instruments such as a ruler so powerful (e.g., the ruler measures in the same way from group to group, there is a clear understanding of what is being measured by the ruler (length)). Rasch measurement can be viewed as allowing for the construction of robust instruments to measure human traits, and these instruments function similarly to instruments in science. Do we suggest that measuring a human trait is as "easy" as one might argue it is to measure length? First, we suggest that at least for those conceiving of measuring length, it was not an easy matter to do so, even though it may appear easy today. Second, although measuring humans is indeed difficult, our perspective is that to accurately advance our understanding in all sorts of human traits (e.g., learning, rehabilitation following an injury, attitude toward products), one must indeed strive to construct measurement instruments that are as valid and reliable as metersticks. In this book we will help readers apply Rasch measurement theory through thinking as well as through using Winsteps Rasch software, which allows the mathematical expression of the Rasch measurement model to be applied to the analysis of a data set.

Goal, Organization, and Scope

We have written this book with the goal of outlining and explaining in step-by-step fashion the concepts and analyses to be considered when one begins to use Rasch measurement in research. Whereas many of our examples come from the field of science education, these examples represent theory, concepts, and processes that are relevant and useful to a broad spectrum of research fields within and beyond science education. Readers will note that we present theory, some real-world data, and appropriate Winsteps (Linacre, 2012a) analyses in each chapter. Whereas there are a number of Rasch programs that one can utilize, we find Winsteps to be quite user-friendly. Winsteps provides many diagnostic features, in particular for the investigation of specific test takers and survey respondents. The techniques that we detail can be used when other software is used.

We have designed end-of-chapter activities for readers that require the application of chapter concepts to data sets that have not generally been used as the core data set of the chapter. In the final section of each chapter, we address a common question that workshop participants have repeatedly posed pertaining to how to communicate Rasch analysis results. Also at the end of each chapter, we provide a sample write-up of the chapter topic as it might appear in a journal or in a conference paper. These write-ups target the very specific topic that has been discussed in the chapter. Some are presented as part of a "Results" section in papers, while other samples provided are presented as part of the "Methods" sections of papers.

Each chapter begins and ends with a dialog that recreates discussions we have had with workshop participants. [The dialog idea draws upon the creative flair of our wonderful Indiana University colleague Alan Bell, as he introduced readers to chapter topics in a similar manner in his thoughtful book (Bell, 1997).] Quite often and with prepared workshop materials and PowerPoint presentations at the ready, individuals approach and ask us: "Can Rasch help me do this?" "I have a problem with my 4th grade science test data that no one can help me solve," "I have brought this really weird data set a colleague is having problems with," and so on. Therefore, we first present a dialog that summarizes a problem that has been brought to us over and over and/or we present a nuance of measurement that has a great impact upon the work of researchers. The end-of-chapter dialog makes use of "aha" comments participants have shared with us and debriefing comments we have overheard as we take workshop coffee breaks. Readers will note that we have attempted to make use of many theories in education that help enhance learning. For example, in most chapters we include formative assessment checkpoints, and most of our activities are "hands-on" activities, which will help readers practice chapter techniques.

Formative Assessment Checkpoint #1

Question: Is Rasch statistics?

Answer: We believe Rasch is really learning and applying the science of developing, examining, and analyzing the performance and quality of measurement instruments (e.g., tests, surveys) that are completed by individuals. Such endeavors belong to the field of psychometrics. Briefly, psychometrics is the study of measuring psychological constructs and processes (knowledge, attitudes, etc.) through the development and validation of surveys, tests, and other assessments. Statistics are used in Rasch measurement. For instance, probability tests can be used to investigate the quality of measurement with an instrument. When you are using Rasch techniques, you are really carrying out a psychometric analysis. Using Rasch, as will be shown in every chapter of this book, involves thinking, as well as using Rasch software.

Road Map Tips and Caveats

Our road map for this book includes some tips and caveats for readers. Our writing is guided not only by our interest in sharing why Rasch techniques must be used within and beyond science education research, but also the book's organization is greatly influenced by our work with preservice and in-service science teachers, as well as our personal collaborations with researchers in the fields of education, psychology, medicine, and business. Many excellent books on Rasch measurement start with a heavy dose of theory. Theory is definitely important; it is a core concept in Rasch measurement that keeps us focused. However, we have organized this book around a sample of applied measurement issues that researchers within and beyond science education often face. In each chapter we do present some theory, but only a limited amount of theory that is relevant to the concepts and processes at hand. We firmly believe in encouraging readers to think and practice from the beginning. As a result of our perspective, we have organized this book in a manner which reflects what results we have found in greater comprehension by workshop participants. A few books exist that introduce readers to Rasch; we suggest readers digest our book, practice with our data sets, and then move onto other books. We have intentionally limited the mathematics in our book. Understanding the mathematics is important, but in this book we wish to present some theory and focus readers on how Rasch can be used to confront common measurement problems. We feel readers will then be able to move onto other books and dive into some of the *hows* and *whys* of the mathematics. In later chapters do provide an overview of how the mathematics of the Rasch model is used to compute the varied indices, measures, and plots commonly presented in Rasch analyses.

Formative Assessment Checkpoint #2

Question: Must one have an extensive background in mathematics to use and understand Rasch?

Answer: All one really needs is some understanding of algebra to start with Rasch.

Each chapter presents an overview and some specific techniques applied to instruments common to education, psychology, medicine, and market research (e.g., tests, attitude instruments, frequency reports). For most chapter topics, we could have written an entire book about the topic, so it is important to remember we provide an overview herein. Readers will note that some topics are presented more than once. This type of organization – present a topic at 9 AM and then revisit the topic at 3 PM during a workshop – is a teaching–learning strategy that has worked well for us when we work personally with colleagues, and we therefore employ that teaching–learning technique in this book.

Our colleagues are, for the most part, interested in being able to confidently apply Rasch measurement, but they are not interested in all of the theory and details of Rasch measurement. As a result of our writing perspective, some specialists in Rasch may be disappointed in topics we have chosen to skip or quickly summarize. However, we cannot emphasize enough that there are so many researchers in fields within and beyond science education who will continue to use raw data and compute only an internal consistency coefficient (e.g., KR-20 (Andrich, 1982) or Cronbach's alpha) to "prove" reliability unless they are quickly engaged in thinking and applying Rasch. Readers will find each chapter easy to understand and will be able to replicate the techniques for their own data sets.

A Sample of Common Problems That Researchers Face and the Application of Rasch Measurement Helps Solve

Survey Data Problems (Ordinal Data Problems)

Q1 I ask open-ended questions when I teach.				
Very Often	**Often**	**Sometimes**	**Seldom**	**Never**
Q2 I use technology when I teach.				
Very Often	**Often**	**Sometimes**	**Seldom**	**Never**

As we review studies in science education that involve the collection and analysis of quantitative data, we note that survey and questionnaire data are very common. If you work in other fields, you will have also found the same situation. Surveys are administered to many types of people, often called respondents or participants. Some examples of respondents are teachers attending a summer K-4 life science inquiry workshop, administrators leading a K-12 statewide grant to integrate the teaching of science and mathematics, parents required to conduct hands-on experiments at home, or businessmen and women answering a survey to gauge their needs or satisfaction in their work. Survey response formats come in many forms. Respondents may be asked to rank a list of attributes (i.e., from 1 to 7, 1 being most helpful and 7 being least helpful). They may be asked to rate using a Likert scale (Strongly Agree, Agree, Neither Agree nor Disagree, Disagree, Strongly Disagree) or a frequency scale (Very Often, Often, Sometimes, Seldom, Never).

What are some common problems researchers must confront with survey data? To whet readers' intellectual appetites, we introduce now but resolve later herein five common problems: First, survey data are ordinal. What do we mean by ordinal data? Suppose that four high school students are using a 4-point agreement scale to respond to the statement, "I like chemistry." John circles *Strongly Agree*, Susan circles *Agree*, Micah circles *Disagree*, and Emily circles *Strongly Disagree*. Is the change in the amount or level of agreement constant from Emily to Micah to Susan to John? For ordinal data, the answer is no. All we know is that John agrees more than Susan, who agrees more than Micah, who agrees more than Emily. With

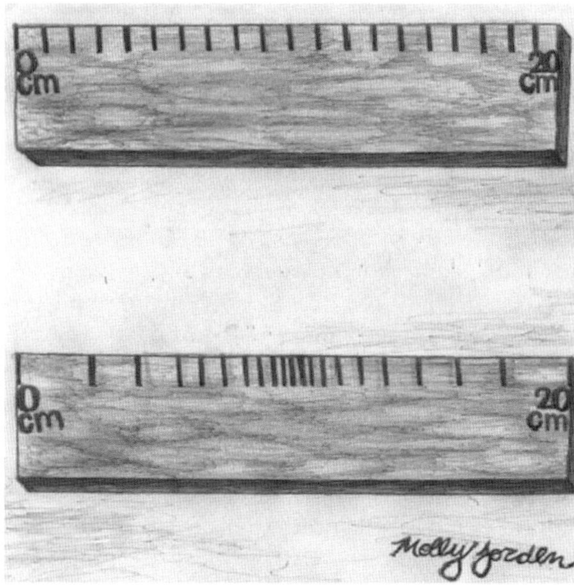

Fig. 1.1 The meterstick (*top*) researchers may think they are using when they immediately utilize raw test data, but the reality may be quite different (*below*) (Figure created by Molly Jorden for this book)

ordinal data, we do not know if the three intervals (Emily–Micah, Micah–Susan, Susan–John) are equal in size. Another way to describe the problem is to say that we do not know if Susan's level of agreement is halfway between Micah's and John's levels of agreement. In this book readers will learn how Rasch measurement helps researchers confidently confront the ordinal (non-equal interval) nature of all rating scale data. One way in which Rasch confronts the ordinal nature of data is to help researchers compute equal interval (linear) measures of respondents that are not impacted by non-equal interval (nonlinear) rating scales. Of great importance for readers is the perhaps surprising nonlinear nature of raw data from tests that include partial credit items or tests that consist of just multiple-choice data. Rasch measurement should be used with all test data in which items will be pooled to describe the performance of a test taker. In this book we primarily utilize rating scale data sets to help readers better understand Rasch. However, it is exceedingly important to note that Rasch should also be used with test data as basic as multiple-choice tests in which answers are scored as right or wrong. Researchers will often treat raw test data as if it marks a nice equal-interval scale, but in reality the meterstick marked by the test items may be warped (Fig. 1.1).

Rating scale data may appear linear (equal interval) as the result of the coding of responses in a spreadsheet (1 corresponds to *Strongly Agree*, 2 to *Agree*, 3 to *Disagree*, and 4 to *Strongly Disagree*). But the problem is that one cannot assume the data are really linear. All one knows is that selection of *Strongly Agree* means more agreement than selection of *Agree*, and all one knows is that selection of *Agree* means more agreement than selection of *Disagree*. After this type of data (numbers 1, 2, 3, 4) are entered into a spreadsheet, researchers can use the Rasch model (a mathematical expression) and a software program such as Winsteps to convert ordinal data to linear measures. We devote entire chapters to these issues. If parametric tests such as ANOVA are used on raw data, a researcher may be violating requirements of parametric tests. Ignoring the parametric requirement of utilizing linear measures can result in incorrect statistical conclusions (a medical researcher may think a treatment has not had a significant impact upon patients when it has impacted them at a statistical level of significance).

There are other roadblocks that researchers often must confront with rating scale data: If respondents fail to answer all items on a survey, must researchers remove such respondents from the study? Are the pre-surveys and post-surveys really equivalent so that respondents can be confidently compared? If judges are used to evaluate piles of essays, can differences among judges' ratings be resolved so that student essays can be fairly compared? How best to present quantitative results so that stakeholders can make informed decisions? Ordinal data problems as well as these problems are just a sample of roadblocks that can impact a data analysis. However, as readers will learn through the use of this book, Rasch measurement will provide not only physical tools (such as software) to confront these issues, but Rasch measurement will also provide a cognitive tool, a way of thinking.

Missing Data Problems

So, what should researchers do when some respondents do not answer all items on a survey? Stated another way, what should researchers do when data are missing? One type of missing data is the case in which a respondent skips one or more items. There are many reasons for skipping an item: the item may be hard to understand, the item may not pertain to the respondent, or sometimes an entire page of survey items may not have been photocopied. Typically, researchers will know that something needs to be done, but they have no idea what to do; thus, in the end the data may be discarded, or researchers may insert the "typical" response that a respondent has answered for non-skipped survey items. In large data sets (such as the collection of data from all 8th grade students in a state (e.g., Florida, USA), a small amount of missing data may not influence the results of data analyses. In small data sets (e.g., data collected from 30 9th grade teachers who attended a 1-week summer workshop), however, and in cases where particular groups of respondents are to be compared, removal of some respondents can strongly influence the results of data analyses. Rasch measurement does not require that all items of a survey be answered, and respondents can still be compared on a single, equal-interval scale. We devote an entire chapter to this issue, and we will show how and why Rasch analysis is not impacted by missing data. Being able to work with missing data provides great flexibility to researchers. Also, since respondents must not complete all items of an instrument, it is possible to create a number of versions of a test and, through a technique named multimatrix design, still compare all respondents on the same scale. The theory and mathematics of Rasch allows persons to not complete all items of a test or a survey yet still be expressed as if they had completed all items.

Problems with Equating Pre-surveys and Post-surveys

A Pre-survey

Q1	I am confident in my ability to teach.
.	
.	
.	
Q15	I am well organized in the classroom.

A Post-survey

Q1	I am confident in my ability to teach.
.	
.	
.	
Q15	I am well organized in preparing for lab experiments.

Of course surveys, questionnaires, and tests are commonly used within and beyond science education research, often in a "pretest/posttest" research design. Our colleagues frequently wonder how Rasch measurement might help them with a problem often confronted, the pretest/posttest equivalency problem. Suppose we give a pretest to a group of students in early September to collect baseline data. We plan to administer a posttest in mid-May, but how should we handle a posttest at the end of the school year? If we administer the identical test, the students may have an advantage in some way, thus inflating their posttest scores. It seems fairer to administer a different test, but how, then, should we compare the two tests? Surely, even if a teacher or researcher tries very hard to ensure the mix of items from easy to hard is the same on the pretest and posttest, his or her attempt will likely not be perfect. How, then, should we deal with this issue?

Researchers have used Rasch measurement successfully to confront this issue for many years. Using Rasch measurement, researchers can develop and evaluate different forms of tests and questionnaires and confidently compare student performance on a single common scale. As long as care is taken to measure on a single variable, it is possible to add new survey items (or test items) to a post-survey (or posttest) and still be able to confidently compare respondents on a single scale. This is one reason why many medical boards (e.g., American Board of Family Medicine) and high-stakes test developers (e.g., PISA) now routinely use Rasch measurement for their test development and analysis. This will also be one of the topics we present in detail to readers of this book.

Formative Assessment Checkpoint #3

Question: Two forms of a survey are developed (Form A and Form B). Both surveys use the same rating scale and both surveys have ten items. Five items are identical on both forms, but the remaining five items on each form are unique to that form. Why could you not compare the overall attitude of a person (Doris) who answered Form A in the fall of a school year to her Form B responses in the spring of a school year?

Answer: Different items can be easier to agree or disagree with. This means that when a different set of items is answered, even if there are some common items presented to the student, the set of "pre" and "post" responses of Doris cannot be immediately compared. A respondent's raw score total for all items may go up or down, but one will not know how much of the change is the result of a change in the respondent over time or the result of a different mix of items presented in each survey form.

Problems with Utilizing Judges, Examinees, and Tasks

In many research projects within and beyond science education, training and use of judges are a part of data collection. To quickly recall such studies, consider when judges evaluate examinees (e.g., students or teachers) with respect to tasks (e.g., essays, teaching performance). There exists, however, a major problem with this commonly used technique of using judges in research: (1) assuming the rating scale is linear (equal interval), (2) assuming the judges can behave in a similar manner, and (3) assuming all the judges should be trained to use a judging scale identically. Using Rasch measurement, agencies that administer high-stakes medical certification board exams and employ judges to score these exams have discovered it is better to have each judge be consistent in his or her severity (how easy or hard he or she scores) rather than to encourage (and supposedly train) all judges to act as identical "robots." (See Looney, 2004, for an application of Rasch measurement to the understanding of judge behavior in sports.) Rasch measurement techniques can be used to help take into consideration the specific mix of judges' severity and leniency. In the case of medical credentialing, this helps ensure that test takers are not penalized for having a tough judge evaluate aspects of their performance. Furthermore, the public is protected from the chance that a test taker might have been very lucky and may have had a number of easy judges (therefore suggesting that the candidate knew more than she or he really did). Another advantage of using Rasch techniques with judge data is that not as many judges need to be utilized (e.g., every judge does not have to evaluate every response to a test item). The use of Rasch techniques can save time and money for researchers as a result of judges not having to evaluate as many respondents and/or items. In our chapter involving the use of judges, we will show how all of the Rasch techniques we share with regard to the analysis of test items, survey items, and respondents can be used to address issues associated with the use of judge data.

Problems Presenting (and Communicating) Research Results

How best to present quantitative results? An often overlooked issue within and beyond the field of education concerns the manner in which quantitative research results are presented in articles, in reports, and at conferences. The old adage "a picture tells a 1,000 words" applies. Workshop participants often approach us with a computer output of their data analysis and ask us how best to explain or present a point in a manuscript. We will share several visual techniques throughout this book that can be used to clearly and simply communicate very complex psychometrics. The Wright Map is one example. Pioneered by Rasch experts such as Benjamin Wright and Mike Linacre, Wright Maps allow Rasch results to be shared and, most importantly, allow those unfamiliar with Rasch in particular, or psychometrics in general, to digest results and make sound decisions using complex data (Bond, 2003). Whereas a complex equation here and there might help a manuscript to be accepted, we must work toward clearly and succinctly communicating results to parents, teachers, school administrators, business executives, medical researchers, community leaders, and legislators if we desire to make broad impacts on the purposes, policies, programs, and practices of many fields. Rasch measurement techniques (e.g., Wright Maps) allow us to do so in a clear and concise manner.

Resources

Question: When you try to learn a new technique such as Rasch, are you on your own?

Answer: In Rasch measurement many individuals such as Mike Linacre, the authors of this book, and many others are very interested in helping others understand, no matter the level of questions. Two websites that are very good starting points for finding individuals to contact and resources are the site hosted by the Institute for Objective Measurement (www.rasch.org) and the site that hosts Rasch Winsteps® software (www.winsteps.com). The authors of this book feel that the fuel that provides energy to Rasch specialists to help others is the view that Rasch provides a critical development that will help many fields of research (e.g., medicine, education) advance.

Software

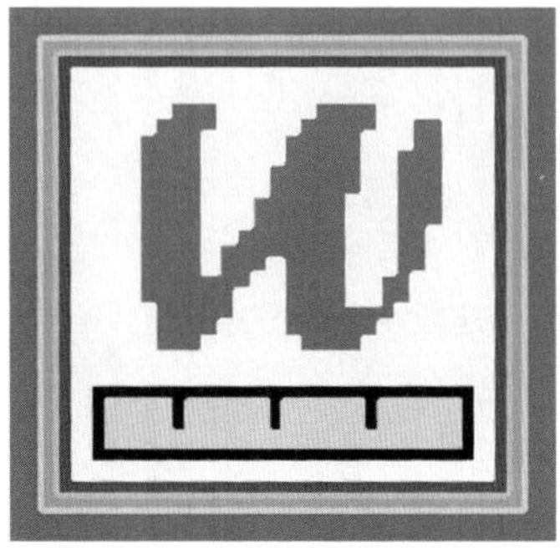

The icon for the Winsteps program by Mike Linacre (2012a)

We conduct workshops using Winsteps, which is Rasch software created and constantly improved by Mike Linacre (2012a). We selected this software many years ago because it is exceedingly user-friendly, the user's manual is very detailed, and there is almost instantaneous online support. If one goes to the Winsteps website, he or she can download a free version of Winsteps called Ministeps (Linacre, 2012b). Ministeps is limited in the number of persons and items that can be evaluated, but this free version of Winsteps is perfect for someone who is new to Rasch. Winsteps is perfect for those conducting an analysis for a thesis or a project of any sort, and it is relatively inexpensive. For this book we have authored most text and end-of-chapter exercises so that the computer files work for both Winsteps and Ministeps. This means we have provided data sets with 75 or fewer respondents and 25 or fewer items. (These are the maximum for Ministeps. Winsteps can evaluate data sets with 30,000 items and 10,000,000 respondents.) The data sets we provide are typical of those evaluated in education, medicine, psychology, and business (e.g., rating scale survey data, partial credit data, test data from multiple-choice tests in which there is only one right answer). The purpose of these small data sets is not to make research conclusions but to teach readers how to use Rasch. When Winsteps is used for an analysis, one can do many things – for example, evaluate the reliability and validity of an instrument in many ways. Additionally, measures of respondents can be easily computed, and it is these measures that must be used for any parametric statistical tests. By the time readers complete this book, they will be able to confidently conduct and interpret a Rasch analysis of both test and survey data.

Formative Assessment Checkpoint #4

Question: Is Rasch only for large education data sets, such as PISA (Programme for International Student Assessment), TIMSS (Trends in International Mathematics and Science Study), and large-scale medical data sets (e.g., validation of the Rasch-based depression screening in a large-scale German general population sample; Forkmann et al. 2010)?

Answer: Many useful analyses can be completed with small data sets. It all depends upon what you hope to learn. John Michael Linacre reports in the article "Sample Size and Item Calibration [or Person Measure] Stability" that 30 items administered to 30 respondents should "provide statistically stable measures" (1994, p. 328). As we will show in subsequent chapters, much can be learned from data sets with a small number of items (<25) and respondents (<75). Also, we will help readers think through different research questions to help them appreciate that "small" depends upon the issue being investigated. There are certainly cases when less than 30 respondents and/or 30 items can be used in a study.

Teaching Techniques

Readers will note some differences in our book compared to other books on Rasch, as well as many education books. We employ a number of research-based teaching techniques that science education researchers recommend to pre-service and in-service teachers. For example, we do not try to present everything, also known as "covering the entire textbook of Rasch." Instead, we focus on a number of important concepts and applications of Rasch measurement and analysis. In particular, we try to address misconceptions common to our workshop participants as we explain concepts and processes. We will revisit previous chapters and topics as a technique to introduce and enhance readers' understanding of new topics.

Isabelle and Ted: Two Colleagues Conversing

Ted: Okay, I am ready for my quiz. Let's go.

Isabelle: Well honestly to begin with, just off the top of your head, tell me a little bit of where you think Rasch measurement came from, and why it is important.

Ted: Rasch measurement is named after George Rasch, a brilliant Danish mathematician who was able to see (and understand) that there were some massive problems with the way in which data such as test data were used in analyses. Anyway, he made this major breakthrough and the University of Chicago's Ben Wright (as well as others who worked with Ben) took the idea and really ran with it.

Isabelle: What do you mean "ran with it"?

Ted: Well it looks to me that Rasch analysis really started with interest in educational testing, but it has now been extended to many fields and to rating scales, partial credit, and data sets that might include judges.

Isabelle: Question #2. Why is it important?

Ted: In research one has to do many things, such as design an instrument, prove an instrument is reliable and valid, revise an instrument, and compute scale scores that are used for parametric statistical tests. Rasch measurement allows one to do all these things. If you do not use Rasch measurement for such tasks, it could easily be that you end up with data that are worthless.

Isabelle: Question #3. Ted can you tell me what might have been some problems with past quantitative work and how knowledge of Rasch might help you correct for those problems.

Ted: Well there are lots of issues. What struck me is that often there has been no guiding theory used when instruments are developed. People do not seem to have asked, "what are we measuring?" and "what might be a variety of items that will allow us to measure all parts of a trait?" The other issue that really is amazing is that in many data analyses raw data are treated as useful for immediate statistical analysis. And in many cases it looks to me that questionable data have been used for parametric tests.

Isabelle: Well one more, this is a bonus question... in your brief look at the book, how might Rasch measurement theory guide your own analysis?

Ted: Actually this is really the fun part. It looks to me as if Rasch provides an organizational framework for me. There are certain things that I need to remember to think about when I design any instrument. Also, after the design of the instrument, a Rasch analysis program such as Winsteps can really help me confidently improve the instrument and also compute the scale scores (Rasch measures) which are the types of numbers which can then be used for parametric statistical tests.

Isabelle: There are some new terms presented in this chapter. Well, maybe not completely new terms, but words that seem to be used in an unusual way. What can you tell me about some of the terms you noticed?

Ted (sipping a cup of coffee): One key word is "instrument." This seems to be a word that can be used for whatever device is used for data collection, a survey or a test. "Rating scale" is another term, although I knew that one. "Rating scale" refers to the way in which respondents can answer an attitude item. So a rating scale is Strongly Agree, Agree, Neither Agree nor Disagree, Disagree, Strongly Disagree, but another rating scale can be Very Often, Often, Sometimes, Seldom, Never.

Isabelle: A+so far, now tell me about judges, examinees, and items.

Ted: I am having some problems with those terms. I can see that later in the book there is more information on them, but here is my best guess. Sometimes statewide tests require students to write essay responses to questions. For example, "Explain why water boils at a different temperature on top of a 5,000 meter mountain than it boils at sea level." Let's pretend that each student had to answer 5 essay items, and each essay was worth between 0 and 10 points. After thousands of students complete the 5 essays, the test booklets are packed up and mailed to a testing company. At the testing company there is a room where trained graders evaluate the essays. Since there are so many essays from students, not all essays can be graded by each grader. So, graders are assigned to a mix of questions. Using this example, the graders are the "judges," the "examinees" are the students, and the "items" are the 5 essays.

Keywords and Phrases

Equal interval data and non-equal interval data
Examinee

Items
Judges
Linear data and nonlinear data
Measurement
Ministeps
Ordinal data
Psychometrics
Rasch, George
Rating scale
Respondents
Tasks
Winsteps
Wright, Benjamin

Potential Article Text

The development of the project's pedagogical content knowledge (PCK) test and the subsequent analysis of collected test data were guided by the application of Rasch theory. Application of Rasch theory provided guidance for the development of two linked PCK instruments used for pre- and post-test comparisions. The first instrument was used to collect the baseline data at the start of the school year; the second instrument was used to collect the post-intervention data at the end of the school year. Linking these two instruments via Rasch measurement techniques allowed student performance at both time points – pre and post – to be expressed on a single measurement scale, even though a different mix of test items was presented on each instrument.

Quick Tips

Raw data from a test (e.g., multiple choice, partial credit) or a survey (e.g., rating, ranking) should not be immediately used for parametric statistical tests. When the data fit the Rasch model, one must prepare data using Rasch measurement techniques and compute, among many things, person "measures" which are expressed on an equal-interval scale.

Data Sets: (go to http://extras.springer.com)

cf used for Chapter 1 activity

Activities

Activity #1

We want you to practice finding resources for Rasch measurement on the internet. Sometimes a basic search can quickly clarify a question that you have. Use a search engine of your choice to find a research article that applies Rasch measurement and analysis. Read part of the article. You should consider some of the keywords we provide at the end of this chapter.

Activity #2

Rasch measurement is used in many fields of research, and this activity is a little more specific than Activity #1 in that readers are asked to find out more about a specific large-scale study which uses Rasch measurement. One very well-known international research effort that uses Rasch measurement is the PISA (Programme for International Student Assessment). Even individuals who are not working within education will be aware that there are large international studies that compare the performance of students in many countries. For this activity, use a search engine to find a Rasch article/report/book that pertains to the use of Rasch in PISA.

Hint: Type in the words "Rasch" and "PISA" to find a multitude of resources.

Activity #3

There are many applications of Rasch measurement in the field of medicine that transfer directly to science educators and other education researchers with an interest in learning about Rasch measurement. Find and read one or two articles in the field of medical research that contain applications of Rasch measurement. Look for some of this chapter's ideas in the article.

Hint: Some of the keywords that you might try include "Rasch," an area of interest within the field of medicine (e.g., "quality of life"), "credentialing," "validity," and "measurement." Finding Rasch articles in other fields can be a great help to researchers in all fields. For example, many medical research articles that involve Rasch are written for readers who are not experts in Rasch. This means that the presentation techniques can provide a model for writing an article. Rasch is used to guide the entire process of instrument development in many of the medical articles. Perhaps the initial Rasch paragraphs of the article will provide an added perspective to the topics raised in our introductory chapter.

Activity #4

Go to the Winsteps website (http://www.winsteps.com) or the Institute for Objective Measurement website (http://rasch.org) and then proceed to the *Rasch Measurement Transactions* (abbreviated RMT). Rasch Measurement Transactions is a publication that provides many useful Rasch articles. For this activity, see if you can find three RMT articles that involve three different words, concepts, and/or phrases that are present in this chapter.

Activity #5

Among the key individuals who have helped researchers appreciate and understand the need for quality measurement is Benjamin Wright. Look up and read the entry for "Benjamin Drake Wright" in Wikipedia to learn a little more about "Ben."

Activity #6

Draw a concept map of what you think has been presented in the chapter. How are topics related? After you have drawn your concept map, write a paragraph summarizing what you think are the key points of this chapter.

Activity #7

Download the *free* version of Winsteps that is available for download. That version of Winsteps is named Ministeps. The only differences between Ministeps and Winsteps are the number of items and persons that can be evaluated with Ministeps. Winsteps (and Ministeps) are extremely user-friendly. After you have installed the program, you can test whether the program runs by using a control file which we have prepared for you. In later chapters we will show you how to run the program; just double-click on Ministeps and click on the word "No," and then read in the file we provide (cf used for Chapter 1 activity).

Activity #8

One of the problems with some analyses of survey data and test data is that data are treated as if they are linear (equal interval) when they are not linear. Some common types of data are nominal data, ordinal data, and equal interval data. Make a list of some examples of nominal data and how those data might be coded in a spreadsheet. Make a list of some types of ordinal data and how those data might be coded in a spreadsheet, and finally make a list of some types of equal-interval data and how those data might be coded in a spreadsheet.

Answer: Some examples of nominal data are gender, race, and school type. In a spreadsheet gender might be coded as a "0" (for male) and a "1" (for female). Gender could also be coded as M and F. Race (African American, White, Hispanic, Asian) might be coded as 1, 2, 3, and 4. It might also be coded as AA, WH, HI, and AS. An advantage of coding nominal data with letters is that it might help a researcher remember that this type of data cannot be evaluated through mathematical steps such as the computation of a mean. Examples of ordinal data might be a rating scale of Agree, Neutral, Disagree in which an Agree is coded with a "2," Neutral is coded with a "1," and Disagree is coded with a "0." This rating scale could also be coded using 3, 2, and 1. Another example of ordinal data might be a scale such as "Very Often," "Often," "Sometimes," "Seldom," and "Never." This scale could be coded as 5, 4, 3, 2, and 1. Also this scale could be coded as 1, 2, 3, 4, and 5. Remember, the coding of rating scale data as we have done here is absolutely fine. It is just that numerical calculations with such ordinal data cannot be conducted. Rasch analysis must first be conducted when data fit the model. Equal-interval data are those data that represent a linear scale. Examples of linear data are the person measures from PISA. Other examples of linear data are the measurement of length with a meterstick or the measurement of time with a stopwatch.

Activity #9

Make a list of instruments that are used to collect data in the fields of education, medicine, and business. Of the data collected from those measurement instruments, predict which provide linear measures (ready for statistical analysis).

Answer: In medicine, data are commonly collected from patients that involve their pulse, weight, temperature, and their views toward different medical options that might be proposed to them to address a medical condition. The data involving pulse rate, weight, and temperature are "measures" that can be used for data analysis. However, survey data such as from an attitude survey (e.g., Strongly Agree, Agree, Disagree, Strongly Agree) cannot be immediately used for statistical analyses.

Activity #10

What does Rasch measurement mean for you? Who needs to worry about Rasch measurement?

Answer: In our own work, we are continually surprised by the wide range of fields that are impacted by poor measurement (and are in need of high-quality measurement of individuals). Anytime sets of items are authored (or used) to provide an overall measure of a person's knowledge (learning of physics concepts in middle school), beliefs (views toward different cars), and actions (what type of unhealthy habits do they conduct), Rasch measurement must be used when the data fit the model.

Activity #11

A test is administered to a group of 1,000 patients at a hospital. The test involves their knowledge of healthy eating habits. Rasch analysis is used to compute "person measures" and a subsequent statistical analysis suggests that female patients have statistically higher person measures than male patients (the females apparently know more than the males with regard to the issue). What is a possible next step in the analysis that should be discussed?

Answer: Determining whether there is a statistical difference is only part of what you should do for a research study. You must also take steps to understand the "meaning" of the difference. It is critical to be able to document in what manner the females, in this case, know more than the males. Application of Rasch measurement techniques as presented in later chapters will allow one to determine in what way respondents are different (in this example, "what do females know that males do not know?"). Knowing in what way individuals differ allows informed decisions to be reached.

References

Andrich, D. (1982). An index of person separation in latent trait theory, the traditional KR-20 index, and the Guttman scale response pattern. *Educational Research and Perspectives, 9*, 95–104.

Bell, A. (1997). *The mind and heart in human sexual behavior: Owning and sharing our personal truths*. Lanham, MD: Arnoson.

Bond, T. G. (2003). Validity and assessment: A Rasch measurement perspective. *Metodología de las Ciencias del Comportamiento, 5*(2), 179–194.

Forkmann, T., Boecker, M., Wirtz, M., Glaesmer, H., Brähler, E., Norra, C., & Gauggel, S. (2010). Validation of the Rasch-based depression screening in a large scale German general population sample. *Health and Quality of Life Outcomes, 8*, 105.

Linacre, J. M. (1994). Sample size and item calibration stability. *Rasch Measurement Transactions, 7*(4), 328.

Linacre, J. M. (2012a) Winsteps (Version 3.74) [Software]. Available from http://www.winsteps.com/index.html

Linacre, J. M. (2012b) Ministeps [Software]. Available from http://www.winsteps.com/ministep.html

Looney, M. A. (2004). Evaluating judge performance in sport. *Journal of Applied Measurement, 5*(1), 31–47.

Chapter 2
Rating Scale Surveys

Ted: Why start by analyzing a rating scale and not a multiple-choice test? This is strange.

Isabelle: Most Rasch analysis books begin with right/wrong tests, such as multiple-choice tests, but that logic is often difficult to follow. The logic is absolutely correct, but sometimes it is hard to accept. It is hard to believe on a 50-point test that the 1-point difference between Bob (25/50) and Dennis (26/50) is not necessarily the same as the difference between Janet (49/50) and Jane (50/50).

Ted: Why not start with theory? Theory's role is to explain and predict.

Isabelle: Because the theory is abstract and mathematical. Rating scales are common in science education research (and in fields such as medicine, market research, and psychology) and more clearly illustrate some applications of Rasch measurement. The authors' strategy is to introduce the theory as they present the applications. Later in the book, there are chapters on theory, but the groundwork of theory lies within the application chapters. This really helped me understand how to apply the theory.

Ted: That makes sense. Besides, the data set I have is a rating scale survey, so this information will help me right away.

Isabelle: You know what?

Ted: What?

Isabelle: I have to confess something. When I went to my first Rasch workshop, there was a lot of theory the first three hours. I could follow it, but I had no clue how it related to my analysis of rating scale data.

Ted: My guess, Isabelle, is that after one has practiced Rasch theory and Rasch analysis, then maybe such an organization might work?

Isabelle: Yes, I think you are right Ted.

Ted: Maybe the way to think of this book is it's one total newbies might start with, and then they would move on to other books.

Isabelle: Ted one more question, ok? I see that this chapter used 13 rating scale items to define a respondent's "self-efficacy measure." Tell me what this means.

Ted: Alright...when we want to know what a person's attitude is, we must have a set of items for one issue, one trait, and one variable. In this chapter the authors make use of a data set

W.J. Boone et al., *Rasch Analysis in the Human Sciences*,
DOI 10.1007/978-94-007-6857-4_2, © Springer Science+Business Media Dordrecht 2014

collected from preservice teachers using 13 self-efficacy items. The goal of that data collection is to compute a single measure for each preservice teacher, which summarizes their overall "self-efficacy" with regard to the teaching of science. This then would allow one to compare the respondents as in "Stefanie has a higher self-efficacy measure than Katie and Amy." This means that Stefanie is more confident, based upon her answers to the 13 survey items, in her ability to teach science than Katie and Amy.

Isabelle: Ted, do you think it really matters in learning Rasch that we are using a science education instrument to evaluate preservice teachers?

Ted: No I do not think it really matters. The self-efficacy measure of the STEBI just provides a very familiar example of a set of rating scale items which is then used to compute a single "measure" for each person. Individuals in education, medicine, and business will be able to quickly adapt what we say to their needs.

Introduction to Rating Scale Surveys

Chapter 1 presented some basics of the Rasch model and some theory. In this chapter, additional Rasch theory will be introduced, as well as an initial Rasch analysis. In later chapters, we will consider multiple-choice tests and revisit the topic of surveys through the presentation of additional introductory analysis techniques as well as intermediate level Rasch analysis for rating scales. For the moment, however, we will consider a rating scale survey commonly used in science education research. Readers who work in other fields will have no difficulty thinking of a survey in their field that uses a rating scale to evaluate respondents (e.g., customer market research, patient medicine). The survey is the 13-item self-efficacy subscale of the *Science Teaching Efficacy Belief Instrument Version B* (STEBI-B) (Enochs & Riggs, 1990). In this case, the word "subscale" refers (perhaps not transparently) to the fact that the STEBI-B consists of items that are used to evaluate "self-efficacy" and other items that are used to evaluate the "outcome-expectancy" of respondents. Most of the issues discussed in regard to Rasch and the STEBI-B self-efficacy scale are issues that researchers in and beyond science education must confront when they collect and analyze rating scale data.

For the STEBI-B, 13 of the 23 items involve the trait of self-efficacy and the other 10 items involve the trait of outcome-expectancy. This means that researchers use the set of 13 items to ultimately compute a person's self-efficacy measure. In the case of the STEBI-B, which was designed for science education research, this means a "measure" of how confident respondents are in their future teaching of science. The ten items of the STEBI-B that are not used for computation of a self-efficacy measure are used for the computation of what is termed an "outcome-expectancy measure" (this measure can be thought of as a measure of respondents' overall view of students' capabilities in terms of learning science). So the STEBI-B really collects data on two variables, "self-efficacy" and "outcome-expectancy." In essence the STEBI-B is really two surveys, with the items for the two variables presented in one survey. However, it is important to mention the data for the two variables are never combined. To understand this, consider a test of 20 items which might be presented to 16-year-old students in which 10 items involve mathematics and 10 items involve history. In this case it only makes sense to compute a measure of the students' mathematics ability and to compute a separate measure with respect to their history ability.

In this chapter (and many chapters of this book), we will focus upon the analysis of only the self-efficacy survey items of the STEBI-B. The self-efficacy scale consists of 13 rating scale items that measure preservice teachers' self-efficacy. The 13 items consist of eight negatively worded items (e.g., *Even if I try very hard I will not teach science as I will most subjects*) and 5 positively worded items (e.g., *I will continually find better ways to teach science*). Often researchers will include so-called negative survey items in an instrument to force respondents to carefully read survey items. The published version of the STEBI-B uses a rating scale of *Strongly Agree*, *Agree*, *Uncertain*, *Disagree*, and *Strongly Disagree*.

Formative Assessment Checkpoint #1

Question: Is the STEBI-B a single scale?

Answer: The STEBI-B is a 23-item instrument in which all items are presented in one survey. However, 13 items are self-efficacy items, and 10 items are outcome-expectancy items. This means when you collect data with the instrument, you do not use all 23 items to compute a single measure. Rather, you will use 13 items to compute a self-efficacy measure and the 10 remaining items will be used to compute an outcome-expectancy measure. This situation is similar to a 30-item multiple-choice test in which 20 items are math items and the remaining 10 items are English items. If you were to report the performance of a student on the test, a math measure should be computed and an English measure should be computed.

Entering Rating Scale Data into a Spreadsheet

The issues addressed in this chapter concern topics such as how to code data, understanding the implications of utilizing ordinal (rating scale) data, and the importance of equal-interval scaling.

Whenever researchers collect any type of data, they must use some sort of coding technique to keep track of the data. The most common technique for entering rating scale data is to "code" each rating category level with a number. For example, a researcher has collected data using all 13 STEBI-B self-efficacy items and the rating scale as presented in the original instrument. Figure 2.1 presents one STEBI-B item and the rating scale suggested by Enochs and Riggs (1990). One possible coding scheme is to code a response selection using the number "1" for the selection of

I will continually find better ways to teach science

Strongly Agree	**Agree**	**Uncertain**	**Disagree**	**Strongly Disagree**

Fig. 2.1 An item from the STEBI-B (Enochs & Riggs, 1990) self-efficacy scale

Strongly Agree, the number "2" for the selection of *Agree*, and so on. Alternatively, the researcher could easily have selected the number "5" to indicate the selection of *Strongly Agree*, a "4" to indicate the selection of *Agree*, and so on. For a single analysis, it makes no difference which number is used to code, for instance, *Strongly Agree* (it could be a "1" or a "5" for a 5-step scale), but researchers must always keep track of what nomenclature they use. Moreover, researchers must remember what the phrases "go up the scale" and "go down the scale" mean numerically. We have seen analyses in which researchers lost track of the directions – up and down – of their rating scale. Depending upon the goals of an analysis, it is sometimes helpful to select coding directions that will make it easier to communicate findings. For example, consider a 100-item test. Usually the number of correct responses is reported, since parents, patients, clients, and researchers are used to the idea that a higher number is somehow better. With regard to attitude scales, it helps an analysis if the higher number for coding items is the "better" response. Thus, if the positive items of a rating scale are what one would like to see, it is helpful to code the highest part of the rating scale using the greatest number to be used for coding. For instance, if a 10-item constructivist teaching scale is presented to preservice teachers (e.g., *I rarely lecture to students*) and the rating scale is *Strongly Agree*, *Agree*, *Disagree*, and *Strongly Disagree*, then it is preferable to code the *Strongly Agree* response with the highest of the numbers used to code responses. Thus, if the analyst has selected the coding scheme using the numbers 0, 1, 2, and 3, the number 3 will be used to code the *Strongly Agree* answer and so on. Other numerical coding can be used (0 for *Strongly Agree*, 1 for *Agree*, 2 for *Disagree*, 3 for *Strongly Disagree*), but our experience is that such coding causes confusion in later parts of an analysis. Our tip is, when possible, to code with the highest number being the best response you want to observe. That means if a scale of *Never*, *Sometimes*, *Often*, and *Always* is presented to respondents and you hope to see respondents selecting *Never* or *Sometimes*, then you would choose a coding scheme of *Never* (4), *Sometimes* (3), *Often* (2), and *Always* (1).

Why do researchers use this type of coding? By coding in this manner, a researcher hopes to communicate that selecting *Strongly Agree* means a higher level response than selecting *Agree*. And, the researcher by "coding a response" is also trying to keep track of data, although he or she might forget this. The correct part of this data entry and coding technique is that such numbers (1, 2, 3, 4, 5, 6) can be used to label a person's responses. Remember the word "label." The problem is that researchers make an unfortunate leap when they assume that the distance from any one rating category to the next rating category (e.g., *Strongly Disagree* to *Disagree* and *Disagree* to *Barely Disagree*) is exactly known and exactly the same. Most researchers would agree that all they really know is that *Strongly Agree* (SA) represents more agreement with a statement than *Agree* (A), that *Agree* means more agreement than *Barely Agree* (BA), that *Barely Agree* means more agreement than *Barely Disagree* (BD), that *Barely Disagree* means more agreement than *Disagree* (D), and, finally, that *Disagree* means more agreement than *Strongly Disagree* (SD) (SA>A>BA>BD>D>SD). When the order of the categories is known but the intervals or distances between categories are not equal, the scale is called an ordinal scale (Stevens, 1959). One of the key issues all researchers using rating scale data must remember is that the data are ordinal, and one cannot magically assume the data are linear (interval) data that can be immediately used for parametric statistics. As readers will learn, the Rasch model will allow a researcher to take ordinal data (from a set of items which define a single trait) and confidently compute a linear (equal-interval) measure for respondents. This linear measure is a value that can then be used for parametric statistical calculations even though one began an analysis with ordinal data.

Figure 2.2 presents a sample of 26 respondents to the 13 self-efficacy items of the STEBI-B. The coding of 6 (*Strongly Agree*), 5 (*Agree*), 4 (*Barely Agree*), 3 (*Barely Disagree*), 2 (*Disagree*), and 1 (*Strongly Disagree*) was used.[1]

Formative Assessment Checkpoint #2

Question: If a rating scale has, for instance, five categories, how do you decide which category should be coded with the highest number? And what exactly do "going up the scale" and "going down the scale" mean?

Answer: A 5-category scale (e.g., *Very Often*, *Often*, *Sometimes*, *Seldom*, *Never*) does not have to be coded as a 5 for *Very Often*, a 4 for *Often*, a 3 for *Sometimes*, a 2 for *Seldom*, and 1 for *Never*. One could just as easily and correctly code the data as 1 for *Very Often*, 2 for *Often*, 3 for *Sometimes*, 4 for *Seldom*, and 5 for *Never*. All that is important when coding is to express the ordinal nature of the rating scale. We have found that coding so that the highest rating scale number is what you would like to observe helps one avoid confusion later in an analysis, for example, if a smoking cessation survey included the three step rating scale of *Never*, *Sometimes*, and *Always*. If the phrasing of the items was such that an answer of *Never* was the best answer, then a rating scale of 1, 2, and 3 might be used, in which *Never* is coded as a "3." In the case, for example, with the 13-item self-efficacy subscale of the STEBI, one would want individuals to have strong self-efficacy, so in the case

[1] Our purpose in using the STEBI-B is to present useful examples for colleagues in and beyond science education. The number of rating categories is not central. Our past work suggests that removal of the original middle category of *Uncertain* (and the addition of the *Barely Agree* and *Barely Disagree* categories) yields added measurement information (see Bradley, Cunningham, Akers, & Knutson, 2011). As a result, readers will note that our STEBI-B examples present a six-category scale.

of positively worded items of the 13-item scale, one would want to code *Strongly Agree* with a higher number than *Agree*.

We use the phrase "going up the scale" to mean moving toward the type of response one would like to see when comparing respondents or comparing a respondent over time. Suppose a male patient in the hospital is administered a pain scale when he is admitted; he then receives physical therapy for a period of time, and then he is readministered the pain scale. In this case one would hope to see lessened pain with time. Thus, in this case we think of "going up" the scale as "lessened pain." So in this example, if the pain scale consisted of 4 rating scale categories, we would choose to code the rating scale steps with the numbers 1, 2, 3, and 4 with the difference of moving from a 1 at *admission* to a 2 *following therapy* as indicating lessened pain.

Entering Negatively Phrased Items

Another key issue in rating scales is the idea of negatively phrased items, and the STEBI-B self-efficacy scale is a good example. Negatively phrased items are typically meant to keep respondents attentive. The rationale is that respondents will not be tempted to read the first item, answer it, and then repeat the same answer for all remaining items. For instance, the STEBI-B self-efficacy item 3 –*Even if I try very hard, I will not teach science as well as I will most subjects* – is answered using the same rating scale as that presented for all the other self-efficacy items. However, the "better" response for this item would be *Strongly Disagree* instead of *Strongly Agree*. Therefore, as data are entered into a spreadsheet, all *Strongly Agree* responses for item 3 are entered as a "1," not as "6"; all *Agree* responses are entered into the spreadsheet as a "2," not as "5"; and so on. This reversing, or flipping, of the rating scale numbers used to code negatively phrased items is a common technique to facilitate the analysis of a set of survey items that involve one trait (e.g., self-efficacy). This is because one wants the meaning of movement regarding the trait, from say a "2" to a "3" on a rating scale, to mean the same direction of movement along the trait. For instance, suppose a survey included the items Q1-"I like Biology," Q2-"I like Chemistry," Q3-"I do not like Physics," Q4-"I like Geology" all with the rating scale of *Strongly Agree*, *Agree*, *Disagree*, and *Strongly Disagree*. Suppose further that a researcher decided to code the rating scale with the numbers 4 (*Strongly Agree*), 3 (*Agree*), 2 (*Disagree*), and 1 (*Strongly Disagree*). Before the set of four items might be used together to learn something about each respondent, the data for Q3 must be flipped (reverse coded, recoded). This is because without flipping the data, the meaning of a student moving from selecting a "2" as opposed to a "3" for item Q3 means the opposite of what it means for a respondent to select a "2" as opposed to a "3" for items Q1, Q2, and Q4. Our personal view is to not create surveys with some items which must be flipped.

If data are collected electronically or on a bubble sheet (scantron) that is fed into an optical scanner, the analyst will need to flip item responses after data have been placed into a data manager such as SPSS or Excel. The recode option for SPSS can be used to reverse code (flip) the answers for all items that need to be recoded. The Rasch program Winsteps also provides an easy to use option for recoding selected items. Once we introduce you to the file that you need to run a Winsteps/Ministeps Rasch analysis, we will explain this recoding technique. This means one can enter the raw data; then before the final Rasch analysis takes place, the program will internally recode items as long as you remember to tell the program which items need to be reverse coded (flipped).

A Sample Spreadsheet with Survey Data and Basic Non-Rasch Calculations

In Figure 2.2 a unique five-digit ID is presented for each student. Following entry of these data is "PR" to remind the researcher these data are from the pre-data collection. Then 13 entries are made to indicate the answer, or lack of answer ("x"), for each respondent to the 13 self-efficacy items. Recall that the self-efficacy items are part of a larger scale and therefore are not numbered as items 1–13. The first column of item data contains each respondent's answer to item 2, which is the first self-efficacy item presented in the 23-item STEBI-B. In scanning the data, 25 of the 26 students answered either *Strongly Agree* (labeled by a "6") or *Agree* (labeled by a "5") for item 2, found in the second column of the spreadsheet. The third column pertains to the second self-efficacy item, item 3 presented on the STEBI-B. This item is a flipped item; student ID = 21141 (second row) answered *Disagree* to item 3. Because the item is a negatively phrased item for teaching science self-efficacy, the labeled responses must be flipped. Therefore for student 21141, a response of *Disagree* is coded as a "5" not a "2." This means that the data are presented in this spreadsheet after flipping of data for the appropriate items.

Formative Assessment Checkpoint #3

Question: Can the numbers entered into a spreadsheet to indicate which rating scale category was answered by a respondent be used to compute means, standard deviations, and conduct parametric tests, such as *t*-tests? (A *t*-test allows one to compare the means of two groups of respondents. For example, to compare the mean self-efficacy measure of 100 female preservice high school teachers to the mean of 115 male preservice high school teachers, the *t*-test allows one to determine if the difference in the means is a significant difference.)

Answer: The numbers you enter into a spreadsheet are simply labels that indicate which rating scale categories were answered by a respondent. Those labels should

	A	B	C	D	E	F	G	H	I	J	K	L	M	N	O	P	Q
1	ID	Item 2	Item 3	Item 5	Item 6	Item 8	Item 12	Item 17	Item 18	Item 19	Item 20	Item 21	Item 22	Item 23	Raw Score Total	Items Answered	Raw Score Mean
2	21141PR	6	5	2	6	5	2	5	5	4	5	5	5	5	60	13	4.62
3	91052PR	6	4	2	5	5	5	x	x	x	x	x	x	x	27	6	4.50
4	95793PR	6	6	5	5	6	5	x	x	x	x	x	x	x	33	6	5.50
5	08453PR	5	3	2	5	5	4	4	4	2	3	3	5	2	47	13	3.62
6	36281PR	6	3	2	4	5	4	4	3	2	4	5	5	4	51	13	3.92
7	85453PR	6	6	3	5	6	5	4	5	3	5	5	6	5	64	13	4.92
8	46328PR	5	3	3	3	4	3	x	x	x	x	x	x	x	21	6	3.50
9	41024PR	5	2	2	5	5	2	2	2	2	2	3	5	2	39	13	3.00
10	08746PR	5	5	4	3	5	4	2	4	1	2	2	5	2	44	13	3.38
11	09132PR	5	2	3	5	5	5	4	5	3	5	5	5	4	56	13	4.31
12	28100PR	5	5	2	5	5	2	2	4	1	1	2	5	3	42	13	3.23
13	43532PR	6	6	4	5	5	6	6	5	4	5	5	5	4	66	13	5.08
14	36754PR	4	3	4	4	5	4	5	3	2	5	5	6	4	54	13	4.15
15	53695PR	5	3	2	2	3	2	3	2	2	4	4	5	2	39	13	3.00
16	65759PR	6	3	5	4	5	4	3	5	2	3	3	5	3	51	13	3.92
17	40166PR	5	5	3	6	5	4	5	5	4	4	4	5	4	59	13	4.54
18	55959PR	6	6	5	5	6	5	6	5	1	5	6	6	6	68	13	5.23
19	97766PR	6	6	4	5	6	5	5	4	4	5	5	6	6	67	13	5.15
20	78880PR	5	4	2	4	2	4	2	4	2	3	4	4	1	41	13	3.15
21	33573PR	6	3	2	4	3	2	2	2	2	2	2	5	2	37	13	2.85
22	99365PR	6	5	3	5	6	4	6	5	4	6	5	6	5	66	13	5.08
23	18489PR	6	5	4	5	5	4	5	4	2	4	4	5	4	57	13	4.38
24	96468PR	6	4	5	4	5	5	4	4	3	2	3	6	2	53	13	4.08
25	37854PR	5	4	3	5	5	5	4	5	4	4	5	6	5	60	13	4.62
26	71215PR	5	4	3	5	5	4	5	5	4	3	4	5	4	56	13	4.31
27	87610PR	5	5	2	4	4	3	5	4	1	5	5	4	5	52	13	4.00
28	# of Responses	26	26	26	26	26	26	23	23	23	23	23	23	23			
29	Item Mean	5.46	4.23	3.12	4.54	4.85	3.92	4.04	4.09	2.57	3.78	4.09	5.22	3.65			

Fig. 2.2 A spreadsheet of raw data (*after flipping of appropriate items*) for 26 respondents to the 13 self-efficacy items of the STEBI-B. A six-step scale of SA, A, BA, BD, D, and SD was utilized. In addition, computations are provided for the total raw score, total items answered, and raw mean score for each respondent, and number of responses and mean for each item. Readers need to note that such calculations can help one double-check one's understanding of data coding, but due to the nonlinearity of rating scale data, the Rasch measures of items and Rasch measures of persons must be used for any calculations and interpretations. As we will explain to readers, raw score totals should not be mindlessly used in an analysis

not be used to immediately compute values such as means and standard deviations. Rasch analysis must be used to compute person "measures" that are values that express where each respondent falls on a linear scale.

One practice of scoring rating scale data is to sum the raw values of all the items and calculate a mean item score. Again, this practice assumes that all intervals between the categories of the rating scale (*Strongly Agree*, *Agree*, etc.) are equal (i.e., 1 to 2 = 2 to 3). Figure 2.2 presents the mean score of each student (last column) and the mean response for each item (last row). For student ID = 21141, a mean of 4.62 was computed. This value can be verified by adding all of student 21141's responses and then dividing that sum by the number of survey items answered (6 + 5 + 2 + 6 + 5 + 2 + 5 + 5 + 4 + 5 + 5 + 5 + 5 = 60; 60/13 = 4.62).

The mean of a set of values can be very misleading. This mean suggests that this respondent (respondent 21141) could have selected a "4" (*Barely Agree*) or a "5" (*Agree*) for each item. Clearly for this respondent, this mean only tells the broadest of stories (another reason why raw data cannot be immediately evaluated with parametric statistics) because this person also rated one item very highly (6) and gave another item a low rating (2).

Formative Assessment Checkpoint #4

Question: Does it make a difference if you use raw item data to compute person raw score totals or if you conduct a Rasch analysis and compute Rasch person measures?

Answer: Raw data are not always linear. This means conclusions you draw that are based on parametric statistical analyses (e.g., *t*-tests, ANOVA) of raw data may be incorrect. By conducting a Rasch analysis, you are able to evaluate the validity and reliability of a measurement device much more thoroughly than can be done in a traditional analysis. And then you can use Rasch person measures and item measures in parametric statistical tests and communicate findings making use of the fact that items are on the same scale as persons.

Additionally, some researchers will compute a number to show how difficult or easy it was to agree with an item. Researchers do this by adding all the responses to a single item and then dividing the sum by the total number of respondents. Reviewing Fig. 2.2, a researcher might note the sum total for item 2 as 142 and the item mean as 5.46 (6+6+6+5+6+6+5+5+5+5+5+6+4+5+6+5+6+6+5+6+6+6+6+5+5+5=142/26=5.46). As was the case for the student data, this computation of an average requires a very big leap of faith, namely, that the step between rating categories is linear.

Flaws in the Use of Non-Rasch Techniques to Confront Missing Data

A further step, often carried out but rarely discussed, is how researchers confront missing data. Any researcher realizes that missing data are a way of life in research projects. We present and discuss typical techniques for handling missing data and help readers see that past techniques used for missing data are questionable.

When data are missing, for example, the responses to the second page of a survey, a researcher will sometimes throw out the data. If a data set is massive and the missing data seem random, then this might be acceptable. However, data are often very difficult and costly to collect. For small data sets, each piece of data is potentially helpful. For example, in a study involving multiple schools, perhaps at a particularly important school to study (the school district might have pumped a lot of money into this one school), the second page of the survey was not photocopied

Fig. 2.3 Difference in mean responses based on 13 items and first six items for two respondents. The items selected for computation of mean responses greatly impact the mean response and also perceptions of the similarities and differences in overall attitude of the two respondents

ID	21141PR	08453PR
Q2	6	5
Q3	5	3
Q5	2	2
Q6	6	5
Q8	5	5
Q12	2	4
Q17	5	4
Q18	5	4
Q19	4	2
Q20	5	3
Q21	5	3
Q22	5	5
Q23	5	2
Mean Response of all 13 items	4.62	3.62
Mean Response of first 6 items	4.33	4.00

and was not even presented to respondents. In this case it is certainly preferable to keep and use all the data (no matter how much data is missing).

Prior to the advent of Rasch measurement, techniques that researchers used with good intentions could be very problematic in terms of yielding high quality results. One technique that researchers sometimes use when data are missing is to compute a person's raw mean based only upon the answered items. Let's consider this technique for person 91052 in Fig. 2.2 (91052PR: 6+4+2+5+5+5=27; 27/6=4.50). What are the problems with this number? First, in calculating the mean the researcher commits the major error of confusing numbers used as labels for an ordinal rating scale with an equal-interval scale. Second, one is assuming that all items should share an equal weight and are created equal. Computing a mean with a smaller set of items likely will not yield the same mean as if all items were answered. Figure 2.3 illustrates this problem. Using the first and fourth rows of data in Fig. 2.2, we have calculated two means for student 21141 and student 95793. The first mean is based on all 13 items, and the second mean is based on only the first six items (Q2, Q3, Q5, Q6, Q8, Q12). When the students' respective means are based on only the first six items, student 21141's mean decreases by 0.29, yet student 95793's mean increases by 0.38.

When a rating scale is ordinal – which it is for the STEBI-B – computing a mean based upon the items answered (in an effort to consider missing data) often results in different means for the same persons as a function of items included in calculating those means. This, of course, is attributable to the differences across survey items. All items do not measure self-efficacy in the same manner; one wants survey items that measure different aspects of a single trait. In fact, if all items measured self-efficacy in exactly the same way, then a researcher would not need to administer a survey longer than one item to a respondent.

Researchers sometimes attempt to correct for missing data by reporting an item mean that was computed from just the scores of other respondents who answered that item. For example, to compute the mean response for the 26 people presented in Fig. 2.2 for item 2 of the survey, a researcher would make the following computation:

$$6+6+6+5+6+6+5+5+5+5+5+6+4+5+6+5+6+$$
$$6+5+6+6+6+6+5+5+5=142/26=5.46$$

But, for item 17 of the survey, a researcher would make the following computation:

$$5+x+x+4+4+4+x+2+2+4+2+6+5+3+3+5+6+5+2+2+$$
$$6+5+4+4+5+5=93/23=4.04$$

What is the problem? Again, the researcher commits an error by using the rating scale as if the data are equal interval, not ordinal. Also, the sample of 26 respondents, who will exhibit a range of self-efficacy, will not be duplicated by the sample of 23 respondents. Perhaps the three respondents with missing data for item 17 are individuals who would have entered a very low self-efficacy rating? If so, then the mean will be impacted by the increased variance in the data. The bottom line is that since Rasch analysis does not require every item to be answered by a respondent, then it is possible to calculate a person measure as if the respondent had answered all survey items, even if the respondent did not answer an item (or items). As we teach readers how to conduct a Rasch analysis, we will present an entire chapter on the issue of missing data (Chap. 18).

Action and Consequence of Just Entering Data and Not Conducting a Rasch Analysis

Why do researchers commit an error when they (1) compute a mean for each respondent and then use each person's mean for later parametric statistical tests or (2) compute a mean item score? First, the error itself is that one does not know that *Strongly Agree* is the same distance away from *Agree* as, for example, the distance from *Disagree* to *Strongly Disagree*. All that is known is that the rating scale represents an ordinal scale. Second, the consequence of the error is massive because it can impact all later parametric statistical analyses. When researchers immediately conduct parametric statistical procedures on ordinal data, they ignore or forget that the numbers are only labels for the responses circled. All parametric statistical procedures that would be conducted on such raw data would rest on the assumption that the data are equal interval.

Formative Assessment Checkpoint #5

Question: Does telling respondents that the "jump" from one rating scale category to the next is identical (Teacher to class: "Students when you are answering the STEBI-B, I want you to imagine the change in attitude from *Strongly Agree* to *Agree* is the same change in attitude from *Agree* to *Barely Agree*, and so on. OK?") mean you do not have to conduct a Rasch analysis?

Answer: You could say something like this to respondents, but it is highly unlikely the respondents will really understand what you are trying to say. Also, even if they might understand what you are saying, a non-Rasch analysis will result in an analysis that does not take into consideration that items are not equally agreeable. Furthermore, as we will see in later chapters, Rasch analysis provides sophisticated techniques for evaluating the validity and reliability of instruments and monitoring the quality of data. Using raw data and computing a mean answer of a respondent results in a number that is flawed. The numbers used for the mean are not equal interval and survey items are not all created equal (not all are as easy to agree with as other items).

We cannot overemphasize the magnitude and importance of this issue. A frequent outcome of ignoring this problem is rejection of a null hypothesis when it is true, a Type 1 error (Glass & Stanley, 1970). For those researchers who might need a few words on this issue, just think of "rejection of a null hypothesis when it is true" as a situation in which you might think there is no difference between the self-efficacy of male preservice teachers ($n=56$) and the self-efficacy of female preservice teachers ($n=47$). If readers consider this problem of needing to use linear measures for parametric statistical tests, they will immediately see that, if Rasch addresses this issue (which it does), then this is one reason they must use Rasch measurement techniques when analyzing rating scale data.

This may lead readers to ask, "How does Rasch measurement resolve this problem?" That is, of course, a focal point of this book. To begin answering how Rasch measurement resolves the unequal-interval problem, one must understand a few Rasch measurement assumptions. Rasch measurement provides a technique by which sample-independent item measures and item-independent respondent measures can be computed. Item independence means that a researcher can collect data from respondents, but each respondent can be presented with a different mix of items. For example, School A completes algebra items 1–10 and algebra items 11–20, School B completes algebra items 1–10 and algebra items 21–30, and School C completes algebra items 1–10 and algebra items 31–40. As long as all items 1–40 involve the same trait (and requirements of the Rasch model are met), it does not matter which items any single respondent completes. They (the respondents) can be measured upon the same single algebra scale as all respondents. This is what is meant by item-independent respondent measures.

Sample-independent item measures mean if one is interested in understanding in what manner the 40 algebra items define the trait of algebra knowledge (e.g., which items are easy items, which items are middle of the road items, which items are difficult items), then it should not matter which persons complete the algebra items. How one develops an instrument that leads to (1) item-independent person measures and (2) person-independent item measures requires a lot of thinking about what it means to measure and the application of Rasch measurement techniques. The important point is, when you think about what it means to measure, the measurement problems can be confronted and solved, and in the end you can conduct high-quality analyses.

Formative Assessment Checkpoint #6

Question: A math test is to be administered to 10-year-old students. What would be an example of test items that would involve the same trait (the same variable), but the items would tap different parts of the same trait?

Answer: If one considers a math test that could be completed by 10-year-olds, items involving addition, subtraction, multiplication, and division could be administered. These items all involve the trait of mathematics as taught to these students, but the items differ in which part of the trait is defined. Division items and multiplication items would define the more difficult part of the trait, while addition and subtraction items, generally, would define the easier (less difficult) portion of the trait.

Isabelle and Ted: Two Colleagues Conversing

In this scenario we are able to eavesdrop on Isabelle and Ted as they talk to a measurement class they are teaching.

Isabelle: Good morning class, I hope you had a good weekend. Today Ted and I are going to return to the issue that we should not simply enter rating scale data and then use that data to conduct a parametric statistical analysis in SPSS or SAS.

Ted: Class, this whole idea was really confusing to me. Isabelle and I want to show you what we are talking about by using some props.

Isabelle: Alright class, I have five signs that I wrote on paper file folders. You will see five folders, each with the words of one of the five rating scale steps. Ted, do you mind showing the class the five signs?

Ted lifts up each sign one at a time. One folder has only the words "Strongly Agree."

Isabelle: Now I need 5 volunteers; Kim, John, Charlie, Molly, and Carolyn, why don't you come on up. Each of you, please take a single sign. Kim, you can take Strongly Agree. Get in the order of the rating scale, but make sure that you have a distance of 1 meter between yourselves.

(The students move to their places.)

Kim	Carolyn	Molly	John	Charlie
I	I	I	I	I

Isabelle: Kim, please use this meterstick to check all spacing between adjacent students, including you. (Kim does so and replies that there is an equal spacing between each student.)

Isabelle: Class, would you agree that our 5 students are organized to illustrate what is implied if we label each category with a whole number from 1 to 5?

Ted: If it helps, I have just written down the numbers 1, 2, 3, 4, and 5 on post-it notes. And now let me put one post-it note on each of the five signs to remind us what numbers we are using to code the possible responses of Strongly Agree, Agree, Uncertain, Disagree, and Strongly Disagree.

Isabelle: Okay, does everyone agree that our five students are in the right order?

Class: Yes.

Isabelle: Now, does everyone agree that the post-it notes show what coding we are using for each rating scale?

Class: Yes.

Isabelle: Now I am going to modify this line of people, but notice I am not going to touch the post-it notes Ted has placed on the signs!

Isabelle moves the students to varying distances apart. Kim, who is holding Strongly Agree, is moved 1 meter away from Carolyn, who is holding Agree. Molly, who is holding Uncertain, is placed 2 meters away from Carolyn. John, holding Disagree, is 2.5 meters away from Molly, and Charlie, holding Strongly Disagree, is positioned 0.5 meters away from John.

Ted: Everyone, please notice that we have changed the spacing of the rating scale, but we have not changed the post-it notes. Do you agree that this example with unequal spacing of rating scale categories is just as reasonable as the previous spacing when labels 1–5 are used?

Isabelle: Class, note we have not changed the numbers we use to represent the rating categories! This unequal spacing is a great visualization of an ordinal scale. More often than not, there will be a problem with immediately computing a mean response for an item and/or computing a total raw score for a person.

As a final example of comparing the use of raw scores and Rasch measurement scores, we present a results table (Fig. 2.4) from an introductory Rasch article by Boone, Townsend, and Staver (2011) that employed Rasch analysis techniques on a set of STEBI-B self-efficacy scale data. The table displays the findings of the change of student self-efficacy over time, from the start of a course to the end of a course. For this study, a *t*-test was conducted on the difference between the students' pre- and post-raw mean scores. The top section of the table presents the raw pre-mean score, the raw post-mean score, and the result of a paired sample *t*-test using the raw mean scores. The bottom section of the table presents the pre- and post-Rasch measure scores for the two groups and the result of a paired sample *t*-test comparing pre- and post-Rasch measure scores. In this instance, there is a difference in the conclusion, depending upon whether raw mean scores or Rasch person measures were used for

	Time Point	Mean Score
Raw Score	Pre	55.5
	Post	59.2
	t-test sig.	$p < .05$
	Time Point	Measure
Rasch Measures	Pre	573
	Post	599
	t-test non-sig.	$p > .05$

sig. = significant
non-sig = non-significant

Fig. 2.4 A comparison of respondents (pre versus post) using the STEBI-B. When raw scores are used to compare pre to post, a significant change in attitude is suggested. But when Rasch measures are used, no significant difference is suggested (*sig* significant, *non-sig* non-significant)

the analysis. Researchers will not always reach a different conclusion from the results of a *t*-test computed with raw mean scores or Rasch person measures; however, this example shows that different conclusions can be made based on the type of scores used in a parametric statistical analysis. Using the incorrect type of scores increases the probability of error, Type 1 and Type 2 (Glass & Stanley, 1970). The techniques of a basic Rasch analysis can be easily undertaken to compute person measures and then confidently conduct parametric statistical analyses.

Formative Assessment Checkpoint #7

Question: Can the numbers you enter into a spreadsheet to indicate which rating scale category was answered by a respondent be used to compute means, standard deviations, and conduct parametric analyses, such as *t*-tests?

Answer: The numbers you enter into a spreadsheet are simply labels that indicate which rating scale categories were answered by respondents. Those raw values should never be used immediately to compute values such as means and standard deviations. First, when data fit the Rasch model, Rasch analysis must be used to compute "person measures" that are values that express students' performance on a linear scale.

The Logit

Readers should readily understand the argument that ordinal rating scale data are not linear and cannot be immediately used for parametric statistical analysis no matter how many previous published studies have done so. We have started our book concentrating upon rating scales since that is a very common type of data that many individuals in business, education, psychology, and health sciences use in their research.

$$B_n - D_i = \ln (P_{ni}/1-P_{ni})$$

Fig. 2.5 The Rasch model for dichotomous data (right/wrong test items, a rating scale with only agree or disagree rating categories). B_n represents the ability of a specific person (n) and D_i represents the difficulty of a specific item (i). P_{ni} represents the probability of person n correctly answering item i. The form of the Rasch model for rating scales in which multiple rating categories are presented is described in detail in the seminal Rasch book *Rating Scale Analysis* (Wright & Masters, 1982)

Below we provide a brief introduction to the unit of measurement in Rasch measurement; that unit is the logit. Later herein we present a chapter concerning the logit, and in almost every chapter, readers will be provided with practice in reading Rasch analysis results presented in logits. However, at this early point in the book, we will provide an introduction to the "unit" which is used in Rasch measurement to express "person measures" and "item difficulties." We present our explanation using data for a test in which items can be scored as right or wrong. Readers can think of a right/wrong test as a survey in which respondents can either *Agree* or *Disagree* with an item, so only one of two rating scale categories can be selected.

Figure 2.5 presents the Rasch model equation. Readers should note that this basic equation is the one used for right/wrong test data, as well as (although rarely found) the case in which a survey might have a 2-category rating scale of *Agree* and *Disagree* (or *Yes* and *No*). The application of the concepts expressed in this equation when one runs a Rasch analysis program such as Winsteps is how one eventually ends up with person measures of all respondents expressed in logit units and the item measures, which are also expressed in logits.

The important aspect of the left side of the equation is the subtraction of D_i from B_n. This subtraction is important on many fronts. First, because we are doing a subtraction, both B_n and D_i must be expressed in the same units. Only if two variables are expressed in the same units can there be a subtraction. For example, 5 cm are subtracted from 75 cm (75–5 cm), and it makes perfect sense that the units for subtraction must involve the same variable. It would be nonsensical to subtract 5 °C from 75 cm (75 cm – 5 °C). As readers will see in later chapters, there are many amazing aspects of the left side of this equation involving two variables with the same units. However, to begin, readers are encouraged first to note that the left side of the equation involves two variables that are expressed in the same unit (logits). What exactly does $B_n - D_i$ mean? For us in our work, it means that we understand that if one variable is used to express where a person is on a single trait, and where an item is on the same single trait, then there will be a difference between where the person and item are, unless both the person and the item are at the exact same spot of the trait. In Fig. 2.6 we plot a person Amy and one item (Q8) from a survey in which items can be answered only as *Agree* or *Disagree* (note if you wish, you can think of *Disagree* as *Not Agree*). Readers should be able to see that both Amy and item 8 are expressed in logits and there is a difference between the location of Amy and item 8. That is exactly what is being presented in the left side of the equation. The right side of the equation is an expression that involves the probability of Amy

Amy (+.25 logits)

<-->

-2 logits **Q8 (-1.2 logits)** **+2 logits**

Less Agreement **More Agreement**

Fig. 2.6 A plot of a respondent and item along a trait. The respondent "Amy" is greater in measure than the measure of item 8. This means that when Amy answers item 8, there is a greater than 50/50 probability that Amy will answer the item correctly (if it is a right/wrong item). If we think of a dichotomous rating scale item (agree/disagree), then this plot could express the greater than 50/50 chance that Amy agrees with item 8

(denoted with the letter "n") agreeing with item 8 (denoted with the letter "i") divided by the probability of Amy not agreeing with item 8.

We do not present the derivation of the equation, nor will we present, for now, the Rasch equation used when rating scale surveys with multiple rating scale categories are used, or when tests might involve partial credit. The important thing for readers to note is that the Rasch model equations for those more sophisticated types of analyses have this equation, and thinking, at their core. Also, when we conduct an analysis with a program such as Winsteps, there is a calculation of person measures and item measures, and those values are expressed in units of logits. Logits is short for log odds units. This should make sense in that the mathematical term "ln" is in the equation, and also an "odds" (the fraction that has the probability of – in our example – Amy agreeing with item 8 divided by the probability of Amy not agreeing with the item).

Before we finish the chapter, we will discuss briefly one aspect of how the terms of the Rasch equation are determined (e.g., if Amy takes a survey in which items can be rated as either agree or disagree, how can Amy's person measure (B_n) be computed as +.25 logits, and how can the difficulty of survey item 8 (D_i) be computed as −1.2 logits?) To begin, readers should remember that Rasch measurement is based on measuring one variable or trait. This means that in the survey Amy completes, there will be items that involve one trait (e.g., belief in one's ability to teach). The items will not be identical and items will mark different parts of the trait. Some items will be easy to agree with, some harder to agree with, and some very hard to agree with. One might visualize the items as notches on a meterstick, some items at the bottom end (from 0 to 30 cm) while other items are located at the top end (70–100 cm). Once data are collected from Amy as well as other respondents, if a variable has been well defined by the survey items, then one would expect to see a pattern of responses similar to that provided in Fig. 2.7 when items are organized by their location along the trait (are items easy to agree with, hard to agree with, etc.) and when persons are organized by their overall level of agreement (Amy is overall more agreeable to survey items than Joe). The response pattern in Fig. 2.7 is a perfect response pattern, which in the real world we would not expect to see. As readers think about the Rasch equation, it should make sense to readers that there must be probabilities involved when each person responds to an item, for in

	Q2	Q5	Q7	Q8	Q11	Q13	Q1	Q6	Q3	Q9	Q12	Q4	Q10
Lisa	1	1	1	1	1	1	1	1	1	1	1	1	1
Heidi	1	1	1	1	1	1	1	1	1	1	1	1	0
Tina	1	1	1	1	1	1	1	1	1	1	0	0	0
Rose	1	1	1	1	1	1	1	1	0	0	0	0	0
Amy	1	1	1	1	1	1	1	1	0	0	0	0	0
Katie	1	1	1	1	1	0	0	0	0	0	0	0	0
Stef	1	1	1	1	0	0	0	0	0	0	0	0	0
Dave	1	1	1	0	0	0	0	0	0	0	0	0	0
Joe	1	0	0	0	0	0	0	0	0	0	0	0	0
Jay	0	0	0	0	0	0	0	0	0	0	0	0	0

Fig. 2.7 A table of 10 respondents' answers to the 13 self-efficacy items when respondents are only presented with two rating scale answers (*Agree* = 1 or *Disagree* = 0). Items are organized from easiest to agree with (Q2) to hardest to agree with (Q10). Persons are organized by most agreeable (Tina) to least agreeable (Jay). The distinctive pattern of 0s and 1s is what one expects to see if a variable has been defined by a set of items. Think of the Rasch model as utilizing this data pattern to compute the person measures and item measures expressed in the Rasch model. Also if one carefully thinks about what it means to measure, one is setting the stage for the type of data pattern and usefulness of data, which will facilitate measurement

Figure 2.7, one can see that there are differences in how easy it is to agree with the 13 survey items, and there are also differences in the agreeability of respondents. Also, it should make sense there will be a unique probability of an agreement as a function of respondents' overall location on the trait (where the person is located along the line in Fig. 2.6) and as a function of the location of the item on the trait (where the item is located along the line in Fig. 2.6).

By using Fig. 2.7, one can begin to quickly identify persons who are agreeable (or not very agreeable) and items that are easy to agree with (or not easy to agree with). Also, it is easier to understand what is expressed by the left side of the Rasch equation. Consider "Amy" who is very agreeable and item Q8, which is fairly easy to agree with. In this case Amy (B_n) has a higher logit measure than the logit measure of item Q8 (D_i) – just think of our plot in Fig. 2.6. Amy's logit measure being more agreeable than the logit measure of Q8 can also be seen by using the data pattern and looking at both the right side and left side of the Rasch equation. Given that Amy, when answering the set of items, agreed to some harder to agree with items (items 11, 13, 1, 6), then it should make sense that the probability of Amy answering *Agree* to item Q8 is greater than .50 (50/50). When the probability of answering *Agree* to an item by a person (the P_{ni} in the numerator of the Rasch equation in our example) is greater than .50, it means the denominator of the right side of the Rasch equation is less than .50. This results in a number which is positive when *ln* is computed for $P_{ni}/(1-P_{ni})$ and the only way to have a positive value on the left side of the equation is if $B_n > D_i$ (this means that the logit measure of the person is greater than the item measure).

Isabelle: Ted, are you ready for your quiz?

Ted: Go ahead!

Isabelle: Explain to me how to understand an analysis I conducted. I collected data from 200 15-year-old students regarding their interest in science from three school types (rural schools, urban schools, and suburban schools). I created a 25-item rating scale survey with questions such as "I would like to work as a scientist some day" and the students could answer using a rating scale of Strongly Agree, Agree, Disagree, and Strongly Disagree. No items needed to be flipped. I evaluated the data by first computing person measures with my Rasch program Winsteps, and then I conducted some statistical analyses. Those analyses suggested that the rural students were statistically more interested in science than the urban students. And those parametric calculations suggested the urban students were statistically more interested in science than the suburban students. But when I did not use the Rasch program, my statistical analyses suggested no differences in attitudes of the three groups. How can this be?

Ted: That is easy, Isabelle. Do you remember what you have been reminding me about for weeks, namely, that a rating scale is ordinal and not linear? Well to me it makes sense that if you throw ordinal data into SPSS, and conduct statistical analyses that assume the data are linear, then it is not surprising you will end up with results that might be different.

Isabelle: Good going, Ted. Okay a few more questions, the first two questions are related to each other. What does it mean when people talk of items being independent for measurement in a Rasch analysis? And what does it mean when people talk of sample independence?

Ted: This is actually pretty cool. When one uses Rasch measurement, and one is very careful about building a measurement instrument with items that involve one trait, then it does not matter which items a person completes. That person can be measured on the same scale as the scale used to measure other persons who take items for that trait. So "item independence" means, in my mind, different people can take different combinations of items but everyone can be expressed (measured) on the same scale.

"Sample independence" is very similar. What this phrase refers to is that if one wants to know where items for a trait fall on the trait (for instance, for the self-efficacy scale of the STEBI-B), then it should not matter which persons complete the items. One should still be able to figure out where items lie along the trait. It all has to do with the mathematics and the logic of the Rasch model, but the cool thing is that one can have "item independence" and "sample independence." For most of my work, it will be item independence that will be most helpful. I will be able to create different forms of a test, but I'll still be able to confidently put all of my data into a single spreadsheet and compute Rasch person measures as if everyone completed the same set of items! One more thing is that when I try to remember about sample independence, I just think of a meterstick. That meterstick will work very well both for lengths of leaves that range from 25 cm to 35 cm in length as it does to leaves of 5 cm to 15 cm.

Isabelle: Okay one last question. The Rasch software you use for your analysis uses, perhaps not surprisingly, the Rasch model to compute person measures and item measures. Can you explain to me, in a very simple way, how the program applies the Rasch model and makes sense out of the data?

Ted: Well Isabelle, the key is really starting with the requirement that when we collect data with an instrument, we are collecting data to measure one trait. If we are measuring one trait, then there should be a predictable pattern in the response of each person to each item depending upon where the person is on the trait and where the item is on the trait. I know I am working with a rating scale survey, but I really got the idea by thinking about a 5-item test in which items are right or wrong. It made sense to me that if I lined up the test items in terms of their difficulty, say from easy to hard, for each person, I should see an answering pattern from right (a 1 in my coding) to wrong (a 0 in my coding). Also if I then organized my data from the highest performers (highest ability students) to lowest ability students, I would see a very interesting pattern in their responses which looks like this:

	Q2	Q5	Q3	Q4	Q1
Doris	1	1	1	1	1
Sam	1	1	1	1	0
Jack	1	1	1	0	0
Billy	1	1	1	0	0
Tommy	1	1	0	0	0
Clemens	1	0	0	0	0
Elisabeth	1	0	0	0	0

Then when I looked at the Rasch model formula for dichotomous data, I could see that the left side of the equation made sense… that there would be a difference between each respondent and each item, and that difference would determine what I should see in terms of each person's response to each test item. Also, it made sense to me that by using the data in the rows and the data in the columns, I would be able to begin to compute some probabilities that would express the chances of someone getting an item right or wrong.

I also then did one other thing, I wrote out what a matrix of data might look like, if respondents completed the 13-item self-efficacy scale using a 6-step rating scale (6-Strongly Agree, 5-Agree, 4-Barely Agree, 3-Barely Disagree, 2-Disagree, 1-Strongly Disagree). This really helped me begin to gain a feel for the whole issue of what one gains when a single trait is measured. In my example I only used 5 items.

	Q2	Q5	Q3	Q4	Q1
John	6	6	6	6	5
George	6	6	6	5	5
Joe	6	5	5	4	4
Pete	6	5	5	4	3
Sue	5	5	4	4	3
Tony	5	5	4	3	3
Ken	5	4	4	3	3
Rich	5	4	4	3	3
Peggy	4	4	3	3	2

Keywords and Phrases

Equal interval
Flipping
Item independence
Label
Linear
Logits
Odds
Ordinal
Outcome-expectancy
Parametric statistics
Person independence
Rasch person measures

Rasch item measures
Raw score
Respondents
Self-efficacy
STEBI
Trait
Variables

Any number entered into a spreadsheet to indicate what rating scale category was selected by a respondent is only a label to indicate what rating scale category was selected by the respondent.

Ordinal data are not equal-interval data; ordinal data are not linear data. Ordinal data should not be immediately evaluated using parametric analyses (e.g., *t*-tests, ANOVA).

An assumption of parametric statistics is that data are expressed using an equal-interval scale. Since raw test data and raw rating scale data are not linear, then all analyses conducted with raw data may violate assumptions of parametric tests. This means that statistical analyses conducted with raw data may be wrong. When using a set of survey items (or a set of test items) to provide an assessment of a respondent along a trait, Rasch measurement must be used before parametric statistical tests are conducted (those statistical tests will use the person Rasch measures).

It is important, before entering survey data, to evaluate if any items need to be flipped. Often researchers will present wording in survey items to supposedly keep respondents alert. This unique wording would cause someone who selects *Strongly Agree* for most survey items to select *Strongly Disagree*. Data for such items need to be recoded before an analysis is conducted.

Potential Article Text

Data were collected at the University of XYZ from 237 preservice biology teachers using the preservice version of the STEBI-B. The original rating scale of Enochs and Riggs (1990) was altered to provide additional measurement precision. In this study a 6-category rating scale of *Strongly Agree* (6), *Agree*(5), *Barely Agree* (4), *Barely Disagree* (3), *Disagree* (2), and *Strongly Disagree* (1) was utilized. Two Rasch analyses were completed using the Winsteps (Linacre, 2012) computer program. Two linear measures (a self-efficacy person measure and an outcome-expectancy person measure) were computed for each respondent. These measures were then used to conduct parametric statistical analyses. Some respondents ($n=45$) did not complete the second sheet of the survey (items 17–23), which included both outcome-expectancy and self-efficacy items. Since Rasch analysis does not require all respondents to answer all items, person measures could be confidently computed only utilizing the survey items that were completed by respondents.

Quick Tips

To Show that a 4-Step Rating Scale May Not Be Linear. Have four colleagues get in a row. One person is Strongly Agree, one person is Agree, one person is Disagree, and one person is Strongly Disagree. First have them stand 50 cm from each other in a row. Point out that this could be the rating scale. Then ask the colleagues to make different spaces between each other. Point out that this new spacing could be the rating scale, and that the numbers used to code responses are just labels that help show the ordinal nature of the data.

The Rasch Model for Dichotomous Data. $B_n–D_i = \ln (P_{ni}/1 - P_{ni})$. B_n represents the ability of a respondent along the trait. *Di* represents the difficulty of an item along the trait. The relationship between the person ability and an item difficulty is described by a probability. For the case of dichotomous items, the probability of a person correctly answering an item is expressed by P_{ni}. The probability of the same person not correctly answering the same item is given by $1 - P_{ni}$. If one looks carefully at the model, it really summarizes the key ingredients of Rasch, and one can see that the Rasch model really emphasizes the importance of a single trait.

Look in the Winsteps manual under the keyword "recode," and you will be able to see how to reverse code items before a Rasch analysis.

Data Sets: (go to http://extras.springer.com)

No data sets

Activities

Activity #1

Find a rating scale that is different than the one discussed in this chapter. Decide how you would code data collected with this rating scale (e.g., "What numbers will you enter to indicate what a person answered?"). Often you can go to the search engine of your choice and simply type in keywords such as "rating scale," "survey," and "questionnaire." You might also consider looking at articles in journals such as the *Journal of Applied Measurement*. Many of the articles in this publication involve the use of rating scales. The rating scale might be presented in the article, or it will at least be referenced and easy to find.

Answer: Different rating scales will mean different coding. A three-category scale (e.g., Yes, Maybe, No) could be coded as 1, 2, 3 or 3, 2, 1 or 0, 1, 2 or 2, 1, 0. The important aspect of how you code is to remember the numbers are labels expressing the ordinal nature of data. Also remember our tips on the direction of the scale.

Activity #2

Pass out a copy of the STEBI-B to 10 friends. After they have answered the survey, set up two Excel spreadsheets, one for the outcome-expectancy items (1, 4, 7, 9, 10, 11, 13–16) and one for the self-efficacy items (2, 3, 5, 6, 8, 12, 17–23). Decide upon a coding scheme for each rating scale category. Be sure to code negatively phrased items correctly. (You could enter all the data for each scale first and then recode the negative items or you could recode negative items as you enter data.) The negative self-efficacy items are 3, 6, 8, 17, 19–21, and 23. The negative outcome-expectancy items are 10 and 13.

Answer: For each spreadsheet, you will have 10-item columns (OE) and 13-item columns (SE). Each row will be a respondent. To further practice data entry, you can add a column with a respondent ID or a column with a code for gender (perhaps, M and F or 0 and 1).

Activity #3

Look in your field's top research journals and also on the Internet. Find 2 or 3 rating scale surveys with items that must be reverse-coded.

Answer: Sometimes you have to dig to find surveys that contain negative items. Articles should have some explanation as to which items, if any, need to be recoded (flipped). It is very helpful, even if such guidance is provided, to try to figure out the recoding on your own and then check it with what authors may have suggested. This will help you think about the survey items.

Activity #4

Create your own 10-item survey for consumer ratings of the amount of product usage. Use a scale of *Often* (3), *Sometimes* (2), and *Seldom* (1). After you have completed the survey, repeat Activity 2 using your survey.

Activity #5

You have been asked by a high school principal to talk to teachers about how their state (in Germany "Land," in Australia "Territory") evaluates survey data. Prepare a brief script outlining how you would explain the main points of this chapter. Also prepare potential questions the teachers might have, and prepare your response to their questions.

Answer: Our tip to you is: Keep it simple. Yes, you might want to impress your audience with what you know, but communicate in a way so they can understand.

Activity #6

What is the difference between "counting" and "measuring"?

Answer: In this chapter we learned that numbers can be used to indicate what rating scale category was selected by respondents for survey items. In this chapter we hope that readers have learned that survey data can be entered into a spreadsheet, but because of the nonlinear aspect of survey data, one must take care to compute linear Rasch measures using the rating scale data as a starting point. Also, try reading "A History of Social Science Measurement" (Wright, 1997).

Activity #7

Pretend that you are presented with a sample of data in a spreadsheet in which students are in rows and the columns represent the 13 STEBI-B self-efficacy items. How might you check whether items were flipped or not by scanning the data set with your eyes?

Answer: The answer by each student to each of the 13 self-efficacy items will of course depend upon each student's assessment of his or her self-efficacy. Also, the response provided by each student for each item will depend upon if the item describes a component of self-efficacy that is generally easier to agree with or less easy to agree with. What we do when we scan data is scan a line of data for respondents. If we find an item that is quite different than the typical response of a single respondent, we then look for the same pattern for other respondents. Below we provide some sample self-efficacy data and will talk readers through this checking procedure. The rating scale is *Strongly Agree* (6), *Agree* (5), *Barely Agree* (4), *Barely Disagree* (3), *Disagree* (2), and *Strongly Disagree* (1).

ID	Q2	Q3	Q5	Q6	Q8	Q12	Q17	Q18	Q19	Q20	Q21	Q22	Q23
90	6	2	6	2	2	6	2	6	2	2	2	6	2
91	4	1	5	1	1	5	1	6	2	1	1	5	2
92	5	1	5	2	1	4	1	5	1	1	2	5	1

Looking through the data for person 90, we see some items for which a "6" is entered and also some items for which a "2" is entered. We would expect to see some similar answers for most items for this respondent. This suggests that the items for person 90 which have a "2" are items that need to be flipped. As we try to see if our guess is correct, we can look at the other two respondents (persons 91 and 92). Although the numbers entered are not exactly the same, the same general pattern is present. Review of the published scale reveals that the negatively worded items are 3, 6, 8, 17, 19–21, and 23. These are exactly the items we have identified as being different in the coding. This means that these items need to be flipped before an analysis of the data is completed.

Activity #8

One way to improve one's ability to identify items that need to be flipped is to author items for a scale and to write versions of an item with both positive and negative wording. Below we provide the text for each of the 13 self-efficacy items of the STEBI-B. Author the non-flipped version of the items that are negatively worded. Also author a new version of the five self-efficacy items that do not need reverse coding. This means, for example, for item 2, write a version of that item so it will need to be flipped for an analysis.

Q2 I will continually find better ways to teach science.
Q3 Even if I try very hard, I will not teach science as well as I will most subjects.
Q5 I know the steps necessary to teach science concepts effectively.
Q6 I will not be very effective in monitoring science experiments.
Q8 I will generally teach science ineffectively.
Q12 I understand science concepts well to be effective in teaching elementary science.
Q17 I will find it difficult to explain to students why science experiments work.
Q18 I will typically be able to answer students' science questions.
Q19 I wonder if I will have the necessary skills to teach science.
Q20 Given a choice, I would not invite the principal to evaluate my science teaching.
Q21 When a student has difficulty understanding a science concept, I will usually be at a loss as to how to help the student understand it better.
Q22 When teaching science, I will usually welcome student questions.
Q23 I do not know what to do to turn students on to science.

Answer: There are many potential ways of changing item wording so that the items will need to be flipped later, and there are many ways of un-flipping an item with words. An example of alternate wording for Q23 is "I do know what to do to turn students onto science."

Activity #9

Draw a meterstick from 0 to 20 cm and mark the units for each centimeter (e.g., 11, 12 cm). This meterstick provides linear (equal-interval) units of measurement. Second, imagine a second meterstick is made of rubber, squish parts of the meterstick, and expand parts of the meter stick. Draw that second meterstick that has markings of 0 to 20 cm.

References

Boone, W., Townsend, S., & Staver, J. (2011). Using Rasch theory to guide the practice of survey development and survey data analysis in science education and to inform science reform efforts: An exemplar utilizing STEBI self-efficacy data. *Science Education, 95*(2), 258–280.

Bradley, K., Cunningham, J., Akers, K., Knutson, N. (2011, April). *Middle category or survey pitfall: Using Rasch modeling to illustrate the middle category measurement flaw*. Paper presented at the annual meeting of the American Educational Research Association.

Enochs, L. G., & Riggs, I. M. (1990). Further development of an elementary science teaching efficacy belief instrument: A pre-service elementary scale. *School Science and Mathematics, 90*(8), 694–706. doi:10.1111/j.1949-8594.1990.tb12048.x.

Glass, G. V., & Stanley, J. C. (1970). *Statistical methods in education and psychology*. Englewood Cliffs, NJ: Prentice-Hall, Inc.

Linacre, J. M. (2012) Winsteps (Version 3.74) [Software]. Available from http://www.winsteps.com/index.html

Stevens, S. S. (1959). Chap. 2: Measurement, psychophysics and utility. In C. W. Churchman & P. Ratoosh (Eds.), *Measurement: Definitions and theories*. New York: Wiley.

Wright, B. D., & Masters, G. N. (1982). *Rating scale analysis*. Chicago: MESA Press.

Additional Readings

A classic text that specifically considers the ins and outs of Rasch analysis for rating scales.

Wright, B. D., & Masters, G. N. (1982). *Rating scale analysis*. Chicago: MESA Press.

In this article Ben Wright presents a summary of the work of Stevens and discusses in what manner Rasch measurement addresses the issue of "measurement".

Wright, B. D. (1997) S. S. Stevens revisited. *Rasch Measurement Transactions 11*(1), 552–553. http://www.rasch.org/rmt/rmt111n.htm

A brief sample article that provides readers with an exemplar of Rasch measurement applications in the field of medicine.

Gothwal, V. K., Wright, T. A., Lamoureux, E. L., & Pesudovs, K. (2009, October). Rasch analysis of visual function and quality of life questionnaires. *Optometry and Vision Science, 86*(10), 1160–1168.

Chapter 3
A Rasch Rating Scale Analysis (Step I): Reading Data and Running an Analysis

Isabelle and Ted: Two Colleagues Conversing

Ted: Okay Isabelle, I am here at 8 AM as promised. I have all of the STEBI self-efficacy data entered into a spreadsheet, and I am ready to go.

Isabelle: Ready to go? Are you now going to leave?

Ted (ignoring Isabelle's humor): I have a data set from 75 students. So I have 76 rows in my spreadsheet; the top row is where I named each column. I have 15 columns of data; each column is either a survey item (there were 13 of them), an ID column, or a gender column.

Isabelle: Okay, you have checked your data for errors?

Ted: Yes I did.

Isabelle: Okay here we go. You will see that we can construct a simple Winsteps control file and do a quick analysis. We might discover some things we need to change, but we can do an initial analysis in just a few minutes.

Ted: So I might be able to run this and include it in my conference proposal that is due in 20 hours?

Isabelle: Potentially! But I do have an initial question for you. When you use a set of survey items, such as these 13 survey items to "measure" a person, what does that mean?

Ted: That is easy. First think of a test in which students can get an item right or wrong. When we give a test to students, we hope that by administering a number of test items involving one trait, we can get a good idea of what the students do or do not know. By administering items with a range of difficulty, we also hope to be able to differentiate students, to figure out how students compare. The same is true with a survey that involves one trait and a rating scale. In this case we are also administering a number of items that will allow us to compare respondents and determine each respondent's overall attitude. I had mostly worked with tests in the past, but I now understand that just as one might want to compare the test performance of two subgroups (say boys and girls), there are many situations in which one might want to compare the attitudes of groups of respondents at one time point (treatment patients and control patients at a hospital) or over time (how did the attitudes of treatment patients change over time).

W.J. Boone et al., *Rasch Analysis in the Human Sciences*,
DOI 10.1007/978-94-007-6857-4_3, © Springer Science+Business Media Dordrecht 2014

Introduction

In Chap. 2, we explained that rating scale data are ordinal, and that one should not immediately compute raw score totals for respondents, nor should one compute and use person raw score means for analyses. We also emphasized that techniques such as ignoring missing data can cause inaccurate calculations of raw scores. Missing data rarely cause problems in a well-constructed Rasch analysis; this is, in part, because one is working with a single variable. By conducting a Rasch analysis of an appropriate data set, one can compute equal-interval scale person measures that are appropriate for *t*-tests and other parametric statistical analyses. A person measure is defined as a quantitative measure of a person's ability, attitude, self-efficacy, etc., on a unidimensional scale. In this chapter, readers will be introduced to Winsteps software through the entry of data and the construction of a simple Winsteps control file, which then allows the computation of person measures. Examples of person measures would include, but of course would not be limited to, student performance on a multiple-choice test, the medical condition of a patient being treated for anxiety, and the overall views of a consumer about a type of product.

A Rasch analysis, of course, provides much more than person measures, but we will start with person measures, for these are the data most often used in science education studies as well as in other fields of educational research, in medical research and market research.

Preparing Your Data for a Rasch Analysis

Spreadsheet Data Entry

To conduct a Rasch analysis of rating scale data using Winsteps, one needs to first enter the raw data into a file. We will use an example in which a researcher begins an analysis by entering data into an Excel spreadsheet. Excel, of course, is a software program familiar to most researchers. Certainly there are many other programs that can be used to enter, save, and organize data (e.g., SPSS is a software package that facilitates statistical analysis of data and also provides a spreadsheet format for data entry and organization). Figure 3.1 presents our now familiar STEBI (Enochs & Riggs, 1990) self-efficacy scale data, in which the collected data have been entered into a spreadsheet using numbers as labels to indicate which rating scale category was selected by each and every respondent. In addition to each respondent's answers to the set of 13 self-efficacy items, the Excel spreadsheet includes a column for a student ID, a column noting whether the data were collected as "pre" (at the start of a semester) or "post" (at the end of the semester), a column with a letter to indicate the gender of a respondent, and a column indicating which school the respondent attended. Only the first three lines of data are presented in Fig. 3.1, but the entire Excel file can be found in the Chapter 3 data file named "Chap. 3 Excel

	A	B	C	D	E	F	G	H	I	J	K	L	M	N	O	P	Q
1	ID	PR	Gender	Schl	Q2	Q3	Q5	Q6	Q8	Q12	Q17	Q18	Q19	Q20	Q21	Q22	Q23
2	21141	PR	M	A	6	5	2	6	5	2	5	5	4	5	5	5	5
3	91052	PR	M	A	6	4	2	5	5	5	x	x	x	x	x	x	x
4	95793	PR	M	A	6	6	5	5	6	5	x	x	x	x	x	x	x

Fig. 3.1 A sample Excel spreadsheet following the entry of data for three students who completed the STEBI. The first row provides column headers for student IDs, a code to indicate whether student data were pre (at the start of the semester data, denoted with the letters PR), gender, a code letter to identify the school attended by students, and student responses for the 13 self-efficacy items

Self-Efficacy Data". Only the 13 self-efficacy items are entered. Columns are labeled with the original item numbering of the STEBI. Data have been flipped, when needed, prior to entry into the spreadsheet.

Creating a Winsteps Control File

To run Winsteps and compute, among other things, Rasch person measures, a researcher must create a Winsteps control file. A control file is simply a file that helps the program understand the form of the data (e.g., how many rating scale categories are used for each item, what numbers have been used to code/label each of the rating scale categories). The control file does two other things as well; the control file includes "names" (descriptors) of items and also includes only the data to be evaluated (your spreadsheet may have lots of information that is unrelated to an analysis of a survey or test).

Creating a control file can be completed in a few simple steps. Researchers frequently edit a control file that has already been created for running Winsteps, but as a first step, we think it is easier if readers utilize Winsteps' ability to quickly create a control file. Below we explain the steps a researcher can use to create a control file using the file "Chapter 3 Excel Self-efficacy Data." [Remember you can use the free Ministeps program (limited to $n=25$ items, $n=75$ respondents) or the Winsteps program for your analyses.] Once the control file is created, it takes just a few simple steps to run Winsteps and compute person measures. As you learn how to create a control file, remember there are three parts to the control file. Part 1 helps the Winsteps program know how the data are organized and also provides some nuances germane to a Rasch analysis of the data. Part 2 contains the descriptors or labels of each item (e.g., test item, survey item). These labels (descriptors) will allow you to quickly remember the core meaning of each item as you look through the Rasch Winsteps output (e.g., you could name the 4th item on the survey "Q4," but it may be better to name it "Q4-Like Lab Group Work"). Part 3 of the control file simply contains the data to be evaluated in a Rasch analysis.

Step 1: Double-click on the Winsteps icon to open the program. The following introductory window will appear:

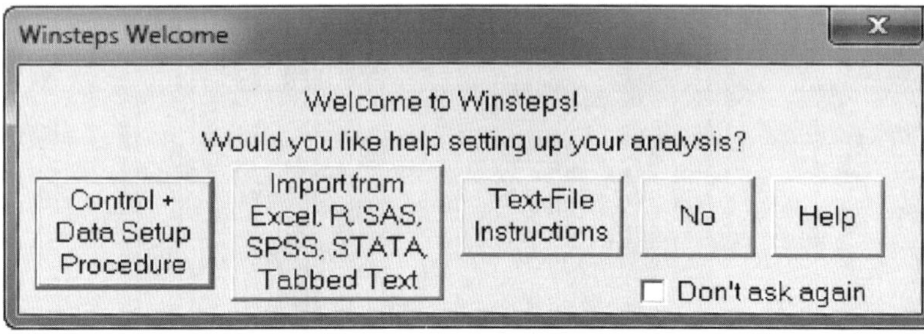

When the program opens, a gray box appears with the question: "Would you like help setting up your analysis?" Since we want to import data from another program, we will click the button concerning importing data.

Step 2: Click on **Import from Excel**, **R**, **SAS**, **SPSS**, **STATA**, **Tabbed Text** to import data and create a control file. The following box with colored squares will then appear (e.g., in color the box entitled "Excel" is a light green color and the box entitled "SPSS" is a lavender color):

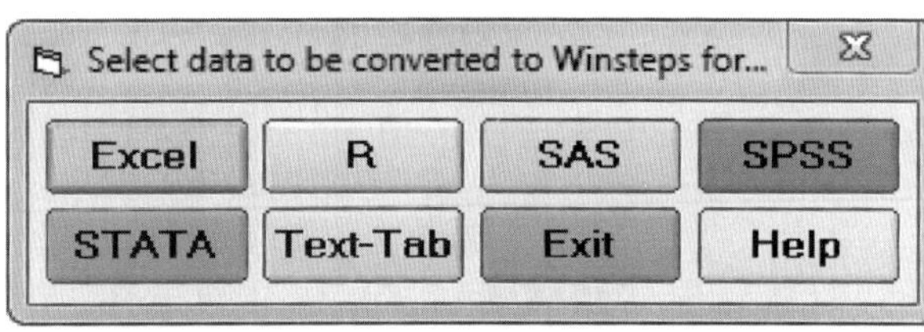

Each colored square lists a possible data analysis program that is compatible with importing data into Winsteps.

Step 3: Click on the square marked **Excel**, since Excel is the program used for the data of this example.

Now a new box will appear:

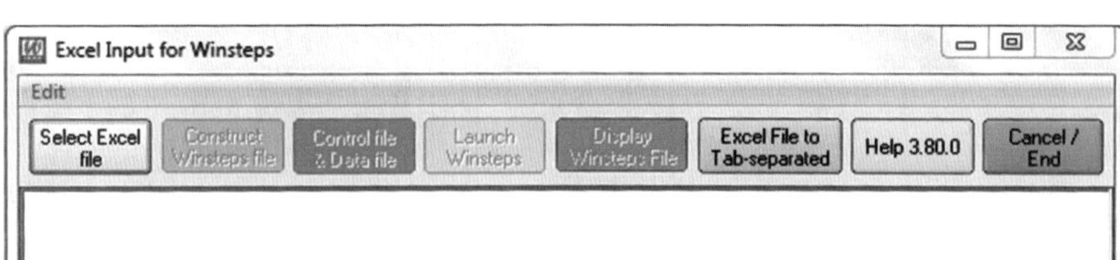

Step 4: Click on the square marked **Select Excel File**. This step allows the researcher to provide the location of the Excel file. Now a screen familiar to those who use Windows will appear:

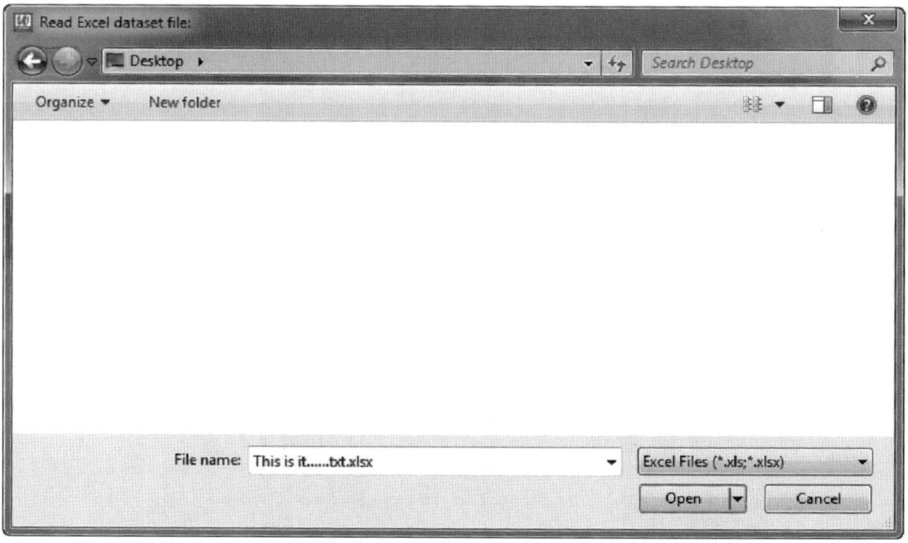

For our analysis, the sample Excel file was saved to the desktop, so we will select the proper file from our desktop. We have named our file "This is it…..txt.xlsx". The file you will select will be "Chapter 3 Excel self-efficacy data."

Step 5: Click on the Excel file you have selected. A message will appear on the screen that says "Processing…." If your data set is small, this message will appear only briefly. Then you will see the following on your screen. Please note the five critical lines that appear in red on your screen. For the text below, we put those five red lines in **bold** and underlined text. Although there are several lines in the text below, the most important lines for setting up the control file for Winsteps are simply (1) the lines we have underlined and bolded and (2) the lines presented below the underlined and bolded lines (the first of which is the line ";Variable (First Cell Value)" and the last of which is the line "@Row number ; row in Excel spreadsheet").

; Excel File: C:\Users\Desktop\This is it......txt.xlsx
; Dataset name: This is it.....
; Number of Excel Cases: 75
; Number of Excel Variables: 17

; Choose the variables listed below under "Other Variables" that you want to form part of the Winsteps person labels.

(continued)

(continued)

; Copy-and-paste those variables under "Person Label Variables" in the order you want.
; There will be one space between the variables in the person labels.

; Choose the variables listed below under "Other Variables" that you want to be Winsteps item response-level data.
; Copy-and-paste those variables under "Item Response Variables" in the order you want.
; Numeric item variables are truncated to integers.

; The same variables can be placed in both sections and in any order.
; Constant values may be included in the "Person" and "Item" variable lists with " ", e.g., "1"

; Click on "Construct Winsteps file" when completed

! Item Response Variables. (Do not delete this line - item variables on left-side if this line before "Person Label Variables")

! Person Label Variables. (Do not delete this line - person variables on left-side if before "Item Response Variables")

! Other Variables (ignored) - if this looks wrong, save the Excel file as .xls and rerun.
;Variable Label (First Cell Value)
A ; ID (21141 PR 465526555425445554555 ; spring 2008 PRE)
B ; PR (PR)
C ; Gender (M)
D ; Schl A or B (A)
E ; Q2 (6)
F ; Q3 (5)
G ; Q5 (2)
H ; Q6 (6)
I ; Q8 (5)
J ; Q12 (2)
K ; Q17 (5)
L ; Q18 (5)
M ; Q19 (4)
N ; Q20 (5)
O ; Q21 (5)
P ; Q22 (5)
Q ; Q23 (5)
@Case number ; Person entry number
@Row number ; Row in Excel spreadsheet

Note the letters A through Q presented at the bottom of the screen. Each line represents an Excel column and includes the column heading. Columns A, B, C, and D are the demographic variables id, pre or post, gender, and school, respectively. Columns E through Q present each of the self-efficacy items. The information between the "( )" simply presents the actual data for the first person in the data set.

Step 6: Copy and paste all of lines A, B, C, and D under the red lines (on your screen, bold and underlined in our book) that start with "Person Label Variables." You could "cut and paste" if you wanted to, but by copying and pasting, if you get confused and/or make an error, it is easier to backtrack and start over.

! Person Label Variables. (Do not delete this line - person variables on left-side if before "Item Response Variables")
A ; ID (21141 PR 4655265555425445554555 ; spring 2008 PRE)
B ; PR (PR)
C ; Gender (M)
D ; Schl A or B (A)

This step tells the program which data lines describe the respondents. In other words, columns A, B, C, and D explain the respondents' demographic characteristics, not the rating scale data for the 13 self-efficacy items. The next step tells the program which data lines contain the item responses.

Step 7: Copy and paste all of lines E–Q under the red lines (again in our book noted in bold and through underlining) that start with "! Item Response Variables."

! Item Response Variables. (Do not delete this line - item variables on left-side if this line before "Person Label Variables")
E ; Q2 (6)
F ; Q3 (5)
G ; Q5 (2)
H ; Q6 (6)
I ; Q8 (5)
J ; Q12 (2)
K ; Q17 (5)
L ; Q18 (5)
M ; Q19 (4)
N ; Q20 (5)
O ; Q21 (5)
P ; Q22 (5)
Q ; Q23 (5)

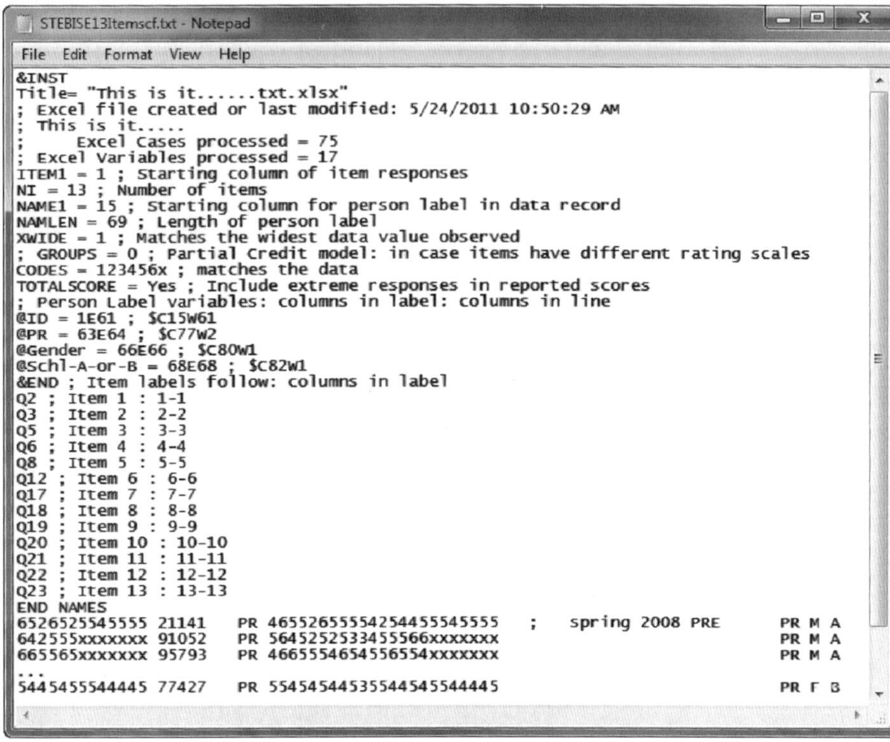

```
&INST
Title= "This is it......txt.xlsx"
; Excel file created or last modified: 5/24/2011 10:50:29 AM
; This is it.....
;     Excel Cases processed = 75
; Excel Variables processed = 17
ITEM1 = 1 ; Starting column of item responses
NI = 13 ; Number of items
NAME1 = 15 ; Starting column for person label in data record
NAMLEN = 69 ; Length of person label
XWIDE = 1 ; Matches the widest data value observed
; GROUPS = 0 ; Partial Credit model: in case items have different rating scales
CODES = 123456x ; matches the data
TOTALSCORE = Yes ; Include extreme responses in reported scores
; Person Label variables: columns in label: columns in line
@ID = 1E61 ; $C15W61
@PR = 63E64 ; $C77W2
@Gender = 66E66 ; $C80W1
@Schl-A-or-B = 68E68 ; $C82W1
&END ; Item labels follow: columns in label
Q2 ; Item 1 : 1-1
Q3 ; Item 2 : 2-2
Q5 ; Item 3 : 3-3
Q6 ; Item 4 : 4-4
Q8 ; Item 5 : 5-5
Q12 ; Item 6 : 6-6
Q17 ; Item 7 : 7-7
Q18 ; Item 8 : 8-8
Q19 ; Item 9 : 9-9
Q20 ; Item 10 : 10-10
Q21 ; Item 11 : 11-11
Q22 ; Item 12 : 12-12
Q23 ; Item 13 : 13-13
END NAMES
6526525545555 21141    PR 46552655554254455545555   ;   spring 2008 PRE       PR M A
642555xxxxxxx 91052    PR 5645252533455566xxxxxxx                         PR M A
665565xxxxxxx 95793    PR 4665554654556554xxxxxxx                         PR M A
...
5445455544445 77427    PR 55454544535544545544445                         PR F B
```

Fig. 3.2 The control file created by Winsteps (Linacre, 2012)

Step 8: Now click on the button marked **Construct Winsteps file**.

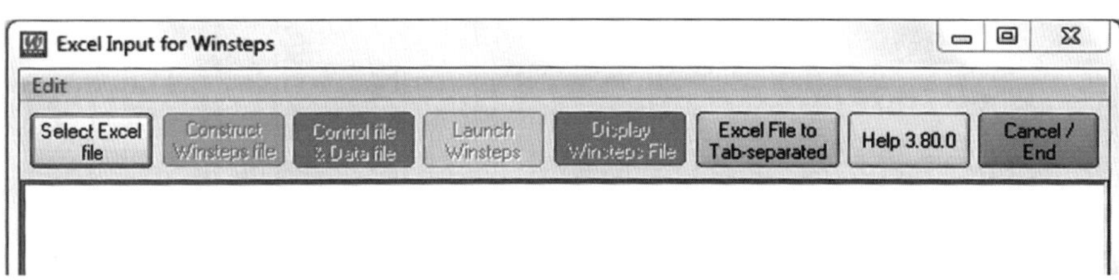

Once step 8 is completed, the program will ask where to save the control file and what name you want to give the file. To remember that the file is a control file, it is helpful to include the letters "cf" in the control file name. It is also helpful to add a name that will remind you about the data set. In this case, we will name the file "STEBISE13Itemscf."

Step 9: Enter "STEBISE13Itemscf" or your name for the control file, and then click **Save**. The control file should appear on the screen; you are now ready to conduct a Rasch analysis of the STEBI self-efficacy rating scale data. In Figure 3.2 almost all lines of the control file are presented. To save space, we removed all but the first three lines of person responses and the last line of the person measures.

Step 10: For now exit the program and close all windows. Then double click on Ministeps (Winsteps). Then click "No" in the gray box. Then push the "Enter" key on your keyboard. Select your control file. Then click on "Open." Then press the Enter key. Then press the Enter key again. Then the program will run.

Within the control file, we usually make one small edit before we run the analysis. We typically add descriptions of each item to the control file so each item can be easily recalled in the Winsteps output. These descriptions help one stay organized and save time (look at the item descriptors between the lines "&END" and "END NAMES"; these descriptors can be edited to provide clarity to the researchers who will later have to plow through data tables). Item descriptions can be added easily to lines of the control file. The next section provides examples of potential item descriptions for the analysis. Moreover, for the first one-third of the control file, note that the program ignores any text after a semicolon (;) along a single line; thus, the line "Q23 ; Item 13 : 13-13" could be edited to simply "Q23" and the program would run in an identical manner. Finally it is important to note that the discussion of how to name an item is a different discussion than our earlier chapter discussion of the importance of not confusing the use of numbers to code rating scale data with measures. When we discuss how to describe an item, we are describing the importance of using letters and numbers to summarize an item. This is quite different than the issue of what numbers do and do not represent when data are coded in a spreadsheet. Why might you want to describe each item, and why might you wish to describe an item in shorthand? When you conduct your Rasch analysis and attempt to synthesize the results, a number of tables and plots become available that will include a listing of instrument items. Your interpretation of data will be much faster if you develop descriptors for items. We suggest a shorthand naming of items because long item names can slow down your interpretation of results. At any point in an analysis, one can write in a longer item name in a control file, but short, thoughtful item names are very useful.

Formative Assessment Checkpoint #1

Question: Is it hard to construct a control file?

Answer: No. When your data are in a spreadsheet, it is quite easy to create a control file. Spreadsheets can be in many forms, for instance, EXCEL, SPSS, and SAS.

Naming Survey Items

One can use any symbol or set of symbols to describe a survey item. In our constructed control file above, we described the 13 survey items based upon the headers presented in the Excel file (e.g., the Excel column with the data for Q2 of the survey had the header "Q2"). Generally, it is advantageous to create an item identifier that is a shorthand summary of the item. For simplicity, the description should be short but

Item 7 - Original Wording
If students are underachieving in science, it is most likely due to ineffective science teaching.

Item 7 – Identifier Options
7StuUnderachSciDueIneffSciT OR Q7 Stu UnAch S Due Ineff STeach

Item 10 - Original Wording
The low science achievement of some students cannot be blamed on their teachers.

Item 10 – Identifier Options
10FlipLowSciachCanBeBlamedonT OR Q10 Flip_Low_S_Ach_CAN_Blamed_OnTeach

Fig. 3.3 Two examples of shorthand summaries for items. It is a personal choice to use spaces or underline to make a clearer descriptor

informative. Short item identifiers will allow you to quickly scan Rasch analysis outputs for patterns. Below we provide some tips on naming survey and test items. When you wish to rename an item, you need only to edit the control file; this can be done through an "edit" option in Winsteps or by simply editing the control file in a text editor.

In Fig. 3.3 we present some different examples of shorthand text for two STEBI self-efficacy items. The second item is one that was flipped with the numerical labels of the respondent rating scale category selections.

Succinct wording for item identifiers is helpful in terms of efficiently using Rasch analysis tables (e.g., quickly being assured that the third line of data of the STEBI items refers to item Q5). However, there is a second reason for adding item identifiers. Sometimes a critical error is made in that just as the coding of student answers are flipped before analysis, one must also adjust the wording of flipped items to identify such items. It is important to adjust the wording of how an item is named so that the item name reflects the flipped item. For instance, if an item originally was "I do not have the confidence needed to teach science," the new item name (after flipping) should be "I have the confidence needed to teach science." An easy way to check the wording of the item identifiers is to imagine a respondent who has an extremely confident view of himself or herself as a science teacher, read through the item identifier names for all items in the control file, and then answer the item. Using the new wording for each survey item, all responses should be at one end of the rating scale.

Formative Assessment Checkpoint #2

Question: Is it important to worry about items that are worded in a negative manner when the majority of items are worded in a positive manner?

Answer: When you have some negatively worded items, it is very important to keep track of those items to make sure data are entered correctly for those items (or at least later recoded to reflect the negative nature of the items). Moreover, you should name the items in such a manner to reflect the impact of recoding.

Reverse coding data ensures that increasing a label value for a rating scale means the same thing for all items. Thus, when negative item data are entered, they are flipped because the direction of a negatively stated survey item's wording does not match other survey items. Just as this reverse coding must be done when data are entered, there must also be recoding in the descriptions of negatively stated survey items. Such flipping of item wording is usually accomplished by adding or removing words such as "not" and prefixes. For example, remove the prefix "un-" appearing in the word "unsatisfactory," so that the phrase "satisfactory" is presented.

To finish up this discussion, we emphasize that when data are entered, you could enter all data from the beginning and make sure to flip the negative items. An alternative is to use some options in Winsteps to "flip" those items that you will look at during your Rasch analysis (the option to flip before a Winsteps analysis is detailed in the Winsteps manual – see RECODE). For readers interested in immediately reading more on how to reverse code with Winsteps, simply use the Winsteps manual and look up the term "RECODE=."

The important thing is to remember to reverse code appropriate items. Those just beginning to use Rasch for the analysis of data should not fret too much over trying to find the "best" or "correct" item identifier. Remember, it is important that you provide enough of a description so that when you are reviewing Winsteps output, items can be quickly identified. In our workshops and classes, we emphasize that writing an item descriptor for a Winsteps control file is similar to authoring a word or phrase to describe data in an Excel file. Figure 3.4 contains the set of descriptors that we often use when we are evaluating data sets with these 13 self-efficacy items.

This chapter detailed how a control file can be quickly created and edited to run a Rasch analysis of a rating scale data set with Winsteps. Part 1 of the control file (from "&INST" to "&END") contains several lines of code that tell the program how to read the data. Part 2 of the control file consists of item descriptions. Part 3 of the control file contains that data you will be evaluating. By following the steps outlined, researchers can quickly put a control file in a form so that resulting outputs and plots can, if needed, be placed immediately in publications and reports. When you practice constructing a Winsteps control file, you will see red lines appear on your screen. These are the lines that we underline and make bold in our book. For better descriptors for items, recall that the headers in your spreadsheet will be read in as item names. However, at a later point, you can also improve your naming of items. To do so, in the control file, remove the old item name by removing the text and type in a modified descriptor on a single line. Finally, remember to save your changes.

Isabelle and Ted: Two Colleagues Conversing

Ted: *So, if I have this right Isabelle, I can enter my survey data into an Excel sheet, one line per person and one column per item. Then Winsteps can create a control file for me, correct?*

Isabelle: *Yes indeed, you are right. Winsteps is user-friendly for converting other data formats, such as Excel, to the necessary control file format.*

```
&INST
Title= "This is it.....txt.xlsx"
; Excel file created or last modified: 5/24/2011 10:50:29 AM
; This is it....
;   Excel Cases processed = 75
; Excel Variables processed = 17
ITEM1 = 1 ; Starting column of item responses
NI = 13 ; Number of items
NAME1 = 15 ; Starting column for person label in data record
NAMLEN = 69 ; Length of person label
XWIDE = 1 ; Matches the widest data value observed
; GROUPS = 0 ; Partial Credit model: in case items have different rating scales
CODES = 123456x ; matches the data
TOTALSCORE = Yes ; Include extreme responses in reported scores
; Person Label variables: columns in label: columns in line
@ID = 1E61 ; $C15W61
@PR = 63E64 ; $C77W2
@Gender = 66E66 ; $C80W1
@Schl-A-or-B = 68E68 ; $C82W1
&END ; Item labels follow: columns in label
Q2IWILLContFindBetWays2TeachS
Q3FLIPITryHrdIWillTchSAsWelOthSubj
Q5IKnowStepsNecc2TeachSEff
Q6FLIPIWillBeEffInMonitSExp
Q8FLIPIWillTchSEff
Q12IUnderSConceptWell2TchS
Q17FLIPIWilNOTFindDiff2ExplWhySExp
Q18IWILLBeAble2AnsStudSQ
Q19FLIPIDoNOTWonderIfISkills2TchS
Q20FLIPIWILLInvitPrinc2EvalSTch
Q21FLIPIWILLNOTBAtLoss2HelpStudUSCon
Q22WhenTchSIWillWelcomeStuQ
Q23FLIPIDoKnowWhat2DotoTurnStuOntoS
END NAMES
```

Fig. 3.4 Shorthand summaries for all 13 self-efficacy items

Ted: *Let's do a final fact check, okay?*

Isabelle: *Okay.*

Ted: *The control file has three parts. The first part is the part of the code that tells Winsteps how to read the data. But this part of the control file also does some other things. For example, this part of the code guides the program as to what should be done in an analysis. The second part of the control file contains the names for each survey item. So, if there are nine survey items, we will see nine lines, where each line describes an item. The last part of the file contains the data. If we look at the first line of data, we should be able to find that data in our original Excel spreadsheet.*

Isabelle: *Excellent! But one more question Ted. You have been learning how to create a control file for Winsteps that allows a Rasch analysis of data. Last week you and I spoke*

about the problems with "raw rating scale data." Can you tell me in a general way how you think Winsteps allows us to compute "measures" that are linear?

Ted: Okay, this is where I am in my understanding. George Rasch developed an equation (and also a theory) that involves what it means to "measure." He started with tests in which items are right or wrong, but it makes sense that it would not be hard to extend his work to rating scales. After all a test in which students either get an item right or wrong is a lot like a survey in which someone can either agree or disagree with an item. When I look at the equation of the Rasch model for a survey with just agree/disagree (let me write it below),

$$B_n - D_i = \ln\left(P_{ni} / 1 - P_{ni}\right),$$

it makes sense to me that the Winsteps program will attempt to see if it is possible to compute a person attitude, B_n (in the case of surveys), and also compute an item difficulty, D_i (think of this as how easy it is to agree with an attitudinal item). It also makes sense to me that if you have many responses from one person, you should be able to figure out what a person's attitude is, and if you have numerous responses to one item (by many people), you should be able to figure out how easy or hard it was to agree with an item. I see that the Rasch model has probabilities, and that is a distinctive feature of the model, at least for me. The probabilities help remind me that Rasch analysis uses the model to see if the data can fit the model, and if so, how well. Anyway, in a very basic way, that is where I am with the Winsteps program and the Rasch model and what must take place when the program runs a data set.

Keywords and Phrases

Control file
Item response variables
Item names
Negatively worded items
Person label variables

The control file tells the Rasch analysis program what data to evaluate and provides guidance to the program with regard to many issues (e.g., what type of data will be evaluated, which items need to be reverse coded, the location of information that involves variables such as gender and race).

Quick Tips

Enter your survey data into a spreadsheet. Recode any items that are negatively worded when you enter your data into the spreadsheet. Make sure to label each column of the spreadsheet with a code that indicates which item is presented in the given column. Then follow the steps that our book provides to create a control file. Remember to copy and paste the entire line you wish to copy and paste under the two red lines that are provided on your screen as you create the control file (see Steps 6 and 7 of this chapter).

There are three parts to the control file. The first part of the file begins with the line "&INST" and ends with the line "&END." The second part of the control file consists of the names of all items to be evaluated. The last line of the second part of the control file ends with the line "END NAMES." The third and final part of the control file consists of the data to be evaluated in the analysis. In this section, you will see the respondents' answers to the survey (or test), demographic data (e.g., gender), and respondents' IDs that you have included in your analysis when you were constructing the control file.

The Winsteps manual provides extensive details on a variety of commands that can be added to the control file. Do not be overwhelmed by the length of the manual; the steps we describe herein will allow you to conduct a Rasch analysis of your test and/or survey data. And as you advance and practice, you will learn how to quickly find the guidance you need for an analysis of most all conceivable data sets.

Data Sets: (go to http://extras.springer.com)

Chapter 3 Excel Self-Efficacy Data
Excel Data Outcome-Expectancy – Negative Items Already Flipped In Data Set Do NOT Flip These Items

Activities

Our colleague Naz Bautista at Miami University has kindly provided a nonrandom sample of outcome-expectancy data to us for use in this book. Readers should recall that the 23-item STEBI of Enochs and Riggs contains 13 self-efficacy items and 10 outcome-expectancy items. Much of the text in this book makes use of the 13 self-efficacy items. Bautista's student data are provided to readers in an Excel spreadsheet. For those items that are negatively worded, the items were reverse coded prior to entering the data; thus, data are ready to be used for the creation of a control file, without having to later (in the control file) flip the response. Fictitious genders have been added to the data set. This gender data will be used for other chapter activities, such as activities that focus on differential item functioning (DIF).

Activity #1

The spreadsheet with the very long name "Excel Data Outcome-Expectancy – Negative Items Already Flipped In Data Set Do NOT Flip These Items" contains the responses of 74 students to the 10 items of the STEBI outcome-expectancy scale. Find a copy of the STEBI, identify the 10 outcome-expectancy items, and then identify what rating scale coding Bautista most likely used. The original STEBI used a 5-step scale; therefore, which code was used for each rating scale step?

Answer: Review of the data entered into the spreadsheet suggests that the coding used for the data was 5 (Strongly Agree), 4 (Agree), 3 (Uncertain), 2 (Disagree), and 1 (Strongly Disagree).

Activity #2

Readers should remember that the data presented in the file are already reverse coded. Take a copy of the original STEBI and circle the answers that would have been selected by respondents 401 and 402 (the first and second persons in our data set) for the outcome-expectancy items.

Answer: Since we are considering only the outcome-expectancy items, not all items of the STEBI should be circled. The items that should be circled for both persons 401 and 402 are items Q1, Q4, Q7, Q9, Q10, Q11, Q13, Q14, Q15, and Q16. Recall that (1) the flipped outcome-expectancy items are items Q10 and Q13 (these two items were denoted with shorthand Q10-rc and Q13-rc in the Excel sheet headings) and (2) the original coding of the data was 5 (Strongly Agree), 4 (Agree), 3 (Uncertain), 2 (Disagree), and 1 (Strongly Disagree). This means that if in the spreadsheet we see the following coding for persons 401 and 402 (401,4,4,4,4,3,4,4,4,4,4,F; 402,2,2,2,3,3,2,4,2,4,2,F), it means that person 401 originally circled *Agree, Agree, Agree, Agree, Uncertain, Agree, Disagree, Agree, Agree, Agree*. The "3" entered for person 401's answer to item Q10 (the 5th outcome-expectancy item) was originally a "3" for uncertain, and that entry does not change with a recoding since the recoding of the middle category will remain a "3." The "4" that was entered into the spreadsheet as the flipped response for item Q13 (the seventh outcome-expectancy item) indicates that the original answer to item Q13 was *Disagree*. Person 402 data entry for item Q10 is a "3." This means that this person also answered *Uncertain* for item Q10. Person 402 must have originally circled a *Disagree* for item Q13, and the reverse coded response entered in the spreadsheet was "4."

Activity #3

Take three blank copies of the STEBI. Cross out all self-efficacy items. When this is done, only the 10 outcome-expectancy items remain. Put the name "Ms. Confident" on one survey, and then answer the 10 outcome-expectancy items as if you were very confident. Put the name "Mr. Not So Confident" on another survey and then answer the 10 outcome-expectancy items as if you were not very confident. Then, put the name "Mr. Middle of the Road" on a third survey, and answer the 10 outcome-expectancy items as if you had some confidence in what students can accomplish, but you were not as positive as Ms. Confident nor not as unconfident as Mr. Not So Confident.

Answer: To fill out Ms. Confident and Mr. Not So Confident, you will have to first figure out which items are positively worded and which items are flipped. This will determine for each person if you are circling a *Strongly Agree* (SA) answer or a *Strongly Disagree* (SD) answer for an item for these two individuals. If you carefully read the survey items, you should be able to identify the positive items (Q1, Q4, Q7, Q9, Q11, Q14, Q15, Q16) and the negative items (Q10, Q13). This means that Ms. Confident will answer in the following manner to the outcome-expectancy items (Q1-SA, Q4-SA, Q7-SA, Q9-SA, Q10-SD, Q11-SA, Q13-SD, Q14-SA, Q15-SA, Q16-SA). Mr. Not So Confident will answer in the following manner to the outcome-expectancy items (Q1-SD, Q4-SD, Q7-SD, Q9-SD, Q10-SA, Q11-SD, Q13-SA, Q14-SD, Q15-SD, Q16-SD). Mr. Middle of the Road will have a mix of answers depending upon your prediction.

Activity #4

It is a helpful technique to temporarily insert fictitious people who are at the extreme part of the scale when you work with any type of data. In activity #3 you figured out two extreme respondents. For this activity, you should hand enter the data for the three respondents that you created for activity #3. Then if you are using Ministeps, you will need to remove 3 students (since the free Ministeps program has a limit of 75 respondents it can analyze). Then save the Excel file with a new name. After you have completed the activities in this chapter, you may want to go back and repeat many of the activities with your new Excel file, which includes your three fictitious people.

Answer: Following a Winsteps analysis, your two extreme people should be at the top and bottom of plots that we will talk about in later chapters. Again, get in the habit, with rating scales and tests, of entering sample people whose attitude (or performance) you know to be extreme. This allows you to double check the coding, your understanding of the answer key to a test, and the direction of a rating scale.

Activity #5

Take the outcome-expectancy Excel file supplied for this chapter and create a control file.

Answer: Below we present the "pasting" that you must carry out to create the Winsteps control file. Remember, you open Winsteps and then indicate that you will read data from an Excel file. Next, copy and paste particular lines under each of the two red lines (the red line for person label variables and the red line for item response variables). In our black and white text below, we underline and bold the red lines.

```
; Click on "Construct Winsteps file" when completed
! Item Response Variables. (Do not delete this line - item variables on left-side if this line before "Person Label Variables")
B ; Q1OE (4)
C ; Q4OE (4)
```

```
D ; Q7OE (4)
E ; Q9OE (4)
F ; Q10OE-RC (3)
G ; Q11OE (4)
H ; Q13OE-RC (4)
I ; Q14OE (4)
J ; Q15OE (4)
K ; Q16OE (4)
! Person Label Variables. (Do not delete this line - person
variables on left-side if before "Item Response Variables")
A ; Stud ID (401)
L ; Gender (F)
```

Following the steps to create a Winsteps control file, you should get the following control file. Below we provide the file as created by Winsteps, but we only include the first person in the data portion of the control file.

```
&INST
Title="ExcelDataOutcome-Expectancy(NegativeItems
AlreadyFlippedinDataSetDoNOTFlipTheseData).xls"
; Excel file created or last modified: 8/16/2011 9:00:30 AM
; Dash comma text naz oe
;     Excel Cases processed = 74
; Excel Variables processed = 12
ITEM1 = 1 ; Starting column of item responses
NI = 10 ; Number of items
NAME1 = 12 ; Starting column for person label in data record
NAMLEN = 6 ; Length of person label
XWIDE = 1 ; Matches the widest data value observed
; GROUPS = 0 ; Partial Credit model: in case items have
; different rating scales
CODES = 12345 ; matches the data
TOTALSCORE = Yes ; Include extreme responses in reported
; scores
; Person Label variables: columns in label: columns in line
@Stud-ID = 1E3 ; $C12W3
@Gender = 5E5 ; $C16W1
&END ; Item labels follow: columns in label
Q1OE ; Item 1 : 1-1
Q4OE ; Item 2 : 2-2
Q7OE ; Item 3 : 3-3
Q9OE ; Item 4 : 4-4
Q10OE-RC ; Item 5 : 5-5
Q11OE ; Item 6 : 6-6
Q13OE-RC ; Item 7 : 7-7
Q14OE ; Item 8 : 8-8
```

```
Q15OE ; Item 9 : 9-9
Q16OE ; Item 10 : 10-10
END NAMES
4444344444 401 F
```

Activity #6

Take the control file that you have created for activity #5 and create short item descriptors for each of the ten outcome-expectancy items.

Answer: You can abbreviate the items however you wish. The important thing to remember is that when you write the words for a negative item, you must enter an abbreviation for the item as if it had not been presented as a negative item.

Activity #7

Following your authoring of new descriptors and insertion of those new descriptors into your control file, run the control file.

Answer: If you created the control file with Winsteps and carefully inserted your descriptors so that you have a single line descriptor for each item, the program will run.

Activity #8

Find a survey of your choice. Print ten copies of the survey and enter answers on paper copies for ten different people. Play the role of different "types" of students (that means do not answer haphazardly). Carefully enter the data into a spreadsheet. Create a control file and run the data with Winsteps. Make sure to reverse code items that need to be flipped. Also make sure to reword the item descriptors as needed. Tip: When we have a new data set (entered or not entered into a spreadsheet), sometimes we take a small amount of the data to create the control file. Then we can make sure we are reading the data correctly. Once one is sure that the data are entered correctly and one understands the small data set, it is simple to repeat the procedure for a larger data set.

Activity #9

We have completed this chapter using an Excel data set. If you use a statistical package such as SPSS, open the desired statistical package, read in the Excel data for outcome-expectancy, and then save the spreadsheet in your statistical package. Create a Winsteps control file using the new SPSS file or other desired statistical package.

Answer: The same steps are taken to create the control file using a data file other than Excel.

Activity #10

Below we provide a fictitious health survey in which respondents are asked to indicate how often they participate in the following activities. Identify which items need to be reverse coded.

1. I use tobacco products.
Very Often Often Sometimes Seldom Never

2. I exercise at least 60 min per day.
Very Often Often Sometimes Seldom Never

3. I eat high-fat fast food.
Very Often Often Sometimes Seldom Never

4. I eat salads.
Very Often Often Sometimes Seldom Never

5. I watch television more than 30 min a day.
Very Often Often Sometimes Seldom Never

6. I do not take medicines "as prescribed" by my doctor.
Very Often Often Sometimes Seldom Never

7. I meet with friends.
Very Often Often Sometimes Seldom Never

Answer: To help students understand what items need to be flipped in a survey, we ask students to answer the survey in an extreme way. In this case it could be someone with an extremely healthy lifestyle (let's call her Frau Gesund). In that case the answer would be Never, Very Often, Never, Very Often, Never, Never, Very Often. This suggests that items Q1, Q3, Q5, and Q6 need to be recoded.

Activity #11

Think of techniques that you could use to spot items that need to be flipped and techniques that could be used to see if you have accurately flipped items that needed to be flipped.

Answer: In this chapter (text and activities), we have suggested some techniques that would allow you to spot items to flip and also techniques that would allow you to check if any mistakes have been made. A technique that will become clearer in a few chapters is looking at the "measures" of the survey items from "easiest to agree with" to "hardest to agree with." If you compare your predictions to the results of the analysis of data, it is easy to spot an item that has not been flipped (when it should have been flipped) or to spot an item that has been flipped (when it should not have been flipped). Those items will be "out of place" when you compare your predictions to the result of the analysis.

Activity #12

It is easy to create a control file and conduct a Rasch analysis of survey data (and compute person measures and item measures). Can you think of situations when you would not create a control file for survey data and compute a Rasch person measure and a Rasch item measure?

Answer: A set of test items or survey items can be a very useful way of determining what a student knows or a respondent views. However, using a set of survey items (or test items) together is only useful when a single trait is being measured with test or survey items. If you are not able to make the case, from a theoretical perspective, that the items involve different aspects of a single trait, you should not conduct a Rasch analysis. Also as readers will learn (and appreciate why), the data must "fit" the Rasch model.

References

Enochs, L. G., & Riggs, I. M. (1990). Further development of an elementary science teaching efficacy belief instrument: A pre-service elementary scale. *School Science and Mathematics, 90*(8), 694–706. doi:10.1111/j.1949-8594.1990.tb12048.x.

Linacre, J. M. (2012). Winsteps (Version 3.74) [Software]. Available from http://www.winsteps.com/index.html

Additional Readings

A technical report provided by a US state that uses Rasch and Winsteps for the analysis of thousands of students who are tested annually.

Ohio Achievement Test Technical Report (2008, May) Administration, Ohio Department of Education, Columbus, OH and American Institute for Research. Retrieved April 22, 2013, from http://www.ohiodocs.org/Technical%20Docs/TR%202008%20-%2015%20OAT%20Technical%20Report%20SP08.pdf

An article that presents some of the pitfalls of not correctly recoding negatively worded items.

Hughes, G. D. (2009). The impact of incorrect responses to reverse-coded survey items. *Research in the Schools, 16*(2), 76–88.

An article that discusses some techniques of authoring item descriptions in Winsteps control files.

Boone, W. (1991). Naming elements for understanding. *Rasch Measurement Transactions*, *5*(1), 130. http://www.rasch.org/rmt/rmt51d.htm

An article that provides examples of numerous US states which use Rasch analysis for the evaluation of high-stakes tests.

Boone, W. J. (2009). Explaining Rasch measurement in different ways. *Rasch Measurement Transactions, 23*(1), 1198.

The website and user manual for Winsteps provides introductory as well as more advanced guidance for those interested in using Rasch measurement.

Linacre, J. M. (2012). A user's guide to Winsteps Ministeps Rasch-model computer programs [version 3.74.0]. Retrieved from http://www.winsteps.com/winsteps.htm

Chapter 4
Understanding Person Measures

Isabelle and Ted: Two Colleagues Conversing

Ted: *Isabelle, wait a second; I am a little confused. I usually administer a survey and compute the total raw score of each person. Then, I use the total scores for many of the parametric statistical tests that I want to perform.*

Isabelle: *Please tell me a little more; I'm not sure what you are saying.*

Ted: *Okay, here is what I mean. Let's pretend I have a 13-item survey, and each item can be answered using one of six rating scale steps (categories). This means that a person could have a low score of 13 if he or she selected the lowest rating scale step, which is coded with a "1," for all 13 items. The highest score a person could get would be a 78. This would be the case in which a person selected a "6," the highest step of the 6-step scale for all 13 items. Usually, I put those total scores in a spreadsheet and then conduct statistical tests. What do I do if I am doing a Rasch analysis?*

Isabelle: *When you conduct a Rasch analysis, you will get a number, usually called a measure, for each person. And, that number is what you must use for subsequent statistical analyses. The numbers might look a little odd; sometimes they range from a negative number to a positive number, but these are the numbers you want to use.*

Ted: *So, if I have this right, those person measures I get from a Rasch analysis take the place of the total raw score I have normally used?*

Isabelle: *Correct. And, most important, those numbers are linear/equal-interval numbers. You need to remember you should not use the raw data.*

Ted: *That makes sense. One more thing, sometimes I read that some researchers have used Rasch, but when I look at their person measures, the numbers do not look like logits; by this I mean I see that the person measures might vary from 0 to 1,000. Can you tell me what is going on?*

W.J. Boone et al., *Rasch Analysis in the Human Sciences*,
DOI 10.1007/978-94-007-6857-4_4,

Isabelle: Easy! In Rasch we always compute Rasch person measures. These are the values we must use for statistical analyses. However, often people will rescale, which means they will just make a linear transformation of the measures. This is how one might only report measures that are positive values. This is just like converting from Celsius to Fahrenheit. To make this change, we will use a line in our control file that starts with the word UMEAN, and we will have a second line in our control file that starts with the word USCALE.

Introduction

Using Winsteps, we have created a control file and fine-tuned it by adding improved item identifiers. The control file is now ready for the initial computation of Rasch person measures, which will be used for many types of statistical analyses frequently reported in the science education literature, the broad education research literature, and in fields beyond education such as medicine. There are many important steps that can, and should, be used to evaluate the function of the measurement instrument and the quality of the data. However, this chapter will specifically focus on how to compute the person measure values and put those values in a form that is useful (and helpful) for parametric statistical analysis.

Preparing to use Rasch person measures, we remind readers that Rasch measurement, in part, provides person measures that *can* be used for parametric statistical tests because they avoid problems associated with the nonlinearity of rating scales as well as the nonlinearity of raw test data. The numbers reported for person measures might seem odd to researchers because such numbers will most often range from a negative number to a positive number. As we will demonstrate, however, these numbers should not be confusing, and we can easily express them using a linear scale that is all positive and causes less confusion. When we are teaching our classes, we point out to students that it is not uncommon to have temperature data in which there are both positive and negative values. We emphasize to our students that one could take the negative and positive temperature data and successfully conduct an analysis. The same is true if one uses the initial person measures that result from a Rasch analysis of data.

The units of Rasch measurement are named "logits" (Linacre & Wright, 1989), and these units are used to express both person measures and item measures. The important point for readers to note at this point in your learning is that logits express where an item is on the single variable being measured, AND logit units also express where each person is located on that same variable. Because persons and items have the same unit, and because logits are equal-interval units, not only can persons be compared to other persons (Charlie has a more positive attitude than John) and items can be compared to other items (item 6 was easier to agree with than item 12) but also items and persons can be compared (Charlie has a high likelihood of agreeing in some manner with item 6; Charlie has a high likelihood of disagreeing with item 12 in some manner).

Formative Assessment Checkpoint #1

Question: Why is it that when I run a Rasch analysis, I will have some persons (e.g., Billy is −1.32 logits, Tommy is −.32 logits) or items (e.g., Q2 is −3.56 logits, Q13 is −2.45 logits) with negative measures?

Answer: When Winsteps is used to run a Rasch analysis of data, all persons and all items are expressed on the same linear scale. When the analysis is run, the mean item measure is set to 0.00 logits. This means that one will have items with a positive measure and a negative measure. How persons respond to items determines how many persons have a negative person measure and how many persons have a positive person measure.

Running the Data

A very common situation is one where a researcher has created an initial control file, made minor revisions, saved the control file, and then set it aside before running the Winsteps program.[1] The following steps allow the researcher to access the control file and run it again.

1. Open Ministeps by clicking on the program icon. The introductory gray box will appear with the statement "Welcome to Ministeps! Would you like help setting up your analysis?"

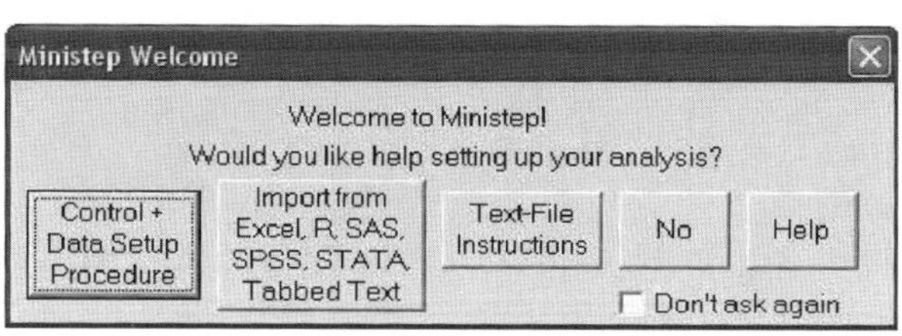

2. Since the control file has already been developed, click "**No**." The following line will appear on the screen: "Control file name? (e.g., exam1.txt). Press Enter for Dialog Box:" At this point press the Enter key on your keyboard.

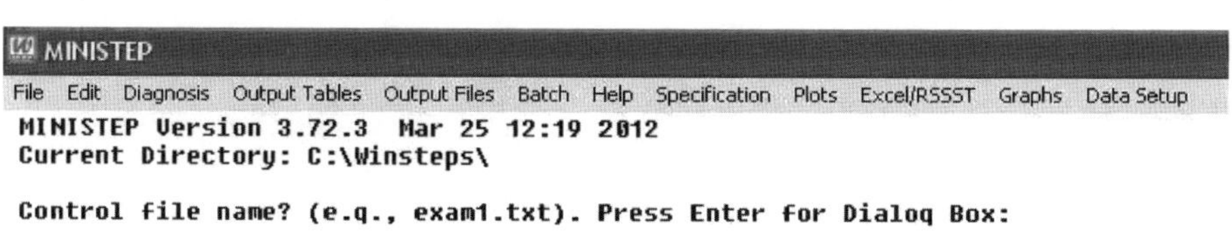

[1] n.b. Winsteps and Ministeps are the identical program, the only difference is since Ministeps is free there is a limit to the number of items and persons which can be evaluated. This means almost all that one sees on a screen will be identical for Ministeps and Winsteps.

3. To select the control file from its saved location, go to "**File | Open File**" and find and select the proper control file. After you have completed this step, the following line will appear: "Report output file name (or press Enter for temporary file, Ctrl+O for Dialog Box):"

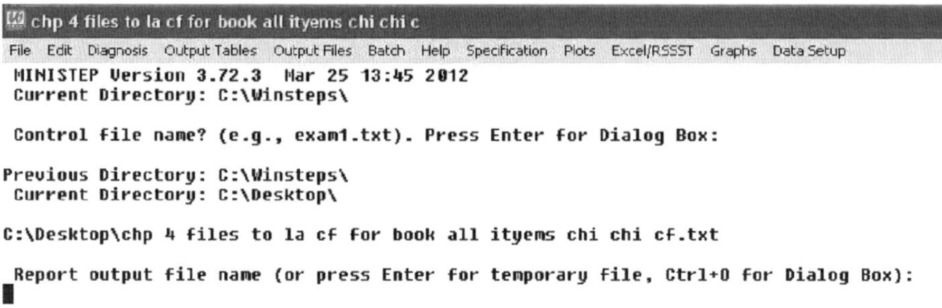

4. Press the "**Enter**" key on the keyboard. (For this example, only a temporary file is being created; therefore, no information needs to be entered). Then a new line will appear on the screen: "Extra specification (if any). Press Enter to analyze:"

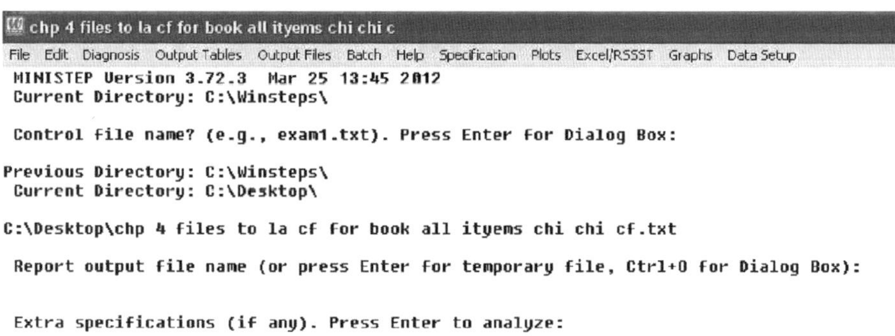

5. Press the "**Enter**" key on the keyboard again. For this example, there are no extra specifications to be entered. Now the program will run.

Formative Assessment Checkpoint #2

Question: Once you have constructed a control file, is it easy to run an initial Rasch analysis?

Answer: There are many nuances to conducting a Rasch analysis and interpreting its results. However, a number of menu-driven clicks will enable you to run the program and conduct an initial analysis of data.

Understanding the Output Tables: Person Measures

Following a successful run of the program on the data set, a gray bar will appear at the top of the screen.

Select the option "**Output Tables**," then all the possible output tables that can be selected will be presented. First, select "**Table 20**," the Score Table which appears immediately below as Fig. 4.1.

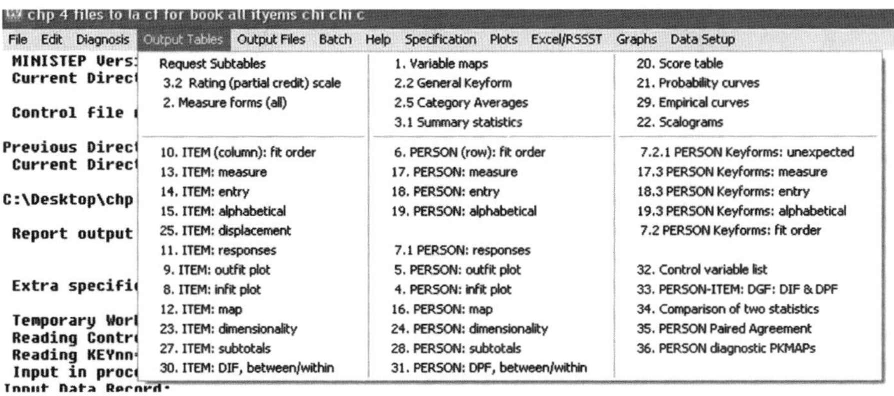

This is the table for the analysis of data in which there are 13 self-efficacy items and a 6-category rating scale. Remember, this is a table that was created after appropriate items were reverse coded.

The "Table of Measures on Test of 13 Item" provides the Rasch person measure for each and every possible raw score total on the 13-item self-efficacy scale, starting with the lowest raw score. Notice that the lowest raw score (top of left column SCORE) is 13, which corresponds with all 13 items being rated with the lowest rating category (labeled with a "1"). This would yield a score of 13 by multiplying 13×1. Also, note that the highest possible raw score is 78, which corresponds with all 13 items receiving the highest rating category, labeled "6." To check, multiply 6 by 13 to obtain the maximum raw score a person could reach on the survey ($6 \times 13 = 78$).

```
TABLE 20.1 SCIENCE TEACHER EFFICACY BELIEFS ZOU325WS.TXT  Jan 23 12:11 2012
INPUT: 75 PERSON  13 ITEM  REPORTED: 75 PERSON  13 ITEM  6 CATS     WINSTEPS 3.73
--------------------------------------------------------------------------------

                     TABLE OF MEASURES ON TEST OF 13 ITEM
--------------------------------------------------------------------------------
| SCORE  MEASURE    S.E. | SCORE  MEASURE    S.E. | SCORE  MEASURE    S.E. |
|------------------------+------------------------+------------------------|
|   13    -7.28E    1.88 |   35    -1.05     .33  |   57     1.00      .33 |
|   14    -5.93     1.09 |   36     -.94     .33  |   58     1.11      .33 |
|   15    -5.06      .81 |   37     -.84     .32  |   59     1.22      .34 |
|   16    -4.51      .69 |   38     -.74     .32  |   60     1.34      .35 |
|   17    -4.09      .60 |   39     -.64     .31  |   61     1.47      .36 |
|   18    -3.77      .55 |   40     -.54     .31  |   62     1.61      .38 |
|   19    -3.49      .50 |   41     -.45     .30  |   63     1.76      .39 |
|   20    -3.25      .47 |   42     -.36     .30  |   64     1.92      .41 |
|   21    -3.04      .45 |   43     -.27     .30  |   65     2.09      .43 |
|   22    -2.85      .43 |   44     -.18     .30  |   66     2.28      .45 |
|   23    -2.68      .41 |   45     -.09     .29  |   67     2.49      .47 |
|   24    -2.51      .40 |   46     -.01     .29  |   68     2.72      .48 |
|   25    -2.35      .39 |   47      .08     .29  |   69     2.96      .50 |
|   26    -2.20      .39 |   48      .17     .29  |   70     3.22      .52 |
|   27    -2.05      .38 |   49      .25     .29  |   71     3.50      .54 |
|   28    -1.91      .37 |   50      .34     .30  |   72     3.79      .56 |
|   29    -1.78      .37 |   51      .43     .30  |   73     4.12      .58 |
|   30    -1.64      .36 |   52      .52     .30  |   74     4.47      .62 |
|   31    -1.52      .35 |   53      .61     .30  |   75     4.89      .68 |
|   32    -1.39      .35 |   54      .70     .31  |   76     5.43      .79 |
|   33    -1.27      .34 |   55      .80     .31  |   77     6.24     1.06 |
|   34    -1.16      .34 |   56      .90     .32  |   78     7.53E    1.86 |
--------------------------------------------------------------------------------
CURRENT VALUES, UMEAN=.0000 USCALE=1.0000
TO SET MEASURE RANGE AS 0-100, UMEAN=49.1339 USCALE=6.7515
TO SET MEASURE RANGE TO MATCH RAW SCORE RANGE, UMEAN=44.9371 USCALE=4.3885
Predicting Score from Measure: Score = Measure * 6.4964 + 32.4868
Predicting Measure from Score: Measure = Score * .1436 + -4.6647
```

Fig. 4.1 (Winsteps Table 20.1): The Score Table which is created for each potential score earned by someone answering the 13 self-efficacy items using a 6-category rating scale (1, 2, 3, 4, 5, 6). All possible raw score totals are presented as well as the linear logit measures for each possible raw score. If a respondent answered in a way that expressed the highest self-efficacy to each of the 13 rating scale items, she or he would be noted as having 78 raw score points (13×6=78) and a measure of 7.53 logits. In your analysis you may not see all possible measures. For example you may have no students who had a raw score of 56, and thus you will not see any students with a measure of .90 logits

---

Formative Assessment Checkpoint #3

Question: A student answers all 13 items of the STEBI using a 6-step scale in which SD is a 1, D is a 2, BA is a 3, BD is a 4, D is a 5, and SD is a 6. The student's raw score (after correcting for items that need to be flipped) is 18, but the logit measure is −3.77. Is something wrong with the analysis when one has a negative value computed for a person measure? The raw score can only be positive, so is it the case that the person measures in logits must be positive?

Answer: No. When you run a Rasch analysis, the program is set to compute an average (mean) item measure of 0.00 logits. Since persons and items are expressed on the same scale, the location of the average item on the logit scale will, in part, determine the range of logit values computed for respondents. Also, you must not assume that a student who has a negative logit person measure is someone who was very disagreeable to a survey such as the STEBI. To understand what the student's person measure represents, one can get a ballpark feel for the student's attitude by looking at the measure-score table and investigating what a typical raw score was for a particular measure. Negative person measures do not mean that you have made a mistake in your analysis. Also negative person measures cannot be assumed to be respondents who are disagreeable (or in the case of a test did not do well on the test).

Now look at the column titled "MEASURE"; this column contains the person measures. The values for all possible person measures vary from a low of –7.28 to a high of 7.53 logits. Two obvious questions are: How can there be negative values for the person measures, and why do the values vary (in this case) from approximately –7 to 7? When the Rasch measurement model is used in any Rasch software, linear/equal-interval person measures are computed. These values will be expressed in "logits," which is short for "log odds units," the unit of measurement in Rasch measurement. Negative values occur because the mean item measure has a default logit value of 0.0. Therefore, a negative value simply implies that the respondent has a person measure less than the mean item difficulty. Although a negative value for a person measure may cause confusion for some analysts, the most important point to remember is that all of the person measures can be expressed on a scale with only positive numbers. Such person measures are more palatable for some analysts as well as non-analysts who might be presented with the measures. We will now explain how to transform the logit scale to include only positive person measures. But you can conduct your analysis with person measures that are both positive and negative and conduct accurate statistical analyses.

Formative Assessment Checkpoint #4

Question: Can negative person measures computed from a Rasch analysis be immediately used for statistical analysis of data?

Answer: The negative (and positive) person measures that result from a Rasch analysis *can* be used for the statistical analysis of data. These values are linear measures and are thus the types of numbers (scales) that are fair game for parametric tests. The use of negative numbers has no bearing on the conclusions that a researcher would reach with regard to an analysis of data.

Transforming (Rescaling) Person Measures

The Rasch Winsteps control file can be edited so that a new scaling can be used to express person measures (think of this as writing a computer program so that Fahrenheit temperature data are immediately converted to the Celsius scale before you do any work with the temperature data set). However, it can be very informative

SCORE	MEASURE	MEASURE+ 10 logits	SCORE	MEASURE	MEASURE + 10 logits	SCORE	MEASURE	MEASURE+ 10 logits
13	-7.28E	2.72E	35	-1.05	8.95	57	1.00	11.00
14	-5.93	4.07	36	-.94	9.06	58	1.11	11.11
15	-5.06	4.94	37	-.84	9.16	59	1.22	11.22
16	-4.51	5.49	38	-.74	9.26	60	1.34	11.34
17	-4.09	5.91	39	-.64	9.36	61	1.47	11.47
18	-3.77	6.23	40	-.54	9.46	62	1.61	11.61
19	-3.49	6.51	41	-.45	9.55	63	1.76	11.76
20	-3.25	6.75	42	-.36	9.64	64	1.92	11.92
21	-3.04	6.96	43	-.27	9.73	65	2.09	12.09
22	-2.85	7.15	44	-.18	9.82	66	2.28	12.28
23	-2.68	7.32	45	-.09	9.91	67	2.49	12.49
24	-2.51	7.49	46	-.01	9.99	68	2.72	12.72
25	-2.35	7.65	47	.08	10.08	69	2.96	12.96
26	-2.20	7.8	48	.17	10.17	70	3.22	13.22
27	-2.05	7.95	49	.25	10.25	71	3.50	13.50
28	-1.91	8.09	50	.34	10.34	72	3.79	13.79
29	-1.78	8.22	51	.43	10.43	73	4.12	14.12
30	-1.64	8.36	52	.52	10.52	74	4.47	14.47
31	-1.52	8.48	53	.61	10.61	75	4.89	14.89
32	-1.39	8.61	54	.70	10.70	76	5.43	15.43
33	-1.27	8.73	55	.80	10.80	77	6.24	16.24
34	-1.16	8.84	56	.90	10.90	78	7.53E	17.53

Fig. 4.2 (Winsteps Table 20.1): An edited version of Table 20.1. This table presents all possible raw scores and all possible person measures for the analysis of a 75-person data set which was created through the collection of data using the 13-item STEBI self-efficacy scale. An additional column has been added which reflects the addition of 10 logits to each logit measure presented in Table 20.1

to calculate the conversion, which is a linear transformation, by hand at least once, in order to better understand the automatic conversion conducted by Winsteps. Let's begin by adding the number 10 to each logit person measure in Fig. 4.1. The person measure values now range from 2.72 to 17.53 instead of –7.28 to 7.53. The original person measures and converted person measures are displayed in Fig. 4.2. The original Fig. 4.1 (Winsteps Table 20.1) is presented, but with an extra column, MEASURE +10. This extra column contains the converted or transformed person measures, where 10 was added to all of the original person measures (logits). This table was created by the authors by manually adding the number ten to the original values.

This conversion, also called a linear transformation, does not alter the scale distribution. Using the scale ranging from –7.28 to 7.53 or the scale ranging from 2.72 to 17.53 will produce identical statistical results. To illustrate this issue, we direct readers' attention to Fig. 4.3. Figure 4.3 presents the descriptive statistics of the original logits for a single group of 75 students who completed the STEBI. We also present the mean and standard deviation of the person measures after the value 10 has been added for the same 75 students. Readers will note that the standard deviation is unchanged, and the means differ exactly by 10.

Fig. 4.3 Means and standard deviations of the same group of 75 students before and after a linear transformation of the data

Descriptive Statistics			
	N	Mean	Std. Deviation
Original Logits	75	1.045	1.295
Logits plus 10	75	11.045	1.295

Hopefully, this brief example helps show that a scale ranging from –7 to 7 is uncommon but not all that odd. Moreover, we do have everyday experiences with negative numbers on a scale, for instance, in degrees Celsius or degrees Fahrenheit.

Formative Assessment Checkpoint #5

Question: One has conducted a Winsteps analysis of data and conducted a *t*-test comparing male and female person measures. The *p* value is 0.03, what will be the *p* value when a comparison of males and females is made with rescaled male and female measures?

Answer: The *p* value will be the same. Rescaling the original measures using the techniques we present still means that accurate person measures are being used for analyses, and, as a result, the same parametric statistical tests will produce the same results.

Above, we demonstrated a conversion from a scale with negative and positive person measures to a scale with all positive person measures. However, another type of conversion is very useful that conversion presents values ranging from 0 to 100 or from 0 to 1,000. In most people's experience, there is an understanding that a higher value is somehow better. Because so many tests in schools might be based upon a maximum performance of 100 possible points, many stakeholders seem comfortable interpreting results when data are presented on such a scale.

When possible, we find another alternative to be even better. That is to present data using a scale from 0 to 1,000. As is the case with the 0 to 100 scale, there is an understanding that a higher value is better. The advantage of using a 0 to 1,000 scale, we have found, is that it does not create the impression that 100 items were administered. And by avoiding numbers from 0 to 100, we are reminding stakeholders not to make assumptions based upon percentiles and/or raw scores.

There is an easy way to convert the initial logit values of any Rasch analysis to a range from 0 to 100 or 0 to 1,000. In Fig. 4.4, we have underlined the text that allows us to change the logit scale to an alternative logit-based scale. The underlined lines

```
&INST                  ; shows this is a control file (optional)
TITLE = 'SCIENCE TEACHER EFFICACY BELIEFS' ; Report title: data from paper version
STEBI
NAME1 = 1              ; First column of person label
NAMELENGTH = 10        ; Length of person label
ITEM1 = 11             ; First column of responses in data file
NI = 23                ; Number of items
CODES = "123456"       ; Valid response codes in the data file
```

```
&INST                  ; shows this is a control file (optional)
TITLE = 'SCIENCE TEACHER EFFICACY BELIEFS' ; Report title: data from paper version
STEBI
NAME1 = 1              ; First column of person label
NAMELENGTH = 10        ; Length of person label
ITEM1 = 11             ; First column of responses in data file
NI = 23                ; Number of items
UMEAN=49.134
USCALE=6.751
CODES = "123456"       ; Valid response codes in the data file
```

Fig. 4.4 The first 7 lines of the original control file (used to create Winsteps Table 3.1) and the first 9 lines of the edited control file which rescales person measures to a scale from 0 to 100. The line for UMEAN and the line for USCALE can be placed anywhere after the first line of the control file, but before the line "& END." When you add these two lines to the control file, the lines will not be underlined; we have just done so to highlight the lines. In Fig. 4.1, you can find the values for UMEAN and USCALE. When you look at Winsteps Table 20, you will want to use UMEAN and USCALE from the line that has the phrase "0-100". Please note that when you run other analyses (say a different survey), you will need to use a different UMEAN and USCALE which you can look up

are values for two control variables that can be added to the control file. By adding these control variables to the control file, all person measures can be effortlessly converted to a new range that remains linear/equal interval. In Fig. 4.4 we present the first few lines of our original control file, and we present an edited control file with two lines added. One new line begins with the phrase "UMEAN," and the other new line begins with the phrase "USCALE." When we add these two lines (and the information following the insertion of those two words), the program automatically converts all person measures to a scale ranging from 0 to 100. Note carefully that the numbers that follow UMEAN and USCALE are provided in Fig. 4.1. Finally, it is important to comment that a researcher would first conduct a Winsteps Rasch analysis using the control file constructed through Winsteps, which we described in Chap. 3. Then the researcher would look at Winsteps Table 20 of that initial analysis and note the values of UMEAN and USCALE to make a conversion from the initial logit values to a scale ranging from 0 to 100. Where do the values of UMEAN and USCALE come from? We feel the best way to understand these numbers is to first think of Winsteps computing the person and item measures using the logit scale with both negative and positive numbers. Then understand that Winsteps computes what the conversion would be for a rescaling that maintains the exact linear nature of the data. Imagine that you had collected data concerning the distances 100 cars traveled in a year, but you collected the data using the metric of miles. Then you

TABLE OF MEASURES ON COMPLETE TEST

SCORE	MEASURE	SCORE	MEASURE	SCORE	MEASURE
13	.00E	35	42.06	57	55.90
14	9.07	36	42.79	58	56.63
15	14.95	37	43.49	59	57.40
16	18.70	38	44.17	60	58.21
17	21.49	39	44.83	61	59.08
18	23.71	40	45.48	62	60.00
19	25.57	41	46.11	63	61.00
20	27.16	42	46.72	64	62.08
21	28.58	43	47.33	65	63.26
22	29.88	44	47.93	66	64.54
23	31.07	45	48.51	67	65.94
24	32.20	46	49.10	68	67.46
25	33.27	47	49.68	69	69.10
26	34.29	48	50.26	70	70.86
27	35.27	49	50.84	71	72.73
28	36.22	50	51.43	72	74.74
29	37.14	51	52.03	73	76.91
30	38.03	52	52.63	74	79.34
31	38.89	53	53.25	75	82.17
32	39.73	54	53.88	76	85.77
33	40.53	55	54.53	77	91.29
34	41.31	56	55.20	78	100.00E

Fig. 4.5 (Winsteps Table 20): The Score Table created when the analysis was conducted through the use of UMEAN and USCALE control lines in the control file. Use of the exact values results in a rescaling for the STEBI that has a lowest measure of 0 and a maximum measure of 100

needed to express those distances in kilometers. Remember, when you collected the distance data you started your work with a linear metric, so you did not have to confront the rubber ruler that Rasch measurement helps us confront! Then to convert to kilometers, you simply need a transformation.

Formative Assessment Checkpoint #6

Question: Where exactly do you find the values for UMEAN and USCALE that allow you to correctly rescale so that you do not have negative person measures?

Answer: The Winsteps Table 20 provides a variety of data for a number of purposes. The first part of this table provides all possible raw scores for respondents completing the instrument, and all possible measures are reported. Immediately below that table are a number of horizontal lines of information, one of which presents the correct UMEAN and USCALE values if one were to rescale the logits to an equal-interval scale that begins with a minimum of 0 and proceeds to a maximum measure of 100 (Fig. 4.5).

Finally, how does one use the UMEAN and USCLALE values to create scales from 0 to 1,000 or other scales? To convert the original scale (−7.28 to 7.53) to a scale of 0 to 1,000, one needs only to move the decimal point for USCALE and UMEAN by one column to the right. Thus, use of UMEAN = 491.34 and USCALE = 67.51 will provide logit measures on a scale of 0–1,000. If one wanted to use a scale ranging from 200 to 1,200, then a value of UMEAN = 691.34 and USCALE = 67.51 would be used, adding 200 to *only* the UMEAN value. Thus, to change the 0–1,000 scale by a set amount (e.g., to a scale from 200 to 1,200 or to change the scale from 300 to 1,300), one only adds a value to the UMEAN number and one keeps USCALE "as is." There are other conversions with UMEAN and USCALE that are used in some research, but in the vast majority of research we have been involved with, the conversions that we describe above should be useful for readers.

Following the editing of a control file to rescale the linear measures of items and persons, a researcher can cut and paste the data (the Rasch person measures) into SPSS [and other spreadsheets]. After the running of Winsteps, the gray bar at the top of the screen provides a number of options. As previously shown, the option "Output Tables" provides many key Rasch tables. Another option, "Output Files," is immediately to the right of "Output Tables." Clicking the "Output Files" displays a long list of options for saving particular output tables to separate files. Selecting the button "Person File PFILE=" requests a file that contains the Rasch person measures, which can be used for a parametric statistical analysis. Numerous statistical program file types can be selected to store the person measure data. Moreover, an analyst can create either a temporary or permanent file. For this example, we want to create an output file in Excel with only the first 5 columns (Entry Number, Measures, Status, Count of Observations, and Raw Score) and the last column (Name or Label) of the possible output. To select specific fields, click on "Select fields+other options." Figure 4.6 presents the screens which show how one creates such a spreadsheet. Figure 4.7 presents the Excel file that results from the creation of a spreadsheet that contains the six columns.

Formative Assessment Checkpoint #7

Question: It is commonplace to rescale test results; however, one rarely sees rescaled science attitude survey results. Is this because attitude survey results cannot be rescaled; only test results can be rescaled?

Answer: No. The same type of techniques that are used to rescale logits to a different but mathematically equivalent scale can be used for survey data.

To conduct a parametric statistical analysis one must use the data in the MEASURE column, not the SCORE column, because the measures are the linear/equal-interval values and the results of the Rasch analysis. Scores are the raw scores. If a researcher wanted to extract only the MEASURE column, he or she would

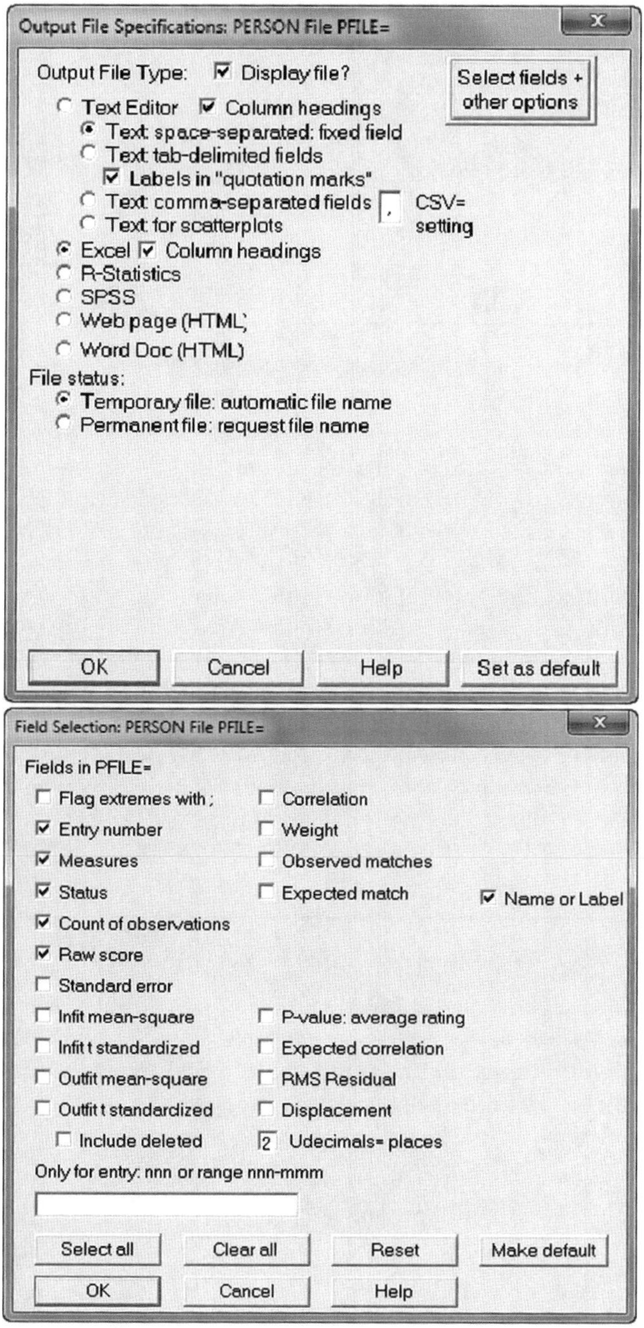

Fig. 4.6 The options that are provided when selecting a file (such as Excel) that will contain person measures computed following a Winsteps analysis

	A	B	C	D	E	F
1	ENTRY	MEASURE	STATUS	COUNT	SCORE	NAME
2	1	58.2	1	13	60	21141PR
3	2	54.78	1	6	27	91052PR
4	3	71.31	1	6	33	95793PR
5	4	49.68	1	13	47	08453PR
6	5	52.02	1	13	51	36281PR
7	6	62.07	1	13	64	85453PR
8	7	46.32	1	6	21	46328PR

Fig. 4.7 A component of the Excel spreadsheet that resulted from the use of the procedures discussed in the text

select only that field for the output table in the "Field Selection: PERSON File PFILE=" window. A tip to researchers, before you conduct a statistical analysis using person measures, scan through the measures for each person. Make sure that if someone did not answer any items (probably a person who did not take the survey/test) that a measure value is not reported for that person. If you see a value of "0" in the count column, that means a person did not take the survey/test. Please note we said a value of 0 in the COUNT column. If you see a value of 0 in the measure column, depending on the scaling, that person may indeed have a measure and therefore would have answered survey or test items. Another tip is when you are just starting to conduct a Rasch analysis, you might want to select all the output columns we have selected above (and more), but then at a later time, you will just cut and paste the measure column into the spreadsheet (e.g., SPSS) that you use for your statistical analysis.

Formative Assessment Checkpoint #8

Question: You have conducted a Rasch analysis of survey data from 500 students. You can see in your Winsteps person entry table that you have indeed computed a person measure for each respondent. How can you quickly place the person measures in a spreadsheet for subsequent data analysis?

Answer: There is a gray bar at the top of the Winsteps analysis screen that includes the option "Output Files." By clicking on the "Output Files" button, selecting PFILE, and following the prompts you are presented, you can create a spreadsheet with the person measures in a few seconds. If you wish, these values can then be pasted into a spreadsheet that you may already have created.

We would like to make two final points: First, a comment about person measures. When a person "maxes out" (marks the highest answer to each item on a survey, assuming there are no items that must be recoded due to survey item wording), you will see that Winsteps computes a person measure, but you will also see text in the Winsteps tables which list respondent measures. For persons who "max out," you will see a comment ("MAXIMUM MEASURE"). From a measurement perspective, the important aspect of the comment is that we learn very little about such respondents. We know they may care (or know) a lot about the surveyed (or tested) concept, but we do not know the limits of their knowledge or views. Second, after the initial analysis is conducted, and perhaps the person measures and item measures are rescaled, one should not immediately decide that the data are ready for statistical analysis. As we will show in subsequent chapters herein, Rasch measurement provides great insight into the quality of the measures, and in many cases, there will be items that are best removed from an analysis and also respondents who might need to be removed prior to additional analysis. Generally we recommend doing a rescaling later in an analysis to make sure you understand the higher and lower measures for respondents and items.

Isabelle and Ted: Two Colleagues Conversing

Isabelle: Ted, tell me about UMEAN and USCALE; what is the point?

Ted: Okay, when we run our data without these two lines in the control file, we get person measures that typically range from numbers that are negative to positive numbers. These numbers could be used for a statistical analysis, but people are often confused by negative numbers. For example, I could compare the mean (1.04 logits) of all males completing the self-efficacy part of the STEBI to the mean (–0.17) of all females completing the self-efficacy of the STEBI.

Isabelle: I am not confused when I see it is –2 °C outside, I just grab my hat.

Ted: Well, not everybody is like you. So, what we do is similar to converting from Fahrenheit to Celsius or from Celsius to Fahrenheit. We can conduct a linear conversion of the data.

Isabelle: What are some good selections for conversions?

Ted: Well, I find reporting measures from 0 to 100, 0 to 1,000, or 200 to 1,200 seems to work. People seem to be more familiar with these first two ranges of measure values. The 200 to 1,200 is a range I use, for in this case the lowest value on a test or survey does not look as bad as a "0." One advantage of using the scale from 0 to 1,000 and 200 to 1,200 is that I remain organized. It would be hard to make the mistake of reading raw data and evaluating that data. That would be a waste because we want to use the Rasch measures.

Isabelle: What is this plot of yours?

Ted: Well, I did a little experiment. For a group of 75 students, I computed their self-efficacy measures. Next, I reran the analysis but set UMEAN and USCALE to give me the range of measures from 0 to 100. Then I did an analysis with UMEAN and USCALE set to give me person measures from 0 to 1,000. And then I rescaled from 200 to 1,200 and computed student measures using that scale. Then I plotted the persons against one another. For example, on one plot I plotted each person's measure using the original scale from –1 to 8 logits for the person measures against the person measures computed using the scale from 0 to 100. I did other plots, such as person measures on the 0–100 scale against person measures for 200–1,200. In every case I got a straight line when all the persons were plotted. That proved to me that when USCALE and UMEAN are used, then one is simply making a change from one system of measurement (meters) to another systems of measurement (centimeters). Here is a picture of one of the plots; this one shows the measures I computed from my initial Winsteps run and the analysis in which the scale is set to range from 200 to 1,200. So, each point is located as a function of its person measure on the initial logit scale and the person measure on the rescaled metric.

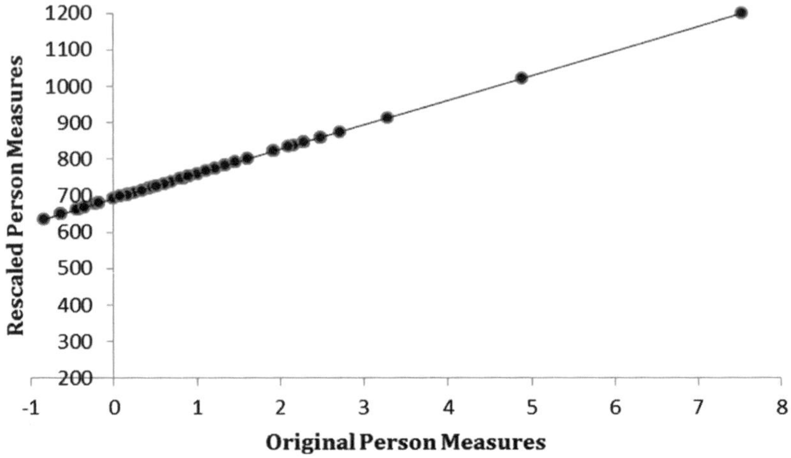

Isabelle: *That is very cool. I'll bet that you could also teach students about this by plotting some experimental temperature data in degrees Celsius and then plotting that same experimental temperature data in degrees Fahrenheit and then in degrees Kelvin. Also, you would be able to show that it does not matter what scale you use, as long as it is a linear scale and as long as you do not goof things up when you convert from one scale to another.*

Ted: *I feel as if I might have taught you something today*!

Keywords and Phrases

Person measures
Rescaling
Logit
Score
Measure
Linear conversion
Equal-interval
Count
UMEAN
USCALE
John Michael Linacre
Benjamin Wright

The Rasch person measures were rescaled to a scale that ranged from 0 to 1,000. These measures are expressed on an equal-interval scale. One can think of this conversion as converting temperature data from Celsius to Fahrenheit. Statistical analyses using the original logit values will result in the same conclusions as statistical analyses conducted with the rescaled person measures.

Potential Article Text

Data were collected from a sample of 75 students who were administered the STEBI of Enochs and Riggs (1990). Rasch Winsteps software (Linacre, 2012) was utilized to compute person measures. Initially, person measures were expressed using a linear logit scale that ranged from a low of −2.5 to a high of 3.0. Because the initial Winsteps analysis is simply expressed on a scale where the 0 value is the location of the mean item difficulty, it is possible and acceptable to rescale person measures. Person measure data for this project were rescaled from the original logit scale to a user-friendly, but still linear, scale ranging from 200 to 1,200.

Quick Tips

To rescale from logits to another linear scale, use USCALE and UMEAN. Winsteps provides the conversion from logits to another equal-interval Rasch scale of 0 to 100. This conversion is provided in the Score Table of Winsteps. By rescaling, you retain the Rasch measures of persons and items, but you have the ability to report results without the use of negative values for persons and items. This often enhances communication of results to stakeholders. We strongly encourage researchers to rescale but to also avoid a rescaling of 0–100. We have found that too often there is confusion with stakeholders that values on a 0 to 100 scale are raw scores or percentiles. We often use a low value of 200 and a top value of 1,200.

A person who has a negative person measure in the case of a test is not necessarily a person who performed very badly on a test. A person who has a negative person measure is not necessarily someone who (when answering a survey with rating categories of *Strongly Disagree*, *Disagree*, *Agree*, and *Strongly Agree*) selected *Disagree* for most survey items.

In algebra one might rescale an equation to better understand its meaning. Pretend the relationship between air temperature (degrees C) and meters above sea level is $T = .006 \times$ (meters above sea level) $+ .012$. One could write the equation as 1,000 $\mathrm{T} = 6 \times$ (meters above sea level) $+ 12$. Both formulas communicate the same information. You can rescale or not rescale. It is up to you!

The Winsteps table entitled “Score Table” provides the values of UMEAN and USCALE to convert your person measures from a lowest possible person measure of 0 to a highest possible person measure of 100. This information can also be used to very quickly create a scale that ranges from, for example, 1 to 1,000 or 200 to 1,200. Just find the line “TO SET MEASURE RANGE AS 0-100” in the “Score Table” and you will find what you need. Then type in the UMEAN and USCALE lines and values in your control file, or copy and paste from the Score Table. Please note one cannot just type any number after UMEAN and USCALE; one must use numbers that retain the linear scale that is presented in the initial analysis that presents the initial logit measures. To help the researcher, Winsteps computes the correct values to rescale from 0 to 100.

Using the option Output Files, one can quickly make a spreadsheet that contains all person measures and item measures. These measures can be pasted into several of other types of spreadsheets.

A quick way to show someone the nonlinear nature of your raw data is to find two raw scores in the middle range of the Score Table and compute the difference in the measures for those two scores. Then pick two raw scores that differ by the same raw score amount in a different part of the Score Table. We suggest picking two raw scores that are extreme. If you compute the difference in the measures for the two sets of items, you will see that the differences in measures are not the same.

You can run an analysis of data, for example, from 75 students who completed a rating scale instrument. You can convert those person measures to a scale from 200 to 1,200. Now you can conduct parametric statistical tests to learn about the respondents. However, if you are asked to compare the 75 students to a data set of 50 students you have been sent, you cannot simply run a Winsteps analysis of the 50 students, convert to a scale from 200 to 1,200, and then compare the values for the group of 75 students and 50 students. This is because you must take steps to ensure that the survey items defined the trait in the same manner for both groups. In later chapters, we will teach you how to "link" a scale so that you *can* compare measures of two different groups of respondents.

Data Sets: (go to http://extras.springer.com)

cf Chem Edc ETSU

Activities

Activity #1

It is very helpful to be able to find how the same people are presented in different output tables and plots. If Dave has a logit value of 2.02 on a test and Stephanie has a value of 3.03, these two individuals will be presented in a number of Winsteps tables. In each of these tables, they will be plotted and/or listed using their "measures." It is helpful for researchers to be able to quickly find the different locations of the same person in different Winsteps tables. Run an analysis of the control file cf Chem Edu ETSU. This is a nonrandom sample of students who completed a multiple-choice chemistry education research test kindly supplied by Chih-Che Tai of East Tennessee State University and Keith Sheppard of SUNY Stony Brook. First, print out a Wright Map (Table 12 of Winsteps) and then find any single person on the Wright Map. This will be an "X" on the left side of the table. After circling that person on the Wright Map, go to any of the Person Measure tables. Can you find the person whom you circled in the person measure table? What is that person's measure value? What raw score did he or she earn on the test? How many test items did he or she attempt? Later in our book, we present two chapters on Wright Maps.

Answer: To complete the exercise, circle the person of your choice and then look at the scale on the Wright Map (Table 12 of Winsteps). This scale expresses both the person measures and the item difficulties using the same metric. If you have not changed the control file to a "user-friendly scaling," then the person measures and item measures will be reported in logits. Usually you will see a scale that ranges from −3.0 to +3.0, but the exact range of the scale will depend in part upon the most extreme items and persons in your data set.

Once you find the approximate measure of the person you circled, go to the Person Measure table, look at the "MEASURE" column, and find the person with that measure. It is okay if several persons have approximately that measure. This has to do with how the Wright Map is printed. If we could stretch out the Wright Map to perhaps 2 m long, we would be able to find most people. Now examine the "score" column and find the person's raw score. Next, look at the "count" column and determine the number of items the person answered. Remember, all individuals do not need to take the same number of items, but all respondents can still be expressed on the same scale! [A multimatrix design such as that used by Dr. Hans Fischer (Physics Education, University of Duisburg-Essen, Germany) is possible because respondents do not need to take an identical set of items]. Instead of simply looking at the columns in the person measure table that say "SCORE" and "MEASURE" to identify how many items a person correctly answered, one could write down a respondent's measure and then look at the Score Table of Winsteps. That table provides the raw score for each possible person measure.

Activity #2

Part I

Conducting a Rasch analysis helps researchers build measurement instruments and provides linear measures of respondents that can and must be used for parametric statistical tests. In this exercise, we will practice conducting a statistical analysis of person measure data. Run a Rasch analysis of the data provided in the file cf Chem Edu ETSU. Use the output file option to create an SPSS or Excel spreadsheet with the person measures of the persons who are in the data set. The last number (a 1 or a 0) in each student's ID indicates gender; in this case, a 0 is a female and a 1 is a male. Using the person measures in your spreadsheet, use the statistical software of your choice to compute the mean person measure and the standard deviation of the person measures. Then compute the mean and standard deviation of the females and the mean and standard deviation of the males. Then conduct a *t*-test to see if there is a statistical difference between the distributions of men and women. Our book is not an introduction to statistical analysis techniques; thus, consult a statistic textbook for more information about standard deviation and *t*-tests.

Answer: The mean measure of all 75 respondents is 1.29 logits, and the standard deviation of the 75 respondent measures is .97 logits. The female (code "0") mean measure is 1.45 logits, and the SD of the female measures is .87 logits. The male (code "1") mean measure is 1.27 logits, and the SD of the male measures is .98 logits. If equal variances are assumed, you should compute a significance level of .522 when you compute your *t*-test.

Part II
You have computed a mean person measure for the entire group of respondents, and you have also computed the standard deviation of the person measures. Examine either the Winsteps "person measure" table or the "person entry" table. In either table you will find reported both the mean and the SD of the person measures. Can you find them?

Answer: A great amount of information is provided in each Winsteps table. For beginning researchers it is very helpful to learn where key information that will help in analysis is located. Look at the very bottom of the Winsteps "person measure" table, and you will see reported the mean and the standard deviation of the persons who were included in your analysis of the data.

Activity #3

One aspect of Rasch measurement that beginners often do not remember is that it is important to use some of the very basic descriptive statistics that are used in the analysis of education, psychology, and health science data to understand the data that have been evaluated with Rasch techniques. The important point for researchers to remember is that no matter the Rasch scale (e.g. from –4.0 to 4.0) for person measures and/or items, or a rescaled set of measures which might range from a low of 200 to a high of 1,200, the same statistical techniques one uses for data analysis in other fields can be used. Using the computations (mean measures of all respondents, mean measures of females, mean measures of males, standard deviation of all respondents' measures, standard deviation of all females' measures, standard deviation of all males' measures), plot the location of the means on your Wright Map from Activity #1. Then plot one standard deviation up and down for each mean measure. Remember, the sample size is small, so the standard deviation will be large!

Answer: Let's carefully plot the mean and standard deviation of all respondents. The procedures will be the same for any subgroup (e.g., mean female measure, standard deviation of female measures). The mean measure of all 75 respondents is 1.29 logits; the standard deviation of the 75 respondent measures is .97 logits. The left side of the Wright Map presents the person measures. Then find the scale that is presented on the Wright Map. If you are doing your plot by hand, take a colored pen and make a dot at the location of the mean person measure. Then carefully draw a line upward from the dot, which is .97 logits long. Then draw a line from the dot downward, which is .97 logits long. Our sample size is not huge, so it is not surprising that the SD is large. Now you will have an idea of how to graph the location of the mean person measures of a group of respondents and the SD of the person measures as well.

Activity #4

There are many ways for researchers to pull up the same data in a Winsteps Rasch analysis. Depending upon the type of analysis you are conducting and the stage of your analysis, you may use different techniques to find data you need to review. Using the chemistry education data set, print out a person entry table. Make sure to rotate the page, put it in courier font, and set your spacing to "1". Now also use the output file option of Winsteps and create a PFILE (a person file) in the spreadsheet of your choice. Print out that spreadsheet and verify that the information provided in the Winsteps table is indeed identical to that presented in the spreadsheet you created. Check to see if the data are organized in the same manner in the spreadsheet and the Winsteps table.

Answer: The same headings are used in the Winsteps table as well as the output files that you create, and those data are provided in the same order.

Activity #5

From time to time, it will be important to place person measures you compute into a spreadsheet that has additional data. For example, you may have 20 different types of demographic data in a spreadsheet for each student in a study. You need to compute Rasch person measures and then place those person measures as an additional column in your data set. Run the chemistry education data set and create a spreadsheet (using the files option in the gray menu bar). Then bring up a blank spreadsheet (pretend this spreadsheet is full of information about each respondent) and paste the person "name" column and the person "measure" column into this blank spreadsheet. Then save the spreadsheet.

Answer: Very often you will have to cut and paste person measures so that the data can be inserted into a spreadsheet. The reason why you must do this often is that all of your statistics and all of your graphing of how "persons" did on your test will make use of the Rasch measures. Remember, the raw data are nonlinear and thus violate requirements of parametric tests. Also remember, if you used multimatrix design, the two identical raw scores may not have the same meaning because of the difficulty of a set of items that a respondent completes. We have also found it useful to make sure to paste a person ID as well as a person measure into other spreadsheets. Often we will insert the column of person IDs from Winsteps and the person measures from Winsteps next to the column that includes IDs in a preexisting spreadsheet. Then we examine the IDs of the inserted Winsteps column and the IDs in a spreadsheet to make sure that the IDs match. It is easy to make little errors (e.g., shifting inserted data down by one row), which result in the wrong person measures being inserted into the wrong spot in the spreadsheet. So be careful and double check respondents' IDs and person measures.

Activity #6

Both Mike Linacre (author of Winsteps) and Ben Wright (one of the pioneers of Rasch Measurement) have always advocated plots. Plots summarize data very well, and the brain can process a plot quicker and easier than a column of numbers. Winsteps provides a very user-friendly "plot" option in which Excel can be effortlessly used to plot almost all data that are provided in almost all output tables. For this activity, use the chemistry education data set, run a Winsteps analysis, then use the Plots option (look at the gray bar at the top of the computer screen following an analysis), select the scatter plot option, and then plot the person measures (X-axis) against the person error (Y-axis). What do you see?

Answer: You should see that person measures farther from the mean have a higher measurement error than person measures nearer the mean. When you set up the plot, make sure to note that you can plot any data from items and persons. Plots such as the one you just made allow you to better understand your data, verify the quality of the measures you make in your research, and also explain what you did for talks and research articles.

Activity #7

A very important aspect of Rasch measurement is experimenting and thinking about what it means to measure. In Dr. Hans Fischer's physics research group in Essen, Germany, a common step is to collect data that will inform the development of a final instrument. This means that a large number of items will be authored, multiple instruments constructed using item links (multimatrix design), and an informed decision made as to which items will be used for a final data collection.

One aspect of empirically establishing which items make up a final instrument involves how items work together to measure a respondent. One technique of evaluating the impact of one set of items being selected over another is to conduct two separate Rasch analyses.

For this activity, pretend that a decision has been made to present only 23 items (not 24 items) to respondents in a final test form. Also, let's pretend that two items (item 1 and item 2) are the two items that are being considered for removal to get to a total of 23 items. One technique of assessing the impact of removing one item over another is to plot the person measures computed with items 2–24 and to plot the person measures computed with items 1, 3–24. Now conduct two analyses with these two combinations of items. Add the command IDELQU=YES to your control file, and you will be able to tell Winsteps when you run your data which items you want to remove!

After each analysis, create a PFILE for your person measures. When you have completed these two analyses (both with 23 items, but not an identical 23 items), use the plot/scatter plot option to graph the person measures from the two analyses. Look at your plot and remember that each dot represents one person, and that the location of the person is based upon his or her computed measure using the two different sets of items. What do you see in the plot? If there is a straight line of points, it means that using the different sets of items did not make much difference in terms of person measures. This means that you must look at other evidence to decide what set of items to use. If there had been a difference, this observation would be used as part of a decision to select one set of items over another. The word "part" is critical to remember; in Rasch analysis, we carry out a wide range of analyses to inform our decisions with respect to issues such as which items to present in an instrument, which people are providing spurious responses, and with what level of assurance we can present arguments supporting the reliability and validity of instruments.

Answer: In order to complete the analysis, we suggest a Rasch analysis with item 1 removed and then the creation of a PFILE. Please note that you will need to save the file as .txt file. You can do so with the options provided for PFILE. Then exit Winsteps and conduct a second analysis with the same data, but this time remove item 2. Then as before, use the PFILE option to save the person data. Again, make sure to save the data for the second run of data as a .txt file. Once you have saved this second file, you can use the plot scatter plot option of Winsteps. By plotting the measures of the 75 respondents using all items but item 1 and the measures of all 75 respondents using all items but item 2, you will be able to see if there are any respondents whose measures changed dramatically. Those respondents whose measures changed will be "off diagonal," which means they will be away from the diagonal line that cuts through the plot of person measures. If this plot were the only piece of evidence that we were consulting to decide which items to remove from a test, we would conclude it did not really matter which of the two items was removed from the test. Later herein, we will discuss the two curved lines which are presented in this plot. For the time being, suffice it to say that, since no respondent was located outside of the curved bands, there was not a significant statistical shift in the person measures using the two different combinations of test items.

Activity #8

Sometimes when one is conducting an analysis, it is helpful to be able to very quickly compute the mean measures and the standard deviation of measures of person subgroups, for example, males and females. In an earlier activity, we conducted an analysis, placed data into an SPSS file, and computed the means and the standard deviations of males and females who completed this instrument. We then computed a *t*-test to compare the mean measures of males and females who completed the instrument. Winsteps provides a very quick way to review the mean person measures and the standard deviation of person measures for subgroups of respondents.

Run a Rasch analysis on the entire data set, then look at the table entitled "PERSON: subtotals." You will see that you can tell Winsteps (by clicking on the small rectangular white box) which way you would like the program to sort your data (e.g., by gender, by school) and then you will be provided with the mean and standard deviation of each subgroup of a variable. Use this table to look at the values provided for gender and school.

Answer: The female (code "0") mean measure is 1.45 logits, and the standard deviation of the female measures is .87 logits. The male (code "1") mean measure is 1.27 logits and the standard deviation of the male measures is .98 logits. School D has a mean of 2.14 logits and a standard deviation of .35 logits. School E has a mean of 1.26 logits and a standard deviation of .99 logits.

References

Enochs, L. G., & Riggs, I. M. (1990). Further development of an elementary science teaching efficacy belief instrument: A pre-service elementary scale. *School Science and Mathematics, 90*(8), 694–706. doi:10.1111/j.1949-8594.1990.tb12048.x.

Linacre, J. M., & Wright, B. D. (1989). The "length" of a logit. *Rasch Measurement Transactions, 3*(2), 54–55.

Additional Readings

A very good discussion of the problems with using raw data.

Wright, B. D. (2001). Counts or measures? Which communicate best? *Rasch Measurement Transactions, 14*(4), 784.

Chapter 5
Item Measures

Isabelle and Ted: Two Colleagues Conversing

Ted: I am looking through the Winsteps tables, and I think I have a good understanding of person measure tables, but there are also these item tables. I am guessing I can use some of the things I have learned about person measure tables to understand them. What do you think? Please tell me this is true!

Isabelle: Yes, Ted, it's true. With these tables you can understand what was used to compute each item measure, and just as with the persons, you can conduct some exceptional diagnoses of the quality of items that you are using in your instrument. One thing that is really important for you to remember is that the persons and items in a Rasch analysis are expressed in the same units. This will allow you to not only compare items on a scale and compare people on a scale but also compare items to persons on the same scale.

Introduction

This chapter takes us a step further in developing knowledge, confidence, and fluency in interpreting Rasch analysis tables. In particular, we will emphasize a number of tables that focus on survey items, again using the STEBI self-efficacy scale as the example. As in prior chapters, we provide targeted guidance on central issues that will help readers build their knowledge with each chapter, but we do not present every minute detail. By the end of this chapter, readers will be able to interpret item measure tables, perform some simple calculations to check their understanding of item measure tables, and learn more about some unique possibilities when using the Rasch model. Two of the most important aspects of this chapter are the following: (1) Our continuing emphasis that items must involve a single trait, a single variable, when one measures respondents with a scale and pools items together for a person measure; and (2) helping readers gain skill and confidence in interpreting what numbers are used to lead to a calculation of both person measures and item measures.

W.J. Boone et al., *Rasch Analysis in the Human Sciences*,
DOI 10.1007/978-94-007-6857-4_5, © Springer Science+Business Media Dordrecht 2014

Item Measures

Researchers can employ Rasch analysis to evaluate the responses of STEBI survey takers. Initially, it is important to see that some of the techniques employed to understand and use person measure tables also can be used to understand and use item measure tables. Below we provide the entire item entry table (Fig. 5.1) for the 13 STEBI self-efficacy items. This table is a product of the second analysis of the STEBI data, in which UMEAN and USCALE were set to provide linear/equal-interval, all positive Rasch scale person measures from 0 to 1,000. Also, some similarities exist in the organization of this item measure table compared to the person measure table. For example, at the top of the table is a header that contains a title, as stated in the control file. Note at the top of the table (in fact all tables) that Winsteps indicates how many people (75) and items (23) were read into the analysis. The "read evaluated data" is indicated with the term "(INPUT)." Moreover, the phrase "(MEASURED)" indicates how many persons (75) and items (13) were evaluated. In this case all persons and 13 of 23 items were evaluated. Reviewing the header, which shows the input and the analyzed results, is a great way to catch errors in the control file. Readers should notice that 23 items were inputted, but only 13 items were measured.

The STEBI includes items that provide two measures. The ten deleted items (1,4,7,9,10,11,13,14,15,16) are the STEBI outcome-expectancy items. The STEBI

```
TABLE 14.1 SCIENCE TEACHER EFFICACY BELIEFS      ZOU136WS.TXT  Sep  5 12:03 2011
INPUT: 75 PERSON  23 ITEM  MEASURED: 75 PERSON  13 ITEM  6 CATS  WINSTEPS 3.70.6
--------------------------------------------------------------------------------
PERSON: REAL SEP.: 2.52  REL.: .86 ... ITEM: REAL SEP.: 7.00  REL.: .98

        ITEM STATISTICS:  ENTRY ORDER
```

ENTRY NUMBER	TOTAL SCORE	TOTAL COUNT	MEASURE	MODEL S.E.	INFIT MNSQ	INFIT ZSTD	OUTFIT MNSQ	OUTFIT ZSTD	PT-MEASURE CORR.	PT-MEASURE EXP.	EXACT MATCH OBS%	EXACT MATCH EXP%	ITEM
1	DELETED												Q1oe
2	410	75	323.03	15.03	1.06	.4	1.08	.5	.31	.41	55.4	64.4	Q2se
3	317	75	506.14	8.93	1.69	3.5	1.64	3.1	.52	.59	33.8	44.5	Q3se-rc
4	DELETED												Q4oe
5	258	75	569.60	8.38	1.28	1.7	1.34	1.9	.54	.67	37.8	41.2	Q5se
6	352	75	459.26	10.27	.96	-.2	.89	-.5	.52	.53	51.4	54.5	Q6se-rc
7	DELETED												Q7oe
8	369	75	429.77	11.43	1.03	.2	.97	-.1	.55	.50	59.5	57.9	Q8se-rc
9	DELETED												Q9oe
10	DELETED												Q10oe-rc
11	DELETED												Q11oe
12	310	75	514.25	8.78	.99	.0	.96	-.2	.56	.60	52.7	44.2	Q12se
13	DELETED												Q13oe-rc
14	DELETED												Q14oe
15	DELETED												Q15oe
16	DELETED												Q16oe
17	277	69	525.27	8.97	.69	-2.1	.72	-1.7	.71	.61	39.7	42.8	Q17se-rc
18	298	69	498.88	9.51	.66	-2.1	.62	-2.3	.62	.57	47.1	46.0	Q18se
19	206	68	603.19	8.99	1.10	.7	1.21	1.2	.67	.69	44.8	40.8	Q19se-rc
20	262	69	542.68	8.76	1.07	.5	1.13	.8	.65	.63	39.7	41.3	Q20se-rc
21	295	69	502.85	9.41	.77	-1.4	.84	-.9	.65	.58	45.6	44.9	Q21se-rc
22	364	69	367.55	14.34	.97	-.1	.93	-.3	.51	.44	64.7	63.7	Q22se
23	260	69	544.95	8.75	.80	-1.3	.86	-.8	.70	.63	44.1	41.1	Q23se-rc
MEAN	306.0	71.7	491.34	10.12	1.00	.0	1.01	.1			47.4	48.2	
S.D.	53.9	3.1	75.80	2.10	.26	1.5	.26	1.4			8.6	8.4	

Fig. 5.1 A sample Winsteps Rasch analysis table presenting the results of evaluating the 13 self-efficacy items of the STEBI. In this analysis, the 10 outcome-expectancy items are not used; those items are removed in the Rasch analysis through use of the command line IDFILE

can be thought of as including 13 items that involve the field of history, and 10 items involve the field of mathematics. Since our analysis focuses only on the self-efficacy items, a line was inserted in the control file (IDFILE=) to conduct a Rasch analysis only on the 13 self-efficacy items. One can also create a control file using only the items of interest, in this case the 13 self-efficacy items, and in that case, one would only see the 13 items listed in Fig. 5.1.

Key columns to understand in Fig. 5.1 are ENTRY NUMBER, TOTAL SCORE, TOTAL COUNT, MEASURE, and ITEM. The ENTRY NUMBER column gives the sequence in which items are read into the program for analysis. In our example, all 23 STEBI items were entered (ENTRY NUMBER). But the 10 outcome-expectancy items were not evaluated. Six self-efficacy items were answered by all 75 respondents, another six items were answered by 69 respondents, and one item was answered by 68 respondents.

A Noteworthy Detail

At this point, we focus on a frequently encountered detail that can cause a problem or even a misconception. Researchers often think the entry number will match the item name, but that is not the case. The entry number is only a tally of the first, second, third (and so on) items read into the program for analysis. For example, a researcher might wish to evaluate a 20-item 6th grade test. Perhaps the researcher has found an error in item 4 and decides to remove that item from an analysis before a control file is made. In this case, the entry numbers and associated item names would look like the following lists in Fig. 5.2.

In Fig. 5.2, recall that two columns to the right of the ENTRY NUMBER column is the column entitled "TOTAL COUNT." This is another place where researchers can use the tables to spot errors in an analysis (and better understand an analysis). The TOTAL COUNT column shows how many people responded to each item. Depending

Entry Number	Item
1	Q1
2	Q2
3	Q3
4	Q5
5	Q6
.	.
.	.
.	.
19	Q20

Fig. 5.2 An example of the "entry number" not always corresponding perfectly to the item name

upon how researchers have coded their data, they should see a number for TOTAL COUNT that is about the same as the number of individuals who took the survey. This will often be a number close to that reported in the header (see Fig. 5.3) under INPUT and MEASURED as in "INPUT: 75 PERSONS" and "MEASURED: 75 PERSONS."

We now return to the column with the heading "TOTAL SCORE" in Fig. 5.1. Look at Item Q2se in the "ITEM" column on the far right of Fig. 5.1. It is the second item of the whole survey and the first self-efficacy item presented to students. Item Q2se has a TOTAL SCORE of 410. This value (410) is the raw score sum total of all responses to this item. Thus, for this six-category rating scale where "6" represents the selection of *Strongly Agree*, "5" represents the selection of *Agree*, "4" represents the selection of *Barely Agree* and so on, the number 410 is computed by adding all the responses to this item by all respondents. For example, 410 could be the result of 37 respondents having answered *Strongly Agree*, 36 respondents having answered *Agree*, and 2 respondents having answered *Barely Agree* $(410=(37\times6)+(36\times5)+(2\times4))$. One could compute this value in a spreadsheet, but these numbers should not be used for any analysis purposes because they are not linear. To see the actual distribution of responses that produced the raw score of 410, a portion of Winsteps Table 14 (Fig. 5.4) provides the relevant information.

```
TABLE 14.1 SCIENCE TEACHER EFFICACY BELIEFS       ZOU136WS.TXT  Sep  5 12:03 2011
INPUT: 75 PERSON  23 ITEM  MEASURED: 75 PERSON  13 ITEM  6 CATS  WINSTEPS 3.70.6
--------------------------------------------------------------------------------

PERSON: REAL SEP.: 2.52  REL.: .86 ... ITEM: REAL SEP.: 7.00  REL.: .98

         ITEM STATISTICS:  ENTRY ORDER

------------------------------------------------------------------------------------------
|ENTRY   TOTAL  TOTAL            MODEL|   INFIT  |  OUTFIT  |PT-MEASURE |EXACT MATCH|         |
|NUMBER  SCORE  COUNT  MEASURE  S.E. |MNSQ  ZSTD|MNSQ  ZSTD|CORR.  EXP.| OBS%  EXP%| ITEM    |
|------------------------------------+----------+----------+-----------+-----------+---------|
```

Fig. 5.3 The header of Winsteps Table 14.1 and the terms used to identify the meaning of key columns

```
TABLE 14.3 SCIENCE TEACHER EFFICACY BELIEFS       ZOU136WS.TXT  Sep  5 12:03 2011
INPUT: 75 PERSON  23 ITEM  MEASURED: 75 PERSON  13 ITEM  6 CATS  WINSTEPS 3.70.6
--------------------------------------------------------------------------------

        ITEM CATEGORY/OPTION/DISTRACTOR FREQUENCIES:  ENTRY ORDER

------------------------------------------------------------------------
|ENTRY   DATA  SCORE |    DATA    | AVERAGE  S.E.  OUTF PTMEA|         |
|NUMBER  CODE  VALUE |  COUNT   % | ABILITY  MEAN  MNSQ CORR.| ITEM    |
|--------------------+------------+--------------------------+---------|
|    2   4         4 |    2    3 |  541.98  3.25  1.4  -.04 |Q2se     |
|        5         5 |   36   48 |  533.69*10.48  1.2  -.31 |         |
|        6         6 |   37   49 |  590.45 16.60  1.1   .32 |         |
TABLE 14.3 SCIENCE TEACHER EFFICACY BELIEFS       ZOU873WS.TXT  Sep  5 10:54 2011
INPUT: 75 Person  23 Item  REPORTED: 75 Person  13 Item  6 CATS  MINISTEP 3.72.2
--------------------------------------------------------------------------------
```

Fig. 5.4 A portion of Table 14.3 from Winsteps. Key information involves the heading ENTRY NUMBER, DATA CODE, SCORE VALUES, DATA COUNT, DATA %, and ITEM. This table can be used to understand the combination of rating scale categories used to calculate the raw score total for a particular item. Although raw score totals cannot be used for analysis of persons and items, the review of rating scale categories selected for items is very important. For example, a rating scale category rarely used for items may suggest a category that could be dropped from a scale

---

Formative Assessment Checkpoint #1

Question: You are reviewing an item entry table containing the following information regarding the 12th item of the STEBI. The 12th item is a self-efficacy item. What do the numbers that are reported above this item mean? (We entered dashes (--) in for columns that we are not discussing at this point.)

ENTRY NUMBER	TOTAL SCORE	TOTAL COUNT	MEASURE	MODEL S.E.	INFIT MNSQ	INFIT ZSTD	OUTFIT MNSQ	OUTFIT ZSTD	PT-MEASURE CORR.	PT-MEASURE EXP.	EXACT MATCH OBS%	EXACT MATCH EXP%	ITEM
12	310	75	514.25	----	----	----	----	----	----	----	----	----	Q12se

Answer: Looking at the far right of the table, one can see that the name used to describe this item in the control file is "Q12se." What about the far left column with the number 12? So if one looked at the entire data set in which each person was a row, the 12th piece of data read in regarding survey items was this item. One can also see that the total of all responses to this item by all respondents who answered this item was 310. In a typical spreadsheet used to organize rating scale data, this number 310 would be the grand total if one added all numbers used to code the responses of all respondents to this one item. The column labeled TOTAL COUNT shows the total number of people ($n = 75$) who answered this item. The MEASURE column lists the Rasch measure (514.25) of the item. The large number enables one to see that UMEAN and USCALE were used for rescaling.

Readers should note that for this raw score of 410, the distribution of responses is exactly that which results in a raw score total of 410. An important tip is that when you look at Fig. 5.4, you see only data for categories 4, 5, and 6 for this item. Does this mean that something is wrong with your data or your analysis? No. Remember, if you do not see a particular value listed for the codes you used to label your rating scale, one possibility is that no one in your data set selected the missing codes for item Q2se. If your data set is not large, we suggest that you scan the column of your original spreadsheet for data entry to check this observation in Fig. 5.4.

Regarding item Q2se, two important columns to look at in Fig. 5.4 are SCORE VALUE and COUNT. This table shows that only two out of 75 respondents selected response BA for this 6-category, SA (*Strongly Agree*), A (*Agree*), BA (*Barely Agree*), BD (*Barely Disagree*), D (*Disagree*), SD (*Strongly Disagree*), survey item. The SCORE VALUE for BA is "4," and the corresponding COUNT is "2." Now we ask: How many people selected SD, D, BD, A, and SA? Since no "score values" of 1, 2, or 3 are presented, no respondents selected SD (coded "1"), D (coded "2"), or BD (coded "3"). Thirty-six people selected A (coded "5") and 37 people selected SA (coded "6"). If a researcher wishes, he or she may use this part of Fig. 5.4 (Winsteps Table 14.3) to better understand the distribution of responses for each item.

The values in the SCORE VALUE and COUNT columns can also be used to calculate the TOTAL SCORE for Item Q2se ($n=410$) in Fig. 5.1 as follows: Multiply the count value of respondents who answered each item category response by the coded score value for the item category response and add the products: 2(people) × 4(BA) + 36(people) × 5 (A) + 37(people) × 6 (SA) = (8 + 180 + 222) = 410. Therefore, the total score is the total of all raw scores coded for each item. Prior to learning about Rasch analysis, you would have calculated this number (410) if you added all the numbers in a spreadsheet entered in the column for this item. As we have mentioned over and over, this raw score value should not be used for parametric statistical comparisons due to the nonlinearity of rating scales. This value is, however, useful for checking the coding of the data and eyeballing in what manner items were harder to agree with and easier to agree with. Readers will also notice that the total value of 410 for this item is the same TOTAL SCORE for item Q2se in Fig. 5.1.

Formative Assessment Checkpoint #2

Question: What are the ENTRY NUMBER, the TOTAL COUNT, and the TOTAL SCORE presented in the item entry table?

Answer: The ENTRY NUMBER indicates the order in which items are read into the Winsteps program. It is important to remember that if the first item in your test is named item Q2, then item Q2 will have an entry number of 1. The TOTAL COUNT simply reports the total number of respondents who answered the item, and the TOTAL SCORE is the sum of the coded raw scores of all the answers to each item.

The next important step is to understand what a higher total score means. Within this understanding lies what it means for an item to have a higher or lower Rasch measure. To explain, look at item Q5se in Fig. 5.5, which displays the pertinent information from Fig. 5.1 for item Q5se.

From Fig. 5.5, we see a total score of 258 reported for all the responses by 75 people to item Q5se. From Fig. 5.6 for this item, we see that 2 people answered SD, 20 people answered D, 13 people answered BD, 24 people answered BA, 15 people answered A, and 1 person answered SA. Comparing the total scores for item Q2se (410) and item Q5se (258), one can understand in general that a higher total score for one item (e.g., Q2se) compared to another item (e.g., Q5se) means that item Q2se was generally easier to agree with. Remember, a higher number was used to code a more agreeable response (e.g., a 6 was used for SA; a 5 was used for A).

```
TABLE 14.1 SCIENCE TEACHER EFFICACY BELIEFS      ZOU136WS.TXT  Sep  5 12:03 2011
INPUT: 75 PERSON  23 ITEM  MEASURED: 75 PERSON  13 ITEM  6 CATS  WINSTEPS 3.70.6
--------------------------------------------------------------------------------
PERSON: REAL SEP.: 2.52  REL.: .86 ... ITEM: REAL SEP.: 7.00  REL.: .98

         ITEM STATISTICS:  ENTRY ORDER

-----------------------------------------------------------------------------------------------
|ENTRY   TOTAL  TOTAL            MODEL|   INFIT   |  OUTFIT   |PT-MEASURE |EXACT MATCH|        |
|NUMBER  SCORE  COUNT  MEASURE  S.E. |MNSQ  ZSTD|MNSQ  ZSTD|CORR.  EXP.| OBS%  EXP%| ITEM    |
|-----------------------------------+----------+----------+-----------+-----------+--------|
|      5   258     75  569.60    8.38|1.28   1.7|1.34   1.9|  .54   .67| 37.8  41.2| Q5se
|
```

Fig. 5.5 The header of Fig. 5.1 and the table data for item Q5se

```
TABLE 14.3 SCIENCE TEACHER EFFICACY BELIEFS      ZOU136WS.TXT  Sep  5 12:03 2011
INPUT: 75 PERSON  23 ITEM  MEASURED: 75 PERSON  13 ITEM  6 CATS  WINSTEPS 3.70.6
--------------------------------------------------------------------------------

          ITEM CATEGORY/OPTION/DISTRACTOR FREQUENCIES:  ENTRY ORDER

------------------------------------------------------------------------
|ENTRY   DATA  SCORE |    DATA     | AVERAGE  S.E.  OUTF PTMEA|         |
|NUMBER  CODE  VALUE |  COUNT    % | ABILITY  MEAN  MNSQ CORR.| ITEM    |
|--------------------+-------------+-------------------------+---------|
|     5  1        1 |    2    3 |  511.70 20.72  1.2  -.10 |Q5se     |
|        2        2 |   20   27 |  509.17*11.44  1.5  -.37 |         |
|        3        3 |   13   17 |  555.59 17.00  2.1  -.03 |         |
|        4        4 |   24   32 |  561.36 10.94  1.2   .00 |         |
|        5        5 |   15   20 |  616.11 22.51  1.5   .31 |         |
|        6        6 |    1    1 |  999.63          .0   .59 |         |
|                   |             |                          |         |
```

Fig. 5.6 Data pertaining to item Q5se provided in Winsteps Table 14.3

Whereas one should not make more specific inferences regarding the differences between items without using Rasch analysis tools, this simple comparison allows us to understand what a higher or lower total raw score means when one evaluates survey items. This step will be used to help beginners to keep track of the meaning of lower and higher item logit measures.

Even more important, this comparison of item total scores allows us to remember confidently what a higher item measure signifies. In this particular case, a higher item measure signifies an item that is harder to agree with. Item measure information found in Fig. 5.1 under column MEASURE shows that item Q2se has an item measure of 323.03 and item Q5se has an item measure of 569.60. Because item Q2se has a higher total score (410) than item Q5se (258), we can see that when data are entered using a higher number for more agreement with a rating scale ("6" is SA, "5" is A, etc.), the item Q2se with the higher total score of 410 is the easier item to agree with. However, when comparing two item measures (Q2se = 323.04 and

Q5se = 569.59), a lower item measure implies (with this rating scale and coding) more agreement.

We continue this example to help readers understand what a higher measure means for people and for items. For the beginner, this can be one of the most difficult concepts. Let's pretend a 10-item rating scale was used to collect data on the level of constructivist teaching taking place in a classroom. For readers who may not be in the field of education, think of constructivist teaching as the type of innovative teaching that will help students learn and apply class material (rote memorization would not be constructivist teaching; encouraging students to develop alternative experimental hypotheses would be good constructivist teaching). The rating scale that was used to collect data is 1 = No sign of constructivist teaching, 2 = mix of constructivist teaching and non-constructivist teaching, and 3 = constructivist teaching. These ten items are presented to teachers to evaluate their self-reported level of constructivist teaching. If this type of coding were used, then a higher raw score for each participant would mean a higher level of constructivist teaching. For example, if John received a raw score of 29 for his answers to all ten items, it might mean he selected a 3 (constructivist teaching) for 9 items and a 2 (mix of teaching) for 1 item (9 items × 3 raw score points per item + 1 item × 2 raw score points = 29). When Rasch analysis of this data (with the coding-higher number is more constructivist teaching) is conducted a higher person measure (in logits) would mean a higher raw score and in turn would mean a higher level of constructivist teaching.

Now consider the meaning of a higher person measure if the following coding were used to code teacher answers to the survey items: 3 = No sign of constructivist teaching, 2 = Mix of constructivist teaching and non-constructivist teaching, and 1 = Constructivist teaching. In this case a higher raw score for each respondent would mean a lower level of constructivist teaching, and a lower raw score would mean a higher level of constructivist teaching. Since Winsteps (and any analysis program) knows only that 3 is larger than 2 and 2 is larger than 1, in this case the higher person measure would mean a lower level of constructivist teaching. We strongly recommend that researchers code data in such a way that a higher person measure means that the person is doing more of what the researcher wants them to do. In education, that might mean teaching in a way that will enhance learning. In medicine, that might mean a patient is farther along in terms of recovery from an illness. To repeat, our advice is: When coding data, code responses such that a higher number means "better." Then memorize that a higher person measure (in logits) will also mean better. If you do not code the data in this way, all is not lost, but you must be careful to remember what it means to have a higher person raw score and thus a higher person measure.

Formative Assessment Checkpoint #3

Question: If you were evaluating a 10-item Likert scale survey in which the scale is coded *Strongly Disagree* (1), *Disagree* (2), *Agree* (3), and *Strongly Agree* (4) and *Strongly Agree* was the best type of answer from respondents, how should you explain the Rasch measures that you would see for the respondents and the items?

Answer: An item that is more difficult to agree with will be higher up (more positive) on the logit scale used to express item measures. Respondents who have higher person measures are more agreeable respondents.

We conclude this section of the chapter by revisiting two important issues we have presented for this data set, and we will explain why what is seen, is seen! Readers will remember that the self-efficacy data were coded such that a higher person measure meant the person had more self-efficacy than a person with a lower person measure. Readers also will remember that a survey item with a higher total raw score than another item will be more negative (have a lower item measure) than the item it is compared to. In Fig. 5.7, we present a Wright Map with three persons and two items.

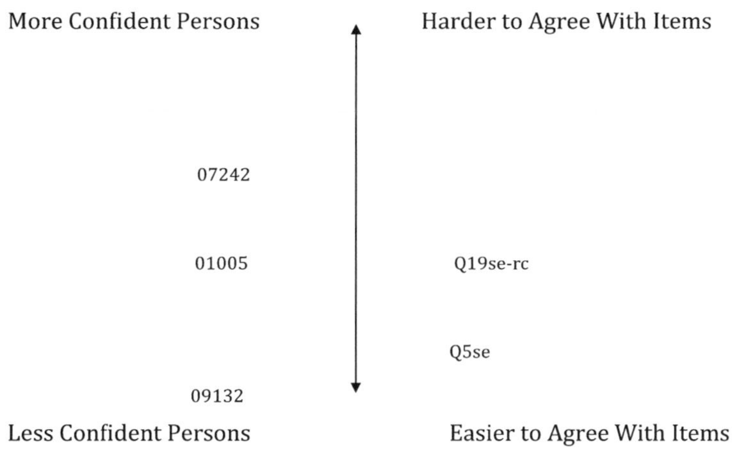

Fig. 5.7 A Wright Map of three persons and two items from the self-efficacy data of the STEBI

Formative Assessment Checkpoint #4

Consider the following sentence: A higher item measure (compared to another item) will always mean "more difficult to agree with in some manner" when a rating scale of SA, A, BA, BD, D, and SD is used for an analysis. Yes or No?

Our response is "No." The meaning of the item measure value depends upon the coding used for your data set. If you choose to use coding of a 6 for SD, 5 for D, 4 for BD, 3 for BA, 2 for A, and 1 for SA, the higher item measure would mean "more difficult to disagree with in some manner."

More Agreement or Less Disagreement?

At this point, we want to introduce an issue that has often confused first time Rasch users. Let's think about the same rating scale and these same survey items. There are several phrases that one can use to express the difference between two items that differ in their "item measure." In the example immediately above, we pointed out that a lower item measure (compared to another item) means easier to agree with in some manner by respondents to the item. It is extremely important to point out that "easier to agree with" may not mean that one sees a higher percentage of respondents marking, for example, *Agree* or *Strongly Agree* to an item. An item "easier to agree with" may in fact be an item in which there is less disagreement compared to another item. For example, if we hypothetically compare two items, one with an item measure of .5 logits and another with an item measure of 1.25 logits, it is quite possible that none of the respondents selected any of the possible agree answers (SA, A, BA) for either item. Thus, all respondent answers for both items used the disagree (SD, D, BD) part of the rating scale. Even though (in our example) the agree rating categories were not used, it remains that the item with item measure of .5 is easier to agree with in comparison to the item with item measure of 1.25. "Easier to agree with" may not necessarily indicate that more agree categories were selected; it may indicate, for instance, that fewer extreme disagree categories were selected. This may seem odd to readers, but the difference between these items can still be expressed by a difference in a level of agreement (which is also an example of a difference in agreement).

A technique we have often used when attempting to interpret item measures is to write or type a quick note in the item entry table to remember what a higher item measure value indicates in terms of a particular survey. For this example, one could note "higher measure = less agreement" or "higher measure is harder to agree with" and "lower measure = more agreement" or "lower measure is easier to agree with." In Fig. 5.8, we present an example of notes we might make in a person table and an item table.

```
TABLE 17.1 SCIENCE TEACHER EFFICACY BELIEFS      ZOU454WS.TXT  Feb 29 11:46 2012
INPUT: 75 Person  23 Item  REPORTED: 75 Person  13 Item  6 CATS    WINSTEPS 3.73
--------------------------------------------------------------------------------
Person: REAL SEP.: 2.52  REL.: .86 ... Item: REAL SEP.: 7.00  REL.: .98

          Person STATISTICS:  MEASURE ORDER

-------------------------------------------------------------------------------------------------
|ENTRY   TOTAL  TOTAL             MODEL|   INFIT  |  OUTFIT  |PT-MEASURE |EXACT MATCH|           |
|NUMBER  SCORE  COUNT  MEASURE  S.E. |MNSQ  ZSTD|MNSQ  ZSTD|CORR.  EXP.| OBS%  EXP%| Person    |
|------------------------------------+----------+----------+-----------+-----------+-----------|
```

HIGHER TOTAL SCORE MEANS MORE AGREEMENT BECAUSE HIGHEST RATING CATEGORY WAS STRONGLY AGREE AND THAT CATEGORY WAS CODED AS A "6." THIS MEANS PERSON 65730 WAS THE PERSON MOST AGREEABLE TO THE 13 SELF-EFFICACY ITEMS. THIS MEANS THEY HAD THE STRONGEST SELF-EFFICACY. THEY ANSWERED ALL 13 ITEMS AND HAD A TOTAL SCORE (RAW SCORE) OF 78.

```
|    46     78     13     7.53     1.85|      MAXIMUM MEASURE|  .00     .00|100.0 100.0| 65730     PR|
|    36     75     13     4.89      .68| .93     .0| .70    -.1|  .37     .32| 69.2  77.8| 65234     PR|
|     3     33      6     3.28      .81| .62    -.5| .57    -.5|  .61     .46| 66.7  66.9| 95793     PR|
|    17     68     13     2.71      .48|3.90    3.6|2.57    2.5|  .53     .50| 61.5  66.2| 55959     PR|
|    18     67     13     2.49      .46|1.00     .2| .98     .1|  .64     .51| 61.5  63.7| 97766     PR|
|    50     67     13     2.49      .46|2.24    2.0|2.07    1.9|  .48     .51| 53.8  63.7| 68028     PR|
|    73     67     13     2.49      .46| .19   -2.3| .19   -2.5|  .82     .51| 92.3  63.7| 81223     PR|
|    12     66     13     2.28      .45| .99     .2|1.05     .3|  .47     .52| 53.8  63.8| 43532     PR|
|    21     66     13     2.28      .45|1.37     .8|1.23     .6|  .62     .52| 61.5  63.8| 99365     PR|

...

     30     22      6     -.22      .44|2.46    2.1|2.12    1.7|  .32     .70| 33.3  41.4| 80392     PR|
|    11     42     13     -.36      .30|1.12     .4|1.00     .1|  .78     .71| 23.1  43.5| 28100     PR|
|    57     42     13     -.36      .30|1.48    1.2|1.34     .9|  .66     .71| 30.8  43.5| 99843     PR|
|     7     21      6     -.42      .44| .27   -1.7| .29   -1.5|  .91     .71| 66.7  42.2| 46328     PR|
|    19     41     13     -.45      .30|1.17     .5|1.09     .4|  .61     .72| 30.8  43.7| 78880     PR|
|     8     39     13     -.64      .31| .52   -1.3| .41   -1.7|  .86     .73| 61.5  45.3| 41024     PR|
|    14     39     13     -.64      .31| .91    -.1| .80    -.3|  .70     .73| 46.2  45.3| 53695     PR|
|    55     39     13     -.64      .31|1.32     .9|1.34     .9|  .51     .73| 30.8  45.3| 12103     PR|
|    20     37     13     -.84      .32| .40   -1.7| .46   -1.4|  .91     .74| 61.5  46.5| 33573     PR|
```

THE PERSON WITH THE LOWEST TOTAL SCORE HAS THE LOWEST SELF-EFFICACY. THIS IS BECAUSE OUR CODING WAS TO USE 1 FOR STRONGLY DISAGREE AND TO USE 6 FOR STORNGLY AGREE. THE WAY IN WHICH THE ITEMS WERE PHRASED WAS IN SUCH A WAY SO THAT A HIGHER NUMBER (A STRONGLY AGREE SELECTION) MEANT MORE SELF-EFFICACY. THE PERSON WITH THE LOWER SELF-EFFICACY WILL HAVE A LOWER TOTAL SCORE, BECAUSE THEY ARE NOT AS AGREEABLE TO THE ITEMS AS OTHER RESPONDENTS.

```
TABLE 13.1 SCIENCE TEACHER EFFICACY BELIEFS      ZOU670WS.TXT  Feb 25 16:31 2013
INPUT: 75 Person  23 Item  REPORTED: 75 Person  13 Item  6 CATS  MINISTEP 3.75.0
--------------------------------------------------------------------------------
Person: REAL SEP.: 2.52  REL.: .86 ... Item: REAL SEP.: 7.00  REL.: .98

          Item STATISTICS:  MEASURE ORDER

-----------------------------------------------------------------------------------------------
|ENTRY   TOTAL  TOTAL             MODEL|   INFIT  |  OUTFIT  |PT-MEASURE |EXACT MATCH|         |
|NUMBER  SCORE  COUNT  MEASURE  S.E. |MNSQ  ZSTD|MNSQ  ZSTD|CORR.  EXP.| OBS%  EXP%| Item    |
|------------------------------------+----------+----------+-----------+-----------+---------|
```

THE HIGHER ITEM MEASURE MEANS THAT THE ITEM WAS HARDER TO AGREE WITH. THIS CAN BE SEEN BY LOOKING AT THE ITEM MEASURE. THE ITEM Q5se HAS A HIGHER ITEM MEASURE THAN ITEM Q12SE. BUT IF ONE LOOKS AT THE TOTAL SCORE, THE REVERSE IS TRUE. SINCE A CODING OF 6 WAS USED FOR THE HIGHEST RATING CATEGORY POSSIBLE, THIS HELPS ONE SEE THAT A HIGHER ITEM MEASURE MEANS THE ITEM WAS HARDER TO AGREE WITH THAN AN ITEM OF LOWER MEASURE.

```
|    19    206     68     1.66      .13|1.10     .7|1.21    1.2|  .67     .69| 44.8  40.8| Q19se-rc|
|     5    258     75     1.16      .12|1.28    1.7|1.34    1.9|  .54     .67| 37.8  41.2| Q5se    |
|    23    260     69      .79      .13| .80   -1.3| .86    -.8|  .70     .63| 44.1  41.1| Q23se-rc|
|    20    262     69      .76      .13|1.07     .5|1.13     .8|  .65     .63| 39.7  41.3| Q20se-rc|
|    17    277     69      .50      .13| .69   -2.1| .72   -1.7|  .71     .61| 39.7  42.8| Q17se-rc|
|    12    310     75      .34      .13| .99     .0| .96    -.2|  .56     .60| 52.7  44.2| Q12se   |
|     3    317     75      .22      .13|1.69    3.5|1.64    3.1|  .52     .59| 33.8  44.5| Q3se-rc |
|    21    295     69      .17      .14| .77   -1.4| .84    -.9|  .65     .58| 45.6  44.9| Q21se-rc|
|    18    298     69      .11      .14| .66   -2.1| .62   -2.3|  .62     .57| 47.1  46.0| Q18se   |
|     6    352     75     -.48      .15| .96    -.2| .89    -.5|  .52     .53| 51.4  54.5| Q6se-rc |
|     8    369     75     -.91      .17|1.03     .2| .97    -.1|  .55     .50| 59.5  57.9| Q8se-rc |
|    22    364     69    -1.83      .21| .97    -.1| .93    -.3|  .51     .44| 64.7  63.7| Q22se   |
|     2    410     75    -2.49      .22|1.06     .4|1.08     .5|  .31     .41| 55.4  64.4| Q2se    |
|------------------------------------+----------+----------+-----------+-----------+---------|
| MEAN   306.0   71.7      .00      .15|1.00     .0|1.01     .1|            | 47.4  48.2|         |
| S.D.    53.9    3.1     1.12      .03| .26    1.5| .26    1.4|            |  8.6   8.4|         |
-----------------------------------------------------------------------------------------------
```

Fig. 5.8 An example of the notation one can make in an output table so that one can remember meaning of what it means for a person and item to have a higher or lower logit measure

---

Formative Assessment Checkpoint #5

Question: If you administer a 15-item multiple-choice test to 100 students, and you code a correct answer by a student to an item with a 1 and an incorrect answer with a 0, what will you expect to observe in the TOTAL SCORE and MEASURE columns for respondents and for items?

Answer: A student who does well on the test will have a TOTAL SCORE that is higher than a student who does not perform as well. The Rasch person measure of the student who did better will be higher than the Rasch person measure of the student who did not do as well.

With respect to items (when you code correct answers with a 1, incorrect answers with a 0), the items with a lower TOTAL SCORE are harder items. These harder items will have higher item measures than items which were easier for respondents.

Isabelle and Ted: Two Colleagues Conversing

Ted: There certainly are some new things for us to think about with Rasch. Some of it takes some time, but once I get it, it really helps me see all sorts of errors, and lots of possibilities, too.

Isabelle: Well, give me some examples.

Ted: I can see that when I run Winsteps with my rating scale data, unless I think from the beginning that all items involve one trait, I think it might be a waste of time to try to do too much with the data.

Isabelle: What do you mean?

Ted: Thinking about the measures, the total scores, and so on helps me understand that measures computed from the raw data really have no meaning unless there is a single trait. Even though it is quite easy to make a control file and run a Rasch analysis, I need to ensure the measures mean something. I could easily enter data for a class of 30 students into a spreadsheet. I could enter the responses of students to their level of agreement with respect to ten survey items that concerned their interest in studying science, for example, using a rating scale of Strongly Agree, Agree, Barely Agree, Barely Disagree, Disagree, and Strongly Disagree to indicate a response to "I would like to take a physics class when I am in high school," and I could also enter data on the students' study habits using the same rating scale to indicate a response to "I usually study at my own desk." When the data are run, the program does not know if I have or have not thought about whether or not the survey items involve a single trait. The program will compute person and item measures as if all the items involve the same single trait.

Isabelle: What is your point?

Ted: One really has to think before one pushes a button before computing a person measure or an item measure.

Isabelle: What else have you noticed?

Ted: I think after a while I probably will get really fast with understanding what a higher item measure or person measure denotes, but for now I think it is really important that I look through the chapter figures and make sure, at least for a few items, that I understand what a higher or lower measure really denotes.

There is one other thing that has been really tough. It is this idea that when I compare two items, one item might be easier to agree with than another item. However, it could be that for the two items I am comparing, no one selected an agree rating category at all! It is really relative. If one item has all Strongly Disagree and Disagree, while another item had only Barely Agree, then the item with only Barely Agree selections was easier to agree with than the other item.

Isabelle: Great thinking Ted. That last part will help you with some other things later on, too!

Keywords and Phrases

Total score
Total count
Item measure

If coding test results using 1 for correct and 0 for incorrect, then when comparing two test item measures, the item with the higher logit measure will be the item that is harder to correctly answer.

Potential Article Text

Data collected for the K Project included the use of the STEBI (Enochs & Riggs, 1990). One main project goal was to improve the self-efficacy of future biology teachers in Germany. The complete cohort of preservice teachers ($n = 75$) completed the instrument, and the data were analyzed using the Rasch Winsteps program (Linacre, 2012). The quality of data entry, coding, and analysis were in part verified through a review of the Winsteps item entry table (Table 14). One of the STEBI items was randomly selected, and a tally of rating scale categories responses were noted using the original paper copy surveys. This tally (as a function of rating scale category) was then compared to information provided in Table 14. Analysis of the sample STEBI item, both raw data and the results, suggested that data were entered, coded, and evaluated correctly.

Quick Tips

To understand the meaning of moving from a lower item measure to a higher measure, find two items that have been answered by the same number of respondents (have the same COUNT value). Then look at the coding rules that were used to code the initial answers into a spreadsheet. What numbers were used for the "best" type of response and the "worst" type of response? Maybe you are collecting self-efficacy data with the STEBI, and *Strongly Agree* is coded as a "6," and the answer *Strongly Agree* is the type of answer that you would like to observe from respondents. In this example, the "worst" response would be a "1." Now look at the TOTAL SCORE for the two items you have selected. Which item has the highest TOTAL SCORE? Now write down the measure of that item. Now look at the item with the lowest TOTAL SCORE and write down the item measure of that item. Now you will be able to quickly see the meaning of an item having a higher item measure. In this case, higher logit measure of items means the item was harder to agree with.

In the data set of this chapter, item Q18se and item Q22se were answered by 69 respondents. Since a coding of 6 *Strongly Agree*, 5 *Agree*, and so on was used for the responses of students to these two items, it can be seen that item Q22se was easier to agree with than item Q18se (the raw score of Q22se is 364; the raw score of Q18se is 298). Since the item measure of Q18se was 498.88 and the item measure of Q22se was 367.55, a lower item measure, in this example, indicates that item was easier to agree with.

Do not assume you know what it means for an item to have a higher measure than another item. Be sure to carry out the steps we describe above.

Also, remember an item for one scale (say attitude toward studying) cannot be compared to an item from another scale (say attitude toward teaching techniques). The values for the items may be expressed using similar numbers, but the items cannot be compared. This means that if an item of the attitude toward studying scale has a measure of 235.67 and an item of attitude toward teaching techniques scale has a measure of 280.98, one cannot compare these two numbers. The items represent two different variables; thus, a one-to-one comparison of item measures has no meaning. You would not compare a 20kg rock to a 20 meter high tree!

Do not assume the entry number will always match the item number of an instrument. Look at the item name to identify items.

When you compare people, compare items, or compare items to people, you know only the manner in which they differ. For example, when comparing two items, a researcher may be able to say that one item is harder to agree with than the other item. But this does not mean (e.g., with the use of a scale of *Strongly Agree*, *Agree*, *Disagree*, and *Strongly Disagree*) one item typically was answered with a disagree rating and the other item was typically answered with an agree rating. It could be that one item was rated with *Strongly Agree* more often by respondents and another item was more often rated with *Agree* by respondents. You can look at details of the item measure table to find these details (e.g., Winsteps Table 14.3).

Data Sets: (go to http://extras.springer.com)

cf for SE for Chp 5 not rescaled
cf for SE for Chp 5 1 to 100
cf subset for SE Chp 5 not rescaled

Activities

Activity #1

Stakeholders and reviewers alike sometimes have difficulties understanding how a person or item could have a negative number. It is therefore a good skill to at least know how to convert person and item measures to positive values in such a way to retain the linear aspect of Rasch measures.

For this chapter we have included a control file for the 13 self-efficacy items, which is named "cf for SE for Chp 5 not rescaled." This control file is identical to the file used for the text of this chapter, except that the Rasch logit scale for items and persons has not been rescaled. That means you will see positive and negative numbers for logits. Run the file and look at Table 14 and in particular item Q5. As shown in this chapter, are harder to agree with items still harder to agree with?

Answer: The rescaling of a Rasch measurement scale does not change anything. You will notice that you get the same values for TOTAL COUNT and TOTAL SCORE. Also notice that the meaning of going up the measurement scale (to more positive numbers for items) has the same meaning as what was observed in the main chapter analysis. Also remember rescaling will not change the results of parametric statistical tests.

Activity #2

A second control file is provided with the self-efficacy data (cf for SE for Chp 51 to 100). This control file is almost identical to that presented in the Activity 1 control file and the chapter control file. The only difference is this control file is set to provide Rasch person measures that vary from a low of 0 to a high of 100. Repeat Activity 1 using this control file.

Answer: As was true for the Activity 1 control file, it does not make any difference if there is a rescaling from 0 to 100. The researcher will get the same results when looking in detail at Table 14 for this data.

Activity #3

Using the control file "cf subset for SE Chp 5 not rescaled," identify an item that was hardest for respondents to agree with in comparison to the other items. After a few Rasch analyses of your own, you will be able to quickly understand how to identify items that are (in the case of a rating scale with agreement) easiest to agree with and hardest to agree with.

Answer: Readers should remember that the coding used for this self-efficacy data was SA (6), A (5), BA (4), BD (3), D (2), and SD (1). This means (when all respondents have answered all items) that the items that were hard for respondents to agree with will be the items with the lowest TOTAL SCORE. Item Q5 was the item which was the second hardest for respondents to agree with. That item, in this data set, has a TOTAL SCORE of 195. The most difficult to agree with item was item Q19se, which was answered by 49 students.

Activity #4

The Rasch Winsteps tables and plots provide a range of data that can help you save time. For example, you can avoid cutting and pasting item measures into a spreadsheet to compute an item mean.

Using the control file "cf subset for SE Chp 5 not rescaled," find the line in Table 14 that reports the mean measure for all items that were evaluated.

Answer: That value is 0.00. Unless told otherwise, Winsteps will always set the mean item difficulty of a set of items at 0.00 logits. Look for the word "Mean" in the table.

Activity #5

Run an analysis using the file "cf for SE for Chp 5 not rescaled" and find Table 14.1. How can it be that the item measures for item Q5se and item Q23se are so different when the TOTAL SCORE is almost identical (258, 260)?

Answer: When one is collecting data using a single trait and conducting a careful Rasch analysis, then it does not make a difference which mix of items along a trait a respondent completes. If readers look at the column that is entitled TOTAL COUNT, you will see that 75 respondents answered item Q5se, but only 69 respondents answered item Q23se. Only when the TOTAL COUNT is the same for items you are comparing can you make immediate (rough) raw score assessments of how respondents' answers may have defined an item along the trait.

Activity #6

Find a survey instrument that requires respondents to indicate how often something takes place, for example, using a rating scale of *Very Often*, *Often*, *Sometimes*, *Seldom*, and *Never*. To ensure an analysis that is similar to that presented in this chapter, first ensure that the selection of *Very Often* is the type of response that you want for the analysis (if it is not, then remember you will need to reverse code the item). Enter your data into a spreadsheet using the coding scheme 5 for *Very Often*, 4 for *Often*, 3 for *Sometimes*, 2 for *Seldom*, and 1 for *Never*. Then use earlier chapters to create a control file and conduct a Rasch analysis. After you have run the data, work through this chapter, but with your data set. Practicing in this manner will not only help you understand this chapter's text (as well as other chapters), but you will then not have to memorize all the steps that we present in this book.

Answer: Correctly following the steps we have outlined will allow this activity to be completed. Remember to reverse code, also remember to alter the item names for the items that had to be recoded.

Activity #7

Review the control file "cf for SE for Chp 5 not rescaled." Where can you find the coding used for the data? Where can you find the meaning of the numbers used to code the responses?

Answer: The numbers used for the Rasch analysis are contained in the line that starts with the phrase "CODES." The numbers that follow are the numbers of the codes used to code responses. The meaning of the numbers is not in the control file. The analyst is responsible for keeping track of what numbers were used to code responses.

Activity #8

In each chapter of this book, we provide readers with details of how to run a Rasch analysis using Winsteps. Go to the Winsteps website and download the manual. Then search for the terms UMEAN and USCALE to see added details regarding these two terms. Also, scan the Winsteps Table of Contents to obtain an overview of the material contained (as of this date) in the 677 page manual.

Activity #9

Discuss with a colleague when it might be useful to use USCALE and UMEAN to create a scale that does not have negative numbers. If you plan to transform the scale of negative and positive values to a scale which is only positive, what might be the range you choose to use? Why?

Activity #10

The STEBI, which is used for much of this book, contains 23 items. Thirteen items measure the trait of self-efficacy, and 10 items measure the trait of outcome-expectancy. The 23 items of the STEBI in the paper presentation of the survey are not presented as a function of trait (the self-efficacy items are not all presented first and then followed by all the outcome-expectancy items). Can you think of pros and cons of presenting items on a survey in this manner?

Answer: Some survey developers will mix the order of items in a survey. The idea is to keep the respondent alert. We think it is preferable to present items from one trait first, then to another trait. Ordering items in this manner might make a survey less taxing for a respondent to complete and will lessen the chances of coding errors by the analyst. Our personal view is also to skip avoid using negatively worded items in a survey.

References

Enochs, L. G., & Riggs, I. M. (1990). Further development of an elementary science teaching efficacy belief instrument: A pre-service elementary scale. *School Science and Mathematics, 90*(8), 694–706.

Additional Readings

Review the discussion in the Winsteps user manual pertaining to Table 14.

Linacre, J. M. (2011). *WINSTEPS user manual*. Chicago: MESA Press.

This article presents an early Rasch analysis by Boone for a science education rating scale data set in an effort to better understand students' views toward a science methods curriculum.

Boone, W. J., & Andersen, H. O. (1994). Designing, evaluating, and reacting to a secondary science methods class. *Journal of Science Teacher Education, 5*(1), 15–22.

An excellent, brief, easy to read article concerning the problems with raw scores.

Wright, B. D. (1993). Thinking with raw scores. *Rasch Measurement Transactions, 7*(2), 299–300.

Chapter 6
Wright Maps: First Steps

Isabelle and Ted: Two Colleagues Conversing

Ted: Okay. Well, we have looked through the person measure table, and I think I understand the basics of such tables. I can find the total score of the persons, and I understand that the person measures are what I need to use for further statistical analysis. I should not use the raw data.

Isabelle: What else?

Ted: Well, when I code my rating scale where a higher number is better, then the person measures represent people who are doing more of what I would like to see. So, if I give a survey to teachers where Strongly Agree is what I hope to see for the best teachers, and if I code SA with a higher number, say a 6, if the scale is SD, D, BD, BA, A, and SA, I can use the same techniques to understand the item tables.

Isabelle: How?

Ted: Well…if I code data in the manner I just mentioned, I now know that a lower item measure means that an item is easier to agree with compared to an item with a higher measure. Now I want to understand how the item measures and person measures fit together.

Isabelle: Ahh, you're talking about Wright Maps.

W.J. Boone et al., *Rasch Analysis in the Human Sciences*,
DOI 10.1007/978-94-007-6857-4_6, © Springer Science+Business Media Dordrecht 2014

```
           Person - MAP - Item

  3           T+
             X |
               |
           XXX |
       XXXXXXX |T
  2      XXXX S+
               |
            XX | Q19se-rc
         XXXXX |
          XXXX |S Q5se
  1        XXX M+
XXXXXXXXXXXX | Q20se-rc Q23se-rc
       XXXXX | Q17se-rc
    XXXXXXXX | Q12se
         XXX | Q18se  Q21se-rc Q3se-rc
  0      XXX S+M
       XXXXX |
        XXXX | Q6se-rc
         XXX |
           X |
 -1          + Q8se-rc
            T|S
             |
             |
             | Q22se
 -2          +
             |T
             | Q2se
             |
```

Introduction

Wright Maps (person-item maps) are a revolutionary technique for displaying very complex rating scale data and test data. In this chapter, we will present the particulars of Wright Maps, so that researchers can construct and incorporate Wright Maps into their publications. The Wright Map was, for a long period of time, named a person-item map, but more often now the map is referred to as a Wright Map (Wilson & Draney, 2000).

We begin with some theory and then, using a key Winsteps table, turn to the construction and interpretation of a simple Wright Map. As readers should note, we are believers that one understands by doing. It is possible to click a button and generate a Wright Map with Winsteps; however, by constructing a Wright Map "by hand," readers will construct a deeper level of understanding.

In previous chapters, we discussed the importance of conceptualizing the single variable, the single trait that is to be measured by a set of survey items. When a researcher wants to administer a set of items to a group of respondents in order to begin pondering the use of total scores to differentiate between respondents, then a requirement is that all items of a survey involve only one general issue. Stated another way, the instrument must be unidimensional. If unidimensionality is not met, then it is meaningless to compute a total measure and "compare" respondents or items. We have also introduced the importance of conceptualizing a variable. You can conceptualize a variable when you draw a line. This is when you draw a horizontal line and you describe what it means to be at different parts of a variable. In the case of self-efficacy, you predict some examples of self-efficacy that would be easy for someone to achieve and some examples of self-efficacy that would be hard for someone to achieve. By conceptualizing a variable, one is able to better define what it means to be at different parts of a trait. In the case of self-efficacy, the various parts of the trait are defined by the terms low self-efficacy, medium self-efficacy, high self-efficacy, and so on. When one uses theory to define a trait with items, one helps improve the possibility that a range of a trait is defined. This helps avoid presenting respondents with redundant items, which not only waste respondents' time but also decrease the quality of their remaining responses, as they often become frustrated, tired, or lose interest. Of course, the reality of the matter is that one can attempt to predict how items define a trait, but reality can be different than what one thinks. As readers will see, Wright Maps allow researchers to quickly and thoughtfully evaluate how items of a survey (and tests) define a trait. Being able to see how items define a trait helps researchers in many important ways. First, researchers can assess an instrument's strengths and weaknesses. Second, researchers can use the Wright Map to document the hierarchy of survey items as expressed by the surveyed group of respondents. Third, with a Wright Map, one can quickly compare theory with what was observed in the data set. To help readers learn how to use Wright Maps in these three ways, we will first walk through the construction, by hand, of a Wright Map. Whereas a Wright Map can be obtained by simply pushing a button, we urge readers just beginning Rasch not to do so. The prize of constructing one by hand is deeper understanding.

Formative Assessment Checkpoint #1

Question: If you imagine the self-efficacy trait of the STEBI provides person measures and item measures that mark parts of a meterstick, can you make a drawing that would show the self-efficacy items and some of the persons who take the STEBI on the meterstick?

Answer: You can imagine that the 13 self-efficacy items of the STEBI define a variable of self-efficacy. Some items are easier to agree with than other items. Those items that are easier to agree with exhibit lower item measures than items that are harder to agree with. Persons higher in self-efficacy measure have higher person measures; they have higher person measures because they are more agreeable to more items. The meterstick immediately below in Fig. 6.1 could be used to schematically show the manner in which persons fall on the trait of self-efficacy and the manner in which items define the trait.

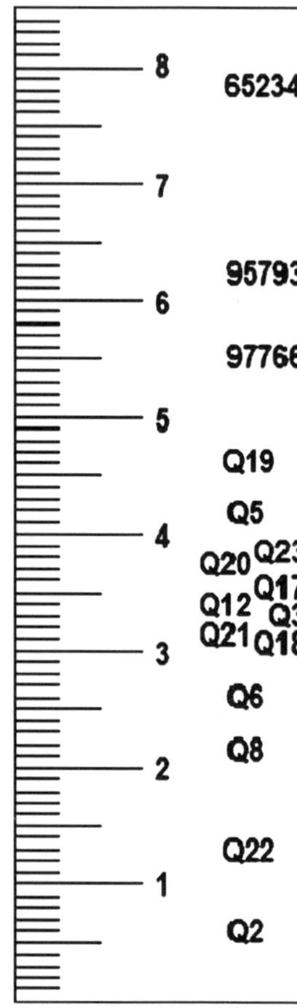

Fig. 6.1 STEBI self-efficacy items and persons transposed onto a meterstick. The three respondents are plotted at 5.5, 6.2, and 7.9 on the meterstick; the items are plotted below the three respondents

Constructing a Wright Map by Hand

Wright Maps display both people and items along the unidimensional logit scale used in Rasch measurement. To begin, rerun the self-efficacy data (cf for SE for Chp5 not rescaled) that were used for Chap. 5. We provide this control file again as a part of this chapter. After rerunning these data, select table "13. Item: measure" from the *Output Tables* list. This table, which is presented in Fig. 6.2, gives the calibrations or measures of items in measure order. For example, the 19th item (Q19) of the survey, the self-efficacy item that was flipped, is calibrated at 1.66 logits. The item listed immediately below item Q19 is item Q5. This item has a calibration of 1.16 logits. Careful review of the item measure values reveals that all 13 self-efficacy items are presented from highest measure to lowest measure (item Q2 exhibits a value of −2.49 logits).

To begin construction of a Wright Map, first determine the range of measure units, the number between the highest item measure and the lowest item measure.

```
TABLE 13.1 SCIENCE TEACHER EFFICACY BELIEFS      ZOU938WS.TXT  Sep 16 11:33 2011
INPUT: 75 Person  23 Item  REPORTED: 75 Person  13 Item  6 CATS  MINISTEP 3.72.3
--------------------------------------------------------------------------------
Person: REAL SEP.: 2.52  REL.: .86 ... Item: REAL SEP.: 7.00  REL.: .98

        Item STATISTICS:  MEASURE ORDER
```

ENTRY NUMBER	TOTAL SCORE	TOTAL COUNT	MEASURE	MODEL S.E.	INFIT MNSQ	INFIT ZSTD	OUTFIT MNSQ	OUTFIT ZSTD	PT-MEASURE CORR.	PT-MEASURE EXP.	EXACT MATCH OBS%	EXACT MATCH EXP%	Item
19	206	68	1.66	.13	1.10	.7	1.21	1.2	.67	.69	44.8	40.8	Q19se-rc
5	258	75	1.16	.12	1.28	1.7	1.34	1.9	.54	.67	37.8	41.2	Q5se
23	260	69	.79	.13	.80	-1.3	.86	-.8	.70	.63	44.1	41.1	Q23se-rc
20	262	69	.76	.13	1.07	.5	1.13	.8	.65	.63	39.7	41.3	Q20se-rc
17	277	69	.50	.13	.69	-2.1	.72	-1.7	.71	.61	39.7	42.8	Q17se-rc
12	310	75	.34	.13	.99	.0	.96	-.2	.56	.60	52.7	44.2	Q12se
3	317	75	.22	.13	1.69	3.5	1.64	3.1	.52	.59	33.8	44.5	Q3se-rc
21	295	69	.17	.14	.77	-1.4	.84	-.9	.65	.58	45.6	44.9	Q21se-rc
18	298	69	.11	.14	.66	-2.1	.62	-2.3	.62	.57	47.1	46.0	Q18se
6	352	75	-.48	.15	.96	-.2	.89	-.5	.52	.53	51.4	54.5	Q6se-rc
8	369	75	-.91	.17	1.03	.2	.97	-.1	.55	.50	59.5	57.9	Q8se-rc
22	364	69	-1.83	.21	.97	-.1	.93	-.3	.51	.44	64.7	63.7	Q22se
2	410	75	-2.49	.22	1.06	.4	1.08	.5	.31	.41	55.4	64.4	Q2se
MEAN	306.0	71.7	.00	.15	1.00	.0	1.01	.1			47.4	48.2	
S.D.	53.9	3.1	1.12	.03	.26	1.5	.26	1.4			8.6	8.4	

Fig. 6.2 STEBI data analyzed and presented in Winsteps Table 13

In this case, the range is 4.15 ((1.66–(–2.49))=4.15). Later we may revise the range of our map due to wishing to plot all respondents, but for now let's start with the items! To construct our Wright Map, we need a scale that runs from a maximum of 1.66 to a minimum of –2.49. When we conduct this exercise in our classes, we use large chart paper, which makes graphing easier by minimizing the time spent on computing how many centimeters represent how many logits. Here, we will use a single sheet of paper and calculate the conversion for readers. With a sheet of paper typically used in a computer printer, use a scale of 1 cm=.15 logits to quickly create our Wright Map. The next step is to place a ruler from top to bottom in the center of the page, use the edge of the ruler to draw a single vertical line through the center of the page, and then mark a horizontal line at each centimeter. The single vertical line is approximately 28 cm long.

After drawing this line, label the scale starting at 1 cm at the base of the page with the number –2.50 (we pick a round number to start the bottom of the scale, but a number that is a little more negative than the logit value of the most negative item). Then, use the centimeter marks up the vertical line to mark an increase of 0.15 logits. The result should be a drawing like a thermometer, where the highest mark is at least 1.66 and the lowest mark is at least as negative as –2.49. We could easily have used increases of 0.20 logits or 0.25 logits. However, having 1 cm be as small a logit increase as possible allows us to provide more detail in the Wright Map.

The next step is to carefully label what it means for an item to be placed toward the top or toward the base of the Wright Map. Considering a thermometer, a reading higher on the thermometer means a higher temperature – a higher measure of heat than a reading lower on a thermometer, a lower temperature – a lower measure of heat. When we construct our Wright Map, we need to make sure we understand what it means to be higher and lower on the Wright Map.

To understand what going up or going down means, we first plot the location of the survey item with the highest item measure (Q19 at 1.66 logits). We then plot the location of the survey item with the lowest item measure (Q2 at –2.49 logits). After plotting these two items, look at the total score of each item and also check how respondents' answers were coded. In the case of the self-efficacy survey, a rating scale of SA, A, BA, BD, D, and SD was used for the rating scale categories. Recall that the number 6 was used for SA, 5 for A, and so on. When students and workshop participants are reminded of this coding (and the rating scale words the numbers represent), they quickly realize that in this particular data set, a higher item measure means an item was harder to agree with compared to an item with a lower item measure that was easier to agree with. Following the identification of what it means to have a high item measure and a low item measure, readers should label the top of their Wright Map. Regarding this rating scale, helpful labels are "easier to agree with" at the bottom of the Wright Map and "harder to agree with" at the top. Regarding other rating scales, the high and low labels could be "more frequent" and "less frequent" for a frequency scale, or "more supportive" and "less supportive" for an attitude scale. We encourage readers to put (in this example) "harder to agree with" at the top of the plot and "easier to agree with" at the base of the plot. This is because all we know at this point is the meaning of moving up or down the map.

Moving forward in our construction project, readers should now mark the location of each item on the vertical line and initially use the shorthand coding we used to identify items. This shorthand can be found in the far right-hand column of our Fig. 6.2 (Winsteps Table 13.1). Our Wright Map with items is now nearly completed – nearly completed because in Rasch measurement, error is always considered. This is true in all measurements in science. To show the error of measurement for each survey item, one can plot at least one standard error above each item and at least one standard error below each item. In Fig. 6.2 (Winsteps Table 13.1), the standard errors of the measures are found in the "Model S.E." column. For example, the standard error for item 18 is 0.14 logits. Generally the more people who complete an item and provide information regarding an item, the less measurement error an item exhibits.

The next step is to include the text of each survey item along the Wright Map. Remember to include revised wording of items that were reverse coded prior to analysis. For example, item 19 might be rephrased to "I do not wonder if I have the necessary skills to teach science" from "I wonder if I have the necessary skills to teach science."

Once you have completed your Wright Map with items, compare your map to the map created by Winsteps (Winsteps Table 12). If you have plotted correctly, you should see a similar pattern of items (from easier to harder). Why is the pattern not identical? If you did not make a plotting error, any differences have are due to scaling (which for us means how much room there is vertically for plotting persons and items). If your plot provides more room for plotting items, you may see some items that are not plotted at the same part of the vertical axis spread out. This is simply because one can plot the location of items more exactly with the more detailed scale. The same issue is present for plotting persons. An important aspect of Wright Maps is plotting person measures. For now we will discuss what can be learned just from a Wright Map with items.

Formative Assessment Checkpoint #2

Question: Should the Wright Map constructed by hand match the Wright Map created by Winsteps?

Answer: Yes and No. Yes, if you pick an identical scale to the Winsteps map for the length of a logit on your paper. No, if you pick a different scale (most likely your scale will have a unit of logits that are longer in centimeters than in the Winsteps map); there will not be an exact lineup of your map and the Winsteps map. The important thing to remember is any differences between the Wright Map you construct by hand and the Wright Map constructed by Winsteps are a matter of scaling of the vertical line that is used to plot the person measures and the item measures.

Informing Instrument Design from the Plotting of Item Difficulty

Developers of robust instruments that accurately measure respondents' responses must address a number of important issues. One issue focuses on the manner in which a set of survey items defines a trait. A review of the self-constructed Wright Map reveals gaps in the distribution of survey items (e.g., between Q8: –0.91 and Q22: –1.83). Also, the pattern of items from "less easy to agree" with toward "easier to agree with" reveals that some portions of the trait are oversampled. For example, there are four items (Q3, Q18, Q12, Q17) along the trait between Q21: 0.17 logits and Q20: 0.76 logits.

To understand the implications of large and uneven gaps between items, we will utilize an analogy that two of this book's authors have used in an article (Boone, Townsend, & Staver, 2011). Imagine that you are running a meterstick factory and you have lots of blank pieces of wood one meter long. Also, imagine that you have a machine that can make "cuts" in the meterstick, but only a limited number of cuts can be made on each piece of wood. If we say that only five cuts can be made in the meterstick, it should be evident to readers that if we have no idea of the length of objects we may be measuring, then a meterstick with large gaps between cuts and duplicate cuts (cuts on top of one another or so closely cut to another cut) will do a poorer job of measuring a range of item lengths. It would be better to have a meterstick with a more evenly spaced (more optimized) distribution of cuts.

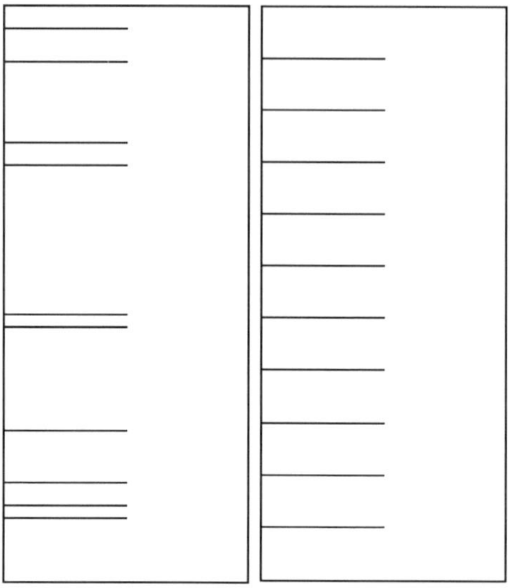

Fig. 6.3 Two metersticks with differing distribution of "cuts." The distribution of cuts greatly impacts the quality of measurement that can be made with a meterstick

Figure 6.3 displays two metersticks. One meterstick has ten cuts, but there are some gaps and also some overlaps of cuts. Wooden blocks of very similar length would not be accurately measured with this ruler. The second meterstick of the same length has ten equally spaced cuts. Just as the evenly spaced cuts of a meter stick provide a better measure of blocks of unknown length, a maximized distribution of items (i.e., no overlap of items) is the "better" meterstick when the location of persons along the trait is unknown.

Wright Maps can inform several aspects of instrument design. In subsequent chapters, we will introduce additional nuances of how researchers can use Wright Maps to evaluate and refine measurement instruments. For the time being, simply think of items as making cuts on a meterstick. Generally, it is most often best to have a nice-sized, consistent distribution of cuts. However, it is not always best to conclude "too many items at a difficulty level, let's get rid of some of these items to improve the instrument." One example is when a researcher wants to ensure a test taker has exhibited a particular competence level. In this case, one might want to have several cuts near a particular logit value. The reason for this is one does not really care if the respondent can correctly answer items that are quite a bit harder than the specific competence level of interest. For rating scales, gaps in cuts often most importantly suggest a possible misconception of the variable on the part of the instrument developer. This of course assumes the developer was thinking about creating items which mark different parts of a single trait.

Formative Assessment Checkpoint #3

Question: Are more items always better than fewer items?

Answer: No. Not always. It depends on where your items fall on the variable and what your measurement goal is.

The Hierarchy of Items

The ordering and spacing of items are important for the assessment of a test's (and a survey's) measurement qualities, but the Wright Map also has profound implications for research in many disciplines (e.g., Wright Maps are now providing great guidance to researchers in fields of medicine). To explain this application of Wright Maps, we will go back in time, over 20 years ago, to an institution (the University of Chicago) and a professor (Benjamin Wright) who was a mentor to one of the authors when he was a graduate student. In psychometric classes, Professor Wright would often hold up a Wright Map that was constructed with elementary school math items that were presented in a test named KeyMath (Wright, 2012). When Professor Wright presented the Wright Map to the class, he talked about how the ordering and spacing presented a roadmap of sorts to teachers. Items at the easy end of the Wright Map represent aspects of the trait that should be mastered by students before items at higher and higher – harder and harder – levels representing more difficult aspects of the trait. He suggested that teachers should consider the order of items as they present a curriculum to students. In recent years, "learning progressions" have been set forth and examined in numerous science education studies (e.g., Liu, 2010). In our opinion, at the core of these learning progressions is the same conceptual idea that Professor Wright presented more than two decades earlier.

To understand how a Wright Map of survey data (presenting a hierarchy of items) could inform research, we will turn briefly to a number of examples of Wright Maps constructed with survey data. We then present a number of observations that could be made in a conference talk or a manuscript submitted for publication. The number and the order of steps can, of course, be altered to suit a particular study.

Since the Wright Map of the STEBI self-efficacy data has already been constructed, let's review the ordering and spacing of items. Some particular questions to contemplate are: Is the ordering surprising? Does it match what one would predict from theory? (Perhaps some groups of items match theory, while other items do not match.) Why might that be? Do items need to be improved in some manner? Do items need to be added? Removed? Does the theory need to be revised? We pose such questions to illustrate the value of Wright Maps. Item ordering and spacing can, for example, be reviewed following a Winsteps Rasch analysis of survey rating scale data, multiple-choice tests, or partial credit tests. Once one masters how to think about and use Wright Maps for instrument development, instrument refinement, and curricular guidance, then readers will discover a multitude of advances that they can make in their field of choice (e.g., education, psychology, medicine, business).

As one reviews the STEBI data collected from the group of 75 pre-service science teachers, one does see a general pattern from "easy to agree with" to "more difficult to agree with." For these respondents, it is quite easy to agree with the statements concerning their interest in finding better ways to teach science (Q2) and welcoming student questions (Q22). However, when responding to questions such as Q18, the students have less self-efficacy, and they indicate less self-efficacy with respect to their understanding of science (Q12). One item that respondents find extremely hard, relative to other items, to agree with is their self-efficacy in knowing the steps to teach science (Q5). Perhaps not surprisingly, respondents found it difficult, relative to the other items, to agree with item Q19 (*I do not wonder if I have the skills necessary to teach science effectively*). As we present this ordering of a handful of the self-efficacy items, readers who prepare science teachers may say "of course," but we have found that the ordering is not always self-evident. If it is evident (at some level), being able to think about (and see) the item ordering and spacing reminds one of issues that might need to be addressed in a class and/or in a science teacher preparation program that might consist of numerous classes.

Formative Assessment Checkpoint #4

Question: Is being enamored with a single variable a waste of time?

Answer: No. Thinking about a single variable helps build measurement instruments that are reliable, valid, robust, and trustworthy. Working with a single variable also allows researchers to look at a hierarchy of items that possess valid meaning and guidance.

Plotting Persons on Wright Maps

Wright Maps are constructed not only to show the hierarchy of survey items but also to show simultaneously the hierarchies of both persons and items. We chose to limit the amount of material to be digested in one sitting by beginning with only a plot of items. We will now add people to a Wright Map and explain how researchers can examine persons in a Wright Map, and examine both items and persons simultaneously in a Wright Map – all to understand a data set and improve a measurement instrument!

Adding person measures follows a process similar to adding item measures to the Wright Map. First, we need to find the calculated person measures from the Winsteps analysis. Figure 6.4 (Winsteps Table 18.1) provides the logit measures for the 75 respondents evaluated in the STEBI self-efficacy data. To add persons to the Wright Map, we use the MEASURE column, which is the 4th column in Fig. 6.4 (Winsteps Table 18.1). Before plotting the data, we must understand what higher and lower person measures indicate. We employ a technique that parallels the process we used to better understand the item measure table. To understand the meaning of the person measures, first find two persons who answered the same number of items. The 1st person in the data set (Person 21141) and the 4th person in the data set (Person 08543) are good candidates, for they both answered all 13 items, as

```
TABLE 18.1 SCIENCE TEACHER EFFICACY BELIEFS       ZOU938WS.TXT  Sep 16 11:33 2011
INPUT: 75 Person  23 Item  REPORTED: 75 Person  13 Item  6 CATS  MINISTEP 3.72.3
--------------------------------------------------------------------------------
Person: REAL SEP.: 2.52  REL.: .86 ... Item: REAL SEP.: 7.00  REL.: .98

        Person STATISTICS:  ENTRY ORDER
```

ENTRY NUMBER	TOTAL SCORE	TOTAL COUNT	MEASURE	MODEL S.E.	INFIT MNSQ	INFIT ZSTD	OUTFIT MNSQ	OUTFIT ZSTD	PT-MEASURE CORR.	PT-MEASURE EXP.	EXACT MATCH OBS%	EXACT MATCH EXP%	Person	
1	60	13	1.34	.35	1.71	1.5	1.63	1.4	.52	.57	46.2	50.2	21141	PR
2	27	6	.84	.50	.71	-.3	.57	-.6	.83	.60	50.0	46.1	91052	PR
3	33	6	3.28	.81	.62	-.5	.57	-.5	.61	.46	66.7	66.9	95793	PR
4	47	13	.08	.29	.31	-2.6	.28	-2.5	.84	.67	46.2	40.0	08453	PR
5	51	13	.43	.30	.52	-1.5	.53	-1.4	.81	.64	38.5	42.7	36281	PR
6	64	13	1.92	.41	.97	.1	.86	-.2	.77	.53	61.5	60.3	85453	PR
7	21	6	-.42	.44	.27	-1.7	.29	-1.5	.91	.71	66.7	42.2	46328	PR
8	39	13	-.64	.31	.52	-1.3	.41	-1.7	.86	.73	61.5	45.3	41024	PR
9	44	13	-.18	.30	1.13	.5	1.08	.3	.69	.70	15.4	39.5	08746	PR
10	56	13	.90	.32	.88	-.2	.87	-.2	.52	.60	61.5	40.9	09132	PR
11	42	13	-.36	.30	1.12	.4	1.00	.1	.78	.71	23.1	43.5	28100	PR
12	66	13	2.28	.45	.99	.2	1.05	.3	.47	.52	53.8	63.8	43532	PR
13	54	13	.70	.31	1.05	.3	1.36	1.0	.45	.62	23.1	41.5	36754	PR
14	39	13	-.64	.31	.91	-.1	.80	-.3	.70	.73	46.2	45.3	53695	PR
15	51	13	.43	.30	.79	-.5	.75	-.6	.73	.64	46.2	42.7	65759	PR
16	59	13	1.22	.34	.50	-1.3	.60	-.9	.60	.57	46.2	49.6	40166	PR
17	68	13	2.71	.48	3.90	3.6	2.57	2.5	.53	.50	61.5	66.2	55959	PR
18	67	13	2.49	.46	1.00	.2	.98	.1	.64	.51	61.5	63.7	97766	PR
19	41	13	-.45	.30	1.17	.5	1.09	.4	.61	.72	30.8	43.7	78880	PR
20	37	13	-.84	.32	.40	-1.7	.46	-1.4	.91	.74	61.5	46.5	33573	PR
21	66	13	2.28	.45	1.37	.8	1.23	.6	.62	.52	61.5	63.8	99365	PR
22	57	13	1.00	.33	.32	-2.2	.33	-2.1	.80	.59	69.2	43.2	18489	PR
23	53	13	.61	.30	1.15	.5	1.07	.3	.70	.62	38.5	41.0	96468	PR
24	60	13	1.34	.35	.45	-1.4	.48	-1.3	.68	.57	53.8	50.2	37854	PR
25	56	13	.90	.32	.41	-1.8	.38	-1.9	.67	.60	69.2	40.9	71215	PR
26	52	13	.52	.30	1.35	1.0	1.40	1.1	.42	.63	15.4	42.6	87610	PR
27	64	13	1.92	.41	.45	-1.2	.48	-1.2	.54	.53	76.9	60.3	00103	PR
28	60	13	1.34	.35	.52	-1.2	.70	-.6	.73	.57	53.8	50.2	66990	PR
29	44	13	-.18	.30	1.29	.8	1.33	.9	.64	.70	38.5	39.5	97427	PR
30	22	6	-.22	.44	2.46	2.1	2.12	1.7	.32	.70	33.3	41.4	80392	PR
31	31	6	2.17	.68	1.28	.6	1.24	.6	.70	.52	33.3	65.1	78255	PR
32	24	6	.17	.45	.77	-.3	.63	-.6	.68	.66	66.7	41.1	19984	PR
33	59	13	1.22	.34	2.10	2.1	1.97	1.9	.41	.57	46.2	49.6	90384	PR
34	51	13	.43	.30	1.42	1.1	1.87	2.0	.32	.64	46.2	42.7	96639	PR
35	59	13	1.22	.34	.29	-2.2	.34	-1.9	.63	.57	76.9	49.6	32253	PR
36	75	13	4.89	.68	.93	.0	.70	-.1	.37	.32	69.2	77.8	65234	PR
37	66	13	2.28	.45	.68	-.5	.85	-.2	.31	.52	69.2	63.8	76868	PR
38	52	13	.52	.30	1.20	.6	1.44	1.1	.58	.63	53.8	42.6	66462	PR
39	62	13	1.61	.38	.63	-.7	.74	-.5	.64	.55	69.2	53.5	13684	PR
40	50	13	.34	.30	1.09	.4	.99	.1	.71	.65	38.5	40.9	84422	PR
41	49	13	.25	.29	.99	.1	.92	-.1	.61	.66	46.2	40.6	41426	PR
42	62	13	1.61	.38	.28	-2.0	.29	-2.1	.86	.55	69.2	53.5	01005	PR
43	56	13	.90	.32	.73	-.6	.69	-.7	.68	.60	53.8	40.9	45208	PR
44	66	13	2.28	.45	1.07	.3	1.15	.5	.58	.52	61.5	63.8	94129	PR
45	51	13	.43	.30	1.68	1.7	1.78	1.8	.79	.64	7.7	42.7	65040	PR
46	78	13	7.53	1.85	MAXIMUM MEASURE				.00	.00	100.0	100.0	65730	PR
47	65	13	2.09	.43	2.16	1.9	2.16	2.0	.23	.53	46.2	63.1	07242	PR
48	51	13	.43	.30	.76	-.6	.70	-.7	.77	.64	46.2	42.7	95626	PR
49	55	13	.80	.31	2.25	2.5	1.99	2.1	.42	.61	15.4	41.4	78221	PR
50	67	13	2.49	.46	2.24	2.0	2.07	1.9	.48	.51	53.8	63.7	68028	PR
51	49	12	.47	.31	1.18	.6	1.27	.8	.50	.59	25.0	41.9	94827	PR
52	44	13	-.18	.30	.94	.0	.90	-.1	.80	.70	23.1	39.5	36206	PR
53	61	13	1.47	.36	.81	-.3	.88	-.1	.46	.56	38.5	52.6	94880	PR
54	64	13	1.92	.41	.34	-1.6	.49	-1.2	.70	.53	76.9	60.3	89570	PR
55	39	13	-.64	.31	1.32	.9	1.34	.9	.51	.73	30.8	45.3	12103	PR
56	53	13	.61	.30	1.17	.6	1.20	.6	.76	.62	38.5	41.0	98375	PR
57	42	13	-.36	.30	1.48	1.2	1.34	.9	.66	.71	30.8	43.5	99843	PR
58	55	13	.80	.31	.28	-2.5	.28	-2.5	.75	.61	76.9	41.4	20408	PR
59	52	13	.52	.30	1.33	.9	1.21	.6	.60	.63	53.8	42.6	01849	PR
60	66	13	2.28	.45	2.40	2.1	2.10	1.9	.25	.52	53.8	63.8	67921	PR
61	56	13	.90	.32	.90	-.1	1.07	.3	.79	.60	23.1	40.9	57052	PR
62	50	13	.34	.30	.63	-1.1	.64	-1.0	.78	.65	46.2	40.9	38019	PR
63	66	13	2.28	.45	.28	-1.8	.31	-1.9	.64	.52	84.6	63.8	37339	PR
64	46	13	-.01	.29	.68	-.9	.69	-.8	.81	.68	61.5	41.5	64842	PR
65	56	13	.90	.32	.49	-1.5	.63	-.9	.70	.60	38.5	40.9	54983	PR
66	48	13	.17	.29	1.50	1.3	1.36	1.0	.67	.67	23.1	40.6	49712	PR
67	47	13	.08	.29	1.04	.2	1.02	.2	.61	.67	38.5	40.0	24639	PR
68	57	13	1.00	.33	.79	-.4	.65	-.8	.60	.59	30.8	43.2	81997	PR
69	57	13	1.00	.33	.97	.1	.95	.0	.73	.59	23.1	43.2	85410	PR
70	55	13	.80	.31	1.02	.2	1.29	.8	.33	.61	53.8	41.4	51994	PR
71	56	13	.90	.32	.78	-.5	.87	-.2	.72	.60	38.5	40.9	31122	PR
72	61	13	1.47	.36	2.71	2.8	2.42	2.4	.50	.56	38.5	52.6	96739	PR
73	67	13	2.49	.46	.19	-2.3	.19	-2.5	.82	.51	92.3	63.7	81223	PR
74	44	13	-.18	.30	.99	.1	1.08	.3	.65	.70	7.7	39.5	99055	PR
75	58	13	1.11	.33	.74	-.6	1.07	.3	.17	.58	30.8	47.3	77427	PR
MEAN	53.0	12.4	1.05	.38	1.04	-.1	1.01	.0			47.7	48.5		
S.D.	11.7	1.9	1.29	.20	.65	1.3	.55	1.2			18.7	9.5		

Fig. 6.4 A Winsteps output table that presents the person measures of 75 respondents presented with the 13 self-efficacy items of the STEBI

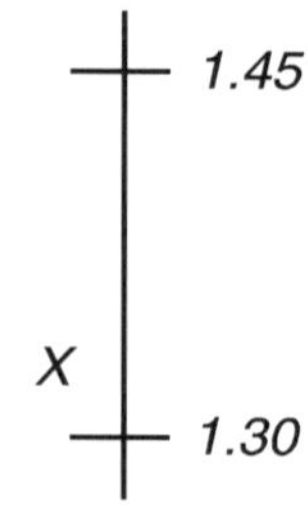

Fig. 6.5 Plot of a person (using a symbol of an X) who has a person measure of 1.34

shown in the TOTAL COUNT column for these two individuals. Next, look at the TOTAL SCORE for these two people and their associated Rasch logit measure (Person 21141, Total Score 60, Measure 1.34; Person 08543, Total Score 47, Measure 0.08). Recall that the coding of respondent answers was 6 for SA, 5 for A, and so on. This enables one to quickly realize that a higher logit value for a person measure means more agreement (and thus more confidence in teaching science). Person 21141 has a higher total score than Person 08543; Person 21141 has a higher person measure in logits than Person 08543.

An equally quick technique to better understand the meaning of a person's measure is to insert two fictitious persons after the last line of the data set in a control file. We find it helpful to provide a person ID, such as "Mr. Positive" or "Mr. Agreeable," and to add answers for this person that would be provided by someone who was extremely agreeable (had strong self-efficacy). After "Mr. Agreeable," we add a "Mr. Disagreeable". That person would be someone who had very low self-efficacy. Such temporary placement of fictitious persons in the control file is one way to quickly clarify or double-check the meaning of the person measures. In this case, the higher a person measure on the Wright Map, the more agreeable the respondent and thus the respondent who has more self-efficacy in teaching science.

To continue, we return to our paper plot of the calibrations of items in logit units. To expand our plot to include person measures, we must of course find the measure of each person and then mark the location of the person on the Wright Map. For example, for the first person (person 21141), mark an "X" at approximately "1.34" on the plot. Figure 6.5 shows the marking of the X for the person who has a value of 1.34. Hint: This will take some time to mark all 75 respondents, so find a friend to help. He or she can read each person's measure and you can plot it.

For this initial plot, the locations of persons need not be 99.99 % accurate. Remember, we are doing this exercise to understand how a Wright Map is automatically made. Since your paper Wright Map may not be very long, there exists a limit to the level of plotting accuracy. Also, recall that some error exists for every item and every person.

In a Wright Map, we look for trends. When you have plotted all 75 persons, you should have a plot that looks like the one presented below (Fig. 6.6). This is the Wright Map that is presented by simply selecting Table 12.2 ("12. Item: map") in Winsteps. Of importance for you is to note when you create similar plots, as we have detailed, you will see that sometimes you will need to lengthen the plot because

```
TABLE 12.2 SCIENCE TEACHER EFFICACY BELIEFS      ZOU938WS.TXT  Sep 16 11:33 2011
INPUT: 75 Person  23 Item  REPORTED: 75 Person  13 Item  6 CATS  MINISTEP 3.72.3
--------------------------------------------------------------------------------

         Person - MAP - Item
             <more>|<rare>
    6           X  +
                   |
                   |
                   |
                   |
    5              +
                X  |
                   |
                   |
                   |
    4              +
                   |
                   |
                   |
                X  |
    3             T+
                X  |
                   |
              XXX  |
          XXXXXXX  |T
    2        XXXX S+
                   |
               XX  |  Q19se-rc
            XXXXX  |
             XXXX  |S Q5se
    1         XXX M+
      XXXXXXXXXXX  |  Q20se-rc  Q23se-rc
            XXXXX  |  Q17se-rc
         XXXXXXXX  |  Q12se
              XXX  |  Q18se     Q21se-rc  Q3se-rc
    0         XXX S+M
            XXXXX  |
             XXXX  |  Q6se-rc
              XXX  |
                X  |
   -1              +  Q8se-rc
                  T|S
                   |
                   |
                   |  Q22se
   -2              +
                   |T
                   |  Q2se
                   |
                   |
   -3              +
             <less>|<frequ>
```

Fig. 6.6 The Wright Map that is constructed in Winsteps (Table 12.2)

there can be people above the highest item measure or below the lowest item measure. The next time that you make a Wright Map by hand, look at the person measure table and the item measure table at the same time. Find the highest person-item, and find the lowest person-item. This will help you plot all persons and items easily.

Now that you have constructed the Wright Map, it is important to understand the meaning of going up and down the Wright Map when you are looking at other data sets.

How do you know what the meaning of going up and down is for both people and items on the Wright Map? Think back to some of the explanations in Chap. 5. We encourage our students to find the two extreme items on the Wright Map and also to find the two extreme persons. Open the item measure table (Table 13.1) and the person measure table (Table 17.1) in Winsteps. The first and last person noted in the person measure table will be the maximum and minimum person measures identified in the Wright Map. By following the same procedure, you can also find the items with the maximum and minimum item measures. Now look up the coding used to code the rating scale for this data set. Then write down the TOTAL SCORE for the two persons and two items. Below we provide some fictitious TOTAL SCORES for two items and two persons on a Wright Map that was created following analysis of data in which respondents could answer *Yes* (2), *Maybe* (1), and *No* (0).

		TOTAL SCORE	MEASURE
Item	17	420	−2.02
Item	4	101	3.31
Person	45220	30	2.98
Person	45208	2	−3.22

By creating this table, we can understand what it means to go up or down either side of the Wright Map. In this case, to go up the Wright Map from item 17 to item 4, one is moving toward items that are harder and harder to answer with a *Yes* (this means items that are harder to give as high ratings as one gave to item 4). Now to understand the meaning of going up in person measure, consider the change in TOTAL SCORE when comparing person 45208 to person 45220. Person 45208 has answered *No* to almost all of the survey items. This can be seen in the low TOTAL SCORE of 2. Now what about person 45220? The most important thing to note is that this person has a higher TOTAL SCORE than person 45208 and person 45220 has a higher measure. Since the rating scale was Yes = 2, one can see that the person 45220 made more positive responses in answering the survey items. Now an analyst can understand, at least with respect to the rating scale, what it means to go up and down the scale. Finally, to understand the meaning of going up and down the scale, one must examine the survey items and, at least in this case, what is meant by a *Yes*. It could be that a *Yes* is the worst type of response one would want. If that were the case, then persons with a negative measure would be the persons you might want to see in your data set.

Now consider what we might learn by reviewing the distribution of persons in the Wright Map (Fig. 6.6). In this case, we see a general, broad distribution of persons in terms of overall self-efficacy as measured by the 13 self-efficacy items of the instrument. One characteristic of the distribution of persons seems to be a skewing of respondents toward the high end of the person measure range. [Remember, the distribution in Winsteps Table 12.2 for persons would not be the distribution observed if you had used raw scores because raw data are nonlinear. A distribution based on raw scores would not exhibit the true spacing between respondents.]

We close our initial consideration of the person component of a Wright Map by presenting some questions an analyst could ask when presented with this distribution of people: Does the distribution make sense? Would it be the predicted distribution? The types of students who are outliers, do they make sense? Are they the ones an instructor would predict would be at ends of the continuum? If concentrations of students with a particular range of measures are observed, is this predictable?

Formative Assessment Checkpoint #5

Question: Is the measurement scale in logits used for person measures different than the item measure scale?

Answer: No. Both person measures and item measures are expressed on the same measurement scale. This fact can be seen every time a Wright Map is reviewed. There is one scale (an equal interval scale) noted on the plot. Both persons and items are plotted using that single scale.

Later herein we present an entire chapter that considers the issue of measurement error. However, here we want to discuss three issues that often arise in our classes as to how to use a Wright Map. Often students are amazed at what they learn about their targeting of items for a survey or test. They see items that oversample a trait in particular spots, and they see gaps in their distribution of items. Often our students will ask, quite logically, "How do I know if a gap is too big, acceptable, or if there truly is an oversampling of a particular portion of a trait?" Responding, we first ask our students to think how the ruler should look if one had no idea of the level of respondents and one was not interested in more information on respondents at a particular location along the trait. Most of our students reach the correct conclusion: If one does not wish to collect detailed information on students near one part of the trait, one would want an even distribution of items marking the trait. Also, students are able to suggest that if one wants particularly detailed information on one part of the trait (e.g., the 1.5 logit value of the self-efficacy trait in the Wright Map presented above), then one would want to have more items near this value. An example of this scenario is the certification of a physician. There is no point in administering potential doctors test items that are exceedingly easy (all doctors should get these items correct), and it makes no sense to administer items that are far too hard for beginning doctors (one would predict that these items would not be correctly answered by most of the doctors at the start of their careers). There is, however, a rule of thumb (and two articles) that we find most useful to supply to our students.

The guidance is provided by DeMars and Linacre (2004) when they state: "Substantively, in many educational situations, one logit approximates 1 year of growth." This means that if there is a gap of more than one logit between items, then there may be major educational growth that is missed. A second article that we suggest students read, if they are in medical research, is an article by Lai and Eton (2002). This article provides details regarding the evaluation of item gaps for medical research purposes.

A second question often asked by our students is as follows: "How do I go about filling the gaps with new items, and how do I decide which items to remove if there are too many marking a part of the trait?" To fill a gap, one has to really know the order and spacing of items in the Wright Map. When teaching we often use this self-efficacy data. We print out very large print copies of each item, and we order the items on the classroom floor. We then have the students discuss in detail what happens to items as one moves from one end of the scale to another. Why are items harder or easier to agree with as one moves in a direction along the variable? The next step is to ask the class in small groups to author items to fill gaps, and to provide a rationale as to why they think the items fill the gaps. Afterward, the students are all asked to present their items and their rationale. Then the class discusses which items they feel, as a group, seem to be ones that from theory would fill the gaps. Then we explain that the next step would be to collect data with the new items, as well as a few of the old items, and conduct an analysis to construct a new Wright Map and evaluate our gap filling skills.

The third question posed by our students is as follows: What targeting of mean item difficulty and mean person ability should I aim for? It seems reasonable to most students that having a targeting of items and persons at the same mean difficulty level/mean ability level should provide the most psychometric power to a test. What role might that goal play as one designs an item in a particular test or survey? In designing a test and looking at targeting of persons and items, we suggest researchers utilize the guidance of Linacre (2000) in a discussion of item targeting and computer adaptive testing. He writes for computer adaptive tests:

> If an optimum-targeting algorithm is employed, then the success rate for all test-takers, of whatever ability, to items will be about 50 % correct. For high ability test-takers, such a low percentage of correct answers is a traumatic experience…Accordingly, testing agencies are suggesting that items be selected to give success rates of 60 %, 70 % or even 80 % by test takers across items … this adjustment in success rate on items is done by administering items to the test-takers about 1 logit less difficult than test takers are able. (p. 27)

The point to make her is, although for a test it might be (from a measurement perspective) best to have a similar average item difficulty and average person abilities, there are also other considerations.

Isabelle and Ted: Two Colleagues Conversing

Isabelle: Hey Ted, how's it going?

Ted: It's going good. Look what I made, a Wright Map. I did it by hand.

Isabelle: What's the point of that? You've got a million things to do as it is?

Ted: Well, I could have just pulled up the table in Winsteps that automatically creates a Wright Map, but I have found at the start of my analysis that it helps A LOT to do a map by hand. In part it warms me up for a lot of the table review I will be doing as part of a project. A second point is that by working through the construction of a Wright Map, sometimes I spot some stupid coding mistakes I have made. Finally, the sheer act of making the map by hand helps me to start some of the higher-level syntheses that I will need to do for my research.

Isabelle: Okay Ted, tell me what you mean by all of this?

Ted: Well, here is my first point. In order to make the Wright Map, I need to be able to pull up the tables that present the item calibrations (the item measures) and the person calibrations (the person measures). I need to know where to find those tables, but also I need to be able to find the person and item measures, and then I need to figure out what those mean. I need to be able to understand what the difference is between a high and a low person measure. I also need to be able to understand what high and low item measures mean. The meaning of the person and item measures will depend upon the coding that was used and of course upon the test or questionnaire that was used for data collection.

Isabelle: Before we move on…what do you mean by "depends upon the coding and the instrument that was used?"

Ted: Do you remember that data set I was working on a few months ago? It involved students' reporting of how often different types of constructivist teaching were used in their classroom. Students had to indicate Very Often, Often, Sometimes, Seldom, and Never. Well if we coded Never as a "0" and worked up to Very Often as a "4," the meaning of moving up in logits would be different than if we did a coding with Never as a "4" and Very Often as a "0." This difference in coding would have the same impact on the person measures as well.

Isabelle: I get it. Now, what did you mean that making the Wright Map helps you spot errors in coding?

Ted: When I am plotting the items on my Wright Map, as I plot I am thinking. I don't even need to remind myself to do so, it just happens – you have to find the measures, find the item name, and think about the words that were in the item. It is human nature to think…. "hmm, it is interesting that that item was so high on the scale…." Well sometimes when I am plotting I see that an item appears to be way off from what I would have predicted. Being way off could mean that I was wrong in an aspect of understanding the construct. But being way off could mean that there was some sort of mistake in coding. Maybe the label naming in an Excel or SPSS variable was wrong, and that error trickled down to this point? It is certainly possible that one could spot an error in a Winsteps constructed Wright Map, but I think by hand construction increases my ability to spot problems.

Isabelle: That makes sense. Can you tell me how you handle big data sets with large numbers of test takers or survey takers? I can't imagine that you spend hours plotting persons.

Ted: You are right. The whole point of the by hand construction is to get into the data and to push myself in a way that will aid me in later parts of the analysis. If I have a LOT of people and/or a LOT of items, I will only plot a subset. Usually what I do is plot a subset throughout the entry table for items and persons.

Isabelle: Why don't you just take the first 10 persons of the person entry table?

Ted: What I have learned is that sometimes mistakes appear in different parts of a data set. I'll give you an example. Perhaps one person entered the first half of a data set, and then another person entered the second half of a data set. Maybe the second person shifted all of her coding by one column. For some reason, she entered the first part of a person's data twice, which in turn meant that all the other information was shifted over one column.

Isabelle: I get it. Now what do you mean by the map-making warming you up for higher-level synthesis?

Ted: You'll remember, Isabelle, that the ordering and spacing of items on the Wright Map are really a test of my theory of what it means to progress along a variable. If there is a disconnect, then maybe my theory was off. Or maybe there is something odd in part of the data? Or a little of both? I've found plotting the items really pushes me to think about the theory when I might have been lazy. The other thing that I have come to appreciate is that the Wright Map helps me see that it is important to look at how items interact with respondents. The Rasch model expressing persons and items on the same scale really opens some additional doors in an analysis.

Keywords and Phrases

Wright Map
Person-item map
Item hierarchy
Person hierarchy
Logits
Person measure
Item measure
Mean item measure
Average Mean person measure

The Wright Map is a graphical, and accurate, representation of the relationship between the measures of persons and items. One can then explain the measure of a respondent in the context of the items and what it means for a respondent to have a particular measure.

Potential Article Text

To evaluate the construct validity of the 15-item measurement device, pilot data were collected utilizing 200 respondents from schools in the Ruhr region of Germany. Following data collection, a Wright Map was constructed with Winsteps Rasch software. Lower person measures and lower item difficulties are presented at the base of the map. Higher performing students and more difficult items are presented at the top of the map. Figure 1 presents this map with all items and respondents.

```
                Person - MAP - Item
                    <more>|<rare>
  6                     X |
                          |
                       XX |  Q9
                          |
                       XX |
  5                     X |
                        X |
                          |
                      XXX |
                          |
  4                       |
                          |
                          |  Q1
                          |
                        X |
  3                       |
                        X |
                       XX |  Q3
                     XXXX |
                XXXXXXXXX |
M> 2          XXXXXXXXXXX |
               XXXXXXXXXX |
                 XXXXXXXX |
                  XXXXXXX |   Q4
                   XXXXXX |                     <M
  1                 XXXXX |
                    XXXXX |   Q8 Q12  Q14 Q15
                      XXX |
                     XXXX |
                       XX |   Q2 Q5 Q6 Q13 Q11
  0                    XX |
                        X |
                          |
                       XX |
                        X |
 -1                       |
                          |    Q7 Q10
                          |
                      XXX |
                       XX |
 -2                    XX |
                       XX |
                          |
                    <less>|<frequ>
```

Analysis of the pilot data suggested some changes in a subsequent version of the instrument. Items Q8, Q12, Q14, and Q15 appear to measure similar portions of the trait and therefore, from a measurement perspective, are redundant. This appears to also be the case for items Q2, Q5, Q6, Q13, and Q11 and items Q7 and Q10. Within these groups of items, individual items can be removed with little measurement precision lost. The Wright Map also shows the need for items to fill the measurement gaps, for example, between items Q1 and Q3 and between items Q3 and Q4.

Quick Tips

On a Wright Map, items that overlap can be viewed as items that "cut" the same part of the trait. Items that cut the same part of the trait provide similar measurement information. To have an effective and efficient instrument, it would be better to not have overlapping cuts. Certainly there may be valid reasons to keep overlapping items, but the issue of overlapping items (from a measurement perspective) should at least be an issue you are aware of.

On a Wright Map, one can see gaps in the regions of a trait "cut" by items. To improve the instrument, it would be best to author items that fill the gaps between cuts. A rule of thumb for learning is that a gap of one logit represents a year of learning. Thus, a gap of more than a logit may mean that some major concepts may have been missed by the manner in which items define the trait.

The ordering of items on the Wright Map should match the prediction made by the instrument developer and/or user. If the item ordering AND spacing do not match the prediction made, then the researcher needs at least to reconsider their his or her definition of the trait.

The ordering and spacing of respondents on the Wright Map should match a prediction made by the researcher. If the ordering and spacing do not match, the researcher needs to consider why her/his prediction was not correct. Has she/he misunderstood the variable?

Optimal targeting of an instrument can be when the average persons are at the same measure as the average item. However, psychologically, it may be better to target test items to be one logit below the mean value of the mean person measure.

When you are labeling a Wright Map for a rating scale that involves "agreement" (e.g., Strongly Agree, Agree, Disagree, Strongly Disagree), you need to indicate the meaning of going up and down the scale. This can be done with different sets of phrases (but phrases that have the same meaning). For example, for SA, A, D, and SD, one might use the set of phrases "least easy to agree with" and "most easy to agree with." But one could also use the phrase "easier to disagree with" and "harder to disagree with." What one wants to do is show the meaning of going up or down a scale.

Data Sets: (go to http://extras.springer.com)

cf for SE for Chp 5 not rescaled
Turkish Sci Educ Data
cf Turkish Sci Educ Data For Wright Map

Activities

The data used in the following activities were collected using a Turkish version of the Test of Science Related Attitudes (TOSRA; Fraser, 1981). The nonrandom sample is supplied by our colleague Dr. Sibel Telli.

The following provides the Turkish numbering nomenclature and item text for one subset of TOSRA items. The subset was named *Enjoyment of Science Lessons* and was viewed as being a single variable. The data set is provided as an Excel sheet named "Turkish Sci Educ Data." Data were entered for the positive items using the coding *Strongly Agree* (5), *Agree* (4), *Neither Agree nor Disagree* (3), *Disagree* (2), and *Strongly Disagree* (1). For negative items, the following coding was used: *Strongly Agree* (1), *Agree* (2), *Neither Agree nor Disagree* (3), *Disagree* (4), and *Strongly Disagree* (5) [this means the flipped data were entered in the spreadsheet]. Negative items are 6, 13, 19, and 31.

Turkish TOSRA	Item
2.	Science lessons are fun.
6.	I dislike science lessons.
10.	School should have more science lessons each week.
13.	Science lessons bore me.
17.	Science is one of the most interesting school subjects.
19.	Science lessons are a waste of time.
23.	I really enjoy science lessons.
26.	The material covered in science lessons is uninteresting.
29.	I look forward to science lessons.
31.	I would enjoy school more if there were no science lessons.

Activity #1

Create a control file for this data set. Make sure to include correct item names. Hint: This means for negative items, make sure to alter wording to reflect that data have been entered as if different item text for these items had been presented to respondents.

Answer: Use earlier chapters of this book to create the control file. A potential control file is attached (cf Turkish Sci Educ Data For Wright Map). Parts of your file may be slightly different. The key is that your file and the file we provide result in the same person and item measures. Tip: We will discuss this issue later, but make sure the CODES line in the cf is set to read CODES="01234".

Activity #2

Place your control file (cf) side by side with the one we provided. What are the differences and why?

Answer: The only real differences should be the names that you give items. We have tried to shorten our item names so they might fit completely on Winsteps tables. A short name is not a requirement, but it makes reading some tables much easier. There will also be some differences due to the number of respondent variables you incorporate into your control file. We chose to use only a student ID as a person label. Please note the way our negative items are phrased, make sure to look at the original items, understand our phrasing, and review your own item names for these negative items.

Activity #3

If you were creating a Wright Map for an analysis using the Activity 2 control file, what will be the maximum value and minimum value for your scale? Look at the range of person measures (from highest to lowest) and look the range of item measures (from highest to lowest).

Answer: The item measures run from a maximum of 1.17 to a minimum of −1.46. The person measures range from a maximum of 1.28 to a minimum of −.43. This means that the Wright Map must range from at least 1.28 logits to at least −1.46 logits. This means that the map spans roughly 3 logits (1.28 – (−1.46) = 2.74).

Activity #4

What might be a good maximum and a good minimum for a "by hand" Wright Map so that graphing and plotting can proceed fairly quickly?

Answer: One possibility is a maximum of 1.50 and a minimum of −1.50. This will facilitate the quick labeling of tick marks and the graphing of persons and items on the map.

Activity #5

Begin making a Wright Map for the data. Plot the item measures of all survey items and the person measures of the respondents. As you plot, provide item descriptions that you will be able to understand.

Answer: Your scale will affect the exact ordering of items/persons. In some scales you choose (perhaps a tight scale), some items/persons will appear very much at the same point on the scale. But, if a scale is not as tight, you will be able to see that items/persons are not at the same location.

Activity #6

What is the meaning of going up the scale from the most negative item to the most positive item? What is the meaning of going up the scale from the most negative person to the most positive person?

Answer: You can review the person measure table and the item measure table of Winsteps. The item measure table reveals that the item with the highest item measure is t2 (1.17 logits) and the item with the lowest item measure is item t19 (−1.46 logits). Item t2 was answered by 74 respondents (look at the TOTAL COUNT column), and the total number of raw score points one has from these 74 people is 79 points. This means that if almost all of the respondents answered *Strongly Disagree* (a coding of 1), then one would get a value of about 79 points. For item t19, it has a total count of 75; this means all respondents answered this item. Also, it has a total score of 266. This means that one could get this raw score total from the 75 respondents if they each used a rating scale of *Agree* (4) or a *Neutral* (3). We can see this is possible by dividing 75 into 266. That number will be between 3 and 4. So, in this data set, those items higher up on the scale are items that are harder to agree with. Those items lower on the scale are easier to agree with. What can we understand about the respondents? The 45th person in the data set has a measure of 1.28 logits, has answered all 10 survey items, and has a raw score total of 31. The 13th person in the data set has the lowest measure of −.43. This person also answered all 10 items, but the raw score total is 16. This means that if one were to compute a mean raw score response, the 45th person typically could have answered a "1" (*Strongly Disagree*) or a "2" (*Disagree*) to the 10 items. We can see that by dividing the raw score total (16) by the number of items attempted (10) (16/10 = 1.6). This means a higher person measure means a higher reported enjoyment of science.

Activity #7

Where along the scale is there a possible overabundance of items defining the trait? Where might there be gaps in the definition of items?

Answer: When items appear near each other or at the same point on the plot, then several "cuts" are being made in our meterstick in the same (or very close to the same) location. One area of the trait that seems to be oversampled can be identified by the locations of items t2 and t23. On the other hand, there is a very large gap between t29 and t13. In future versions of this survey, an item (or items) could be added to fill this gap. The likely result of filling the gap is a decrease in the measurement error which would be calculated for respondents. The type of quality control that you have done for this item is far beyond the type of quality control employed for instrument development unless the developers are already using Rasch techniques.

Tip: Readers should see the broken vertical line in the plot of items from Winsteps. When one sees this, it is the result of the program attempting to plot items next to each other (same/similar measure), but because of the length of the item name, there is wraparound in text. This means one has to do a little editing out of blank spaces. Below we provide an example of wraparound for items t2 and t23, which is followed by the correct editing. Sometimes changing the orientation of the paper when you are viewing this table helps correct for this issue.

```
TABLE 12.2 Turkish Sci Educ Data.xls           ZOU663WS.TXT  Sep 18  7:11 2011
INPUT: 75 PERSON  10 ITEM  REPORTED: 75 PERSON  10 ITEM  5 CATS  MINISTEP 3.72.3
--------------------------------------------------------------------------------

           PERSON - MAP - ITEM
               <more>|<rare>
    2               +
                    |
                    |T
                    |
                    |
                    |
                  . |
                    |  t2-Sci.LessonsAreFun
                       t23-IReallyEnjoySciLessons
    1               +
                  . T|S t17-SciIsOneOfTheMostInterestingSchoolSubjects
                 ##  |
```

```
TABLE 12.2 Turkish Sci Educ Data.xls           ZOU663WS.TXT  Sep 18  7:11 2011
INPUT: 75 PERSON  10 ITEM  REPORTED: 75 PERSON  10 ITEM  5 CATS  MINISTEP 3.72.3
--------------------------------------------------------------------------------

           PERSON - MAP - ITEM
               <more>|<rare>
    2               +
                    |
                    |T
                    |
                    |
                    |
                  . |
                    |  t2-Sci.LessonsAreFun   t23-IReallyEnjoySciLessons
    1               +
                  . T|S t17-SciIsOneOfTheMostInterestingSchoolSubjects
                 ##  |
```

Activity #8

What possible theory could you develop from the ordering and spacing of items? In Rasch analysis, you will have thought about item ordering before your analysis. Moreover, you will have thought about the sense of using the set of items together to define a particular trait. For this exercise, pretend that you have done such thinking and now are examining the Wright Map. Recall that the location of the items shows where the items cut the line of the trait. What do you see when you consider the locations (and text) of items?

Answer: The trends you see will depend, in part, upon the work that you have done. The importance of the Wright Map is its ability to help you generate helpful ideas. The Wright Map is not an answer; however, it provides the substance for reaching informed, useful conclusions.

Activity #9

When looking at a Wright Map, how do you interpret the person side of the Wright Map? Why do you sometimes see "X" and sometimes dots (.)?

Answer: Wright Maps typically have persons plotted on the left side and items on the right side of the map. When Winsteps plots the persons, sometimes not all respondents can fit on one page, so some symbols are used to plot different size groups of respondents.

Activity #10

Theory should be used to create instruments. Where does the theory that is used to author items come from? How does the Wright Map allow one to check theory?

Answer: Theory is built using many resources, including past research. It could in particular make use of theoretical models that have been suggested in the literature. The theory can also be supported through experience of individuals and also data collection such as interviews that might be conducted. The Wright Map can be used to check the theory in a number of ways. For instance, does the ordering and spacing of items match that suggested by theory? Are respondents ordered as one might hypothesize? The Wright Map may confirm your theory or suggest revision of your theory. The important step to take is to think.

Activity #11

The Wright Map of this chapter suggests that there are three items of the STEBI self-efficacy scale that have item measures very close to one another. This suggests the three items might overcut the trait at this point. How do you think you might decide which item to remove, if your goal is to remove one of these three items?

```
XXX | Q18se  Q21se-rc Q3se-rc
```

Answer: In surveys and tests, there will be many issues that you will address as you decide which items might be removed. We commonly make a list of strengths and weaknesses of items. They might include considerations such as the quality of wording in an item, to the length of the item, to whether or not an item needed to be flipped.

References

Boone, W. J., Townsend, J. S., & Staver, J. R. (2011). Using Rasch theory to guide the practice of survey development and survey data analysis in science education and to inform science reform efforts: An exemplar utilizing STEBI self-efficacy data. *Science Education, 95*(2), 258–280.

DeMars, C., & Linacre, J. (2004). Mapping multidimensionality. *Rasch Measurement Transactions, 18*(3), 9990–9991.

Fraser, B. J. (1981). *Test of Science Related Attitudes Handbook (TOSRA)*. Melbourne: Australian Council for Educational Research.

Lai, J. S., & Eton, D. T. (2002). Clinically meaningful gaps. *Rasch Measurement Transactions, 15*(4), 850.

Linacre, J. M. (2000). Computer adaptive testing: A methodology whose time has come. In S. Chae, U. Kang, E. Jeon, & J. M. Linacre (Eds.), *Development of computerized middle school achievement test*. Seoul, South Korea: Komesa Press. [in in Korean].

Liu, X. (2010). *Using and developing measurement instruments in science education: A Rasch modeling approach*. Charlotte, NC: Information Age Publishing.

Wilson, M., & Draney, K. (2000, May). *Standard mapping: A technique for setting standards and maintaining them over time*. Paper presented at the international conference on measurement and multivariate analysis, Banff, Canada.

Wright, B. D. (2012). Benjamin D. Wright's annotated KeyMath Diagnostic. *Rasch Measurement Transactions, 25*(4), 1350.

Additional Readings

Two articles that will help readers develop their understanding and use of Wright Maps.

Boone, W. (2008). Teaching students about Rasch maps. *Rasch Measurement Transactions, 22*(2), 1163–1164.

Stelmack, J., Szlyk, J. P., Stelmack, T., Judith Babcock-Parziale, J., Demers-Turco, P., Williams, R. T., et al. (2004). Use of Rasch person-item map in exploratory data analysis: A clinical perspective. *Journal of Rehabilitation Research & Development, 41*(2), 233–242.

Chapter 7
Wright Maps: Second Steps

Isabelle and Ted: Two Colleagues Conversing

Ted: Isabelle, I am looking at something that appears to be built from a Wright Map. It is a plot of physics test items and persons from Hans Fischer's physics education research group in Essen-Duisburg (Germany). This is really interesting because they are using the Wright Map to communicate what students typically can and cannot do.

Isabelle: Show me what you are looking at.

Ted: See this map. We have the items on the right side and the people on the left side of the vertical line. But, we also have a horizontal line that goes across the Wright Map.

Isabelle: What do I see written next to the line? It seems to say "Pre Treatment Student Average Measure in Logits."

Ted: Yes, that line seems to be the mean person measure of all the students at the start of the project. What I found very interesting was the following comment in their paper: A line is provided in the Wright Map for the mean student performance before onset of the project. Those items below the line are items that a typical student would be expected to answer correctly. Those items above the line are items that the student would be expected to answer incorrectly.

Isabelle: I've done such plots, and you are right Ted; being able to draw such lines in this manner is really amazing. Now you can go beyond just reporting the measure of a student or a group of students. You can also explain the meaning of the measure. You now can describe what students (in this example) could and could not do at the start of a project. This is something that was lacking for decades.

Examining Persons and Items Simultaneously

Much can be learned from Wright Maps by examining the location and order of persons. Also, much can be learned by studying the location of items. Whereas these two investigations can be done separately, a third powerful technique is to examine persons and items on the Wright Map at the same time. How does one do this? We begin by reviewing a few aspects of the Wright Map. Let's look at a

W.J. Boone et al., *Rasch Analysis in the Human Sciences*,
DOI 10.1007/978-94-007-6857-4_7, © Springer Science+Business Media Dordrecht 2014

```
TABLE 12.2 SCIENCE TEACHER EFFICACY BELIEFS      ZOU938WS.TXT  Sep 16 11:33 2011
INPUT: 75 Person  23 Item  REPORTED: 75 Person  13 Item  6 CATS  MINISTEP 3.72.3
--------------------------------------------------------------------------------
          Person - MAP - Item
              <more>|<rare>
    6              X  +
                      |
                      |
                      |
                      |
    5                 +
                   X  |
                      |
                      |
                      |
    4                 +
                      |
                      |
                      |
                   X  |
    3                T+
                   X  |
                      |
                 XXX  |
             XXXXXXX  |T
    2           XXXX S+
                      |
                  XX  |  Q19se-rc
               XXXXX  |
                XXXX  |S Q5se
    1            XXX M+
         XXXXXXXXXXX  |  Q20se-rc   Q23se-rc
               XXXXX  |  Q17se-rc
            XXXXXXXX  |  Q12se
                 XXX  |  Q18se      Q21se-rc   Q3se-rc
    0            XXX S+M
               XXXXX  |
                XXXX  |  Q6se-rc
                 XXX  |
                   X  |
   -1                 +  Q8se-rc
                     T|S
                      |
                      |
                      |  Q22se
   -2                 +
                      |T
                      |  Q2se
              <less>|<frequ>
```

Fig. 7.1 A Wright Map of 75 respondents who responded to the 13 self-efficacy items of the STEBI. Persons are plotted on the *left side* of the vertical line, and items are plotted on the *right side* of the vertical line. In this analysis, persons with a higher measure are those persons who were more agreeable to survey items. Items with a higher measure are items that were harder for respondents to agree with

Winsteps Wright Map of our analysis of the 13 self-efficacy items using the responses of 75 preservice teachers (Fig. 7.1).

First note the letters “T,” “S,” and “M” on the person (left) side of the vertical line. Also note the same letters on the item (right) side of the vertical line. The “M” on the left side marks the approximate location of the “Mean” respondent. Thus, the typical respondent has a mean person measure of about 1.0 logits. The location of the “M” on the item side shows the mean item measure, which is set automatically to be 0.00 by Winsteps. The “S” and “T” notations, respectively, can be used to note the distribution of items and persons. “S” marks one standard deviation from the mean, and “T” marks two standard deviations from the mean. Regarding this Wright Map (after making sure to understand the meaning of going up and down the map for persons and items), one can quickly infer that, in general, the survey items were generally agreed with. This inference is based on the relative positions of the “M” for the items and the “M” for the persons. The “M” for the persons is higher along the scale than the “M” for the items. When researchers have conducted an initial Rasch analysis, we suggest that comparing the "M" for items and the "M" for respondents might be one of the first things to do. This comparison often provides more guidance to test and survey developers than many of the techniques employed prior to using Rasch measurement. In our opinion, the graphical nature of the map facilitates immediate understanding of the relative location of persons and items. In this case, the mean for persons is substantially higher than the mean for items. This means that one improvement of the measurement device would be one in which positively worded items and/or negatively worded items (after reverse coding) would be made harder to agree with through alterations of the text. This would help shift the means closer together.

The payoff is that measurement precision is improved when items are targeted to the mean of the persons. We present this comparison of the mean values for items and persons because this simple step with a Wright Map can provide immense guidance to those developing tests. In many cases, researchers can collect pilot data, produce a Wright Map, and quickly evaluate the quality of item targeting.

Let's examine another Wright Map. Figure 7.2 presents a portion of a Wright Map of chemistry education data presented earlier. Seventy-five (75) students responded to 24 multiple-choice test items.

This Wright Map is presented so that higher-performing students have higher measures and harder items have higher measures. Thus, the three persons with a measure above 2.0 logits are the highest-performing respondents shown in this figure, and q10 was the most difficult test item. In this analysis, the students, on average, are performing at a higher level than the typical test item. If this subset of 75 respondents were representative, then improving the test would be accomplished by including additional harder items. Remember, better targeting of test items and survey items improves the quality of measurement possible with an instrument. Improving a measurement device is more than just writing more items (e.g., if you wish to learn more about the views of customers toward a product, more items will not necessarily mean more certainty in detailing each customer's view).

Formative Assessment Checkpoint #1

Question: Will authoring additional test items automatically improve the quality of measurement possible with the test?

Answer: No. The addition of items to a survey or a test does not mean an automatic improvement of the measurement possible with an instrument. Consider a 10-item test for an initial data collection. A researcher decides that students have time to answer more items so he or she adds three items (a total of 13 items now). The added items are all answered correctly by respondents. As a result, the researcher learns only that the three additional items were easy for this group of respondents. No one knows how easy the items were. Moreover, when all respondents correctly answer the items, no added information is gained to differentiate the performance of test takers. This is a little like asking Olympic figure skaters to show judges that they can skate from one end of the ice rink to the other (item 1), skate and stop (item 2), and catch flowers thrown to them without falling down (item 3).

```
            XXX  |
                 |
                S|
2   XXXXXXXXXXX  +
                 |
                 |
         XXXXXX  |
                 |
             XX  |  q10
                 |  q18a    q18b
          XXXXX  |S
           XXXX M|  q3
           XXXX  |
1           XXX  +
            XXX  |
        XXXXXXX  |
              X  |  q11     q7
          XXXXX  |
                 |
            XXX S|  q15b
                 |  q15a
              X  |
              X  |  q16b    q2
0           XXX  +M q16a
             XX  |  q9
                 |
              X  |
                 |  q12b    q5
                T|
                 |  q14a
             XX  |  q8
                 |  q14b
                 |  q17b
-1               +
                 |  q1
                 |
                 |S q6
                 |
```

Fig. 7.2 The Wright Map of the analysis of 75 respondents to the 20 chemistry test items

Communicating Differences Between Groups with Wright Maps

As we mentioned in the first part of this chapter, Wright Maps provide additional exceedingly informative guidance for those developing tests and surveys (e.g., "What items might be removed?" "Are items well targeted?"), but the maps are also highly informative with respect to bringing meaning to measures and statistics. We will return to this topic repeatedly throughout this book. To begin we consider how Wright Maps might be used to communicate the differences between groups of test takers.

Group comparisons (e.g., males and females) are important within and beyond education research and are commonly presented in analyses. For example, "A *t*-test

```
              XXX  |
                   |
                  S|
   2  XXXXXXXXXXX  +
                   |
                   |
           XXXXXX  |
                   |
               XX  |  q10
                   |  q18a    q18b
            XXXXX  |S
             XXXX M|  q3
             XXXX  |
   1          XXX  +
              XXX  |
Male-----XXXXXXX  |----------------------------------------
                X  |  q11     q7
            XXXXX  |
                   |
              XXX S|  q15b
                   |  q15a
Female----------X  |----------------------------------------
                X  |  q16b    q2
   0          XXX  +M q16a
               XX  |  q9
                   |
                X  |
                   |  q12b    q5
                  T|
                   |  q14a
               XX  |  q8
                   |  q14b
                   |  q17b
  -1               +
                   |  q1
                   |
                   |S q6
                   |
```

Fig. 7.3 The Wright Map of the analysis of 75 respondents to the 20 chemistry test items. A potential mean person measure of all males and a potential mean person measure of all females are plotted

of male and female performances on the science test revealed a significant difference in their performances, with males outperforming females at the 0.01 level." "An ANOVA of Quality of Life data suggested that views of patients differed statistically ($p<.05$) as a function of race."

Fig. 7.3 presents the same Wright Map that presented the performance of respondents completing a multiple-choice test, but two notations have been added, one regarding the location of students who represent the mean performance of females and another regarding the location of students who represent the mean performance of males.

The horizontal line for the males marks the boundary between those items that males had greater than a 50 % probability of correctly answering (all items below

TABLE 13.1 Sample Chem Educ Data from Chihche an ZOU432WS.TXT Sep 20 9:41 2011
INPUT: 76 PERSON 24 ITEM REPORTED: 76 PERSON 24 ITEM 2 CATS WINSTEPS 3.72.3

PERSON: REAL SEP.: 1.17 REL.: .58 ... ITEM: REAL SEP.: 3.54 REL.: .93

ITEM STATISTICS: MEASURE ORDER

ENTRY NUMBER	TOTAL SCORE	TOTAL COUNT	MEASURE	MODEL S.E.	INFIT MNSQ	INFIT ZSTD	OUTFIT MNSQ	OUTFIT ZSTD	PT-MEASURE CORR.	PT-MEASURE EXP.	EXACT MATCH OBS%	EXACT MATCH EXP%	ITEM
14	8	76	3.79	.41	1.16	.6	2.85	2.8	.08	.35	90.7	90.6	q13
10	33	74	1.49	.25	1.15	1.6	1.25	2.1	.22	.38	60.3	66.4	q10
23	36	76	1.39	.25	1.15	1.7	1.25	2.2	.22	.38	57.3	66.0	q18a
24	28	56	1.31	.29	.94	-.6	.89	-.9	.43	.37	61.8	65.3	q18b
3	40	76	1.15	.25	.88	-1.5	.87	-1.2	.49	.37	72.0	65.4	q3
11	46	76	.77	.25	.99	.0	.98	-.1	.37	.36	66.7	67.0	q11
7	47	76	.71	.25	1.13	1.3	1.20	1.5	.23	.36	57.3	67.6	q7
18	52	76	.37	.26	.97	-.3	.94	-.3	.38	.34	76.0	71.3	q15b
17	53	76	.30	.27	.94	-.5	.92	-.4	.39	.34	77.3	72.2	q15a
2	55	76	.15	.27	1.10	.9	1.04	.3	.25	.33	68.0	74.0	q2
20	56	76	.08	.28	.99	.0	1.03	.2	.32	.32	76.0	74.9	q16b
19	57	76	.00	.28	.93	-.4	.83	-.8	.40	.32	77.3	75.9	q16a
9	58	76	-.08	.29	.99	.0	.90	-.4	.34	.31	80.0	76.9	q9
5	61	76	-.34	.30	.96	-.2	.91	-.3	.34	.29	80.0	80.5	q5
13	62	76	-.44	.31	.90	-.5	.89	-.3	.38	.29	84.0	81.7	q12b
15	64	76	-.64	.33	.98	.0	.92	-.1	.30	.27	86.7	84.1	q14a
8	64	75	-.74	.34	1.03	.2	1.05	.3	.22	.26	85.1	85.1	q8
16	62	72	-.80	.36	1.05	.3	1.23	.7	.19	.27	85.9	85.9	q14b
22	66	76	-.88	.35	.89	-.4	.70	-.8	.37	.25	86.7	86.7	q17b
1	68	76	-1.15	.39	.97	.0	.72	-.6	.30	.23	89.3	89.3	q1
6	69	76	-1.31	.41	.92	-.2	.84	-.2	.28	.22	90.7	90.7	q6
4	70	76	-1.49	.44	1.02	.2	1.29	.7	.15	.20	92.0	92.0	q4
21	71	76	-1.70	.47	.85	-.3	.54	-.8	.35	.19	93.3	93.3	q17a
12	72	76	-1.94	.52	.85	-.2	.36	-1.1	.37	.17	94.7	94.7	q12a
MEAN	54.1	74.9	.00	.33	.99	.1	1.02	.1			78.7	79.1	
S.D.	15.4	4.0	1.25	.08	.09	.7	.44	1.1			11.5	9.8	

Fig. 7.4 A Winsteps table that presents the measures of items. Note the table is organized from highest measure to lowest measure

the line) and those items that males had less than a 50 % probability of correctly answering (all items above the line). Looking at Fig. 7.4 (Winsteps Table 13.1), we should remind ourselves that item Q6 was an easy item for the respondents to answer, and that item Q10 was a harder item for respondents to correctly answer. This helps us correctly interpret the meaning of an item being below or above the line that shows the mean measure of the male test takers. Now we know that all items below the line for males are those items one would predict the males can answer. And, all items above the line are the items that the typical male in this data set does not correctly answer. This horizontal line adds more meaning to the males' measure. Now we are provided with a picture of what it means for students to perform at a particular level.

Now examine the horizontal line indicating the location of the mean measure of females (remember to locate this line, one computes the mean measure, in logits, of the females in the data set). Based upon where the line is located with respect to the test items, these females would have a greater than 50 % probability of correctly answering all items plotted below the line. Items farther and farther below this line are items for which there is an increasing likelihood of a mean female correctly answering the item.

Why is plotting the mean performance of a group of respondents so powerful in terms of aiding researchers? It is the same reason why plotting even a single person on a Wright Map is exceedingly informative. Instead of simply stating the mean of a group of individuals, one can now bring conceptual meaning to that mean. Using this Wright Map, one can communicate with words which items the group would have most likely answered correctly and which items the group would have most likely answered incorrectly. If readers are not yet convinced of the explanatory power of a Wright Map, let us present an excerpt from a potential science teacher in-service in which an attending teacher asks a common question.

"Joe the Consultant" Presenting the Results of a State's 10th Grade Science Test

Consultant: As you can see, the state mean for the test is 76 items correct out of 100, but the mean of your school was 84 out of 100. Statisticians who work for me conducted statistical analyses, and they tell me that there is a significant difference between your students' mean performance and the mean performance of the entire state.

Principal: Isn't that wonderful!

Julia the Teacher: For me as a teacher, I guess it is good that we are above the state average, but what do these results tell me?

Consultant: They tell you, you are doing a good job.

Julia: Honestly, I would say sort of.

Consultant: Sort of? What do you mean?

Julia: This result doesn't really give me any specific guidance at all. What does this number 84 mean? Maybe my students got the 84, but they did not achieve what I would have predicted in science terms. Maybe an 84 means they have a good grasp of speed and velocity, but they are not doing well with acceleration. This number does not help me know if there are concepts I need to repeat, and it does not help me know what the students are ready for. If I can be frank, I think what all this expensive test did is tell us that our students are doing better than the state average, but we do not really know how much better and how much better on specific concepts. Finally, these results do not help me improve as a teacher.

What is this teacher saying? So often in research, test results are presented and compared to some group. If a significant result is found, then some sort of conclusion is made (e.g., School A is better than School B, the intervention worked). However, this type of conclusion did not provide substantive guidance at any level. Using the Wright Map, one can provide specific guidance and insight for the teacher, the researcher, and almost any group with an interest in the data.

Formative Assessment Checkpoint #2

Question: Can we use statistical techniques we have learned with Rasch measurement and with Wright Maps?

Answer: Yes. The first thing to remember about Rasch measurement is that when the data fit the model, we can confidently compute a person measure and an item measure expressed with the same equal-interval scale. After we carefully do such calculations, we then can use descriptive tools and parametric statistics. For instance, we could present a box plot on the left side of the Wright Map. Of course, this box plot would be created using the person measures. We could also create a box plot of item difficulty.

Before we move on to another slightly more complex use of the Wright Map, consider the following: What if the trait presented in Fig. 7.3 is technical ability in ice skating rather than chemistry. So the items on the right side of the map represent different technical tasks such as one jump, two jumps in a row, skating backward, and skating backward and then jumping once. Also imagine the plot of persons on the left side of the Wright Map presents the performance of competitive male and female skaters from throughout the world. Being able to note the location of the female ice skaters and review the female group measure in light of specific items (skating tasks), one can quickly bring meaning to a particular measure. If one were a coach of a specific skater, one could see how the average female skater compared to a specific female skater in terms of the techniques each skater could and could not do.

There exists another group comparison technique on the Wright Map that is one of the most important aspects of Wright Map usage in education, medicine, and market research. This technique is reviewing the items between the horizontal lines that mark the location of the mean male and the mean female for this data set. In short, if a statistically significant difference is found between the mean measures for males and females, then the items between the two lines in essence describe the "meaning" of the difference between males and females.

In education, psychology, and medicine, statistical tests are quite often used to compare subgroups of respondents. Perhaps a comparison is made between the STEBI attitudes of males and females. And, perhaps a significant difference in attitude is revealed, with males being "more agreeable" or having more self-efficacy than females. This is interesting; however, a massive piece of information is missing. Not to worry, though, because the missing piece of information, the meaning of the difference, is provided by the Wright Map. Researchers can compute linear measures for statistical tests, and when differences are uncovered, we can use Wright Maps to explain what the differences represent. We have found that the most important items to consider are the items between the means. These are the items that differentiate the two samples. In our chemistry example, the items of great interest are those items below the male mean performance and above the female mean performance. These are items q11, q7, q15b, and q15a.

Before we present a reflection on this chapter, we wish to present an activity that helps our students and workshop participants further understand the power of the Wright Map, aids them in their understanding of the Rasch equation, and provides excellent details regarding "the logit." To begin the activity, we simply write the Rasch equation for dichotomous items on the board: $B_n - D_i = \ln[P_{ni}/(1 - P_{ni})]$.

Then we remind our audience that B_n represents a person ability, D_i represents an item difficulty, and P_{ni} is the probability of that same person answering the same item correctly. And we remind our audience that $1-P_{ni}$ is the probability of that same person NOT answering the same item correctly. A final piece of information that we review is that probabilities of answering an item correctly can range from 0 to 1, and that if one adds the probability of a particular person correctly answering an item to the probability of the same person not correctly answering an item, that sum will be 1.

The next step in our explanation is to draw a vertical line on the board and to mark the right top side of the board with the words "More Difficult Items," to mark the bottom right side of the board with the words "Less Difficult Items," to mark the top left side of the board with the words "More Capable Respondents," and to mark the lower left side of the board with the words "Less Capable Respondents." Then on the vertical line, we provide tick marks going from a low of –4 logits to a maximum of 4 logits.

Following the creation of this plot, we move to the far right side of the board a few steps away from our marked vertical line. We then write down the Rasch equation for dichotomous items and also write three questions: (1) What is the logit difference between a student Rose who attempts an item that is exactly at her ability level? (2) What is the logit difference between Rose when she attempts an item that she has an 80 % (.8) chance of successfully answering? (3) What is the logit difference between Rose and an item when she attempts an item for which she has a 20 % (.2) of correctly answering?

Most of our workshop participants are able to see that they must simply use the Rasch equation to solve these problems. They will realize that we have asked them to compute the difference between B_n (which is Rose in this case) and D_i (three items of the test, let's call the items 14, 15, and 16). So, they will need to use the right side of the equation to solve our three questions. Below we provide the calculations that our students will carry out; we also talk readers through the steps:

1. *What is the logit difference between a student Rose who attempts item 14 which is exactly at her ability level?*

$$Rose - Item\ \ 14 = \ln\left[(.5)/(1-.5)\right]$$

If the item is at Rose's exact ability level, then Rose has a .5 chance of solving the item correctly and has a .5 chance of not correctly solving the item since 1–.5 is .5 (remember 1–.5 allows us to calculate the probability of not solving the item correctly).

$$Rose - Item\ \ 14 = \ln\left[(.5)/(.5)\right]$$

Just doing a subtraction on the denominator gives us a fraction of .5/.5

$$Rose - Item\ \ 14 = \ln\left[(1)\right]$$

.5/.5 is "1"

$$Rose - Item \ \ 14 = 0$$

The ln of "1" is zero. And, it makes sense that the difference between Rose and item 14 (which is at her exact ability level) is 0 logits.

2. *What is the logit difference between Rose when she attempts item 15 that she has an 80 % (.8) chance of successfully answering?*

$$Rose - Item \ \ 15 = \ln\left[(.8)/(1-.8)\right]$$

If Rose has a .8 chance of solving the item 15 correctly, she then has a .2 chance of not correctly solving the item since 1–.8 is .2 (remember 1–.8 allows us to calculate the probability of not solving the item correctly).

$$Rose - Item \ \ 15 = \ln\left[(.8)/(.2)\right]$$

Just doing a subtraction on the denominator gives us a fraction of .8/.2

$$Rose - Item \ \ 15 = \ln\left[(4)\right]$$

.8/.2 is "4"

$$Rose - Item \ \ 15 = 1.38$$

The ln of "4" is 1.38, and it makes sense that the difference between Rose and item 15 (which is an item for which she has a high chance of a successful answer) should be at least a positive number. Remember the Rasch equation is Bn – Di, so if Rose has over a 50 % chance of correctly solving an item, she will have a higher measure than the item which she is attempting. This means that when one takes the measure of Rose and subtracts from Rose's measure the measure of the item, a positive number results!

3. *What is the logit difference between Rose when she attempts item 16 that she has a 20 % (.2) chance of successfully answering?*

$$Rose - Item \ \ 15 = \ln\left[(.2)/(1-.2)\right]$$

If Rose has a .2 chance of solving the item 16 correctly, she then has a .8 chance of not correctly solving the item since 1–.2 is .8 (remember 1–.2 allows us to calculate the probability of not solving the item correctly).

$$Rose - Item \ \ 15 = \ln\left[(.2)/(.8)\right]$$

Just doing a subtraction on the denominator gives us a fraction of .2/.8

$$Rose - Item\ \ 16 = \ln\left[\left(.25\right)\right]$$

.2/.8 is ".25"

$$Rose - Item\ \ 16 = -1.38$$

The ln of ".25" is −1.38, and this makes sense that the difference between Rose and item 16 (an item for which she has a low chance of a successful answer) should be at least a negative number. Remember the Rasch equation is Bn – Di, so if Rose has less than a 50 % chance of correctly solving an item, she will have a lower measure than the item that she is attempting. This means when one takes the measure of Rose and subtracts from Rose's measure the measure of the item, you get a negative number!

Following these calculations, we then go to our Wright Map, which is on the board, and ask a student to plot the location of Rose and the location of the three items. Usually, and totally correct, whomever is asked to plot Rose and the three items looks at the board, the calculations, and the Rasch formula, and realizes that nowhere in what we have done is there a specific "Rose" ability level in logits that has been reported and nowhere is there a specific item difficulty value in logits that has been reported for the items. How can this be? The answer lies, in part, in the formula. Looking at the formula, the students realize that our calculation for Rose and her interactions with an item are based upon her probability of a success on the item. This means we can place "Rose" anywhere in terms of person ability, and we can then plot the location of each item by knowing the gap in logits (and the direction of the difference between Rose and an item). If we pretend Rose is quite capable and has a logit person measure of +2.0 logits, this means the location of item 16 is 3.38 logits (the item is 1.38 logits more difficult than Rose is able), the location of item 15 is .62 logits (the item is 1.38 logits easier than Rose is able, which means the item difficulty is 2.0 logits – 1.38 logits), and the location of item 14 is at the measure of Rose (Fig. 7.5).

The important aspects of this exercise and the figure are a number of points. First, for any difference between a person and an item in logits, the probability of the person correctly answering the item can be computed. Second, for a difference of any amount that you pick, the meaning of that difference is maintained. This means that if we plotted Rose at −2 logits, item 14 would be at −2 logits, item 15 would be at −3.38 logits, and item 16 would be at −.62 logits. This means that when you have the measures for persons and items, you can compute the probability of a specific person answering any item. And perhaps most important, by referring to the Rasch equation, you can start to develop a feel for the meaning of logits, develop an understanding of the interconnectedness of items and persons, and better understand the mechanics of the Rasch formula.

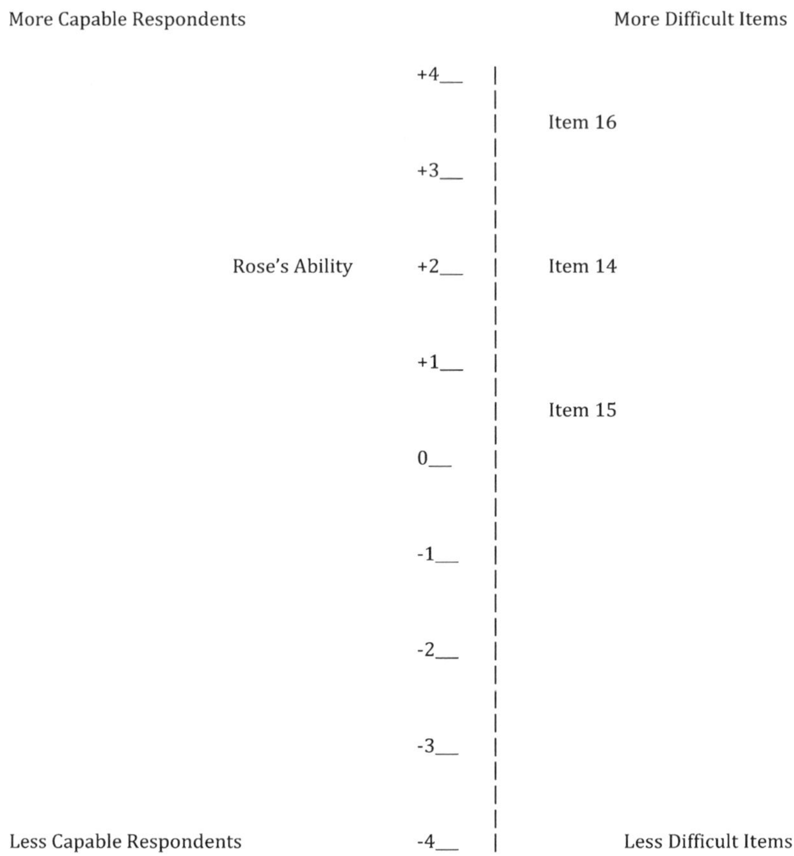

Fig. 7.5 The location of a test taker "Rose" and the location of three test items. One item is one for which there is a .8 probability of Rose successfully answering, one item is .5, and another item is quite a bit harder than Rose's ability level. That item has only a .2 chance of correctly being answered by Rose

By constructing and interpreting Wright Maps, we can describe the specific nature of a statistically significant difference between groups of subjects. For example, national, state, and local policy makers scrutinize international comparisons (e.g., TIMMS, PISA) in science and mathematics. Deeper, more specific understanding of the differences could produce improved policies, higher quality curricula, better teaching, and deeper learning. In fact, Wright Maps are currently being used to inform educational decisions. For example, it is now a common practice to define competency levels by reviewing item difficulty. Student performance, for instance, can be expressed using meaningful words, which result from a synthesis of item content. For example, experts meet to discuss what it means to be competent in 10th grade Biology, and then those definitions are used with Wright Maps to draw the ranges of competency bands. These techniques can be applied in many fields of research.

Formative Assessment Checkpoint #3

Question: Is it difficult to determine where to plot the lines for subgroups on a Wright Map?

Answer: No. Imagine you have conducted a pre-assessment at the start of the school year in two schools (Shiller Elementary and Ramundo Elementary). During the school year, two different interventions were attempted to help the students learn physics. Then at the end of the school year, you collected post data. Now to draw the lines on the Wright Map, all you need to do is compute the person and item mean measures for the two pre-measures and compute the person and item mean measures for the two post-measures. After you have done that, find those means on the Wright Map. Those will be the locations of your lines.

Isabelle and Ted: Two Colleagues Conversing

Isabelle: Well, what do you think of these Wright Maps?

Ted: Initially, I was really overwhelmed, but I reminded myself that I was looking at a thermometer. In the case of a survey on one side, I was seeing the temperature of students, except I was seeing how they compared in terms of the level of their agreement to the set of survey items. When I went back to Winsteps Table 17.1, I could see the person with the top measure of 7.53 had a "raw score" of 78. Since the rating scale was SA (6), A (5), and so on, that told me that this person was the most agreeable person in this sample compared to the other students.

Isabelle: Good, then what next?

Ted: Then I just looked at the items. Also, I reminded myself of what it meant to go up and down with respect to the items. I used Winsteps Table 13.1 to see that item Q2 was the easiest to agree with because it had the highest total raw score and item Q19 (recoded) was the hardest item to agree with because it had the lowest total raw score.

Isabelle: So, even though we have one thermometer, to keep things straight, it seems as if you almost looked at this one Wright Map as if it were two thermometers.

Ted: Yes, and in fact for this article I am writing, I think I may present just the person side of the Wright Map when I am talking about the distribution of person measures because in that part of the paper, it would confuse readers to throw the items at them. I also will have a part of my paper where I am going to have just the item part of the Wright Map. In that part of the paper, I will talk about the measurement quality of the instrument (where items define the trait and so on). Then later on, I will present both persons and items on the plot.

Isabelle: That is really a good way to think about how to use the Wright Map. Okay, you know what I am going to ask you now don't you? Tell me what you think you have learned about the Wright Map when you use persons and items?

Ted: It seems to me that being able to relate a person's performance to the distribution and pattern of items on the Wright Map is really a critical breakthrough. In education we talk about how groups of students did. Now I see so clearly that there is not only a major flaw in using raw data, but also there is something very important that has been missed. I now completely see that we have missed so much, in that when we say there is a statistical difference between groups of students, we might know the direction of the difference (e.g., males are more agreeable than females; females did better on a test than males), but we do not know anything else about the difference. What went into the difference? Are there items that do not differentiate the two groups? What items do differentiate the groups? Those questions are really the most important items for us as researchers. Those items that separate the two groups allow us to understand more about the difference between groups.

Keywords and Phrases

Wright Maps
Bring meaning to measures

If you plot two means (e.g., groups of respondents) on a Wright Map, draw a horizontal line from each mean across the Wright Map. If there is a statistically significant difference in the means, then the items between the two lines help explain meaning of the difference between the two groups.

If you plot a mean (e.g., pre-mean), the items above the mean (in the case of a right/wrong test) are the items the typical pre-student likely could not solve, and the items below the line are the items the typical pre-student likely could solve.

Potential Article Text

Figure 7.6 presents a Wright Map constructed for the STEBI self-efficacy scale data collected from workshop participants at the conclusion of a summer institute. Each survey item is plotted using a Rasch measure. These measures are linear (equal-interval) measures; thus, the location of items is not impacted by the potential nonlinearity of the raw rating scale.

Items are presented from harder to agree with, at the top of the map, to easier to agree with, at the bottom of the map. Items near each other are those items that define the construct in a similar manner. Some items are reverse coded. Those items are presented with the letters "rc." For reverse-coded items, words have been added to present the item as if it had not been presented as a reversed item to respondents.

The ordering of items matches in a general way to the ordering predicted by the course instructors. And, the ordering of items aligns with theories of self-efficacy proposed by numerous researchers. Prior to the analysis of the collected data, the course instructors were asked to provide their predicted ordering and spacing of items based upon their experiences. That predicted ordering matched the general pattern observed in the Wright Map. Items Q5 and Q19 (recoded) were predicted to be among the most difficult for these respondents to agree with.

Over the years, a number of self-efficacy instruments have been developed for data collection from students, preservice teachers, and teachers. These instruments exhibit many similarities, but, of course, many items are unique in terms of item

```
TABLE 12.2 SCIENCE TEACHER EFFICACY BELIEFS      ZOU938WS.TXT  Sep 16 11:33 2011
INPUT: 75 Person  23 Item  REPORTED: 75 Person  13 Item  6 CATS  MINISTEP 3.72.3
--------------------------------------------------------------------------------

         Person - MAP - Item
             <more>|<rare>
    6           X  +
                   |
                   |
                   |
                   |
    5              +
                X  |
                   |
                   |
                   |
    4              +
                   |
                   |
                   |
                X  |
    3             T+
                X  |
                   |
              XXX  |
          XXXXXXX  |T
    2        XXXX S+
                   |
               XX  |  Q19se-rc
            XXXXX  |
             XXXX  |S Q5se
    1         XXX M+
      XXXXXXXXXXX  |  Q20se-rc  Q23se-rc
            XXXXX  |  Q17se-rc
         XXXXXXXX  |  Q12se
              XXX  |  Q18se     Q21se-rc  Q3se-rc
    0         XXX S+M
            XXXXX  |
             XXXX  |  Q6se-rc
              XXX  |
                X  |
   -1              +  Q8se-rc
                  T|S
                   |
                   |
                   |  Q22se
   -2              +
                   |T
                   |  Q2se
                   |
                   |
   -3              +
             <less>|<frequ> TABLE 12.2 SCIENCE TEACHER EFFICACY BELIEFS
```

Fig. 7.6 A Wright Map constructed for the STEBI self-efficacy scale data collected from workshop participants at the conclusion of a summer institute

wording as well as rating scale. The analysis conducted for this study suggests, from a measurement perspective, that some improvements might be made with regard to the measurement precision that is possible with the STEBI. For example, there appear to be some gaps between items (e.g., between items Q8 and Q22 recoded). Also, there appears to be an oversampling of items in the region of the trait covered by items Q18, Q21 (recoded), and Q3 (recoded). A new version of the STEBI might include items that are authored to fill the gap. Also, if the number of items to be completed by respondents is limited, then it would make sense to remove some of the redundant items. Administering items that, from theory, might fill the gap could be administered with some of the original self-efficacy items. The item measures of the new items could be plotted, and one could investigate whether a researcher was successful in filling the gap.

Quick Tips

You have conducted a Rasch analysis of test data (right/wrong) and created a Wright Map. You have also evaluated the person measures of your data set and computed a person measure for all males and all females. You performed a *t*-test and discovered that there is a highly significant difference between the female and male test takers, with the female test takers exhibiting a statistically higher test measure. To show the meaning of the difference, plot the location of the mean male and the mean female on the Wright Map. Draw a horizontal line for the male measure so that the line cuts across the region of the Wright Map where items are plotted. Do the same procedure for the mean female measure. The items between the lines represent the meaning of the difference between male and female test takers. These items between the two lines are those that the females have a higher than 50/50 chance of correctly answering, and these same items are those items which the males have less than a 50/50 chance of answering correctly.

Sometimes when items are "flipped," sorting out the meaning of an item's location with respect to a person can be tricky. The best way to avoid becoming confused is to remember that when an item is flipped, you must think of the text for that item as also flipped. In other words, the survey text that would have been required so that the flipped code would not have been necessary.

Data Sets: (go to http://extras.springer.com)

cf 25 GCKA
cf Turkish Sci Educ Data for Wright Map

Activities

Activity #1

Run the control file (cf Turkish Sci Educ Data for Wright Map) for the "Enjoyment of Science" data collected in Turkey by Dr. Sibel Telli and write what you might assert about the mean person of the data set. Also, edit your Wright Map so that items are plotted correctly. Last, add notes to the map to help you remember the meaning of “going up” in logits for items and persons.

Answer

```
         PERSON - MAP - ITEM
               <more>|<rare>
    2  THESE ARE     +  THESE ITEMS ARE HARDER TO AGEE WITH
       MORE AGREEABLE|
       PERSONS       |T
                     |
                     |
                     |
                  .  |
                     |  t2-Sci.LessonsAreFun t23-IReallyEnjoySciLessons
    1                +
                  . T|S t17-SciIsOneOfTheMostInterestingSchoolSubjects
                 ##  |
                 ## S|
         .#########  |  t10-SchlShouldHaveMoreSciLessonsEachWeekt 29-ILookForwardToSciLessons
----------- ####### M| ------------------------------------------------------------------------
              .####  |
            .######  |
    0           ### S+M
                  .  |
                  . T|
                  #  |
                     |
                     |  t13-SciLessonsDoNotBoreM t26-TheMat t31-IWouldNotEnjoySchoolMoreIf
                     |  t6-IKikeSci.Lessons
                     |S
   -1                +
                     |
                     |
                     |
                     |  t19-SciLessonsAreNotAWasteOfTime
                     |
                     |T
                     |
   -2 THESE ARE LESS + THESE ITEMS ARE EASIER TO AGREE WITH
      AGEEABLE
       PERSONS
```

For this group of 75 students, the mean student (denoted by the letter M on the left side of the vertical line) has a greater than 50 % probability of agreeing at some level to items 13, 26, 31, 6, and 19. Of these items, the item that is by far the easiest to agree with is t19 (*science lessons are not a waste of time*). There is a group of items that an average person of this sample would have less than a 50 % probability of agreeing in some manner to. Those items are 10, 29, 17, 2, and 23.

Activity #2

Using the control file (cf 25 GCKA) of 75 students who completed Kathy Trundle’s (Ohio State University) multiple-choice earth science test, conduct a Rasch analysis and then comment on the item targeting for this group of students. Are there items that are redundant? Are there gaps in how the items define the trait? How would you assess the overall targeting of item difficulty to person performance?

Answer: The Wright Map for the analysis is provided below. Certainly researchers know that looking at the mean of any two things only provides part of an answer to any question. However, a quick comparison of the location of the mean for persons and the mean for items reveals very good item targeting with person ability. In terms of item gaps, a large gap is between two easy items (Q5, Q41). One way to improve this instrument would be to write new items that might land in this gap.

```
TABLE 12.2 GEKA Content only w/o true false item ZOU884WS.TXT  Sep 26  4:34 2011
INPUT: 75 PERSON  25 ITEM  REPORTED: 75 PERSON  21 ITEM  2 CATS  MINISTEP 3.72.3
--------------------------------------------------------------------------------

           PERSON - MAP - ITEM
               <more>|<rare>
   2  Most Capable  +T Most Difficult Items Are Here
      Students       |
      Are Here       |
                     |
                     |
                     |
                     |
                     |
                     |  Q25,
                     |
                    T|  Q39,
                     |  Q19,        Q23,
                     |  Q35,
   1         XXXX   +S
                     |
                     |
           XXXXXX    |  Q45,
                     |
                    S|
         XXXXXXXX    |
                     |  Q3,
                     |  Q21,
        XXXXXXXXX    |
                     |  Q31,
                X    |  Q37,
   0 XXXXXXXXXXXXX  +M Q27,         Q43,
                X  M|  Q7,
                     |
        XXXXXXXXX    |
                     |
                     |  Q13,
    XXXXXXXXXXXXX    |
                     |  Q11,
                    S|
              XXX    |  Q15,
                     |  Q29,
                     |
  -1        XXXXX   +S
                     |
                     |
                     |
                X  T|
                     |
                     |
                     |  Q33,
                     |  Q47,
                     |
                     |
                     |
  -2                +T Q41,
               XX    |
                     |
                     |
                     |
                     |
                     |
                     |
                     |
                     |
                     |
  -3                 +  Q5,

     Least Capable    | Easiest Items Are Here
 Students Are Here

               <less>|<frequ>
```

Activity #3

Pretend that the mean for the group of students who completed an earlier earth science course was 1.0 logits. Also pretend that the average for those students who had not completed an earlier earth science course was –0.25 logits. Show graphically the band that defines the difference between these two groups of students. What do the items within the band tell a teacher?

Answer: A part of the Wright Map from the previous activity is provided. A line is drawn at 1.0 logits. This line marks the average ability level of the students who had completed a previous earth science class. A line is also provided at –0.25. This line marks the ability level of those students who had not completed a previous course. If there is a statically significant difference between the mean logit measures of the two groups, then the most interesting test items for a teacher (and researcher) are those items between the two lines (45, 3, 21, 31, 37, 27, 43, and 7). These are the items that the group who had completed a geology course was likely to have correctly answered (over a 50 % probability). Of importance, those students who had not completed such a course were less likely (less than a 50 % probability) to have correctly answered the items.

```
TABLE 12.2 GEKA Content only w/o true false item ZOU884WS.TXT  Sep 26  4:34 2011
INPUT: 75 PERSON  25 ITEM  REPORTED: 75 PERSON  21 ITEM  2 CATS  MINISTEP 3.72.3
--------------------------------------------------------------------------------

           PERSON - MAP - ITEM
               <more>|<rare>
    2  Most Capable  +T Most Difficult Items Are Here
       Students      |
       Are Here      |
                     |
                     |
                     |
                     |
                     |
                     |  Q25,
                     |
                    T|  Q39,
                     |  Q19,       Q23,
                     |  Q35,
    1          XXXX  +S ----------------------------------------------------
                     |
                     |
             XXXXXX  |  Q45,
                     |
                    S|
           XXXXXXXX  |
                     |  Q3,
                     |  Q21,
          XXXXXXXXX  |
                     |  Q31,
                  X  |  Q37,
    0 XXXXXXXXXXXXX  +M Q27,       Q43,
                  X M|  Q7,
                     |--------------------------------------------
          XXXXXXXXX  |
                     |
                     |  Q13,
      XXXXXXXXXXXXX  |
                     |  Q11,
                    S|
                XXX  |  Q15,
                     |  Q29,
                     |
   -1         XXXXX  +S
                     |
                     |
                     |
                  X T|
                     |
                     |
                     |  Q33,
                     |  Q47,
                     |
                     |
                     |
   -2                +T Q41,
                 XX  |
                     |
                     |
                     |
                     |
                     |
                     |
                     |
                     |
                     |
   -3                +  Q5,

     Least Capable   | Easiest Items Are Here
 Students Are Here
               <less>|<frequ>
```

Activity #4

Pretend that a test was administered to students at the start of a course and the mean person measure of students was 0.00 logits. Also, let's pretend the same test was administered at the end of the course and the mean person measure was .80 logits. Show how you can use a Wright Map to bring meaning to the analysis results.

Answer: First you would conduct a test such as a *t*-test to see if the means are significantly different. If the means are significantly different, you would know there was a significant level of growth in the student performance using logit measures. Then you would plot a line at the 0.00 level (that would be the start of class line). Then you would plot the end of class line at .80 logits. Now the most important items are the ones between the two lines. These items summarize the learning that took place from start to end of the course.

Activity #5

Using guidance provided in this chapter, if a student Rose, attempts a dichotomous test item which is harder than her ability level. The chance of Rose correctly answering this item is .3. If Rose's ability level is 1.0 logits, how far away from Rose, in logits, is the item. Second, what is the logit value of the item?

Answer:

Using the Rasch equation:

$$\begin{aligned} B_n - D_i &= \ln\left[.3/(1-.3)\right] \\ &= \ln\left[.3/(.7)\right] \\ &= \ln[.43] \\ &= -.84 \end{aligned}$$

The item is .84 logits away from Rose. The item has a logit measure which must be more positive than that of Rose, since Rose has less than a .5 chance of correctly answering the item.

If Rose ability level is 1.0, we can then put Rose's measure and the value −.84 in the Rasch equation and compute the difficulty of the item.

$$\begin{aligned} B_n - D_i &= -.84 \\ 1.0 - D_i &= -.84 \\ -D_i &= -.84 - 1.0 \\ -D_i &= -1.84 \\ D_i &= 1.84 \end{aligned}$$

Additional Readings

A very good discussion of the types of validity that can be evaluated with Rasch analysis. Many of the examples relate to the use of Wright Maps.

Baghaei, P. (2008). The Rasch model as a construct validation tool. *Rasch Measurement Transactions, 22*(1), 1145–1146.

An article demonstrating how Wright Maps can be used.

Pesudovs, K., Garamendi, E., Keeves, J. P., & Elliott, D. B. (2003a). Maps for diagnosis. *Rasch Measurement Transactions, 17*(3), 935.

An article in medical research in which a Wright Map was used.

Pesudovs, K., Garamendi, E., Keeves, J. P., & Elliott, D. B. (2003b). The activities of daily vision scale for cataract surgery outcomes: Re-evaluating validity with Rasch analysis. *Investigative Ophthalmology & Visual Science, 44*(7), 2892–2899.

Chapter 8
Fit

Isabelle and Ted: Two Colleagues Conversing

Ted: Isabelle, can you help me with the Rasch term "fit"? Just a little bit, then I think I will be okay.

Isabelle: When I first heard the term, I was confused. In statistics class when I first heard of goodness of fit, I was confused too. Sometimes, thinking of statistics helped me, but at other times thinking about statistics just caused more confusion.

Ted: How about talking a little bit about what you think fit means for Rasch? What's the point of looking at fit in a Rasch analysis? I think that will help me a lot.

Isabelle: Do you remember a few weeks ago when we looked at the Rasch model on the board and I talked about the Rasch perspective? The Rasch model is the only model that conforms to requirements of measurement outlined by people such as Norman Campbell, L.L. Thurstone, and Edward Thorndike. Because the Rasch model is viewed as the definition of measurement, we need to take time to evaluate whether or not the data fit the model. The view is if the data do not fit the model, then you cannot really conduct measurement.

Ted: So, when people say "an item may be misfitting" or "a person may be misfitting," do they mean that an item does not act as the Rasch model would predict? And, does that also mean a person does not act as the Rasch model would predict?

Isabelle: Exactly. But Ted, I also found some other things to be very helpful as I tried to understand fit, and I wish someone had told me before I struggled with it. The idea is quality control. Looking at fit and then considering if a person or item might be removed because of misfit is also quality control, and quality control is so important. When researchers collect data, we must make sure that we implement quality control. Companies make sure that their instruments work correctly on the assembly line. They are forever doing quality control. Also, if they get an odd piece of data from an instrument, they spend quite a bit of time trying to figure out what is going on with their data. Those who conduct work with tests and surveys must do the same. For example, a medical researcher using a published instrument to help measure a patient's recovery from a stroke must also conduct a quality control assessment of data they have collected.

Ted: What are some examples of what someone using a survey or test might do?

Isabelle: They might ask questions such as: Do the results make sense?

W.J. Boone et al., *Rasch Analysis in the Human Sciences*,
DOI 10.1007/978-94-007-6857-4_8, © Springer Science+Business Media Dordrecht 2014

Ted: Okay, back up to fit. What does assembly line production and quality control have to do with fit and research that is conducted in fields such as education and psychology?

Isabelle: In research we need to make sure our instruments are working properly, but we also must do something else. We must make sure we evaluate the quality of the data that we are collecting. That might mean the collection of test data or survey data. So we must conduct a review of the quality of data we collected.

Ted: So, if we find evidence that the survey responses of a teacher are of low quality, that is important to know. If we find evidence that the quality of some survey items is low, that is important to know.

Isabelle: Why do you say this?

Ted: Well, if we have strong evidence that data from some respondents are odd in some manner, it makes sense that we might not use all of those respondents' answers. Using questionable data might invalidate our analysis.

Isabelle: Exactly. If Sir Alexander Fleming had not rechecked his staphylococci bacteria cultures, he would likely not have noticed a culture dish unlike the other culture dishes. This culture appeared to be contaminated with fungus, and the staphylococci surrounding the fungus were dead. This definitely qualifies as odd data. The eventual result of Fleming's observation was penicillin. Of course, Fleming's discovery represents the positive side of detecting odd data. Again, when we use Rasch, we parallel so much of what scientists do in the lab. We continuously examine the quality of our data. Also like scientists, we collect information and we reflect on it; we do not immediately throw things out. And we do not act like robots and keep everything, just because at some point in an analysis, we hypothesized all our items were perfect.

Ted: What else confused you?

Isabelle: These are types of fit – for example, "Person Infit," "Person Outfit," "Item Infit," and "Item Outfit." I remember thinking, why can't there just be one type of fit? But once you get a handle on the concept, you can review all sorts of "fit" quickly, and it becomes second nature in an analysis. And, it really does not take very long. And you start to realize why it is a great help that there are different types of fit.

Ted: Give me an intro to fit to help me with this chapter!

Isabelle: Person fit looks at how a person answered all the items on a survey or test, but those answers are reviewed in light of the person's measure, which is computed using all of the respondent's answers compared to the difficulty level of the items. Person Infit is a statistic that gives more weight to responses on items near a person's measure. Person Outfit is a statistic that gives more weight to responses on items far away from a person's measure.

Ted: Give me an example from science teacher self-efficacy.

Isabelle: Think of someone who is very confident in terms of science teacher self-efficacy. When outfit of this person is computed, the statistic looks at this person's responses to all survey items, but particular attention is paid to answers that are far away from that person's measure. A faraway item for a person with strong self-efficacy would be the item that was easiest for this group of respondents to agree with. Does that help a little?

Ted: Yes it does. So if we were looking at the results of a physics test, for a person who did very well on the test, when Person Outfit is evaluated, there is more emphasis placed in how that person did on very easy test items.

Isabelle: Yes that is exactly right!

Ted: I think I need a little more practice, but it seems to me your main point is that fit allows us to do some quality control. By doing so, we may be able to spot people who are responding in an odd way. Also, we can spot items that are acting in an odd way. By thinking about such issues, we can improve the quality of the measurement we conduct.

Isabelle: And we need to evaluate if data fit the Rasch model.

Introduction to Fit

As researchers, we continually use theory when we evaluate survey and test data via Rasch measurement. Thinking about theory helps keep us grounded, helps organize our analyses, and helps us compute high quality measures. This chapter presents a series of concepts related to fit. After reading this chapter, readers will understand and be able to apply terms such as Person Infit, Person Outfit, Item Infit, Item Outfit, misfitting item, and misfitting person.

Person Fit

As a concept, fit describes how well data conform to the Rasch model. The degree of fit is expressed quantitatively by the results of estimating how well data fit the Rasch model. Remember that the Rasch model has been shown to be a definition of measurement. Being a definition of measurement results in several beneficial by-products. For instance, if an item bank has been created or test forms have been linked, respondents can be administered different combinations of items, but all respondents can be expressed on the same scale. This means that if a respondent does not answer an item, that data need not be discarded from a study. Also, using the Rasch model permits construction of different forms of a test, but all respondents can be expressed on the same scale. Use of the model yields great benefits, but researchers must thoroughly and carefully evaluate whether or not the data fit the model. When data do not appear to fit the model expectations, some sort of divergence exists in respondents' answers to the items and the theory that was used to generate items for the instrument along a single variable. This divergence may be large or small, explainable or unexplainable. The concept of fit helps us identify (and pause and reflect) divergence of data from the Rasch model expectations. When we consider fit, we initially consider respondents/persons, the items of the instrument, and the theory used to predict items and persons along a single trait.

In order to discuss fit, we have created Fig. 8.1. This figure presents the same Wright Map (in terms of self-efficacy item ordering and spacing) that was presented in Chap. 6. However, in this figure we provide three Wright Maps of fictitious respondents. Items at the base of the three Wright Maps are easier to agree with. Items at the top are harder to agree with. For each Wright Map, we provide the rating scale selected by fictitious respondents (Si, Andy, Mel) for each of the 13 self-efficacy

Si		Andy		Mel	
Harder to Agree With		Harder to Agree With		Harder to Agree With	
Q19	Barely Agree	Q19	Strongly Disagree	Q19	Strongly Disagree
Q5	Agree	Q5	Disagree	Q5	Agree
Q20	Agree	Q20	Disagree	Q20	Disagree
Q23	Agree	Q23	Disagree	Q23	Disagree
Q17	Agree	Q17	Disagree	Q17	Disagree
Q12	Agree	Q12	Disagree	Q12	Disagree
Q18	Agree	Q18	Disagree	Q18	Disagree
Q21	Agree	Q21	Disagree	Q21	Disagree
Q3	Agree	Q3	Disagree	Q3	Disagree
Q6	Agree	Q6	Disagree	Q6	Disagree
Q8	Agree	Q8	Disagree	Q8	Disagree
Q22	Agree	Q22	Disagree	Q22	Disagree
Q2	Strongly Agree	Q2	Barely Disagree	Q2	Barely Disagree
Easier to Agree With		Easier to Agree With		Easier to Agree With	

Fig. 8.1 Wright Maps of three fictitious students: Si, Andy, and Mel. In each map, the 13 self-efficacy items (after any needed flipping) are presented. Answers of the three fictitious respondents are presented to the right of each item

items. Let's now look at Si. Scanning from the "easiest to agree with" item (Q2) to the "hardest to agree with" item (Q19), one observes a predictable change in response pattern. As "harder to agree with" items are answered, Si ultimately shifts to a different rating scale category, one that indicates that he is less agreeable with regard to these "harder to agree with" items. This is the expected pattern if the items do a good job of defining a single trait. Moreover, this is the expected pattern if a respondent reacts to the items in a predictable manner. Si would be an example of good fit.

The idea of fit is complex, and we therefore present a second example that also illustrates good fit. The same vertical Wright Map is presented; easier to agree with self-efficacy items are at the base of the map, and harder to agree with items are at the top of the map. Again, we present the responses of a second fictitious respondent, Andy. Andy is not as agreeable as Si. But, as was the case for Si, Andy responds in a predictable manner in light of our theory of self-efficacy. Although the specific categories of responses selected are quite different, in that Andy tends to select rating categories that include *Disagree* and *Barely Disagree*, there exists a similar shift to less agreement as "harder to agree with" items are answered.

Take a moment to review and compare these two respondents (Si and Andy). Yes, their responses are quite different, but the pattern observed as one proceeds from "easier to agree with" items to "harder to agree with" items is the same. This suggests that, at least for these two individuals, the respondents answer the set of items in line with our conception of self-efficacy theory, which is expressed through the text of the 13 self-efficacy items.

We will see shortly that a number of Rasch indices help summarize this predictable pattern of responses, and these indices also help quantify the level of unpredictability. But for now, let's reflect conceptually, without numbers.

What might be a pattern of responses that diverges from our prediction? Figure 8.1 presents a third fictitious person's (Mel) responses to the 13 self-efficacy items. Readers should note that Mel's responses are identical to Andy's responses, with the exception of item Q5, where Mel selected *Agree*. Selection of *Agree* for item Q5 alone does not suggest a misfitting person. Possible misfit is, however, suggested by reviewing Mel's selection of *Agree* for Q5 compared to her responses to the other self-efficacy items in light of the predicted ordering of items (made a priori) from self-efficacy theory. For example, the model predicts that the responses to Q5 should not be vastly different from responses to items Q19 and Q20. This particular divergence of responses from the Rasch model for a single person is an example of a misfitting person.

A myriad of potential reasons for misfit exists; therefore, once identified, a misfitting person need not be removed automatically from an analysis. Whereas the identification of misfit sounds an alarm for consideration and review, it does not identify what caused the misfit. At this point, researchers must consider possible causes for misfit. Perhaps this respondent had a particular unique personal experience that greatly impacted her response to this item. Perhaps this respondent experienced a reading comprehension issue with this item. If these data were hand entered, perhaps a mistake was made in entering the data. As in all things in life, nothing is certain, but through the consideration of fit, one can identify persons and items that do not fit the Rasch model and as a result degrade measurement.

As we conclude this introduction to person misfit, we stress that, throughout the book, readers will continually observe interactions of theory, items expressing theory, and persons reacting to items (which define theory). When an issue such as person fit is considered, there is almost invariably a consideration of items and theory. Can one turn back the clock and talk to a person as she or he completed the apparently misfitting data? Of course not; however, it is possible to try to evaluate the data pattern from a misfitting person and attempt to figure out what might cause the misfit. For example, 10 respondents who completed a survey may misfit, and the researcher notices that, although half of the sample were males and half were females, all of the misfitting responses were females. This might suggest that there was some sort of gender issue that is impacting the manner in which the measurement is taking place with the instrument.

Formative Assessment Check Point #1

Question: Must a misfitting person and/or item be completely removed from an analysis?

Answer: No. An analyst can identify the mathematical reason for the misfit (which item was unexpectedly answered by the respondent). But there are many potential reasons for misfit of a person. Since Rasch analysis does not need respondents to answer all items on a survey such as the STEBI, in most cases it is possible to remove the odd response and keep a respondent as long as that change has brought the person's degree of fit within reasonable bounds.

Item Fit

There exists a continual interplay among items, persons, and the trait in Rasch analysis. As we introduce the idea of item fit to those learning Rasch for the first time, we often find it helpful to shift gears, ever so briefly, to multiple-choice tests, which are typical within and beyond the field of education. In such multiple-choice tests, individual items are right or wrong. To explain the concept of item fit, we ask readers to imagine a 20-item biology test, with items that are conceptualized to define a single trait and range from easy to hard. An example of a "misfitting" item would be a difficult item that is correctly answered by low performing students. Not all low performing students correctly answered this item, but a number of them did. Another type of item which causes misfit is an easy item that is incorrectly answered by respondents who have done very well on the test. Taken together, these two examples of item misfit illustrate a particular kind of fit called outfit. Another type of fit is infit, which is a close relative of outfit. In more technical terms, outfit and infit are chi-square statistics. Chi-square statistics are typically used to measure association between two groups, variables, or criteria. In a Rasch analysis, the criteria is the association between the model and the data, specifically how well the data fit the model. Although, the calculations for outfit and infit indices are more detailed, the general idea of outfit is that it is a fit statistic sensitive to outliers as described above (e.g. guessing or thoughtless errors) and infit focuses less on outliers but more on responses near a given item difficulty (or person ability). For more detailed information on the calculations of outfit and infit, refer to the Winsteps manual ("misfit diagnosis") and other articles (Linacre, 2012; Wright & Masters, 1982).

Person Fit Indices and Item Fit Indices

Several fit indices are provided in a Rasch analysis: Person Infit ZSTD, Person Outfit ZSTD, Person Infit MNSQ, Person Outfit MNSQ, Item Infit ZSTD, Item Outfit ZSTD, Item Infit MNSQ, and Item Outfit MNSQ. Readers should remember

```
TABLE 18.1 se excel used for fit chp item plot s ZOU657WS.TXTd Sep 11 10:41 2011
INPUT: 75 PERSON  13 ITEM  REPORTED: 75 PERSON  13 ITEM  6 CATS  MINISTEP 3.72.3
--------------------------------------------------------------------------------
PERSON: REAL SEP.: 2.52  REL.: .86 ... ITEM: REAL SEP.: 7.00  REL.: .98

        PERSON STATISTICS:  ENTRY ORDER

-------------------------------------------------------------------------------------------
|ENTRY   TOTAL  TOTAL            MODEL|   INFIT  |  OUTFIT  |PT-MEASURE |EXACT MATCH|
|NUMBER  SCORE  COUNT  MEASURE  S.E. |MNSQ  ZSTD|MNSQ  ZSTD|CORR.  EXP.| OBS%  EXP%| PERSON
|------------------------------------+----------+----------+-----------+-----------+------
|     1     60     13     1.34    .35|1.71   1.5|1.63   1.4|  .52   .57| 46.2  50.2| 21141
|     2     27      6      .84    .50| .71   -.3| .57   -.6|  .83   .60| 50.0  46.1| 91052
|     3     33      6     3.28    .81| .62   -.5| .57   -.5|  .61   .46| 66.7  66.9| 95793
|     4     47     13      .08    .29| .31  -2.6| .28  -2.5|  .84   .67| 46.2  40.0| 08453
|     5     51     13      .43    .30| .52  -1.5| .53  -1.4|  .81   .64| 38.5  42.7| 36281
```

Fig. 8.2 A person entry order table: Winsteps Table 18.1

```
TABLE 14.1 se excel used for fit chp item plot s ZOU657WS.TXTd Sep 11 10:41 2011
INPUT: 75 PERSON  13 ITEM  REPORTED: 75 PERSON  13 ITEM  6 CATS  MINISTEP 3.72.3
--------------------------------------------------------------------------------
PERSON: REAL SEP.: 2.52  REL.: .86 ... ITEM: REAL SEP.: 7.00  REL.: .98

        ITEM STATISTICS:  ENTRY ORDER

--------------------------------------------------------------------------------------------
|ENTRY   TOTAL  TOTAL            MODEL|   INFIT  |  OUTFIT  |PT-MEASURE |EXACT MATCH|      |
|NUMBER  SCORE  COUNT  MEASURE  S.E. |MNSQ  ZSTD|MNSQ  ZSTD|CORR.  EXP.| OBS%  EXP%| ITEM |
|------------------------------------+----------+----------+-----------+-----------+------|
|     1    410     75    -2.49    .22|1.06    .4|1.08    .5|  .31   .41| 55.4  64.4| Q2   |
|     2    317     75      .22    .13|1.69   3.5|1.64   3.1|  .52   .59| 33.8  44.5| Q3   |
|     3    258     75     1.16    .12|1.28   1.7|1.34   1.9|  .54   .67| 37.8  41.2| Q5   |
|     4    352     75     -.48    .15| .96   -.2| .89   -.5|  .52   .53| 51.4  54.5| Q6   |
|     5    369     75     -.91    .17|1.03    .2| .97   -.1|  .55   .50| 59.5  57.9| Q8   |
|     6    310     75      .34    .13| .99    .0| .96   -.2|  .56   .60| 52.7  44.2| Q12  |
|     7    277     69      .50    .13| .69  -2.1| .72  -1.7|  .71   .61| 39.7  42.8| Q17  |
|     8    298     69      .11    .14| .66  -2.1| .62  -2.3|  .62   .57| 47.1  46.0| Q18  |
|     9    206     68     1.66    .13|1.10    .7|1.21   1.2|  .67   .69| 44.8  40.8| Q19  |
|    10    262     69      .76    .13|1.07    .5|1.13    .8|  .65   .63| 39.7  41.3| Q20  |
|    11    295     69      .17    .14| .77  -1.4| .84   -.9|  .65   .58| 45.6  44.9| Q21  |
|    12    364     69    -1.83    .21| .97   -.1| .93   -.3|  .51   .44| 64.7  63.7| Q22  |
|    13    260     69      .79    .13| .80  -1.3| .86   -.8|  .70   .63| 44.1  41.1| Q23  |
|------------------------------------+----------+----------+-----------+-----------+------|
```

Fig. 8.3 An item entry order table: Winsteps Table 14.1

that the concept of fit is important for both persons and items. We suggest to our students that when they evaluate the fit of items and the fit of persons, they should initially spend their time identifying items and persons using the fit statistic called outfit and more particularly Item Outfit MNSQ and Person Outfit MNSQ. This suggestion is based on valued guidance provided by M. Linacre in the Winsteps manual (2012) because the outfit statistic is more sensitive to outliers and has a more familiar calculation. The outfit statistic's sensitivity to outliers also makes it easier to identify and correct issues of fit. More so, Linacre (2012, p. 622) states specifically for reporting purposes that only outfit needs to be reported; "unless the data are heavily contaminated with irrelevant outliers," then reporting infit may be appropriate. Below we provide two tables from our now familiar STEBI self-efficacy analysis with a sample of 75 respondents. Figure 8.2 (Winsteps Table 18.1) is a person entry order table that contains a range of information regarding each respondent. We present the first five respondents for readers. Our second table, Fig. 8.3 (Winsteps Table 14.1), is an item entry order table (Item Statistics: Entry Order). This table contains data for all 13 self-efficacy items.

Fig. 8.4 Reasonable ranges for item MNSQ infit and outfit as suggested by Wright and Linacre (1994) (Reprinted with permission)

Reasonable Item Mean-square Ranges for INFIT and OUTFIT	
Type of Test	Range
MCQ (High stakes)	0.8 - 1.2
MCQ (Run of the mill)	0.7 - 1.3
Rating scale (survey)	0.6 - 1.4
Clinical observation	0.5 - 1.7
Judged (agreement encouraged)	0.4 - 1.2

In examining Winsteps Tables 18.1 and 14.1 (Figs. 8.2 and 8.3), identify the two outfit numbers provided for each person and each item. For the first person in Table 18.1, we see the value 1.63 reported for OUTFIT MNSQ and the value 1.4 reported for OUTFIT ZSTD. For the first item (Q2 of the STEBI, the first SE item of the 23 items STEBI) presented in Table 14.1, OUTFIT MNSQ is 1.08 and OUTFIT ZSTD is .5. Briefly, the MNSQ (mean-square) is a chi-square calculation (which measures level of association) for the outfit and infit statistics. The ZSTD (z-standardized) provides a *t*-test statistic measuring the probability of the MNSQ calculation occurring by chance. Since the ZSTD value is based on the MNSQ and in accordance with advice from Linacre (2012), we first examine the MNSQ for evaluating fit. As long as the MNSQ value lies within an acceptable range of fit, we ignore the ZSTD value.

Therefore, we start our evaluations of fit by first looking at the column of data with the header OUTFIT MNSQ for persons and the header OUTFIT MNSQ for items. If we find that the persons in the data set and the items are within acceptable ranges of MNSQ, then we do not investigate ZSTD. What are acceptable ranges for MNSQ? To identify such ranges, we utilize a table provided by Wright and Linacre in their 1994 article entitled "Reasonable Mean-Square Fit Values" (Fig. 8.4). In general, a range between 0.5 and 1.5 suggests a reasonable fit of the data to the model. This is because the calculation of mean-squares produces an average near 1.0 (Wright & Linacre, 1994). Therefore, values greater than 1.0 show underfit meaning there is too much unexplained variance (or noise) in the data, and values less than 1.0 show overfit meaning the model overpredicts the data causing inflated reliability statistics; see Fig. 8.5 for interpretation of mean-square fit statistics based on multiple simulation studies (Linacre, 2012).

Now that we have some guidance as to which fit to look at (outfit), which outfit index to investigate (MNSQ first), and what range of Outfit MNSQ item values is reasonable, then what? Usually, we initially review the Outfit MNSQ values for both persons and items. A scan of Winsteps Table 14.1 reveals that only one item

Interpretation of parameter-level mean-square fit statistics:	
>2.0	Distorts or degrades the measurement system
1.5 - 2.0	Unproductive for construction of measurement, but not degrading
0.5 - 1.5	Productive for measurement
<0.5	Less productive for measurement, but not degrading. May produce misleadingly good reliabilities and separations

Fig. 8.5 Interpretation of mean-square fit statistic values (Reprinted with permission from Wright & Linacre, 1994)

```
TABLE 9.1 se excel used for fit chp item plot sa ZOU618WS.TXTa Sep 11 13:29 2011
INPUT: 75 PERSON  13 ITEM  REPORTED: 75 PERSON  13 ITEM  6 CATS  MINISTEP 3.72.3
--------------------------------------------------------------------------------

         -3       -2       -1        0        1        2        3        4        5
         -+--------+--------+--------+--------+--------+--------+--------+--------+-
 I     2 +                                    |                                  +  2
 T       |                                    |                                  |
 E       |                           A        |                                  |
 M       |                                    |B                                 |
         |                               D    |   C                              |
 O     1 +----E----f-------F---------G----|--------------------------------+  1
 U       |                     e     c  b d   |                                  |
 T       |                           a        |                                  |
 F       |                                    |                                  |
 I       |                                    |                                  |
 T     0 +                                    |                                  +  0
         -+--------+--------+--------+--------+--------+--------+--------+--------+-
         -3       -2       -1        0        1        2        3        4        5
                                         ITEM MEASURE

 PERSON                        1 3131414174247313322 3 26 3 1    1           11
                         T          S          M          S          T
 %TILE                         0  10   20 40 50 60 70 80 90                   99
```

Fig. 8.6 Table 9.1 from Winsteps. Item Outfit MNSQ values are plotted on the vertical axis and item measures are presented on the horizontal axis

should be flagged with regard to Outfit MNSQ – item Q3, with a value of 1.64. The identification of items with high Outfit MNSQ is quite easy to conduct with Table 14.1. However, Winsteps also provides another informative plot (Fig. 8.6, Winsteps Table 9.1).

Figure 8.6 (Winsteps Table 9.1) plots the Item Outfit MNSQ values and the item measures. Table 14.1 provides the exact Item Outfit MNSQ value, but this plot is useful in a synthesis of data. Here we can see that an item identified with the letter "A" has an Outfit MNSQ value above 1.5 (just eyeballing the plot).

Formative Assessment Check Point #2

Questions: Is the range of numbers for Outfit MNSQ and Outfit ZSTD the same?

Answer: No. The scale for MNSQ averages at 1 and is positive. The range for ZSTD is both negative and positive. Use ranges as suggested by Linacre and Wright.

Over many years of analyzing data sets, we have learned that a few odd (unpredictable) responses by individuals can impact the fit of items. So our next step in an analysis is to evaluate the individuals who are misfitting due to their responses to items that have been flagged for possible misfit. Our goal in identifying the persons who acted in an unpredictable manner to specific items is first to note their response (what rating category they selected or for a test was the item right or wrong). We then reflect on their rating. For example, do we have any information that might help us understand why they deviated from our measurement theory? Finally, by identifying the persons who acted in an unpredictable manner to an item, one can experiment by removing only that response from the person (recall that in Rasch measurement, since we are using one trait, not all items have to be answered by a respondent to be expressed on the same scale). Removal of a response does not mean that we forgot what we have done, and removal of a response does not mean that we do not learn from that response. For measurement purposes, this response clouds what we can learn; this odd response distorts our measures. Figure 8.7 (Winsteps Table 11.1) displays the table that we now use in our analyses to quickly identify those individuals who acted in an unexpected manner to the misfitting item (Q3).

The header of Fig. 8.7 (Winsteps Table 11.1) is hopefully familiar to readers at this point. If readers wish to do so, you can compare the values (just below the header) for NUMBER, NAME, MEASURE, INFIT, and OUTFIT to those in our Fig. 8.3 (Winsteps Table 14.1) above. The only additional information in Fig. 8.7 is the letter "A" under the phrase (MNSQ). This letter corresponds to the letter presented in Fig. 8.6 (Winsteps Table 9.1) identifying item Q3.

Figure 8.7 contains the coded responses of all 75 respondents to this one item. Each line that begins with the word RESPONSE presents ten responses, with exception of the last line of RESPONSE, which will be at times shorter if the data set was not a multiple of ten respondents. This data organization means that the response of the first person in the data set was "5" (*Agree*), and the response of the second person in the data set was "4" (*Barely Agree*). The response of the 10th person in the data set was "2" (*Disagree*). The important values for researchers are presented as Z-RESIDUALS. This means that the 45th person of the data set answered *Strongly Agree* (6) for item Q3, and this answer (based upon a z statistic) was unexpected (see the value of 2 beneath the 6?). Remember we can tell that it is the 45th person in the data set in that the first person response in the line: "`41:    2    5    5    6    6    6    5    4    6    5`" is the response for the 41st person. So the

```
TABLE 11.1 se excel used for fit chp item plot s ZOU869WS.TXTd Sep 12  8:32 2011
INPUT: 75 PERSON  13 ITEM  REPORTED: 75 PERSON  13 ITEM  6 CATS  MINISTEP 3.72.3
--------------------------------------------------------------------------------

TABLE OF POORLY FITTING ITEM    (PERSON IN ENTRY ORDER)
NUMBER - NAME -- POSITION ------ MEASURE - INFIT (MNSQ) OUTFIT

     2  Q3                             .22       1.7    A     1.6
   RESPONSE:       1:   5   4   6   3   3    6   3   2   5   2
 Z-RESIDUAL:                                                 -2

   RESPONSE:      11:   5   6   3   3   3    5   6   6   4   3
 Z-RESIDUAL:

   RESPONSE:      21:   5   5   4   4   4    5   5   4   3   1
 Z-RESIDUAL:                                                 -2

   RESPONSE:      31:   6   2   2   5   5    6   5   4   5   3
 Z-RESIDUAL:                   -3

   RESPONSE:      41:   2   5   5   6   6    6   5   4   6   5
 Z-RESIDUAL:                            2    X

   RESPONSE:      51:   5   5   4   5   2    3   6   5   3   5
 Z-RESIDUAL:                                     2

   RESPONSE:      61:   3   3   5   5   5    5   5   5   3   2
 Z-RESIDUAL:                                                 -2

   RESPONSE:      71:   4   2   5   2   4
 Z-RESIDUAL:               -3
```

Fig. 8.7 A portion of Winsteps Table 11.1. This table provides the responses of each respondent to a particular item. In this case, the responses of 75 individuals to Q3 of the STEBI scale. A *z*-residual value is provided for responses that were unexpected given the other responses of the individual to the survey. The 45th and 57th respondents answered a "6" to item Q3. This answer was unexpected (a *z*-residual of "2" is reported for these respondents)

41st person answered a "2" for this item, the 42nd person answered a "5" for the item, and so on. The third "6" to appear in the line is the response for the 46th person. It is easy to look at this line of data and think that all the information pertains to person 41 since the line begins with this number. However, the number 41 only tells you that the first number (a "2") to follow the 41 was the answer to item Q3 for person 41. Importantly, the number 5 which follows the number 2 indicates the answer to this same survey item by the 42nd person in the data set.

Our next step is simply to write down the person entry numbers for all those people who have a *z*-residual of 2 or higher and to write down those individuals who have a *z*-residual more negative than −2, in Table 11.1. To simplify our lengthy chapter, we have developed an example in which we specifically only consider unexpected answers which are more than 2.

Using a value of 2 or higher for a standardized *z*-score in introductory statistics is a common cutoff for statistical tests, and we use these *z*-residuals in a similar manner. For item Q3, respondents 45 and 57 are the two people who responded unexpectedly (given a review of their other responses and the way in which the items define the trait).

Formative Assessment Check Point #3

Questions: How can we identify the persons and items which misfit? And, once we identify those people and/or items, how can we figure out what person responses caused an item to misfit, and how can we figure out what items caused a person to misfit?

Answer: The Winsteps person tables and item tables provide, among many indices, values for Outfit MNSQ. However, to better understand and identify the specific interactions of persons and items which contributed to misfit, it is important to review Winsteps Tables 7 and 11.

Our next action is to investigate the impact of these two people on the Outfit MNSQ value of item Q3. To conduct such an investigation, we take three steps. First, make a copy of the control file that has all the data. Second, name the copy. Third, find the responses of the 45th and 57th persons and replace the responses for Q3 (this is the 2nd item used in our self-efficacy analysis) with a blank or an X. By doing so we are then able to rerun our data and then evaluate the fit of the item after the two odd responses have been removed (an alternative technique is to use the control file command "EDFILE"; this command allows one to edit the data file without permanently removing the data). If this action lowers the fit values, then we have succeeded in not only improving the manner in which the set of 13 self-efficacy items measure the trait, but we have also improved our confidence in the quality of self-efficacy measurement we have computed for persons 45 and 57. Figure 8.8 presents the data for persons 45–57 as originally presented in the control file, as well as the data following the removal of the responses of persons 45 and 57 to Q3 of the self-efficacy part of the STEBI.

The next action in our analysis is to run our new control file and investigate the impact of removing these two responses on our investigation of Q3 item fit. Figure 8.9 presents Winsteps Table 14.1 from our Rasch analysis, having removed the unexpected responses of the 41st and 51st persons. First, we point out that one way to double-check to make sure that the change in the data set was implemented (sometimes you might make a mistake and read the old data set) is to look at the Total Count column for Fig. 8.9 (Winsteps Table 14.1) and pay particular attention to the entry for item Q3. That number is 73, which is two less than you will see above in Fig. 8.3 (Table 14.1) for Q3. This helps the analyst see that the change has been implemented. You could also look at a "person" table and look to see if the "count" for persons 45 and 57 decreased by 1 for each person.

Let us look at the fit value for item Q3. Examining the table from our most recent run, we note that the OUTFIT MNSQ value of item Q3 has dropped from 1.64 to 1.49. Therefore, these two odd responses did impact the fit of item Q3. To summarize these actions, we first focused on Outfit by investigating Item Outfit MNSQ. We next tried to identify the person responses that may have created the misfit of items. We then removed strange responses, noted any change in the

```
6626632322562 65040   PR 5665262653435554232256 2
```

Original Data for persons 45-57.

```
6X26632322562 65040   PR 56652626534355542322562
```

Data following editing. Note the insertion of two "X" symbols.

Fig. 8.8 STEBI self-efficacy responses for persons 45–57

```
 TABLE 14.1 se excel used for fit chp item plot s ZOU339WS.TXTd Sep 12  9:38 2011
INPUT: 75 PERSON  13 ITEM  REPORTED: 75 PERSON  13 ITEM  6 CATS  MINISTEP 3.72.3
--------------------------------------------------------------------------------
PERSON: REAL SEP.: 2.55  REL.: .87 ... ITEM: REAL SEP.: 7.09  REL.: .98

         ITEM STATISTICS:  ENTRY ORDER
```

ENTRY NUMBER	TOTAL SCORE	TOTAL COUNT	MEASURE	MODEL S.E.	INFIT MNSQ	INFIT ZSTD	OUTFIT MNSQ	OUTFIT ZSTD	PT-MEASURE CORR.	PT-MEASURE EXP.	EXACT MATCH OBS%	EXACT MATCH EXP%	ITEM
1	410	75	-2.56	.22	1.08	.5	1.10	.6	.31	.42	55.4	64.8	Q2
2	305	73	.32	.13	1.50	2.7	1.49	2.4	.57	.60	36.1	45.0	Q3
3	258	75	1.18	.13	1.30	1.8	1.36	2.0	.54	.68	37.8	41.5	Q5
4	352	75	-.49	.15	1.00	.1	.93	-.3	.52	.54	50.0	54.6	Q6
5	369	75	-.94	.17	1.08	.4	1.01	.1	.54	.51	59.5	58.4	Q8
6	310	75	.34	.13	1.02	.2	.99	.0	.56	.61	52.7	44.6	Q12
7	277	69	.51	.13	.69	-2.1	.72	-1.7	.72	.61	41.2	43.5	Q17
8	298	69	.11	.14	.65	-2.2	.62	-2.3	.63	.58	47.1	46.5	Q18
9	206	68	1.69	.13	1.12	.7	1.23	1.3	.68	.69	46.3	41.3	Q19
10	262	69	.77	.13	1.09	.6	1.16	.9	.65	.64	39.7	41.5	Q20
11	295	69	.17	.14	.80	-1.2	.87	-.7	.65	.59	45.6	45.3	Q21
12	364	69	-1.89	.21	.99	.0	.94	-.2	.51	.45	64.7	64.2	Q22
13	260	69	.80	.13	.80	-1.3	.87	-.7	.71	.64	44.1	41.8	Q23

Fig. 8.9 Table 14.1 of Winsteps which can be used to double-check whether a change in the data set has been saved correctly

OUTFIT MNSQ values for these items, and checked to see if MNSQ values were within an acceptable range.

Early on in this chapter, we briefly introduced the ZSTD report of fit as it is noted in output such as Winsteps Tables 14.1 and 11.1. Following we provide additional details regarding the use of ZSTD for fit when MNSQ values are not within an acceptable range.

Formative Assessment Check Point #4

Question: With respect to the issue of fit, how do you know when to remove a person from an analysis, when to remove a response of a person from an analysis but not completely remove a person, and when to completely remove an item from an analysis?

Answer: Fit helps identify some instances in which the items and persons are behaving in a manner that does not suggest perfect functioning of the measurement scale. Make a list of persons and items that may misfit. What information can you collect to allow you to assess the misfit. Can you figure out why a person or item misfits? Use the guidance that we provide throughout the chapter to help you think about why an item or person misfits. Then experiment with analyses with and without items, with and without persons, and with and without answers of persons to specific items. Conducting good measurement with humans is an iterative process, very much as scientists in laboratories refine and improve their measurement instruments. We often construct a spreadsheet to keep track of our investigations (fit and otherwise) with the data.

Person Outfit ZSTD

We continue our discussion of fit by turning our attention to Person Outfit ZSTD. In this section we delve into more detail in the event that a researcher must look at ZSTD when MNSQ values did not shift into an acceptable range. Readers will see that the Person Outfit ZSTD index appears in most Winsteps person tables (e.g., Table 17 Person Measure; Table 18 Person Entry). Figure 8.10 presents lines 1–5 of Winsteps Table 18.1 for our original data set. Readers should take note of the OUTFIT ZSTD column (the 9th column of numbers).

What values might one expect to see for misfitting persons? Recall from the last section, the ZSTD (z-standardized) value measures the probability of the MNSQ value occurring by chance when the data fit the Rasch model. Specifically, the numbers reported in the column under the term OUTFIT and ZSTD are standardized values for the fit of each person when more weight is given to the responses of a person to items not near his or her overall attitude measure (in the case of a test, not near his or her overall test performance measure). Readers should recall that the overall measure of a person is computed based upon his or her responses to all items answered. Thus, OUTFIT ZSTD is expressed similar to a z-score with an expected value of 0 and a standard deviation of 1. The level of confidence researchers place in their decisions to identify misfitting individuals determines the value of ZSTD that is used as a "cutoff" (e.g., $\alpha = 0.05$). Typically, we identify respondents with

```
TABLE 18.1 se excel used for fit chp item plot s ZOU450WS.TXTd Sep 12 10:14 2011
INPUT: 75 PERSON  13 ITEM  REPORTED: 75 PERSON  13 ITEM  6 CATS  MINISTEP 3.72.3
--------------------------------------------------------------------------------
PERSON: REAL SEP.: 2.52  REL.: .86 ... ITEM: REAL SEP.: 7.00  REL.: .98

PERSON STATISTICS:  ENTRY ORDER

ENTRY  TOTAL  TOTAL             MODEL|   INFIT   |  OUTFIT   |PT-MEASURE |EXACT MATCH|
NUMBER SCORE  COUNT  MEASURE   S.E. |MNSQ  ZSTD|MNSQ  ZSTD|CORR.  EXP.| OBS%  EXP%| PERSON

1       60     13     1.34      .35 |1.71   1.5|1.63   1.4|  .52   .57| 46.2  50.2| 21141
2       27      6      .84      .50 | .71   -.3| .57   -.6|  .83   .60| 50.0  46.1| 91052
3       33      6     3.28      .81 | .62   -.5| .57   -.5|  .61   .46| 66.7  66.9| 95793
4       47     13      .08      .29 | .31  -2.6| .28  -2.5|  .84   .67| 46.2  40.0| 08453
5       51     13      .43      .30 | .52  -1.5| .53  -1.4|  .81   .64| 38.5  42.7| 36281
```

Fig. 8.10 The first five respondents presented in Winsteps Table 18.1 following the analysis of 75 respondents' answers to the 13 self-efficacy items of the STEBI

Standardized Value	Implication for Measurement
≥ 3	Data very unexpected if they fit the model (perfectly), so they probably do not. But, with large sample size, substantive misfit may be small.
2.0 - 2.9	Data noticeably unpredictable.
-1.9 - 1.9	**Data have reasonable predictability.**
≤ -2	Data are too predictable. Other "dimensions" may be constraining the response patterns.

Fig. 8.11 Guidelines for the interpretation of ZSTD values from Linacre (2002) (Reprinted with permission)

ZSTD values of 2.0 or higher, and those with a ZSTD values of −2 or lower, as worthy of further investigation. In all honesty, we have found that persons with ZSTD values of 3.0 or higher and −3.0 or lower are likely to be those we investigate in detail. Figure 8.11 gives the general guidelines of ZSTD values and implications of measurement as provided in the Winsteps manual (Linacre, 2002, p. 878).

How can individuals with potentially high misfit be identified quickly? As we did for the earlier example, we have developed an activity in which we investigate respondents with high positive ZSTD values to shorten our example. As you explore ZSTD for Person Outfit, you will want to use our techniques for both high positive Person Outfit ZSTD and for very negative Person Outfit ZSTD.

We suggest using Winsteps Table 5.1 (Fig. 8.12) to begin to identify misfitting persons. Certainly, researchers could review Winsteps Table 18 or Table 17 for small data sets; this is a possible technique that is not onerous. However, we usually begin to assess initially the number of misfitting individuals in a Rasch analysis with Winsteps Table 5.1. We have utilized Winsteps Table 5.1 earlier in this chapter, but readers will recall that we were considering MNSQ and that earlier presentation of Table 5.1 had a vertical axis of MNSQ. To alter table 5.1 so that the ZSTD values are presented, we added a simple line of code to the control file to create the plot of Table 5.1 presented below (with the vertical axis with MNSQ values). That line of code to add to your control file is MNSQ=N. By having the line "MNSQ=N" in your control file, Winsteps creates a plot of Outfit ZSTD.

```
TABLE 5.1 se excel used for fit chp item plot sa ZOU116WS.TXTa Sep 12 10:32 2011
INPUT: 75 PERSON  13 ITEM  REPORTED: 75 PERSON  13 ITEM  6 CATS  MINISTEP 3.72.3
--------------------------------------------------------------------------------

      -3      -2      -1       0       1       2       3       4       5
      -+-------+-------+-------+-------+-------+-------+-------+-------+-
    4 +                        |                                        +   4
      |                        |                                        |
      |                        |                                        |
      |                        |                                        |
    3 +                        |                                        +   3
P     |                        |                                        |
E     |                        |           B         A                  |
R     |                        |                                        |
S   2 +------------------------|--I--E---H------GD-F--------------------+   2
O     |                      C |  J                                     |
N     |                        |                                        |
      |                        |   N      K                             |
O   1 +                     M S|L  O Q                                  +   1
U     |                   R    |   TWU                                  |
T     |                        |    X           VPZ                     |
F     |                    Y1 2|1   1 1 1        1                      |
I   0 +-----------------------1|-11----1-----------1------------------1-+   0
T     |                   1    |      2    1  1  1                      |
      |                        |z 1   v     x            q              |
Z     |                        t  y   w1  u                             |
S  -1 +                        |  s   r  p                              +  -1
T     |                        |          1   mk                        |
D     |                 j   d  |  o                                     |
      |                   n    |                                        |
   -2 +------------------------|------I--h--c----e----------------------+  -2
      |                        |       g                                |
      |                        |f    b             a                    |
      |                        |                                        |
   -3 +                        |                                        +  -3
      -+-------+-------+-------+-------+-------+-------+-------+-------+-
      -3      -2      -1       0       1       2       3       4       5
                                 PERSON MEASURE

ITEM       1    1       1  1    2111 2  1   1
             T        S        M        S        T
%TILE      0   10      20 30   40 60 80 90 99
```

Fig. 8.12 Winsteps Table 5.1: A table which facilitates the quick identification of potentially misfitting respondents. A command line was added to the control file, so that this table now displays the OUTFIT ZSTD value for each person as opposed to the OUTFIT MNSQ value

Readers should note that the person measure is on the horizontal axis, which is organized from left, respondents who are least "agreeable" to the set of self-efficacy items (low scale score measure), to right, respondents who are most "agreeable" to this set of self-efficacy items (high scale score measure). To verify this statement about the meaning of measures for each respondent, readers should review Winsteps Table 17 (the person measure table) or Winsteps Table 20 (the raw score to scale score conversion table) to see that lower measures identify respondents with low total raw scores. These respondents more often marked the lower response categories (perhaps D or BD) in contrast to respondents with higher scale scores (perhaps BA or A). If readers need a quick refresher, we suggest rereading Chap. 3.

Formative Assessment Check Point #5

Question: Must all items and all persons exhibit perfect fit before a final analysis is conducted?

Answer: No. The most important first step is to investigate the fit of items. Yes, it is important to investigate the fit of persons, but there comes a point of diminishing measurement returns. This conundrum is like improving the precision with which a building brick is made. For the construction of some buildings, it will not matter if the brick matches other bricks to a nanometer. Much depends upon what is being measured.

The plot in Fig. 8.12 (Winsteps Table 5.1) above includes three prominent dashed horizontal lines, one at −2, one at 0, and one at 2. When investigating Person Outfit ZSTD, recall that a value of 2.0 or higher can be used to spot misfitting respondents. Researchers can therefore focus on the persons plotted above the top dashed line. As mentioned earlier, we often focus on the persons whose misfit ZSTD values are 3.0 or higher; thus, we typically print out this table and then draw a horizontal line at a value of 3.0 or higher.

One additional important point of information remains to be explained, the numbers such as "1" and "2" and letters such as "A" and "B" in the plot. Careful review of this plot reveals that each letter only occurs once, but there are numerous instances of the numbers "1" and "2." The number "1" indicates the location of a single respondent in terms of his/her Outfit ZSTD and measure. Think of this as akin to the x and y coordinates (x, y) of a point on a scatter plot. For example, a person with coordinates (2.0, −.25) has a self-efficacy measure of 2.0 and a Person Outfit ZSTD value of −.25. Review of the figure and the coordinates reveals that this person is plotted as the number "1," just to the right and above the person plotted with the letter "x." The number "2" indicates the Outfit ZSTD and measure coordinates of two people who have identical (or very similar for plotting purposes) Outfit ZSTD and measure values. If three respondents exhibited identical (or similar) measure and Outfit ZSTD values, we would see a "3" plotted in that location.

Readers should recall our suggestion that the persons of immediate interest for investigating are respondents with fit values above acceptable criteria levels. Winsteps provides an identifier in the form of a capital letter that identifies respondents who (for the group being evaluated) have particularly high fit values. The capital letters presented in this plot are identifiers of those individuals who should be investigated in more detail to better understand why they might misfit, as there are many patterns of responses that may not fit predictions of the Rasch model.

We will finish this chapter involving fit with a question often posed in our workshops: When fit is used, is one not just removing data that do not match the theory which you have developed for your measurement scale? Isn't this just somehow stacking the deck? We approach this question that by first commenting that when one is attempting to conduct measurement, one must have a theory as to what it means to measure a single trait. Second, if one wishes to measure a single trait, then the Rasch model, which is a definition of measurement, must be used. Third, if an item or a person does not fit the model, it means that something in some manner is amiss in the way in which the item or person were measured. If one is to have a rigorous measurement of persons and items, then these odd responses degrade the quality of measurement one carries out. Therefore, we always suggest that fit, for persons and items, should be viewed as a quality control step that is similar to the quality control of data that would take place in many settings, be it a factory manufacturing products or in a scientist's laboratory.

Isabelle and Ted: Two Colleagues Conversing

Isabelle: Ted, can you help me with the central point of Winsteps Table 5.1 (Person Outfit ZSTD vs. Person Measures)? Sometimes I get confused.

Ted: The first thing I remind myself about this plot is that it is important to investigate the fit of persons in a data set. I expect to see misfit by chance, but I want to make sure there is not a lot of misfit. If there is a lot of misfit, I need to think about what might be going on. I start with item fit and then go onto person fit. When I am thinking about person fit, I wonder if the data for some respondents were entered incorrectly. Also, I wonder if a survey item had a different meaning for some respondents. So this is one plot I use to gain an overall view for the fit of the data.

Isabelle: What do you look for?

Ted: Well, what I usually do first is remind myself what it means to move from left to right on the horizontal axis. In this case, people to the right are the more agreeable persons for the STEBI data. If I have not looked at the data for a few days, I usually go to the scale score raw score table (Winsteps Table 20) to help me remember what a higher measures means. You can see that in the data in Table 5.1 there are more people plotted on the right side of the plot than on the left side of the plot. This is because we had many people with measures above the average item measure.

Isabelle: What is next?

Ted: Well honestly, since I have looked at many of these plots, I usually just look at the data points plotted above the top horizontal line (the one at +2 ZSTD), and the people who are plotted below the −2 line. Also when I am looking at the plot, I try to remember how many people I would expect to see in total above and below the line by chance. There are 75 people in the sample; if 5 % might misfit by chance, we would expect a total of about 3 misfitting people by chance. I get "3," by multiplying 75 (the number of people) by .05.

Isabelle: Okay, we have these people plotted, but how do you figure out who is who?

Ted: That is something I got confused and frustrated about, but now I know how to find those people very quickly. See those letters for the misfitting people? Take person "B" Isabelle who I underlined and made bold in Winsteps Table 6.1. That person has a measure between 1.0 and 2.0 (actually 1.47) and a misfit above 2.0 ZSTD? Well that person "B" will be identified with the letter "B" in Winsteps tables such as 6.1 and 7.1. With either of these tables I can quickly identify person B and then, if I wish, see the responses for that person and quickly identify the odd responses. Here are edited Tables 6.1 and 7.1 (Fig. 8.13).

Readers will see that Winsteps Table 6.1 looks very similar to Winsteps Table 18, which presented the data for each respondent in "entry order." The difference is that

```
TABLE 6.1 se excel used for fit chp item plot sa ZOU116WS.TXTa Sep 12 10:32 2011
INPUT: 75 PERSON  13 ITEM  REPORTED: 75 PERSON  13 ITEM  6 CATS  MINISTEP 3.72.3
--------------------------------------------------------------------------------
PERSON: REAL SEP.: 2.52  REL.: .86 ... ITEM: REAL SEP.: 7.00  REL.: .98

         PERSON STATISTICS:  MISFIT ORDER

--------------------------------------------------------------------------------------------
|ENTRY   TOTAL  TOTAL           MODEL|   INFIT  |  OUTFIT  |PT-MEASURE |EXACT MATCH|
|
|NUMBER  SCORE  COUNT  MEASURE  S.E. |MNSQ  ZSTD|MNSQ  ZSTD|CORR.  EXP.| OBS%  EXP%| PERSON
|
|------------------------------------+----------+----------+-----------+-----------+-------
|     17     68     13     2.71    .48|3.90   3.6|2.57   2.5|A .53   .50| 61.5  66.2| 55959
|     72     61     13     1.47    .36|2.71   2.8|2.42   2.4|B .50   .56| 38.5  52.6| 96739
|     30     22      6     -.22    .44|2.46   2.1|2.12   1.7|C .32   .70| 33.3  41.4| 80392
|     60     66     13     2.28    .45|2.40   2.1|2.10   1.9|D .25   .52| 53.8  63.8| 67921
|     49     55     13      .80    .31|2.25   2.5|1.99   2.1|E .42   .61| 15.4  41.4| 78221
|     50     67     13     2.49    .46|2.24   2.0|2.07   1.9|F .48   .51| 53.8  63.7| 68028
|     47     65     13     2.09    .43|2.16   1.9|2.16   2.0|G .23   .53| 46.2  63.1| 07242
|     33     59     13     1.22    .34|2.10   2.1|1.97   1.9|H .41   .57| 46.2  49.6| 90384

TABLE 7.1 se excel used for fit chp item plot sa ZOU116WS.TXTa Sep 12 10:32 2011
INPUT: 75 PERSON  13 ITEM  REPORTED: 75 PERSON  13 ITEM  6 CATS  MINISTEP 3.72.3
--------------------------------------------------------------------------------

TABLE OF POORLY FITTING PERSON   (ITEM IN ENTRY ORDER)
NUMBER - NAME -- POSITION ------ MEASURE - INFIT (ZSTD) OUTFIT

    17  55959    PR 666555666565    2.71     3.6   A    2.5
  RESPONSE:      1:   6   6   5   5   6    5   6   5   1   5
Z-RESIDUAL:                                            -4

  RESPONSE:     11:   6   6   6
Z-RESIDUAL:

    72  96739    PR 262555255235    1.47     2.8   B    2.4
  RESPONSE:      1:   6   2   5   5   5    5   5   4   1   6
Z-RESIDUAL:              -3                            -2

  RESPONSE:     11:   6   6   5
Z-RESIDUAL:

    30  80392    PR 551643445355    -.22     2.1   C    1.7
  RESPONSE:      1:   5   1   4   3   4    5   M   M   M   M
Z-RESIDUAL:              -2   2

  RESPONSE:     11:   M   M   M
Z-RESIDUAL:
```

Fig. 8.13 Winsteps Tables 6.1 and 7.1, which are helpful for identification of specific persons and their odd responses

respondents are presented in "misfit order" as noted in the table's title. A second difference is that a letter is presented after the 9th column of data. Let's first look at the letter "A" in Table 6.1. This letter corresponds to the letter "A" that was presented in the Table 5.1. If one writes down the value of this person's measure (2.71) and his or her Outfit ZSTD (2.5), one can plot those two values in table 5.1 and see that the location of this person is, indeed, at the location noted by the letter "A." How does Table 6.1 help? Researchers usually use this table to identify the misfitting person. In the case of including IDs that include embedded information, one can sometimes use this table to identify patterns of misfit as a function of a subgroup. For example, in this data set, if the first two digits identify a specific grade of future teaching, one might investigate quickly if students who planned to teach a particular grade had a propensity for misfit.

```
INPUT: 75 PERSON  13 ITEM  REPORTED: 75 PERSON  13 ITEM  6 CATS  MINISTEP 3.72.3
--------------------------------------------------------------------------------

TABLE OF POORLY FITTING PERSON    (ITEM IN ENTRY ORDER)
NUMBER - NAME -- POSITION ------ MEASURE - INFIT (ZSTD) OUTFIT

     17  55959    PR 666555666565     2.71      3.6   A    2.5
  RESPONSE:       1:    6    6    5    5    6    5    6    5    1    5
Z-RESIDUAL:                                                      -4

  RESPONSE:      11:    6    6    6
Z-RESIDUAL:
```

Fig. 8.14 Excerpt of Winsteps Table 7.1 for a single respondent "A"

Formative Assessment Check Point #6

Question: If a researcher conducts a traditional analysis with raw data and computes a Cronbach's alpha, is that analysis the same as a fit analysis?

Answer: No. Raw data are used to compute a Cronbach's alpha. As a result, the analysis is flawed from the beginning. Second, use of a Cronbach's alpha or KR-20 is a very broad index that does not facilitate the level of data quality analysis that can be conducted with Rasch analysis.

As we explore fit in an analysis, we have found Winsteps Table 7.1 to be particularly beneficial. This table is organized in a manner similar to Winsteps Table 11.1, but Table 7.1 focuses on specific respondents for all items they answered. This table presents the responses of those respondents (e.g., mystery person "A") who are identified with capital letters in Fig. 8.14. We introduce Winsteps Table 7.1 just for the respondent identified with the capital letter "A."

In earlier tables, we learned how to quickly compare the measure and the fit of person "A" compared to the rest of the sample. And, we learned how to spot the specific values for person measure and person fit for respondent "A." As we introduce Winsteps Table 7.1, we first call readers' attention to the number "17" which indicates that this person is the 17th person in the data set. If we were to look at our control file and count to the 17th line of data, this would be our respondent. The numbers and letters "55959 PR" together constitute the person ID, which is read by the program. This information is also provided in our data line in the control file. What follows is the person measure (2.71) and later this person's Outfit ZSTD (2.5). The letter A, which ties to Winsteps Table 5.1.

The next 4 lines take more time to explain, but once they are understood, evaluation of a data set can quickly yield possible causes of person misfit. When reviewing this part of the table, readers will use "sets" of two lines. The first line, which begins with the word "RESPONSE:," gives the actual responses of the 17th person in the data set, the person with the ID 55959 PR. The number and symbol that follows ("1:") indicates that the first response on this line corresponds to the

```
TABLE 18.1 se excel used for fit chp item plot s ZOU153WS.TXTd Sep 14 10:17 2011
INPUT: 75 PERSON  13 ITEM  REPORTED: 75 PERSON  13 ITEM  6 CATS  WINSTEPS 3.72.3
--------------------------------------------------------------------------------
PERSON: REAL SEP.: 2.52  REL.: .86 ... ITEM: REAL SEP.: 7.00  REL.: .98

         PERSON STATISTICS:  ENTRY ORDER

----------------------------------------------------------------------------------------------------
|ENTRY   TOTAL  TOTAL               MODEL|   INFIT  |  OUTFIT  |PT-MEASURE |EXACT MATCH|
|
|NUMBER  SCORE  COUNT  MEASURE  S.E. |MNSQ  ZSTD|MNSQ  ZSTD|CORR.  EXP.| OBS%  EXP%| PERSON
|
|------------------------------------+----------+----------+-----------+-----------+---------
|     1     60     13     1.34    .35|1.71   1.5|1.63   1.4|  .52   .57| 46.2  50.2| 21141
|     2     27      6      .84    .50| .71   -.3| .57   -.6|  .83   .60| 50.0  46.1| 91052
|     3     33      6     3.28    .81| .62   -.5| .57   -.5|  .61   .46| 66.7  66.9| 95793
|     4     47     13      .08    .29| .31  -2.6| .28  -2.5|  .84   .67| 46.2  40.0| 08453
|     5     51     13      .43    .30| .52  -1.5| .53  -1.4|  .81   .64| 38.5  42.7| 36281
```

Fig. 8.15 Respondents 1–5 in Winsteps Table 18.1 (PERSON STATISTICS: ENTRY ORDER). The table presents a range of indices. The 9th column of data presents the Outfit ZSTD value for respondents. The rows are organized by the order in which respondents were presented in the data set

respondent's answer to the first item of the data set. This tells us that this respondent answered "6" (SA) to the first SE item presented on the survey (Q2SE). Continuing to read this line, we see that the respondent also answered "6" (SA) for the 3rd item of the survey (the second self-efficacy item, Q3se). The rest of this line is organized in a similar vein; the responses of this respondent (the 17th person in the data set, person A) are also provided for the 3rd self-efficacy item (a "5" as an answer) to the 10th self-efficacy (a "5" as an answer) item of the data set.

Readers should now focus on and review the line immediately below this line. Identified with the phrase "Z-RESIDUAL," this new line provides information regarding how unexpected the responses to these items listed above the line (q2, q3, q5, q6, q8, etc.) were when the overall measure of the respondent (based upon his/her set of responses to the self-efficacy items) and the overall trait defined by the set of 13 self-efficacy survey items are taken into account. Blanks indicate this response was not unexpected, given the overall measure of the respondent and the location of each specific item along the trait. The third line of this table provides the responses to the 11th to 13th self-efficacy items ("6," "6," "6"). This time, we see a value entered for one of the items for Z-RESIDUAL. Large values (both positive and negative) serve as "flags" of recipient responses to individual items as likely causes of this person's misfit. Scanning the responses of this person reveals that most of the responses to survey items (after flipping of appropriate items) were 5s and 6s; thus, it should make sense that this person's response of "SD" (coded as a 1) for the 9th self-efficacy item (Q19) is indeed unexpected, given that this person exhibits a strong self-efficacy, in that most of his or her responses after flips were SA and A.

Much of what determines which tables are reviewed in an analysis depends upon the size of a data set and which tables and plots one feels most at ease interpreting. There is extensive flexibility, in that many tables provide identical information, but one can select a table which is most helpful for the exploration of a specific issue.

Winsteps Tables 18.1 and 17.1 are presented below in Figs. 8.15 and 8.16, respectively. Only the first five lines of data are presented for each table. Scanning

```
        TABLE 17.1 se excel used for fit chp item plot s ZOU153WS.TXTd Sep 14 10:17 2011
INPUT: 75 PERSON  13 ITEM  REPORTED: 75 PERSON  13 ITEM  6 CATS  WINSTEPS 3.72.3
--------------------------------------------------------------------------------
PERSON: REAL SEP.: 2.52  REL.: .86 ... ITEM: REAL SEP.: 7.00  REL.: .98

        PERSON STATISTICS:  MEASURE ORDER

--------------------------------------------------------------------------------------------
|ENTRY   TOTAL  TOTAL            MODEL|   INFIT  |  OUTFIT  |PT-MEASURE |EXACT MATCH|
|NUMBER  SCORE  COUNT  MEASURE  S.E. |MNSQ  ZSTD|MNSQ  ZSTD|CORR.  EXP.| OBS%  EXP%| PERSON
|-------------------------------------------------------------------------------------------|
|    46     78     13     7.53    1.85|     MAXIMUM MEASURE|  .00    .00|100.0 100.0| 65730
|    36     75     13     4.89     .68| .93    .0| .70   -.1|  .37    .32| 69.2  77.8| 65234
|     3     33      6     3.28     .81| .62   -.5| .57   -.5|  .61    .46| 66.7  66.9| 95793
|    17     68     13     2.71     .48|3.90   3.6|2.57   2.5|  .53    .50| 61.5  66.2| 55959
|    18     67     13     2.49     .46|1.00    .2| .98    .1|  .64    .51| 61.5  63.7| 97766
```

Fig. 8.16 Respondents 1–5 in measure order in Winsteps Table 17.1 (PERSON STATISTICS: MEASURE ORDER). This table also presents a range of indices. The 9th column of data presents the Outfit ZSTD value for respondents. The rows are organized by respondent measures. For this data set with a coding of 6 for SA and 1 for SD, the first person in the data set has the highest measure (7.53), and this respondent has the strongest self-efficacy with regard to science teaching

the column headings of Table 18.1, readers should see that details regarding OUTFIT ZSTD are provided for the first 5 respondents in the data set. For example, the 4th person in the data set (Person 08453) has an Outfit ZSTD of –2.5. If one wishes to see how this person compares in terms of measure and Outfit ZSTD to the other 75 respondents, simply return to Table 4.1 and locate the person with coordinates "measure .08" (look at the horizontal axis), "OUTFIT ZSTD –2.5" (look at the vertical axis).

Winsteps Table 17.1 (Fig. 8.16) is very similar to Table 18.1. The only difference is that the respondents are arranged in measure order. In this data set, this means the first person listed is the person with the highest measure (measure=7.53). The second person (measure=4.89) exhibits the second highest scale score measure. This person has an OUTFIT ZSTD of –.1. Depending upon the types of issues that are being explored with respect to fit, this is another table that can help clarify an analysis.

Tying it all together, this chapter provided a brief introduction to the concept of fit. In particular we focused on person misfit. We think of person misfit as the degree to which a person does not act in a predictable manner with regard to the difficulty ordering and spacing of items. We also introduced some of the indices that can be used to evaluate fit. Some researchers pay additional attention to persons with large negative ZSTD values. A large negative value is viewed as "too predictable" (Linacre, 2002).

We wish to reiterate a point made earlier in this chapter: The identification of a misfitting person does not mean the person must be removed from an analysis. Also, the identification of a response or responses that more than likely caused the misfit does not tell the researcher why the respondent acted in an unpredictable manner to one or more items. We simply view misfit as a label that is attached to a respondent. It is then up to the researcher as to how he/she will try to understand the misfit and what action is taken. Perhaps the researcher knows the schools for each respondent, identifies all misfitting respondents and discovers that all misfitting respondents attend one classroom in one school. The same is true of item misfit. There are many reasons why an item exhibits misfit. The item may truly misfit, perhaps because it is not part of a trait. An item may appear to misfit for a group of respondents, but the collection of added data may reveal that there was something quite different about the two groups, and the item does a pretty good job of defining a portion of the latent trait. To almost close this chapter, we present the answer we received from Mike Linacre as we attempted to summarize fit and how to prioritize steps in a Rasch analysis:

> From a statistical Rasch perspective, persons and items are exactly the same. They are merely parameters of the Rasch model. So the fit criteria would be exactly the same. But, from a substantive perspective, persons and items differ. We expect the items to be better-behaved than the persons. We also expect item difficulties to continue into the future, but we expect person abilities to change. Also, we expect items to be encountered by many, many persons, but persons to encounter relatively few items. Consequently, we are usually stricter in our application of fit rules to items than to persons. A few maverick persons in a data set don't worry us – they will have negligible impact on anything else. But a few maverick items raise questions about test administration, data entry accuracy, the definition of the latent variable, etc. We will immediately focus our attention on them because they may be symptomatic of a more pervasive problem, such as the wrong key for a multiple-choice test, or reversed-coded items on a survey.

Personal e-mail communication from J. M. Linacre to the authors September 12, 2011

Our review of misfit ties to ideas of theory and the latent trait that are core components of Rasch analysis. A particular concern is the quality of data with respect to defining a variable. Reviewing the quality of data with Rasch techniques is much more detailed than the steps taken in the majority of studies prior to conducting statistical tests. Usually a cursory review of data quality is conducted in most studies. As a result, items may be retained in a scale that do not match theory and thus should not be included in steps to compute a person measure. Also, traditional techniques rarely, if ever, investigate the responses of respondents. As a result there is no certainty of whether respondents answered items in such a way as to provide useful data.

As a grand finale to this chapter on fit, we provide Fig. 8.17, a useful visual overview of potential reasons for misfit from the Winsteps manual (Linacre, 2012, p. 624). There are many issues associated with the use of fit to evaluate aspects of data quality, but investigating fit provides a level of analysis not often conducted when test and survey data are evaluated. When only a few persons misfit, what are the implications? More often than not, you have a measure of the person along the trait that may be suspect. What happens if an item is included in an analysis that misfits? More often than not, that item will impact the measures computed for respondents and warp the person measures you compute.

Isabelle and Ted: Two Colleagues Conversing

Ted: I think I get fit.

Isabelle: Okay, go to it; tell me all you know.

Ted: If I were explaining fit to a class, I would start off by talking about how important quality control of data is. It's great to collect data, but before conducting a final analysis, it is most important to conduct a data quality analysis.

Diagnosing Misfit: Noisy = Underfit. Muted = Overfit				
Classification	**INFIT**	**OUTFIT**	**Explanation**	**Investigation**
	Noisy	Noisy	Lack of convergence Loss of precision Anchoring	Final values in Table 0 large? Many categories? Large logit range? Displacements reported?
Hard Item	Noisy	Noisy	Bad item	Ambiguous or negative wording? Debatable or misleading options?
	Muted	Muted	Only answered by top people	At end of test?
Item	Noisy	Noisy	Qualitatively different item Incompatible anchor value	Different process or content? Anchor value incorrectly applied?
		?	Biased (DIF) item	Stratify residuals by person group?
		Muted	Curriculum interaction	Are there alternative curricula?
	Muted	?	Redundant item	Similar items? One item answers another? Item correlated with other variable?
Rating scale	Noisy	Noisy	Extreme category overuse	Poor category wording? Combine or omit categories? Wrong model for scale?
	Muted	Muted	Middle category overuse	
Person	Noisy	?	Processing error Clerical error Idiosyncratic person	Scanner failure? Form markings misaligned? Qualitatively different person?
High Person	?	Noisy	Careless Sleeping Rushing	Unexpected wrong answers? Unexpected errors at start? Unexpected errors at end?
Low Person	?	Noisy	Guessing Response set "Special" knowledge	Unexpected right answers? Systematic response pattern? Content of unexpected answers?
	Muted	?	Plodding Caution	Did not reach end of test? Only answered easy items?
Person/Judge Rating	Noisy	Noisy	Extreme category overuse	Extremism? Defiance? Misunderstanding the rating scale?
	Muted	Muted	Middle category overuse	Conservatism? Resistance?
Judge Rating	Muted	Muted	Apparent unanimity	Collusion? Hidden constraints?
INFIT: information-weighted mean-square, sensitive to irregular inlying patterns **OUTFIT**: usual unweighted mean-square, sensitive to unexpected rare extremes **Muted**: overfit, un-modeled dependence, redundancy, the data are too predictable **Noisy**: underfit, unexpected unrelated irregularities, the data are too unpredictable.				

Fig. 8.17 A visual overview of expressions of and reasons for misfit (Linacre, 2012) as appearing in the Winsteps manual (Reprinted with permission)

Isabelle: That is what scientists do in their labs, right?

Ted: Exactly! After talking about quality control, I would explain that we have a number of ways to perform quality control. One technique is to use Person Outfit ZSTD and Person Outfit MNSQ indices. I would explain that these indices may be used to "flag" respondents who responded to one or more survey items in an unpredictable manner.

Isabelle: What do you mean exactly by unpredictable?

Ted: By unpredictable I mean that we have a trait we have tried to define with a set of survey items. Some items will be (in our example) harder to agree with than other items. For instance, you might agree that almost all teachers will be more agreeable to the statement "I will encourage student questions in my classroom" in comparison to the statement "I will invite my principal into my classroom to observe my science teaching." I would also explain that, if we are going to compute a measure of a respondent using a set of survey items, it means we think the set of survey items will help us distinguish respondents. And, if we think that, then the only way we will be able to distinguish respondents will be if the survey items map different parts of a single trait from less to more. And people's answers should match our prediction of item ordering.

Isabelle: What does all of that have to do with fit?

Ted: If we are pooling items to measure a person and we want to distinguish across respondents, then the survey items must define a trait in the same manner for everyone. But, if there are cases where a respondent answers an item in a strange manner, we must note this. We should know who the person is, what item or items were answered in an odd way, and what the exact response was.

Isabelle: What do you do then?

Ted: Sometimes we just keep track and monitor a person over time. Sometimes we might go back to the original paper survey to see if there was a mistake in data entry. In some cases we might remove that person's answer to that one item.

Isabelle: I remember talking to a friend at NARST (the National Association of Research in Science Teaching), and she said she had done an analysis of misfit of her data set. She mentioned perhaps one possible bad item and mentioned that, given her data set, she really had few people who misfit.

Ted: Sounds good.

Isabelle: Anything else?

Ted: Yes one more thing…I usually start my analysis of fit by concentrating on item fit. I use outfit, as I do for persons. But I start with items, and I begin with OUTFIT MNSQ. Item Outfit MNSQ helps me identify items that may not define my trait in the way that one would predict. An item with unacceptable Item Outfit MNSQ might be an item that really does not work to define the trait as one would expect and is defined by the Rasch model.

Formative Assessment Check Point #7

Question: When I run a Rasch analysis with a large sample of students ($n = 1{,}403$), I get very high values of ZSTD for persons and items. Does this mean that something is wrong with my data?

Answer: No (Paraphrased from Winsteps software "Help" page (http://www.winsteps.com/winman/index.htm?diagnosingmisfit.htm) to match our example). Your results make sense. Here is what has happened. You have a sample of more than 1,400 people. This gives huge statistical power to your test of the null hypothesis: "These data fit the Rasch model (exactly)." For a sample size of 1,403, even a mean-square of 1.2 (and perhaps 1.1) would be reported as misfitting that is statistically significantly. So your mean-squares tell us: "These data fit the Rasch model usefully" and the *z*-values tell us: "But not exactly." This situation is often encountered when we know in advance that the null hypothesis will be rejected. The Rasch model is a theoretical ideal. Empirical observations never exactly fit the ideal of the Rasch model if we have enough of them. You have more than enough observations, so the null hypothesis of exact model-fit is rejected. It is the same situation with the Pythagorean theorem. No empirical right-angled triangle fits Pythagoras' theorem if we measure it precisely enough. So we would reject the null hypothesis "this is a right-angled triangle" for all triangles that have actually been drawn. But obviously billions of triangles are usefully right angled.

Keywords and Phrases

Fit
Misfit
Outfit
Infit
MNSQ (mean-squared)
Z-standardized (ZSTD)
Person Outfit MNSQ
Person Outfit ZSTD
Item Outfit MNSQ
Item Outfit ZSTD

There are many potential reasons for a person "misfitting." The important aspect of investigating misfitting persons is that these individuals may not only distort your measures of other individuals in your data set, but identification of misfitting persons may also help you understand some important nuances of the topic you are attempting to measure.

Potential Article Text

Following data collection and data entry, an initial Rasch analysis was conducted to monitor data quality in light of measurement requirements of the Rasch model. This analysis of data quality constituted an additional step that provides great rigor and mimics steps taken by scientists in research labs.

Rasch analysis via Winsteps provides a large number of fit statistics that can be used to evaluate data quality. In this analysis, Person Outfit ZSTD statistics were utilized to identify respondents who might have provided idiosyncratic answers to one or more survey items. When unpredictable responses are identified, it is important to seek potential explanations for these responses. For example, were data miscoded and did the student attend one particular classroom?

Review of person misfit revealed a total of 11 misfitting respondents from the sample of $n=1{,}403$. A ZSTD cutoff of 2.0 was utilized. Given that by chance one would expect 70 respondents of this 1,403 person data set to exhibit misfit, the total number of misfitting respondents is not above what would be expected by chance. No pattern in the demographics/schools suggested any common characteristics of misfitting respondents.

An analysis of Item Outfit MNSQ was also conducted. One item was identified as having an MNSQ value above that which is recommended. Analysis of all 1,403 responses to this item suggested that a small number ($n=7$) of students answered this item in an unpredictable manner and may have caused item misfit. Using procedures suggested by Wright, removal of the responses of the seven students to this single item brought the MNSQ value of this item to an acceptable range. Then the research team reviewed these fit indices and compared predicted item difficulty based upon theory and item difficulty determined through an analysis of the data set. The research team concluded that "measures" could be computed with the measurement instrument, and statistics could be used to evaluate project goals.

Quick Tips

Steps for fit analysis:

1. Begin by evaluating Item Outfit MNSQ. Utilize guidance with regard to the appropriate range of MNSQ values as provided by Linacre and Wright (Fig. 8.4).
2. If items misfit, then examine the specific responses of individuals who may have caused the items to misfit. Experiment with the removal of responses that were unexpected (z-residual of 2 or higher) and rerun the analysis. Check to see if the misfitting items now fall within acceptable bounds of MNSQ.
3. If these steps have not brought items within acceptable ranges of MNSQ, then investigate ZSTD values as suggested by Wright and Linacre (Fig. 8.11). Perhaps an item is not part of the trait? Perhaps there has been an error in data entry? Perhaps the item should be dropped?

Think of investigating fit as conducting very high quality, and needed, quality control of data and items.

Data Sets: (go to http://extras.springer.com)

Activity #1– cf for fit chp activity #1
Activity #9– cf for fit chp activity #9

Activities

Activity #1

We provide a control file (Activity #1– cf for fit chp activity #1). This is a version of the control file we created earlier herein using chemistry education data that our colleagues Chih-Che Tai and Keith Sheppard have collected. Run a Rasch analysis of these data. Add appropriate control lines so that item outfit-vs.-item measures are expressed using MNSQ outfit values. Make sure you know what to do in order to create plots in which item outfit values are expressed with values of ZSTD.

Answer: Adding the line "MNSQ=No" will result in an analysis in which tables such as table 9 are presented in terms of ZSTD fit statistics.

Activity #2

Using the control file for Activity #1, identify the item with the highest potential misfit.

Answer: Test item Q13 appears to exhibit the highest misfit (using Outfit and MNSQ).

Activity #3

Item Q13 (the 14th item read into the control file) appears to misfit. What is the next step you might take in order to investigate this misfit?

Answer: Misfit can be caused by a number of issues. One possibility is a problem with the answer key, but for this example let's assume there are no problems with the answer key. Another step is to find those respondents for whom there are unexpected answers. You can do this by reviewing the results of Winsteps Table 7.1. A quick review reveals that the 55th person in the data set, who has not done well on the test, has unexpectedly answered item Q13 correctly. Also, table 7.1 shows the 35th person has unexpectedly gotten this item correct.

Activity #4

The Outfit MNSQ for test item Q13 is 3.17. What is the value of Outfit MNSQ for item Q13 if the responses of the 35th and 55th person are made missing only for item Q13 (the 14th piece of item data for each person)?

Answer: MNSQ Outfit for item 3 drops to 1.19.

Activity #5

Using the control file supplied for activity #1, find an item that exhibits some amount of misfit. Then use the table at the end of the chapter to better understand why that item might misfit.

Answer: What you observe will depend upon the item you select. Remember you will never know, in many cases, exactly why an item exhibits misfit. But you CAN collect information that will enable you to make an informed, thoughtful, reasoned, decision as to what to do with an item.

Activity #6

Repeat the activity as described for Activity #5, but do so for persons.

Answer: Same as for Activity #5, but pertaining to persons

Activity #7

Using the control file for activity #1, create a new control file that is an identical copy. Now, remove the 75th person in the data set (the last line in the control file) and then insert a person who would exhibit misfit as that described in one of the appropriate cells of the table presented at the end of the chapter. Hint: As you decide what items should be correctly or incorrectly answered, it will be important to take into consideration the relative difficulty and spacing of test items. This information can be found in many different Winsteps tables. Perhaps most useful is the Wright Map which presents the ordering of items graphically.

Answer: There are many types of misfitting persons to insert. One possibility is to insert a high-performing person (gets most items correct) but who unexpectedly misses an easy item. Another possibility is the converse, a low performing person who unexpectedly gets a hard item correct. You should also try to experiment with a person who is a high-performing attentive person who, using the words of Ben Wright, snoozes for a period of time and then goes back to concentrating.

Activity #8

Take the control file provided for activity #1 and alter the answer key so that a hard item is incorrectly coded. What do you predict the impact will be on the change you make? What do you think will be the impact upon the Item Outfit MNSQ and Item Outfit ZSTD? Justify your predictions. Then run Winsteps with the incorrect answer key. What do you see?

Answer: The change you see will partially depend upon the item you select. Our "take home" point is problems in answer keys can be spotted through use of fit statistics.

Activity #9

Using the control file entitled "Activity #9 – cf for fit chp activity #9," first run the control file. You will see that this control file is the SE control file that we have used in earlier chapters. And, this file was used for parts of this chapter on fit. After you have run the control file and reacclimate yourself to this file, make a prediction of what the fit of a person would look like if there had been a mistake in data entry. More specifically, when the data were entered, what if the first answer was entered twice and then all other answers were entered. This would mean that the answer for the 2nd SE item was the answer provided for the 1st SE item, and the answer for the 2nd SE item was the answer entered for the 3rd SE item. Below is an example of an original data line and the incorrect line.

6531365463251
66531365463251

Answer: The misfit will partially depend upon what you entered as the fake data line. After you enter your fake data, you could then bring up the Wright Map from the original data set. Next, you could write the fake answers of this fake respondent next to each item on the Wright Map. More than likely you will not see a nice progression of answers from one part of the continuum to another part of the continuum. After you can see how abnormal and in what way your fake person is, look at the end of chapter table to see if the answer pattern of the fake person matches a particular type of respondent.

Activity #10

In your own words write a paragraph for a research article in which you explain how the analysis of item fit and person fit has provided a qualitative aspect to your data analysis. Make sure to consider the shortcomings of many quantitative analyses and how Rasch addresses some shortcomings that qualitative researchers often mention.

Activity #11

Author a very short fit paragraph that could be included in a grant proposal. Explain why and how you will use fit in your analysis of your instrument and in your analysis of your data.

Activity #12

Fit provides an assessment how much a person or item deviates from the predicted pattern when the items define a single trait and when the persons respond predictably to these items. Must one really have a prediction as to how persons will respond to items and how items will define the trait?

Answer: Yes. If one does not have a prediction as to the manner in which items will define a trait, then one does not have an idea of what the measures mean for a respondent using all the items. If one is not able to predict, we argue that one has no business attempting to use subsequent data to make any conclusions.

Activity #13

Figure 8.17 provides a number of descriptions of different causes of misfit for both items and persons. Use Winsteps Table 7, Winsteps Table 11, and a Wright Map to create fictional response patterns that represent examples of those patterns described in Fig. 8.17.

References

Linacre, J. M. (2002). What do infit and outfit, mean-square and standardized mean? *Rasch Measurement Transactions, 16*(2), 878.

Linacre, J. M. (2012). A user's guide to Winsteps Ministeps Rasch-model computer programs [version 3.74.0]. Retrieved from http://www.winsteps.com/winsteps.htm

Wright, B. D., & Linacre, J. M. (1994). Reasonable mean-square fit values. *Rasch Measurement Transactions, 8*(3), 370.

Wright, B. D., & Masters, G. (1982). *Rating scale analysis*. Chicago: Mesa Press.

Additional Readings

An article in which misfitting items were identified using MNSQ criteria.

Smith, A. B., Wright, P., Selby, P. J., & Velikova, G. (2007). A Rasch and factor analysis of the functional assessment of cancer therapy-general (FACT-G). *Health and Quality of Life Outcomes, 20*(5), 19.

An article that provides an example of the consideration of item fit.

Gallini, J. K. (1983). A Rasch analysis of Raven item data. *The Journal of Experimental Education, 52*(1), 27–32.

A number of Rasch researchers have considered a wide range of issues associated with Fit, a good place to start with additional details with regard to the topic of Fit is provided by the work of Richard Smith.

Smith, R. M. (1991). The distributional properties of Rasch item fit statistics. *Educational and Psychological Measurement, 51*, 541–565.

Smith, R. M., & Hedges, L. V. (1982). Comparison of likelihood ratio X^2 and Pearsonian X^2 tests of fit in the Rasch model. *Education Research and Perspectives, 9*(1), 44–54.

Chapter 9
How Well Does That Rating Scale Work? How Do You Know, Too?

Isabelle and Ted: Two Colleagues Conversing

Ted: Sometimes I think my brain is going to explode, Isabelle. There are so many issues that I have not thought of regarding the use of surveys and tests in research. Really, I just feel overwhelmed at times.

Isabelle: If it helps, the first time I started reading about Rasch that happened to me, too. But, over time, I saw how a piece here and there of Rasch could help make all of my projects better. Some of the things I now know can go straight into publications and grants. Other things that I now understand would not go into a publication. Sometimes reviewers want to be impressed with the instruments, but they don't want pages and pages of details on what I did to create a good measurement instrument, unless of course the purpose of the manuscript is to report the development and validation of an instrument.

Introduction

A number of issues are associated with sound social science measurement. One issue is the importance of "thinking before leaping." Let's digress for a few moments to clarify what we mean by "thinking before leaping." Practicing sound measurement requires researchers to confront and solve problems. We embrace a broad definition of "problem" that was set forth over 30 years ago by John R. Hayes, who studied the relationship between creativity and solving problems. Hayes (1981) states, "Whenever there is a gap between where you are now and where you want to be, and you don't know how to find a way to cross that gap, you have a problem" (p. xii). We often use Grayson Wheatley's definition of problem solving, "What you do when you don't know what to do" (personal communication, summer, 1971). Hayes (1981) describes a two-part problem solving strategy: Develop images of the gap, then examine and test ways to cross it. In other words, when facing a problem, thinking before leaping is crucial in the search for sound, viable solutions.

When creating tests or surveys, researchers must conceptualize the trait to be measured. In the case of tests, for example, doing so will improve the likelihood that

W.J. Boone et al., *Rasch Analysis in the Human Sciences*,
DOI 10.1007/978-94-007-6857-4_9, © Springer Science+Business Media Dordrecht 2014

each item will yield useful information on what respondents know and do not know. In the case of surveys, well-constructed items will produce a clearer picture of respondents' views or attitudes. An additional benefit of thinking before leaping is that a better set of items increases the chances of measuring each respondent in a way that group patterns can be identified.

This chapter will introduce some added steps researchers can take to evaluate how well a measurement device is working. In most chapters, steps and ideas are presented that help researchers improve the quality of the measures they compute for respondents (e.g., student attitudes toward science, teacher science content knowledge). This chapter provides theory and techniques that allow researchers to evaluate the strength of the rating scales selected for a rating scale instrument. This chapter is important in that making a hasty generalization that a rating scale that works for one instrument will also work for another instrument can result in low-quality measures. As researchers will see the work of developing, validating, and maintaining a reliable, valid rating instrument is more complicated than simply citing previous use of an instrument, or the computation of some magical index. Fortunately, readers will see that there exist a few easy steps to assess a rating scale.

As we start in earnest, we want to stress that every data set and instrument is unique in some ways. There is not a one-size-fits-all set of steps that researchers can conduct. Moreover, one particular aspect of a Rasch analysis will almost never dictate a decision (how many items, what rating scale to use, etc.). Usually, researchers will need to weigh a number of issues.

To begin review our Fig. 9.1 (Winsteps Table 3.2) from the output of the self-efficacy data analysis, the first four columns display the Category Label, Category Score, Observed Count, and Observed Percentage of counts for each rating scale category (SD, D, BD, BA, A, and SA). The observed count and the observed percentage of counts of the rating scale categories provide information about the level of agreement of respondents for all 13 self-efficacy items of the STEBI.

For this sample, a total of "975" responses are possible for the sum of the OBSERVED COUNT column (13 items × 75 respondents = 975). In our data set, the sum of the observed counts for scores 1, 2, 3, 4, 5, and 6 is 932 (17 + 111 + 113 + 18 8 + 370 + 133 = 932), which is less than 975. This is because there were 43 instances where respondents (975 − 932 = 43) did not answer a survey item. In Table 3.2, the count of nonresponses is reported in the row labeled "MISSING." Notice that the MISSING count is 43, which matches our calculation immediately above.

How might this table be useful? First, it provides a broad view of how the categories presented on the scale were selected. For instance, categories labeled "1" and "2" were selected only 2 and 12 % of the time, respectively. Recall that the response option represented by "1" was *Strongly Disagree*, and the response option for "2" was *Disagree*. Of importance to remember is that Table 3.2 represents the count of responses on all the items, and some of the self-efficacy items were "flipped." One must therefore remember that this table displays the counts of the labeled categories for all items together, but the code after flipping is used as this tally is created (e.g., Q21se-rc is a negative item that had to be "flipped" before the data was evaluated). It is those flipped values that are used for the summary provided in Table 3.2.

```
TABLE 3.2 SCIENCE TEACHER EFFICACY BELIEFS       ZOU840WS.TXT  Oct  8  7:51 2011
INPUT: 75 Person  23 Item  REPORTED: 75 Person  13 Item  6 CATS  WINSTEPS 3.72.3
--------------------------------------------------------------------------------


SUMMARY OF CATEGORY STRUCTURE.  Model="R"
-------------------------------------------------------------------
|CATEGORY   OBSERVED|OBSVD SAMPLE|INFIT OUTFIT||STRUCTURE|CATEGORY|
|LABEL SCORE COUNT %|AVRGE EXPECT|  MNSQ  MNSQ||CALIBRATN| MEASURE|
|-------------------+------------+------------++---------+--------|
|  1   1      17  2| -.97 -1.18|  1.33  1.45||  NONE    |( -3.98)| 1
|  2   2     111 12| -.71  -.70|  1.02  1.19||   -2.81  |  -1.84 | 2
|  3   3     113 12| -.10  -.16|   .97   .86||    -.45  |   -.61 | 3
|  4   4     188 20|  .38   .50|  1.00   .91||    -.35  |    .27 | 4
|  5   5     370 40| 1.52  1.49|   .97   .97||     .28  |   1.91 | 5
|  6   6     133 14| 3.05  3.04|   .97  1.01||    3.34  |(  4.47)| 6
|-------------------+------------+------------++---------+--------|
|MISSING      43  4|  .62      |            ||          |        |
-------------------------------------------------------------------
TABLE 3.2 SCIENCE TEACHER EFFICACY BELIEFS       ZOU976WS.TXT Dec 13 10:49 2010
INPUT: 143 Person  23 Item  MEASURED: 143 Person  13 Item  6 CATS       3.69.1.9
--------------------------------------------------------------------------------
```

Fig. 9.1 A Winsteps table that allows one to begin to investigate the nuances of how the rating scale selected for an instrument functions. The percentages of all responses to each rating category are provided. For example, for all 75 respondents (being able to answer up to 13 items), 14 % of all responses used the *Strongly Agree* rating category (6). In later portions of this chapter, we will discuss how the column "OBSVD AVRGE" can be used to monitor additional aspects of the scale's function

The impact of "flips" can be confusing, as we have explained in our tips on coding of items (e.g., Q21se-rc). In this case, our table presents the selection of rating scale categories after the flips. Thus, if 20 respondents selected SD for item Q21, this selection would be 20 of the 133 tallies reported for the SA category.

What researchers can learn from this table focuses on the overall agreement or disagreement with the trait of interest, in this case, self-efficacy toward teaching science. For this sample, there are more responses of agreement, which means higher self-efficacy. However, we want to stress that looking at this overall usage of categories is only a very broad assessment of how a scale might function. This is because all items are not alike; not all items are equally easy to agree with. What in the end is our general point regarding this component of Table 3.2? Simply that instrument developers and users can investigate the rate of category usage. By evaluating category usage, one may very quickly be able to note which steps of a rating scale are not being used. This very broad analysis suggests that the SD option is really not used that much (after flipping).

We believe readers should also consider using Table 3.2 to provide a quick check of whether or not the data have been correctly read and entered into the program.

Formative Assessment Checkpoint #1

Question: Should researchers be concerned if respondents do not mark certain rating categories?

Answer: Yes. It may be quite important when certain rating categories are not marked, and the issue at least should be investigated. Categories that are not used may waste respondents' time (it takes longer to think about more categories). Also, the lack of use of categories a researcher thinks are appropriate may suggest a misconception on the part of the researcher. There are, to be sure, many issues to be considered before removing a rating scale step from a survey. For example, maybe the category was rarely used for almost all of the survey items. In that case, does a researcher think it is possible and likely that other surveyed respondents might also exhibit a similar pattern in terms of not using a particular response option for many of the survey items? The other issue that has to be remembered is that if respondents are going to be presented with the same rating scale for all items, one would expect for certain items that a rating scale category might not be used by respondents; however, for a different survey item which taps another part of the variable, the rating scale might be used when respondents answer the item.

The Probability of a Rating Scale Step Being Selected

An added step to assess the function of a rating scale is to examine the probability of a particular response category being selected. Using such probabilities provides a more sophisticated technique for evaluating the nuances of how a group of respondents used the rating scale. Our Fig. 9.2 (Winsteps Table 21.1) provides a graphical display of such probability data. At first, this table may appear to be a jumbled mess of information and difficult to interpret, so let's break it down into smaller parts. First, the vertical axis represents the probability of a particular response selection. You will see that the values range from a minimum of 0 to a maximum of 1. This is the range for probabilities. Second, the horizontal axis presents the difference between a respondent's measure and a specific item's measure. Thus, the location of 0 along the horizontal axis presents the probability of a person selecting each of the response options when that person's measure is exactly the same as the item's measure. So if Dave has a logit measure of 1.5 and he answers an item of difficulty 1.5, one would look at the 0 term on the horizontal axis. Another example would be if Sammy S. had a measure of 1.0 and he answered an item of difficulty –2.0, then the value for this particular interaction of Sammy with an item would be located at the 3.0 value of the horizontal axis [1.0 – (2.0)]. This value is indeed the person measure minus the item difficulty measure as noted at the base of the figure as `Person [MINUS] Item MEASURE`. The numbers (1, 2, 3, 4, 5, 6) presented as sequences of repeated numbers (e.g., 4444444) *within* the table correspond to each rating scale category. For example, the trace of 2s marks the location of each probability level for any combination of "person measure–item measure" for the category *Disagree* (*Disagree* was coded with a "2").

Generally researchers should review this plot and note whether or not each category is "most probable" for at least some combinations of person measure–item difficulty.

```
TABLE 21.1 SCIENCE TEACHER EFFICACY BELIEFS      ZOU840WS.TXT  Oct  8  7:51 2011
INPUT: 75 Person  23 Item  REPORTED: 75 Person  13 Item  6 CATS  WINSTEPS 3.72.3
--------------------------------------------------------------------------------

        CATEGORY PROBABILITIES: MODES - Structure measures at intersections
P      -+-----+-----+-----+-----+-----+-----+-----+-----+-----+-
R  1.0 +                                                       +
O      |                                                       |
B      |                                                       |
A      |                                                      6|
B   .8 +                                                    66 +
I      |11                                                66   |
L      |  1                               5555           6     |
I      |   1                            55    55        6      |
T   .6 +    1      2222               55        55     6       +
Y      |     11  22    22            5            55 66        |
    .5 +       12        2          5               *          +
O      |      221         2        5               6 5         |
F   .4 +     2   1         2      5               6   5        +
       |   22     1         2   44*44            6     55      |
R      |  2        1       33**4 5   44        66        5     |
E      |22          1   333  423*      4      6           55   |
S   .2 +             113   44  * 33     44  66              55 +
P      |             331  4   5 2  3      **                  5|
O      |          333   **1 55   22 33  66  444                |
N      |     33333   444 55*11     *****33     44444           |
S   .0 +*****************66666*****111*************************+
E      -+-----+-----+-----+-----+-----+-----+-----+-----+-----+-
       -4    -3    -2    -1     0     1     2     3     4     5
        Person [MINUS] Item MEASURE                    J
```

Fig. 9.2 Table 21 of Winsteps. This table shows the probability of a specific response selection after one considers the item being answered and the overall attitude measure of the respondent. It is important for each response category to be most probable for some combination of person measures and item measures. The measure of John is plotted with the letter "J" along the horizontal axis. When John has a measure about 4 logits higher than the item he is answering, one predicts that he will answer using a *Strongly Agree* (6)

This means there should be some combinations of person ability and item difficulty when *Strongly Agree* (6) is most probable, one combination when *Agree* is the most probable (5), one combination when *Barely Agree* (4) is the most probable, one combination when *Barely Disagree* (3) is the most probable, one combination when *Disagree* (2) is the most probable, and one combination when *Strongly Disagree* (1) is the most probable.

The Hills

A visual representation of the concept in our Fig. 9.2 (Winsteps Table 21.1) above means that the top trace of the "hills" of numbers should, in the perfect case, include a hill of 1s in which for a person measure minus item measure location along the horizontal axis is most probable (that means higher up the probability scale from 0 to 1). And the same was seen for the other rating scale categories of 2, 3, 4, 5, and 6.

```
TABLE 18.1 SCIENCETEACHER EFFICACY BELIEFS      ZOU631WS.TXT  Apr 10 14:56 2012
INPUT: 75 Person  23 Item  REPORTED: 75 Person  13 Item  6 CATS    WINSTEPS 3.73
--------------------------------------------------------------------------------
Person: REAL SEP.: 2.52  REL.: .86 ... Item: REAL SEP.: 7.00  REL.: .98

         Person STATISTICS:  ENTRY ORDER

-------------------------------------------------------------------------------------------------
|ENTRY   TOTAL  TOTAL           MODEL|   INFIT    |  OUTFIT   |PT-MEASURE |EXACT MATCH|         |
|NUMBER  SCORE  COUNT  MEASURE  S.E. |MNSQ  ZSTD|MNSQ  ZSTD|CORR.  EXP.| OBS%  EXP%| Person  |
|------------------------------------+----------+----------+-----------+-----------+---------|
|     1     60     13     1.34     .35|1.71   1.5|1.63   1.4|  .52    .57| 46.2  50.2| 21141  PR|
|     2     27      6      .84     .50| .71   -.3| .57   -.6|  .83    .60| 50.0  46.1| 91052  PR|
|     3     33      6     3.28     .81| .62   -.5| .57   -.5|  .61    .46| 66.7  66.9| 95793  PR|
|     4     47     13      .08     .29| .31  -2.6| .28  -2.5|  .84    .67| 46.2  40.0| 08453  PR|
```

Fig. 9.3 Data concerning the first four respondents of the data set. The first respondent (person 21141) answered all 13 SE items. Their attitude measure is expressed with a measure of 1.34 logits. In the text we name this person "John"

In our Fig. 9.2 (Winsteps Table 21.1), there are large hilltops of 1s, 2s, 5s, and 6s. And, most importantly, the hilltops of 1s, 2s, 5s, and 6s are most probable for some combinations of persons and item. There are hills for categories "3" and "4," but what is important is that there is no place along the horizontal scale in which the rating scale of *Barely Agree* (4) or *Barely Disagree* (3) is most probable.

What does this pattern tell us? In this case there might be lessened measurement information learned from the use of the BA (4) and BD (3) categories. As noted earlier in this chapter, one usually needs multiple pieces of information to improve the reliability and validity of an instrument via a process of revision. So, review of this table suggests that we perhaps consider whether there is sufficient pay off to include categories 3 (BD) and 4 (BA). Some readers may ask: What difference does it make how many categories are presented to respondents? The difference is with each added decision we present, respondents may lose some energy or interest in what we wish for them to do (namely, provide their views to us). Therefore, fewer categories, especially if we do not gain much from the added categories, may be better for an instrument developer!

Winsteps Table 21.1 can also be used to predict a respondent's choice (1, 2, 3, 4, 5, 6) for any single item based on a respondent's overall attitude. To make such predictions with this plot, a researcher first needs a respondent's measure and a specific item's measure. If we consult Winsteps Table 18.1 (Fig. 9.3) under the column MEASURE, we see that John (person 21141) answered all 13 items, and if we then can consult Winsteps Table 14.1 (Fig. 9.4), we can look up the measure for each item. Let's focus on item Q2se, which exhibits an item measure of –2.49. If we then subtract the item measure for Q2se from John's person measure (1.34–(–2.49)), we obtain 3.83, the location of which we have noted with a "J" in Fig. 9.2 (Winsteps Table 21.1).

Our final step to determine the rating scale selection that is most probable for John is to draw a vertical line upward from the "J" to the top of the figure. Find where this line intersects the highest trace of numbers (in this case one of the "6"s which are plotted). Then draw a horizontal line to the left until it crosses the y-axis of the figure (between the .5 and .6 probability level). From these lines, one can predict that John has a between .55 and .6 probability of selecting the rating scale

```
TABLE 14.1 SCIENCE TEACHER EFFICACY BELIEFS      ZOU631WS.TXT  Apr 10 14:56 2012
INPUT: 75 Person  23 Item  REPORTED: 75 Person  13 Item  6 CATS    WINSTEPS 3.73
--------------------------------------------------------------------------------
Person: REAL SEP.: 2.52  REL.: .86 ... Item: REAL SEP.: 7.00  REL.: .98

        Item STATISTICS:  ENTRY ORDER

-----------------------------------------------------------------------------------------------------
|ENTRY   TOTAL  TOTAL              MODEL|   INFIT  |  OUTFIT  |PT-MEASURE |EXACT MATCH|            |
|NUMBER  SCORE  COUNT  MEASURE  S.E. |MNSQ  ZSTD|MNSQ  ZSTD|CORR.  EXP.| OBS%  EXP%| Item       |
|------------------------------------+----------+----------+-----------+-----------+---------|
|     1     DELETED                  |          |          |           |           | Q1oe       |
|     2    410     75    -2.49    .22|1.06    .4|1.08    .5|  .31   .41| 55.4  64.4| Q2se       |
|     3    317     75      .22    .13|1.69   3.5|1.64   3.1|  .52   .59| 33.8  44.5| Q3se-rc    |
```

Fig. 9.4 Data from the Winsteps table which presents details with regard to items answered on an instrument. This edited table presents data for two items of the STEBI (Q2SE, Q3SE-rc) which were answered by 75 respondents

of *Strongly Agree* (after flipping) for this item. But perhaps the more important observation is that John (with this specific attitudinal measure) would be predicted to have selected an SA (after flipping) for this item. This is because it is for category "6" (SA) that John has the highest probability. If one looks at the other probabilities for John with this item, one sees that it is really only category "5" (A) for which there is any significant probability for John answering, but the probability of a "5" can be seen to be below that observed for an answer of a "6" (SA).

Formative Assessment Checkpoint #2

Question: If Billy has a measure of 1.2 logits, and he answers an item of difficulty .7 logits, using Fig. 9.2, what is the approximate probability of his answering that item with a specific rating scale category? What does this information tell you about Billy and this item?

Answer: Since Billy has a measure of 1.2 logits and he answers an item of .7 logits, this means to answer the question one would find the .5 logit part of the horizontal axis (1.2 logits – .7 logits = .5 logits). Then at the .5 logit location, one draws a vertical line upward so that the line cuts through all the traces of 1s, 2s, 3s, 4s, 5s, and 6s above the .5 logit part of the horizontal axis. That vertical line hits the number 2 at about the .08 probability level, about the .18 probability level for the number 3, about the .35 level for the number 4, and about .40 for the number 5. There is probability for both a 1 and a 6, but those values are very small. The values of the probabilities from our approximations add up close to 1 (.08 + .15 + .35 + .40 = .98). This analysis shows us that for this person there is a range of probabilities for his answers. It is most likely that he will answer a 5 (*Agree*) to this item, but there is a chance that he could answer any of the other categories. Most likely if he did not answer a 5, he would answer with a 4 for this item.

Formative Assessment Checkpoint #3

Question: Below we provide Winsteps Table 21.1 for a survey which included 8 rating scale items with a 5 step rating scale. What initial assessment could you make with regard to the functioning of the scale? If the average female student who completed this survey had an average measure of 1.5 logits, what would be the approximate chance of her selecting each of the 5 categories when she answered an item which had a measure of −1.60 logits?

Answer: An initial evaluation of the table suggests that there are at least some combinations of person measures and item measures for which the rating categories of 1, 2, 3, 4, and 5 are most probable. [There may be a very brief portion of the horizontal axis around −1.25 where the rating category of 2 is most probable]. The pattern of Table 21.1 suggests that perhaps rating category 2 might be monitored over time to evaluate if the category should be retained in the survey.

If the average person measure of the females who were surveyed was 1.5, and the average female answered an item of measure −1.25, this would mean one would look at the part of the horizontal axis with a value of 2.75 [1.5 – (−1.27)=2.75]. At that location, by drawing a vertical line upward, one is able to note that the probability of the average female answering this item with an answer of "4" is .4 and the probability of the average female answering this same item with a rating of "5" is about .6.

```
TABLE 21.1 n 40 Fall 2011 Excel Jordan Data for  ZOU636WS.TXT  May 16 19:38 2012
INPUT: 40 PERSON  8 ITEM  REPORTED: 40 PERSON  8 ITEM  5 CATS    WINSTEPS 3.74.0
-------------------------------------------------------------------------------

        CATEGORY PROBABILITIES: MODES - Structure measures at intersections
P      -+------+------+------+------+------+------+------+------+-
R  1.0 +                                                         +
O      |                                                         |
B      |111                                                      |
A      |   111                                                  5|
B   .8 +      11                                              55 +
I      |        1                                           55   |
L      |         11                                       55     |
I      |           1                                     5       |
T   .6 +            1                                  55        +
Y      |             1                     4444444    5          |
    .5 +              1                  44       44 5           +
O      |               1       3333333344           *4           |
F   .4 +                1     3       433         55  44         +
       |              222*22**       4   3       5      44       |
R      |           222    13  22   44     33    5         4      |
E      |         22      3311   224         3 55           44    |
S   .2 +       22       3    1   422         *3              44  +
P      |    222       33      144   22     55  33              44|
O      |2222        33       4411     22555      333             |
N      |       33333     4444    111*55522222       33333        |
S   .0 +*****************55555555555 11111111********************+
E      -+------+------+------+------+------+------+------+------+-
       -4     -3     -2     -1      0      1      2      3      4
        PERSON [MINUS] ITEM MEASURE
```

We have guided readers through the mechanics of using this table when one knows the measures of a person and an item. But why is this table useful? Readers could rightly argue there are an infinite number of seemingly possible person measure and item measure combinations. We explain this plot to our colleagues by pointing out that we rarely look up specific people for many items. Instead, we might compute the mean measure for a subgroup of respondents (e.g., females) and then check the item with the highest and lowest measure. Using the techniques we detailed for John, we would draw two vertical lines and two horizontal lines. Then we could gauge the range of categories that would be most probable for this respondent (the mean female). If we would see that only one category is most probable when considering both the item with the highest measure and the lowest measure, then we might consider altering our rating scale in some manner and/or adding items which are harder or easier to agree with. Doing so would hopefully expand the range of rating scale categories answered by this type of respondent.

The second, and by far easiest, way to use this table is reviewing the numbers on the highest traces of numbers (the tops of the hills as it were). If a rating scale is functioning optimally, there should always be some sort of person–item measure value at which each rating category is "most probable." In our example there is a very small range over which the numbers 3 and 4 are observed. This means that we might not be getting high-quality measurement from these two categories. If we had collected these data as a pilot, we might drop these two categories to simplify the instrument.

Considering the Mean Person Measure as a Function of Item Response Selected

The two previous tables provided information about the interaction of item categories and persons for the entire 13-item scale. The final technique discussed in this chapter provides additional useful information regarding how persons interact with each survey item. The table that we will use is presented, in part, in Fig. 9.5 (Winsteps Table 14.3). This table provides a listing of the distribution of rating scale categories selected by respondents for each survey item and the mean scale score (person) measures of all respondents who selected a particular rating category for each survey item. We repeat for emphasis: This table provides the mean scale score (person) measures of all respondents who selected a particular rating category for each survey item. The portion of Winsteps Table 14.3 (Fig. 9.5) displayed below presents this information for three self-efficacy items (Q21se-rc, Q22se, Q23se-rc).

For item Q21se-rc, six (6) respondents selected *Disagree* (2), and these six respondents represent 9 % of the respondents who answered this item. The mean person measure for these 6 respondents, using all the items they answered, was −.34. Looking at the other lines, we see the number and percent of respondents who selected the other rating category for item Q21se-rc. Notice also that the mean measure is provided for each category. For example, 31 respondents selected rating category 5 (*Agree*), and the mean measure of these 31 respondents was 1.37.

Item CATEGORY/OPTION/DISTRACTOR FREQUENCIES: ENTRY ORDER

ENTRY NUMBER	DATA CODE	SCORE VALUE	DATA COUNT	%	AVERAGE ABILITY	S.E. MEAN	OUTF MNSQ	PTMEA CORR.	Item
21	2	2	6	9	-.34	.15	.5	-.33	Q21se-rc
	3	3	12	17	.43	.23	1.5	-.22	
	4	4	14	20	.42*	.16	.6	-.25	
	5	5	31	45	1.37	.12	.7	.22	
	6	6	6	9	3.53	.93	.9	.60	
	MISSING	***	6	8#	.97	.60		-.02	
22	3	3	1	1	-.64		.6	-.16	Q22se
	4	4	6	9	.19	.25	1.0	-.21	
	5	5	35	51	.62	.13	.7	-.35	
	6	6	27	39	1.87	.29	.9	.51	
	MISSING	***	6	8#	.97	.60		-.02	
23	1	1	2	3	-.40	.05	.8	-.20	Q23se-rc
	2	2	13	19	-.07	.16	.7	-.42	
	3	3	15	22	.61	.14	.8	-.18	
	4	4	13	19	.92	.14	.4	-.05	
	5	5	21	30	1.70	.21	1.0	.33	
	6	6	5	7	3.50	1.01	1.1	.53	
	MISSING	***	6	8#	.97	.60		-.02	

Fig. 9.5 A portion of Winsteps Table 14.3. This table provides similar data as that which was presented in Figure 9.1. In this case data are presented as a function of specific survey items

---

Formative Assessment Checkpoint #4

Question: Does the process of examining rating scale steps (e.g., *Strongly Agree, Agree, Disagree, Strongly Disagree*) transfer to tests in which items are right and wrong?

Answer: Yes. The process and ideas DO transfer. Think of a test in which items are right or wrong as simply a two-step rating scale. You can experiment with this idea by recoding a rating scale data set into only agree (e.g., *Strongly Agree* and *Agree* = 1) and only disagree (e.g., *Strongly Disagree* and *Disagree* = 0). The same techniques that we outlined above can be used!

Winsteps Table 14.3 provides two important pieces of information about the function of the instrument. First, it provides the details of rating scale responses for each individual item. It is important to remember that all 13 survey items are not equal. Some items represent one part of the trait of self-efficacy, and other items represent another part of the trait. This means that the distribution of rating scale steps selected will vary as a function of the survey item. This is why we emphasized that looking at all the responses for all items was only a broad starting point to an analysis of the rating scale. This fact, that all items are not created equal, can be observed in Chap. 6, when we introduced Wright Maps. Different items mark different points along the continuum (line) of the trait, and the impact of this can be demonstrated in items Q22 and item 23 in Table 14.3. Notice that item Q22 exhibits a higher usage of the SA and A categories than does item Q23. This suggests that item Q22 defines the more agreeable portion of the trait than does item Q23.

The second important piece of information focuses on how respondents use rating categories as they answer particular survey (and test) items. To understand how this technique works recall that when Rasch analysis is used to construct instruments, evaluate instrument function, compute linear measures for parametric statistical tests, and create tools such as Wright Maps. The best measurement instruments are those that measure a single variable or trait. Measurement is always uncertain, be it in a physics lab or in collecting data from humans. Regardless of the setting, we must focus our attention on one variable/trait. A valued byproduct of focusing intently on one variable is a generally predictable use of rating scale categories by respondents for each item. One little used and discussed technique in research fields using rating scales is evaluating the mean measure (based upon using all items of a survey or a test) of respondents who select specific rating scale categories as a function of an item (this means for item Q2 determining the overall measure of all respondents who selected *Strongly Agree* for Q2, determining the overall measure of all respondents who selected *Agree*, and so on). Winsteps Table 14.3 provides this information by not only presenting the number and percent of respondents who select each rating scale category as a function of an item but also by including a column that displays the mean measure (named "AVERAGE ABILITY" in the figure) of all respondents who selected a specific rating category for each item (Fig. 9.6).

Closer examination of item Q23se-rc in Fig. 9.6 (Winsteps Table 14.3) allows one to focus on the mean measure score for each rating scale category. Recall that this is *after* flipping. Two respondents selected "1" (SD), and their mean measure is −.40. Thirteen respondents selected "2" (D), and their average measure was a −.07. Data are also provided for the BD, BA, A, and SA selections: 15 respondents (.61), 13 respondents (.92), 21 respondents (1.70), and 5 respondents (3.50), respectively. Note the increase in overall mean measure as a function of rating category. This is the general pattern one wants to observe for all items. Certainly, when a small number of respondents use a rating category, one does not want to make too many decisions. But when roughly ten or more responses are seen for a rating scale category, then the authors generally use that category's mean measure and compare that mean measure to other mean measures for other rating scale steps for an item. When the mean measures do not increase (as would be predicted for a measurement instrument that is optimally functioning), then it is important to track this type of incongruence found in the analysis. Further examination of any such item is necessary to ensure the quality of the measurement instrument. It does not mean that an item will be automatically discarded. Sometimes an item might be removed, but this information is added to what we learn about the instrument, all with the goal of improving the measurement instrument and/or at least taking the best possible steps

ENTRY NUMBER	DATA CODE	SCORE VALUE	DATA COUNT	%	AVERAGE ABILITY	S.E. MEAN	OUTF MNSQ	PTMEA CORR.	Item
23	1	1	2	3	-.40	.05	.8	-.20	Q23se-rc
	2	2	13	19	-.07	.16	.7	-.42	
	3	3	15	22	.61	.14	.8	-.18	
	4	4	13	19	.92	.14	.4	-.05	
	5	5	21	30	1.70	.21	1.0	.33	
	6	6	5	7	3.50	1.01	1.1	.53	

Fig. 9.6 A portion of Winsteps table only for Q23 of the SE survey

to prepare data for statistical tests. What do we do if we do not see a stepwise increase in the mean person measure for each of the groups of respondents selecting a particular rating category (for one item)? First, we ask ourselves if we have enough responses to confidently accept that the mean is a good representation of the mean of the individuals who selected a rating category. If we do have a sufficient number of responses, then we consider a number of steps; one might be to check for a similar pattern in another data collection, and another possibility is to combine rating categories. Combining rating categories would mean that one might, for the 6-step rating scale we have been using in this chapter, combine the *Strongly Agree* and *Agree* category together, retain the *Barely Agree* and *Barely Disagree* categories as distinct categories, and combine the *Disagree* and *Strongly Disagree* categories. So, this would mean perhaps using the number 4 for the *Strongly Agree/Agree* category, the number 3 for *Barely Agree* category, the number 2 for the *Barely Disagree* category, and the number 1 for the *Disagree/Strongly Disagree* category. Then a new Rasch analysis would be run, and for many evaluations of that new analysis, Table 14.3 would be evaluated for each item, and an assessment of the pattern in mean response as a function of rating scale category would again be conducted.

Disordering

To help readers learn how to spot possible problems in a rating scale, let's now focus on Item Q19se-rc in Fig. 9.7. Looking at categories *Strongly Disagree* ("1"), *Disagree* ("2"), *Barely Disagree* ("3"), and *Barely Agree* ("4"), we observe that the mean measure for *Disagree* is less than the mean measure for *Barely Disagree*. This is not the expected mean measure pattern; rather, we would expect to see a number somewhere between .43 (the mean for those who selected SD) and 1.03 (the mean for those who selected BD). This "disordering" in mean values as a function of rating categories might be a sign of unpredictability (in this case the scale is not working as one might predict, and as one might want). As one learns how to use this technique, there is one added nuance that is helpful when reviewing tables: Do not put too much stock in a disordered step if very few respondents are used to compute the mean. Data for item Q19se-rc are provided in

ENTRY NUMBER	DATA CODE	SCORE VALUE	DATA COUNT	%	AVERAGE ABILITY	S.E. MEAN	OUTF MNSQ	PTMEA CORR.	Item
19	1	1	9	13	.43	.34	2.1	-.19	Q19se-rc
	2	2	18	26	.13*	.14	.4	-.43	
	3	3	15	22	1.03	.15	.6	-.01	
	4	4	16	24	1.42	.18	.9	.15	
	5	5	8	12	1.92	.21	1.0	.24	
	6	6	2	3	6.21	1.32	.4	.70	
	MISSING	***	7	9#	.90	.51		-.04	

Fig. 9.7 A portion of Winsteps table only for Q19 of the SE survey

Fig. 9.7. There is technically a disordered step from category 1 to 2 to 3, but only nine people used rating scale step 1. Because the mean is based upon only nine people, it is important not to make any major decisions on the function of a rating scale step. If one more respondent were to mark rating category 1, the mean measure for category 1 could easily decrease to less than .13, which is the mean for rating category 2.

What do we do for data such as item Q19se-rc? If there are no added issues observed for this item (e.g., item outfit is too large), then our general practice is to retain this item and continue to observe how respondents react to it in future samples. The mean of the respondents selecting each rating category (with the exception of category 2) increases with each rating scale step. After reviewing Fig. 9.7, one can see that the disordering of the steps is really not so much the result of category 2, but rather the result of the mean value of category 1. Since so few responses were observed for category 1, then the ordering of average measures does proceed as one expects if the average of category 1 is ignored.

Of course, it is important to remove items that are clearly odd in some manner and appear to degrade the quality of the measurement we are able to conduct with a set of items which define a trait. Unless a range of evidence exists, we have found it is preferable to monitor instrument items and rating scales over multiple samples. The goal is to maximize the reliability and validity of the measurement instrument. We hope that readers now understand that when serious piloting of items occurs, with extra items, then the types of techniques we outline here can be used to select or remove items. Now that readers have completed this chapter, if you wish to learn even more about the nuances of rating scale categories, we suggest, after completing the end of chapter activities, to review an article entitled "Optimizing rating scale category effectiveness" (Linacre, 2002).

Formative Assessment Checkpoint #5

Question: When evaluating the function of a rating scale, does a researcher have only two options, monitor the item or throw the item out?

Answer: No. An added possibility for the researcher is to explore recoding of data. In Chap. 10 herein, we demonstrate how rating scale step recoding might be used to improve reliability of an instrument. One can also recode and review the plots presented in this chapter. An item that exhibits disordering might not exhibit disordering with recoding.

Formative Assessment Checkpoint #6

Question: How do you decide if a strange pattern in a rating scale is a problem with the scale or a problem with how the data were entered?

Answer: If you see a strange pattern in the rating scale data, for example, one item seems to exhibit strange use of a rating category, it is always a good idea to look at the instrument you used to collect data, to look at a few original surveys, then to check to see how the responses were entered into a spreadsheet (very important), to look at how that data appear at the base of the control file, and to double-check any recoding you might have used for "RESCORE=" and "NEWSCORE=."

```
TABLE 14.3 SCIENCE TEACHER EFFICACY BELIEFS       ZOU977WS.TXT  Oct  9  6:43 2011
INPUT: 75 Person  23 Item  REPORTED: 75 Person  13 Item  6 CATS  WINSTEPS 3.72.3
--------------------------------------------------------------------------------

          Item CATEGORY/OPTION/DISTRACTOR FREQUENCIES:  ENTRY ORDER

------------------------------------------------------------------------
|ENTRY   DATA  SCORE |    DATA     | AVERAGE  S.E.  OUTF PTMEA|         |
|NUMBER  CODE  VALUE |  COUNT   %  | ABILITY  MEAN  MNSQ CORR.| Item    |
|--------------------+-------------+--------------------------+---------|
|                    |             |                          |         |
|   21   2         2 |     6    9  |   -.34    .15    .5  -.33 |Q21se-rc |
|        3         3 |    12   17  |    .43    .23   1.5  -.22 |         |
|        4         4 |    14   20  |    .42*   .16    .6  -.25 |         |
|        5         5 |    31   45  |   1.37    .12    .7   .22 |         |
|        6         6 |     6    9  |   3.53    .93    .9   .60 |         |
|        MISSING *** |     6    8# |    .97    .60        -.02 |         |
|                    |             |                          |         |
|   22   3         3 |     1    1  |   -.64           .6  -.16 |Q22se    |
|        4         4 |     6    9  |    .19    .25   1.0  -.21 |         |
|        5         5 |    35   51  |    .62    .13    .7  -.35 |         |
|        6         6 |    27   39  |   1.87    .29    .9   .51 |         |
|        MISSING *** |     6    8# |    .97    .60        -.02 |         |
|                    |             |                          |         |
|   23   1         1 |     2    3  |   -.40    .05    .8  -.20 |Q23se-rc |
|        2         2 |    13   19  |   -.07    .16    .7  -.42 |         |
|        3         3 |    15   22  |    .61    .14    .8  -.18 |         |
|        4         4 |    13   19  |    .92    .14    .4  -.05 |         |
|        5         5 |    21   30  |   1.70    .21   1.0   .33 |         |
|        6         6 |     5    7  |   3.50   1.01   1.1   .53 |         |
|        MISSING *** |     6    8# |    .97    .60        -.02 |         |
------------------------------------------------------------------------
 * Average ability does not ascend with category score
 # Missing % includes all categories. Scored % only of scored categoriesTABLE 14.3
```

Isabelle and Ted: Two Colleagues Discussing Winsteps Table 14.3 Above

Ted: I cannot believe how easy it is to evaluate some aspects of how this instrument functions. Usually, when I read articles about instruments that are created or old ones that are used, only Cronbach's alpha is reported as a measure of the instrument's reliability. That is it.

Isabelle: Okay, it's test time for you....ready? If you were building this instrument from scratch, what might you do differently?

Ted: Well, the first thing is there might be some things that you gain and some things you lose with items that are flipped. I know that some people think that a "flipped" item keeps people honest. I get that. But, there are a lot of other negative (or at least possibly negative) issues. First, which you can see in this instrument, you are partially presenting SD, D, and BD categories so that respondents can answer the negative items. When you do the "tally" of instrument rating usage, you might miss that, the vast majority of responses are BA, A, and SA. But again, that is after "flipping" answers. Generally, for the negative items (before the flips) most responses are BD, D, and SD. And, most responses for items that are not flipped are BA, A, and SA.

Isabelle: I get this, but what is your point?

Ted: Well, as I can now see, and this really isn't even a Rasch observation, we are presenting a three-step scale to respondents. We are presenting BD, D, and SD to respondents for all of the "need to be flipped items," and we are presenting BA, A, and SA to respondents for all the "do not need to be flipped items." I think a better instrument would be one in which we do not do any flips, and we change the wording of items to make them a little harder to agree with. In that case we could work toward seeing more than three rating scale categories being used.

Isabelle: That makes sense. What do you think we gain by having more rating scale steps potentially used for each item?

Ted: I know there are no guarantees, but it seems to me if we can have a bigger range of rating scale responses being used for many survey items, it might help better distinguish the attitudes of respondents. Really increasing how well the instrument works will, in the end, improve the reliability of the instrument.

Isabelle: You get a gold star! Tell me more....

Ted: Another technique that I had never heard of before was looking at the overall mean attitude of individuals who selected a particular rating scale step for each item. It makes perfect sense to me now that we should see an increase in person measures as a function of rating scale step selections for each item. I had never really thought about that before, but it makes sense. Also, thinking about the interplay of individuals, the rating scale, and the difficulty of each item really makes use of Rasch theory. For any item, after flipping, with our STEBI example, we should see a higher mean measure for persons who are selecting the more and more agreeable rating category. For any item, with the STEBI, the general confidence of the persons who select that Strongly Agree should be more confident than those who select Agree. And the general confidence of the persons who select Agree for the same item should be higher in confidence than those who select Barely Agree. It makes sense to me that if the mean person measures with our STEBI data set do not increase with our rating scale, then something might not be working quite right with a specific item and our rating scale.

Isabelle: Now tell me about those, as I call them, "hills."

Ted: When I first saw this plot, I freaked out. There was too much going on, too many numbers going up and down. I'm still trying to comprehend all the details of the plot, but in the end

I figured out one quick thing that I can do. All I do is take a pen and draw a line along the top of the figure, so that I mark only the top of each hill. I know now that that with a perfect instrument I would hope to see each rating scale category as being "most likely" for some part of the horizontal scale. So for a 6-step scale, as we have for this STEBI, I can see that, from a probability standpoint, the categories of BA and BD might not be optimal. Now this does not mean that we just throw them out, but it might mean that if we revise the instrument, we might indeed attempt to change some item wording. Collect a sample of data, and then look at this plot, as well as other data, to see what we see.

Isabelle: What if I had a measurement instrument with a rating scale of very often, often, sometimes, and never. How would the hills look? What would I want to see?

Ted: In that case if we assumed we coded very often as a 4, often as a 3, sometimes as a 2, and never as a 1, we would want to see some region of the top of the hills be composed of "4"s, we would want to see some region of the top of the hills be composed of "3"s, and so on.

Isabelle: One grand finale question. If someone asked you how many persons you needed to evaluate a rating scale what would you tell them?

Ted: This is the deal Isabelle. I do not think you can really say a certain number of people. This is because what you learn about a rating scale will depend upon the people who are taking the scale and also the way in which items define the trait. So you must really take your time and look at the type of information in Table 14 and 21 of Winsteps, but also you need to review the information provided in the other tables of Winsteps that Boone, Staver, and Yale discuss in other chapters.

Keywords and Phrases

Disordered steps
Most probable

Potential Article Text

An important component of the SHAA (Science Helping All Additions) project was to help the project's 100 teachers improve their belief in their own students' ability to learn and understand science. To assess these students the project investigator wished to utilize the 10 outcome-expectancy items of the STEBI. Prior to large-scale data collection, a pilot data collection was conducted with 62 teachers of the city of St. John. Of particular interest to the research team was the functioning of rating scale categories (SD (1), D (2), N (3), A (4), SA (5)).

Tables 9.1, 9.2, and 9.3 provide the results of a Rasch Winsteps (Linacre, 2012) analysis. Table 9.1 revealed that respondents predominately utilized rating steps N (3), A (4), and SA (5). Given this usage pattern, a review of the most probable response as a function of person measure and item difficulty revealed that each response was indeed "most probable" for some combination of item difficulty and person measure (see Table 9.2). A final analysis was conducted to explore rating step functioning that considered the ordering (disordering) of rating steps as a function of

Table 9.1 Summary rating scale data for the administered outcome-expectancy items. Only one respondent selected the rating category SD (coded 1). The largest number of respondents selected rating category A (coded 4)

```
TABLE 3.2 SCIENCE TEACHER NAZ OE 07             ZOU286WS.TXT  Oct  9  8:00 2011
INPUT: 62 PERSON  10 ITEM  REPORTED: 62 PERSON  10 ITEM  5 CATS  WINSTEPS 3.72.3
-------------------------------------------------------------------------------

SUMMARY OF CATEGORY STRUCTURE.  Model="R"
-------------------------------------------------------------------
|CATEGORY   OBSERVED|OBSVD SAMPLE|INFIT OUTFIT||STRUCTURE|CATEGORY|
|LABEL SCORE COUNT %|AVRGE EXPECT|  MNSQ  MNSQ||CALIBRATN| MEASURE|
|-------------------+------------+------------++---------+--------|
|  1   1      1   0|  1.59  -.65|  2.43  3.27||  NONE   |( -5.70)| 1
|  2   2     69  11|   .07* -.03|  1.11  1.19||  -4.58  |  -2.68 | 2
|  3   3    198  32|   .66   .74|   .87   .84||   -.71  |    .00 | 3
|  4   4    320  52|  1.68  1.67|  1.01  1.01||    .71  |   2.68 | 4
|  5   5     32   5|  3.17  2.99|   .86   .97||   4.59  |(  5.70)| 5
-------------------------------------------------------------------
OBSERVED AVERAGE is mean of measures in category. It is not a parameter estimate.
```

Table 9.2 An assessment of the most probable responses as a function of person measure and item difficulty. These results suggest that each response category is most probable at some portion of the "person measure–item measure" axis

```
CATEGORY PROBABILITIES: MODES - Structure measures at intersections
P      -+---------+---------+---------+---------+---------+---------+-
R  1.0 +                                                             +
O      |                                                             |
B      |                                                             |
A      |                                                             |
B   .8 +1                                                           5+
I      | 11          22222222                   44444444          55 |
L      |   1        2        2                 4        4        5   |
I      |    1     22          2               4          44     5    |
T   .6 +     1   2             2             44             4   5     +
Y      |      1 2               2          4                4 5      |
    .5 +       *                 2  33333  4                 *       +
O      |      2 1                 *3     3*                 5 4      |
F   .4 +     2   1               3 2     4 3               5   4     +
       |    2     1             3   2   4   3             5     4    |
R      |   2       1           3     2 4     3           5       4   |
E      | 22         1        33       *       33        5         44 |
S   .2 +2            11     3        4 2        3     55            4+
P      |               11 33        4   2        33 55               |
O      |               33*1       44     22       5*33               |
N      |           3333    111**44         22**555    3333           |
S   .0 +**********************55*************11**********************+
E      -+---------+---------+---------+---------+---------+---------+-
       -6        -4        -2         0         2         4         6
        PERSON [MINUS] ITEM MEASURE
```

item and mean respondent measure (for each rating scale). Table 9.3 presents those results. One item, Q13oe-rc, exhibited step disordering. None of the other 9 items exhibited disordering. A decision was made to monitor this item during the data analysis of the final data set. If similar behavior were exhibited, then this item would possibly be removed before final person measures would be computed.

Table 9.3 An analysis of rating scale steps as a function of person measure and item. Only item Q13oe-rc of the outcome-expectancy scale exhibits disordered steps. However, close inspection reveals that only one person was used to compute the average of 1.96; thus, one really should not consider this item to have a true disordered step problem

TABLE 14.3 SCIENCE TEACHER NAZ OE 07 ZOU286WS.TXT Oct 9 8:00 2011
INPUT: 62 PERSON 10 ITEM REPORTED: 62 PERSON 10 ITEM 5 CATS WINSTEPS 3.72.3

ITEM CATEGORY/OPTION/DISTRACTOR FREQUENCIES: ENTRY ORDER

ENTRY NUMBER	DATA CODE	SCORE VALUE	DATA COUNT	%	AVERAGE ABILITY	S.E. MEAN	OUTF MNSQ	PTMEA CORR.	ITEM
1	2	2	8	13	.38	.32	1.3	-.31	Q1oe
	3	3	18	29	.69	.16	.7	-.34	
	4	4	31	50	1.59	.19	1.1	.32	
	5	5	5	8	2.54	.16	1.2	.36	
2	2	2	7	11	.55	.38	1.5	-.24	Q4oe
	3	3	22	35	.80	.19	1.0	-.31	
	4	4	31	50	1.57	.15	1.0	.30	
	5	5	2	3	3.68	.99	.7	.42	
3	2	2	5	8	-.21	.24	.5	-.41	Q7oe
	3	3	30	48	.98	.18	1.1	-.25	
	4	4	27	44	1.82	.16	.8	.47	
4	2	2	1	2	-.20		.8	-.18	Q9oe
	3	3	10	16	.13	.18	.7	-.46	
	4	4	41	66	1.36	.13	.7	.14	
	5	5	10	16	2.07	.46	1.1	.34	
5	2	2	17	27	.67	.22	1.3	-.33	Q10oe-rc
	3	3	26	42	1.13	.19	1.2	-.10	
	4	4	18	29	1.78	.19	1.1	.32	
	5	5	1	2	4.67		.6	.41	
6	2	2	5	8	.04	.30	.8	-.34	Q11oe
	3	3	22	35	.64	.15	.7	-.43	
	4	4	32	52	1.76	.18	.9	.49	
	5	5	3	5	2.32	.21	1.2	.23	
7	1	1	1	2	1.96		3.2	.09	Q13oe-rc
	2	2	10	16	.62*	.33	1.6	-.26	
	3	3	19	31	1.00*	.17	.9	-.16	
	4	4	29	47	1.36*	.17	1.5	.10	
	5	5	3	5	3.62	.69	.7	.50	
8	2	2	6	10	.08	.33	.9	-.36	Q14oe
	3	3	19	31	.75	.18	.8	-.31	
	4	4	37	60	1.70	.16	.9	.51	
9	2	2	3	5	-.14	.31	.7	-.29	Q15oe
	3	3	19	31	.63	.18	.8	-.39	
	4	4	38	61	1.56	.15	.9	.37	
	5	5	2	3	3.31	1.35	.8	.35	
10	2	2	7	11	.36	.31	1.3	-.30	Q16oe
	3	3	13	21	.61	.20	.8	-.31	
	4	4	36	58	1.39	.14	1.0	.15	
	5	5	6	10	2.85	.48	.9	.49	

* Average ability does not ascend with category score

Quick Tips

Consider piloting your rating scale. Conduct a pilot with a number of potential scales just as you would pilot potential survey and test items.

Winsteps Table 3.2 presents the percentage of all responses as a function of rating scale category. This can help researchers see whether some rating categories are not often used. But recall that when we use this information, we are not remembering that items differ in the ease or difficulty of agreement.

Winsteps Table 21.1, the "hills" as we call them, helps researchers quickly review how the rating scale adds to our measurement precision. Take a marker and color the top part of each hill. Sometimes a rating category is not observed. That means that rating category might not help measure as much as we want.

Winsteps Table 14.3 (one of our favorites) is a table in which the "steps" of a rating scale can be quickly seen, appreciated, and evaluated for each item. One would like to see the mean measures increase. Also, refrain from making too many conclusions when the steps are "disordered" by small amounts and/or small numbers of people are used to compute a mean measure.

For tests, researchers can also use the observation of disordered steps to increase measurement quality. A disordered step in a test may, among other things, reflect an item that is tapping a misconception or an error in scoring of an item.

A lengthy article that provides numerous tips with respect to the usage of rating scale categories is Linacre (2002). The table below appears in that article and provides very clear succinct guidance for the evaluation of rating scales:

	Guideline	Measure Stability	Measure Accuracy (Fit)	Description of this sample	Inference for next sample
Pre.	Scale oriented with latent variable	Essential	Essential	Essential	Essential
1.	At least 10 observations of each category.	Essential	Helpful		Helpful
2.	Regular observation distribution.	Helpful			Helpful
3.	Average measures advance monotonically with category.	Helpful	Essential	Essential	Essential
4.	OUTFIT mean-squares less than 2.0.	Helpful	Essential	Helpful	Helpful
5.	Step calibrations advance.				Helpful
6.	Ratings imply measures, and measures imply ratings.		Helpful		Helpful
7.	Step difficulties advance by at least 1.4 logits.				Helpful
8.	Step difficulties advance by less than 5.0 logits	Helpful			

Linacre's (p. 337) Winsteps manual provides this explanation of how the mean measures of rating scale categories for the entire instrument or particular items can be interpreted. He provides particularly helpful guidance as to the interpretation of the average of respondents who did not answer an item (or items).

> An "*" indicates that the average measure for a higher score value is lower than for a lower score value. This contradicts the hypothesis that "higher score value implies higher measure, and vice versa." The "average ability" for missing data is the average measure of all the persons for whom there is no response to this item. This can be useful. For instance, we may expect the "missing" people to be high or low performers, or to be missing random (and so the average measure would be close to the average of the sample).

Remember, amass information from many chapters, put observations in a spreadsheet, and then make an informed decision based upon many issues as to whether or not to keep items, remove items, change rating scales, and so on!

The data set that we have evaluated has already been flipped for the items that were reverse worded. Flipping can also be completed by Winsteps. The line RESCORE can be used to tell Winsteps which items needed to be recoded. The line NEWSCORE will tell Winsteps what the new values of the responses should be. In the example below the original codes are identified as being 1, 2, 3, 4, 5, 6. The line RESCORE tells the program that of the 5 items presented to respondents, that the second and the fourth item in the data set need to be recoded (the first "1" is in the second column, and the second "1" is in the fourth column). The last command line NEWSCORE tells the program that any "1" will be recoded as a "6," any "2" will be recoded as a "5," and so on. So you would use these lines in your Winsteps file, if you entered all items as if the items did not need to be flipped, and then you had Winsteps flip the items as Winsteps was run.

```
CODES=123456
RESCORE=01010
NEWSCORE=654321
```

Data Sets: (go to http://extras.springer.com)

cf saed_sabah Jordan

Activities

Activity #1

Dr. Saed Sabah of the Hashemite University (Jordan) has kindly provided us with a sample of data that he collected from students in Jordan as part of a study of students' perceptions of inquiry experiences in science laboratories. This, as is the case with almost all data in this book, is a nonrandom sample of data that is useful for learning Rasch. But no research conclusions should be made due to the sample size and the nonrandom sample. Dr. Saed's specialty areas within the field of science education are assessment and technology integration. The data were collected using the scale of Campbell, Abu-Hamid, and Chapman (2010) in which respondents could answer using a frequency scale (1 = *Almost Never*, 2 = *Seldom*, 3 = *Sometimes*, 4 = *Often*, 5 = *Almost Always*). The scale included two items that were reverse coded. The data in the SPSS spreadsheet used to create the control file for this activity have already been corrected for the reverse item wording of the two items.

We supply a control file for readers (cf saed_ sabah Jordan). Please run the data, and evaluate the use of the rating scale categories in Winsteps Table 3.2.

Answer: Below we provide the summary Table 3.2 for this data set.

```
TABLE 3.2 inquiry_dataset-1.sav                          ZOU955WS.TXT  Oct  9  9:49 2011
INPUT: 75 PERSON  20 ITEM  REPORTED: 75 PERSON  20 ITEM  5 CATS  WINSTEPS 3.72.3
--------------------------------------------------------------------------------


SUMMARY OF CATEGORY STRUCTURE.  Model="R"
-------------------------------------------------------------------
|CATEGORY   OBSERVED|OBSVD SAMPLE|INFIT OUTFIT||STRUCTURE|CATEGORY|
|LABEL SCORE COUNT %|AVRGE EXPECT|  MNSQ  MNSQ||CALIBRATN| MEASURE|
|-------------------+------------+------------++---------+--------|
|  1   1     162  11| -1.41 -1.48|  1.12  1.19||  NONE   |( -2.44)| 1
|  2   2     155  10|  -.59  -.46|   .75   .73||   -.92  |  -1.09 | 2
|  3   3     347  23|   .26   .33|   .92  1.07||   -.85  |   -.09 | 3
|  4   4     429  29|  1.07   .96|   .87  1.16||    .44  |   1.04 | 4
|  5   5     392  26|  1.48  1.54|  1.09  1.08||   1.34  |(  2.67)| 5
|-------------------+------------+------------++---------+--------|
|MISSING      15   1|   .92      |            ||         |        |
-------------------------------------------------------------------
OBSERVED AVERAGE is mean of measures in category. It is not a parameter estimate
```

The results show a fairly nice distribution of rating scale selections. Steps 1 and 2 are used 11 and 10 % of the time, respectively. All remaining categories are used 20–30 % of the time. We must also look at the rating scales selected as a function of item, because all items do not measure the same part of the trait.

Activity #2

With the same data set for Activity #1 and the same table, what do you see and what do you not see in terms of disordered steps?

Answer: In the text we only looked at the idea of disordered steps as applied to specific items. This table reveals that we might also make use of our knowledge of looking at disordered steps for the survey as a whole. Notice that the observed average increases from a negative value of −1.41 to a value of 1.48.

Activity #3

Using the data set from Activity 1, investigate the most probable response plot and Table 14.3 where you can investigate the ordering of steps for each item.

Answer: The most probable response plot reveals that two of the five rating scale steps, the first step (almost never, 1) and the fifth step (almost always, 5) have the highest probabilities of being observed. The second step (seldom, 2) is never a most probable response for a combination of a person measure taking a specific item (person measure–item difficulty). The table that presents the mean measure of respondents for each item as a function of rating category reveals no great disordering. Initially, it might appear as if there is disordering, but the number of respondents used to compute some averages is relatively small.

```
CATEGORY PROBABILITIES: MODES - Structure measures at intersections
P      -+---------+---------+---------+---------+---------+---------+-
R  1.0 +                                                             +
O      |                                                             |
B      |1                                                            |
A      | 111                                                        5|
B   .8 +    111                                                  555 +
I      |       11                                              55    |
L      |         11                                          55      |
I      |           11                                      55        |
T   .6 +             1                                   55          +
Y      |              11                                5            |
    .5 +                1                             55             +
O      |                 1                          55               |
F   .4 +                  11     333333333  4444444*444              +
       |                    1  33        4**3    55    444           |
R      |              2222222**22      44    33 5         444        |
E      |          2222     33 11 222 44       5*3            444     |
S   .2 +      2222      333     11 4*22     55   33             444  +
P      |  2222        33        44*1   22 55       333             44|
O      |22         333       444    11 55*222         3333           |
N      |    3333333     44444     5555*1111  22222        333333     |
S   .0 +****************5555555555         1111111*******************+
E      -+---------+---------+---------+---------+---------+---------+-
       -3        -2        -1         0         1         2         3
        PERSON [MINUS] ITEM MEASURE
```

TABLE 14.3 inquiry_dataset-1.sav ZOU955WS.TXT Oct 9 9:49 2011
INPUT: 75 PERSON 20 ITEM REPORTED: 75 PERSON 20 ITEM 5 CATS WINSTEPS 3.72.3

ITEM CATEGORY/OPTION/DISTRACTOR FREQUENCIES: ENTRY ORDER

ENTRY NUMBER	DATA CODE	SCORE VALUE	DATA COUNT	%	AVERAGE ABILITY	S.E. MEAN	OUTF MNSQ	PTMEA CORR.	ITEM
1	1	1	11	15	-.02	.22	1.1	-.37	ITEM1
	2	2	17	23	.24	.12	.7	-.26	
	3	3	29	39	.66	.09	.9	.13	
	4	4	12	16	.76	.09	1.0	.14	
	5	5	6	8	1.54	.22	.7	.46	
2	1	1	11	15	-.06	.18	.9	-.39	ITEM2
	2	2	24	32	.39	.11	.9	-.17	
	3	3	21	28	.58	.12	1.2	.04	
	4	4	17	23	1.06	.13	.8	.44	
	5	5	1	1	1.25		.9	.13	
	MISSING	***	1	1#	1.02			.09	
3	1	1	9	12	-.02	.23	1.2	-.33	ITEM3
	2	2	15	20	.12	.14	.7	-.33	
	3	3	24	32	.58	.09	.7	.03	
	4	4	16	21	.95	.13	.7	.33	
	5	5	11	15	.95*	.18	1.2	.26	
4	1	1	11	15	-.05	.21	1.1	-.39	ITEM4
	2	2	15	20	.19	.14	.8	-.28	
	3	3	24	32	.53	.08	.5	-.02	
	4	4	14	19	.88	.10	.7	.25	
	5	5	11	15	1.27	.16	.9	.47	
5	1	1	55	73	.51	.09	1.1	-.12	ITEM5
	2	2	13	17	.93	.16	.7	.27	
	3	3	6	8	.27*	.26	3.3	-.13	
	4	4	1	1	-.18*		6.1	-.13	
6	1	1	3	4	-.06	.62	1.8	-.19	ITEM6
	2	2	3	4	-.54*	.27	.5	-.35	
	3	3	15	20	.18	.12	.8	-.29	
	4	4	19	25	.65	.13	1.1	.09	
	5	5	35	47	.80	.09	1.0	.37	
7	1	1	3	4	.08	.51	1.9	-.15	ITEM7
	2	2	4	5	-.03*	.18	1.0	-.21	
	3	3	9	12	.33	.19	1.1	-.13	
	4	4	27	36	.47	.08	.4	-.09	
	5	5	31	42	.80	.14	1.2	.33	
	MISSING	***	1	1#	.52			-.01	
8	1	1	8	11	-.29	.17	.9	-.45	ITEM8
	2	2	5	7	.23	.33	1.3	-.14	
	3	3	24	32	.34	.10	.7	-.23	
	4	4	20	27	.77	.07	.5	.21	
	5	5	18	24	1.05	.14	1.0	.44	
9	2	2	1	1	-1.02		.3	-.29	ITEM9
	3	3	7	9	-.05	.23	.8	-.30	
	4	4	23	31	.67	.10	1.4	.12	
	5	5	43	58	.62*	.10	1.2	.13	
	MISSING	***	1	1#	.61			.01	
10	1	1	40	54	.48	.11	1.3	-.12	ITEM10
	2	2	21	28	.63	.15	1.5	.07	
	3	3	9	12	.57*	.15	1.7	.01	
	4	4	2	3	.77	.32	1.6	.06	
	5	5	2	3	.83	.26	1.9	.07	
	MISSING	***	1	1#	.61			.01	

(continued)

(continued)

11	3	3	7	9	.09	.22	1.0	-.23	ITEM11
	4	4	28	38	.55	.10	1.1	.00	
	5	5	39	53	.64	.11	1.2	.14	
	MISSING	***	1	1#	.61			.01	
12	3	3	9	12	.01	.21	.9	-.31	ITEM12
	4	4	25	34	.54	.11	1.1	-.02	
	5	5	40	54	.68	.11	1.1	.22	
	MISSING	***	1	1#	.61			.01	
13	1	1	1	1	-.54		.7	-.20	ITEM13
	2	2	6	8	-.28	.11	.5	-.38	
	3	3	22	30	.30	.13	1.0	-.25	
	4	4	26	35	.79	.10	.8	.28	
	5	5	19	26	.83	.13	1.1	.25	
	MISSING	***	1	1#	.61			.01	
14	1	1	1	1	-.60		.8	-.21	ITEM14
	2	2	4	5	-.22	.47	1.2	-.29	
	3	3	15	20	.31	.16	1.2	-.19	
	4	4	24	32	.61	.09	.6	.07	
	5	5	30	41	.76	.12	1.1	.27	
	MISSING	***	1	1#	.61			.01	
15	2	2	1	1	.64		2.2	.02	ITEM15
	3	3	13	18	-.17*	.15	.5	-.52	
	4	4	31	42	.63*	.09	1.1	.11	
	5	5	29	39	.78	.11	1.1	.29	
	MISSING	***	1	1#	.61			.01	
16	1	1	1	1	-.24		1.1	-.14	ITEM16
	2	2	3	4	-.17	.39	.9	-.23	
	3	3	12	16	-.21*	.15	.4	-.52	
	4	4	35	47	.72	.07	.9	.25	
	5	5	23	31	.81	.13	1.1	.27	
	MISSING	***	1	1#	.61			.01	
17	1	1	2	3	-.66	.12	.6	-.31	ITEM17
	2	2	4	5	-.05	.30	.9	-.22	
	3	3	23	31	.22	.12	.8	-.35	
	4	4	32	43	.74	.09	.7	.26	
	5	5	13	18	1.05	.14	.9	.36	
	MISSING	***	1	1#	.61			.01	
18	1	1	3	4	-.24	.45	1.1	-.25	ITEM18
	2	2	5	7	-.27*	.38	.7	-.34	
	3	3	31	42	.32	.08	.6	-.30	
	4	4	25	34	.84	.10	.7	.33	
	5	5	10	14	1.17	.13	.8	.38	
	MISSING	***	1	1#	.61			.01	
19	1	1	2	3	-.23	.54	1.1	-.20	ITEM19
	2	2	5	7	-.44*	.27	.5	-.42	
	3	3	20	27	.30	.09	.7	-.24	
	4	4	28	38	.62	.12	1.2	.09	
	5	5	18	25	1.08	.09	.8	.46	
	MISSING	***	2	3#	.55	.06		.00	
20	1	1	1	1	-.77		.4	-.24	ITEM20
	2	2	9	12	-.33	.16	.4	-.51	
	3	3	27	36	.42	.08	.7	-.16	
	4	4	24	32	.74	.10	.8	.20	
	5	5	13	18	1.20	.12	.8	.47	
	MISSING	***	1	1#	.61			.01	

* Average ability does not ascend with category score
\# Missing % includes all categories. Scored % only of scored categories

Activity #4

A researcher wants to develop a new rating scale survey. When she is considering a rating scale, she notices that many researchers in her field as well as other fields have used a rating scale of *Strongly Agree, Agree, Neutral, Disagree*, and *Strongly Disagree*. Why is it the case that the researcher cannot assume that rating scale will provide optimal measurement for her research project?

Answer: There are of course many issues that impact the function of a rating scale in a survey. The main point is that just because other researchers have used a particular rating scale, it does not mean that this rating scale is the optimal one to use. The selection of a rating scale should not be the result of what other researchers happened to use. It might be best for the investigator to not only develop different forms of a survey with different items, but to also consider piloting different rating scales and then evaluate the functioning of the different rating scales. This type of experimentation is what scientists do.

Activity #5

Can you explain why the use of negatively worded items might "hide" the fact that parts of a rating scale are being used, but some parts are not being used?

Answer: This is a tough question. Think about the manner in which the 13 self-efficacy items of the STEBI have been answered in the data set that we have evaluated. If one looks at the data at the end of the control file, one sees that many 5s and 6s have been used to answer the survey. This suggests that the two highest rating scales *Strongly Agree* (6) and *Agree* (5) were used by respondents. However, readers should recall that there were a number of items that were negatively worded and were then "flipped" prior to our analysis. So this means that, with this data set and group of respondents, the two ends of the rating scale were heavily used. This means that *Strongly Agree* and *Agree* categories were frequently selected for items that did not need to be flipped. *Strongly Disagree* and *Disagree* were often selected for those items which did need to be flipped.

References

Campbell, T., Abd-Hamid, H., & Chapman, H. (2010). Development of instrument to assess teacher and student perceptions of inquiry experiences in science classrooms. *Journal of Science Teacher Education, 21*, 13–30.

Hayes, J. R. (1981). *The complete problem solver*. Hillsdale: Lawrence Earlbaum.

Linacre, J. M. (2002). Optimizing rating scale category effectiveness. *Journal of Applied Measurement, 3*(1), 85–106.

Linacre, J. M. (2012). Winsteps (Version 3.74) [Software]. Available from http://www.winsteps.com/index.html

Additional Readings

Additional consideration of disordered steps.

Shaw, F., Wright, B., & Linacre, J. M. (1992). Disordered steps? *Rasch Measurement Transactions*, *16*(2), 225.

Analysis of a rating scale using Rasch analysis.

Chien, T., Hsu, S., Tai, C., Guo, H., & Su, S. (2008). Using Rasch analysis to validate the revised PSQI to assess sleep disorders in Taiwan's hi-tech workers. *Community Mental Health Journal*, *44*(6), 417–425.

An introductory article that considers the issue of how well a rating scale works.

Smith, E. V., Conrad, K. M., Chang, K., & Piazza, J. (2002). An introduction to Rasch measurement for scale development and person assessment. *Journal of Nursing Measurement, 10*, 189–206.

Chapter 10
Person Reliability, Item Reliability, and More

Isabelle and Ted: Two Colleagues Conversing

Ted: When I look at a typical education article, there often is very little written about reliability and validity.

Isabelle: I agree. When I sit in on talks, the presenters usually say something like this: "We established the validity of the instrument by having a panel of five experts review items. Following the review of items, the research team revised items in light of reviewers' comments. The revised items were then sent to the expert reviewers. At this point all five reviewers agreed with item changes. Reliability was evaluated empirically by using SPSS to compute Cronbach's alpha. The computed value of alpha was .83." And that's the end of it. Damn the torpedoes, full speed ahead!

Ted: No kidding. It's as if Cronbach's alpha is the only statistic that anyone needs to look at for reliability. The interesting thing is that the Winsteps summary table seems to provide a lot of guidance with respect to reliability that I can use to improve what I do.

Introduction

Cursory treatments of reliability and validity are commonplace in research reports within and beyond education research. To be sure, most authors devote a large percentage of time, effort, and journal space to reviews of past work and narratives that relate their current results to previous research in an effort to extend the literature. However, we believe that good publication practice of any research involving a measurement instrument should report the analysis of reliability and validity evidence for the given sample. Although text space is often limited in a publication not specific to the field of psychometrics, few would argue that simple steps to monitor the fidelity and reliability of measurement instruments should not appear in research articles.

A Rasch analysis of data not only produces linear measures that must be used for parametric statistical tests but it also provides a number of techniques for evaluating

W.J. Boone et al., *Rasch Analysis in the Human Sciences*, 217
DOI 10.1007/978-94-007-6857-4_10, © Springer Science+Business Media Dordrecht 2014

the psychometric properties of the instrument measures. In this chapter, we focus on evidence of reliability that can be obtained in Winsteps output. The types of ideas that we present here, as in most of our other chapters, are not unique to Winsteps and can of course be employed by researchers who use other Rasch analysis programs.

Reliability as a General Concept

To begin, we review the general concept of reliability in the context of developing and using tests and surveys. In Rasch workshops, we ask participants to think about the word "reliable" and about devices they might use or experience in their daily activities that should be "reliable." All of us have likely experienced being "timed" by a police officer using a radar unit as we drove or rode in a vehicle. Hopefully, we were driving within the posted speed limit, and the officer therefore did not stop us and write a citation for speeding. Let's take, for example, a police officer using a radar unit to determine the speed of vehicles. Today our police officer plans to monitor vehicle speeds in several locations: (1) early this morning in a school zone where the speed limit is 20 miles per hour (mph), (2) later this morning in a residential neighborhood where the speed limit is 30 mph, (3) after lunch on a county road where the speed limit is 45 mph, and (4) late this afternoon on a busy state highway where the speed limit is 65 mph. The officer's radar unit is new and was calibrated over a range of 0–120 mph at the factory. The radar unit will provide a vehicle's speed with an uncertainty of +/−0.1 mph at any speed within the range of 0–120 mph. In our work, we start off by reminding ourselves that "good" reliability means that there is empirical evidence that an instrument, be it a survey or test, measures in the same manner from time to time (e.g., Tuesday and Wednesday), and the instrument will measure people consistently no matter their opinion (attitudes) or knowledge (test). An analogy for the STEBI self-efficacy subscale and our police officer who is monitoring vehicle speeds with his radar unit is that a highly reliable STEBI measures teachers with low, medium, or high confidence with the same reliability.

We assert that researchers in all fields can improve the reliability assessment of their instruments by using Rasch techniques to evaluate reliability. Reliability analysis must be a part of any assessment development or use. The *Standards for Educational and Psychological Testing* (1999) describes in detail the necessary documentation of reliability analysis in any study.

A four-pronged outline of reliability documentation requirements is (a) a description of the population or subpopulations of interest, (b) a description of the measurement procedures and research design, (c) a summary of the assumptions (dimensionality, uncorrelated error, and at least congeneric measures) examined or not examined, and (d) the reliability estimate and the standard error of measurement (Meyer, 2010). Although not all of these can be examined with Rasch analysis output, Winsteps does have the capability to help us examine the model requirements of unidimensionality and provides an unbiased reliability estimate. In this chapter, we

```
TABLE 3.1 SCIENCE TEACHER EFFICACY BELIEFS       ZOU976WS.TXT Dec 13 10:49 2010
INPUT: 143 Person  23 Item  MEASURED: 143 Person  13 Item  6 CATS          3.69.1.9
--------------------------------------------------------------------------------

     SUMMARY OF 142 MEASURED (NON-EXTREME) Person
-------------------------------------------------------------------------------
|          TOTAL                         MODEL         INFIT        OUTFIT    |
|          SCORE     COUNT     MEASURE   ERROR      MNSQ   ZSTD   MNSQ   ZSTD |
|-----------------------------------------------------------------------------|
| MEAN       56.1      12.7     584.02   24.63      1.04    -.1   1.02     .0 |
| S.D.       10.4       1.4      75.18    5.71       .68    1.3    .60    1.3 |
| MAX.       75.0      13.0     830.37   52.74      4.23    3.9   3.30    3.3 |
| MIN.       21.0       6.0     438.68   19.62       .21   -2.7    .20   -2.4 |
|-----------------------------------------------------------------------------|
| REAL RMSE  28.57 TRUE SD   69.54  SEPARATION  2.43  Person RELIABILITY  .86 |
|MODEL RMSE  25.29 TRUE SD   70.80  SEPARATION  2.80  Person RELIABILITY  .89 |
| S.E. OF Person MEAN = 6.33                                                  |
-------------------------------------------------------------------------------
  MAXIMUM EXTREME SCORE:       1 Person

     SUMMARY OF 143 MEASURED (EXTREME AND NON-EXTREME) Person
-------------------------------------------------------------------------------
|          TOTAL                         MODEL         INFIT        OUTFIT    |
|          SCORE     COUNT     MEASURE   ERROR      MNSQ   ZSTD   MNSQ   ZSTD |
|-----------------------------------------------------------------------------|
| MEAN       56.2      12.7     586.92   25.28                                |
| S.D.       10.5       1.4      82.54    9.63                                |
| MAX.       78.0      13.0     999.64  117.83                                |
| MIN.       78.0      13.0     438.68   19.62       .21   -2.7    .20   -2.4 |
|-----------------------------------------------------------------------------|
| REAL RMSE  30.13 TRUE SD   76.84  SEPARATION  2.55  Person RELIABILITY  .87 |
|MODEL RMSE  27.06 TRUE SD   77.98  SEPARATION  2.88  Person RELIABILITY  .89 |
| S.E. OF Person MEAN = 6.93                                                  |
-------------------------------------------------------------------------------
Person RAW SCORE-TO-MEASURE CORRELATION = .78
CRONBACH ALPHA (KR-20) Person RAW SCORE RELIABILITY = .95

     SUMMARY OF 13 MEASURED (NON-EXTREME) Item
-------------------------------------------------------------------------------
|          TOTAL                         MODEL         INFIT        OUTFIT    |
|          SCORE     COUNT     MEASURE   ERROR      MNSQ   ZSTD   MNSQ   ZSTD |
|-----------------------------------------------------------------------------|
| MEAN      618.6     139.7     496.36    7.37      1.01     .0   1.01     .0 |
| S.D.       91.8       3.1      77.67    1.51       .23    1.7    .24    1.8 |
| MAX.      794.0     143.0     606.61   10.97      1.61    4.2   1.58    3.8 |
| MIN.      451.0     136.0     321.16    6.05       .69   -2.7    .65   -2.9 |
|-----------------------------------------------------------------------------|
| REAL RMSE   7.77 TRUE SD   77.28  SEPARATION  9.95  Item   RELIABILITY  .99 |
|MODEL RMSE   7.52 TRUE SD   77.30  SEPARATION 10.28  Item   RELIABILITY  .99 |
| S.E. OF Item MEAN = 22.42                                                   |
-------------------------------------------------------------------------------
            DELETED:      10 Item
UMEAN=496.3600 USCALE=63.4800
Item RAW SCORE-TO-MEASURE CORRELATION = -.97
1803 DATA POINTS. LOG-LIKELIHOOD CHI-SQUARE: 3753.17 with 1645 d.f. p=.0000
Global Root-Mean-Square Residual (excluding extreme scores): .7743
```

Fig. 10.1 Winsteps Table 3.1, which provides a wide range of indices that can be used to evaluate the reliability (and additional functioning) of a measurement instrument. Data are presented for (1) cases in which all respondents are used for reliability assessments even when respondents might have a maximum or minimum measure and (2) cases in which only non-extreme positive and non-extreme negative person measures are utilized

will explain where to find the reliability estimates (indices) in Winsteps output, how to interpret the values, and how to make these values part of an analysis. Figure 10.1 (Winsteps Table 3.1) provides a number of key reliability indices as well as other indices for evaluating our STEBI data set. Table 3.1 is composed of three parts.

Person Reliability, Item Reliability

The first part reports reliability information for the persons of a data set as well as added information concerning the respondents. The values in the first third of Table 3.1 are computed using only those respondents who did not respond with a "1" code (e.g., this would mean removing respondents who answered *Strongly Disagree* [SD] to all items after their answers to negative items have been "flipped") for all 13 items or a "5" code (e.g., this would mean removing respondents who answered *Strongly Agree* [SA] to all items after their answers to negative items have been "flipped") for all 13 items on the instrument. Readers should recall from Chap. 4 that when persons obtain the maximum measure when completing an instrument (e.g., all correct on a test, selecting the highest rating scale for all survey items) or they obtain the minimum measure when completing the instrument (e.g., all items wrong on a test, selecting the lowest rating scale for all survey items), the error of the person measure is infinite (when students get a perfect score on a test, they may know a little more than the hardest item on the test, or they may know a lot more than the hardest item on the test). To playfully remind readers of what they read in Chap. 4, consider a person who answers all items correctly on a test. One knows he or she knows a lot, but one has no idea how much more he or she knows. As a result the measurement error of the person's measurement is infinite. The person could know a little more than what is presented on the test or a lot more; we just do not know. So readers should understand that this first third of the table provides reliability information only for those people who were not extreme in all their responses. It may be that in your analysis you decide that you only want to use those respondents who were not extreme (high or low) for your analysis of instrument function.

Formative Assessment Checkpoint #1

Question: When we conduct Rasch measurement to, among other things, create a measurement scale and compute person measures, sometimes we may remove oddly behaving items and persons. Also we may remove, or not use for some parts of an analysis, persons who are extreme in their answers (all perfect on a test, selecting survey answers that result in the highest person measure possible; all incorrect on a test, selecting survey answers that result in the lowest person measure possible). Why is it possibly okay to remove some persons and items from an analysis?

Answer: The Rasch model is a definition of measurement. If persons and/or items do not fit the model, then those items and/or persons are not contributing to useful measurement. In the case of extreme persons, we might remove the persons as we finalize our instrument. One reason to do so is the understanding that someone who is topping out on a scale (who has infinite measurement error), may not contribute to the computation of measures as well as someone who does not have infinite error.

When to Use Extreme People and Extreme Items

The line in the table that says "SUMMARY OF 142 MEASURED (NON-EXTREME) Person" shows that in this analysis, 142 non-extreme respondents were used for the analysis presented in the first part of the summary table.

The next section of Winsteps Table 3.1 also provides reliability information (and additional indices) concerning "persons," but this table includes the "extremes," meaning those persons who achieved the top or bottom measures. The following line in the "SUMMARY OF 143 MEASURED (EXTREME AND NON-EXTREME) Person" states that 143 persons were used for the analysis presented in the second part. This means that one person either topped or bottomed out when he or she answered the STEBI. The comment "MAXIMUM EXTREME SCORE: 1 Person" tells an analyst that a single excluded person topped out, as the raw score total was the "maximum extreme." If one excluded person had bottomed out, the reported phrase would have been "MINIMUM EXTREME SCORE: 1 Person". The final third of the table, in part, presents reliability information for the survey items.

In Fig. 10.1, which part of the table for persons might one use: the summary information that includes all respondents or the part of the table that includes just the non-extreme respondents? Generally, for the work we have done, we have found no major differences in what we learn from the two parts of this table (be it with maximum and minimum respondents, or be it with these two types of respondents excluded). Usually, we use the part of the table that excludes the extreme respondents, since it makes sense to us that persons with infinite measurement error might impact what we say about a measurement device. Certainly we can imagine that there might be data sets in which it does make a difference in terms of whether or not one uses the first third of the table (only non-extreme respondents) or the second third of the table (all respondents are used, including those who were extreme).

We have used a number of analogies herein to clarify our points, and perhaps we can return to an earlier one when we consider incorporating extreme (the highest and the lowest) respondents in an analysis. Our example involves the policeman with the radar gun. It makes sense to us that if one has a number of cars that are traveling very fast (above the maximum value of the radar gun) or one has cars that are standing still, it probably does not help very much to use those cars to evaluate the radar gun's reliability. In our opinion, the same is true for evaluating the reliability of an instrument. We prefer to exclude the extreme (high and low) respondents from analyses of reliability that we conduct. It makes sense to us that if someone has topped out on an instrument (e.g., driven at a speed beyond the maximum reading on the radar gun) or if one has bottomed out (e.g., not moved at all when the policeman is taking a reading), these types of individuals do not provide useful data that help us understand how accurately the instrument is functioning.

Real Reliability and Model Reliability

We have highlighted a few key sections in the two person tables that address reliability. First, we call attention to the term "person reliability" which appears twice in each table. If one takes care to read the full line of text which contains the term "person reliability," then one will see the term "REAL" on one line and the term "MODEL" on the other end of the line. The important aspect of these two lines is that the indices throughout the line (not just person reliability) are computed in two different ways.

According to the Winsteps Manual, "person reliability" can be interpreted similarly to the more traditional reliability indices in classical test theory (i.e., KR-20 and Cronbach's alpha; Linacre, 2012). Meaning that values closer to 1 indicate a more internally consistent measure. The "Model" person reliability gives the upper limit of the consistency, reliability of the person measures. The "Real" person reliability gives the lower limit of the instrument's consistency, reliability of the person measures.

Note that, in this example, there are no large differences in the reliability values reported for the "REAL" line and the "MODEL" line (using only Non-Extreme respondents: REAL Person Reliability .86; MODEL Person Reliability .89). For research typically conducted in education, medicine, and market research, we suggest using the REAL reliability estimate; this value is a more conservative estimate of the person reliability and item reliability. The key issue, as readers will see in later parts of this chapter, is to be consistent in the type of reliability (Real or Model) that is reported and noted as analyses are conducted to explore how the reliability of an instrument might be improved.

Separation, Strata, and Reliability

Moving beyond the lines that begin with the word "Real" and the word "Model," what is important in terms of an analysis of a measurement instrument? Two key pieces of information are reported. First is the person reliability of .86; this is the value that can be reported in papers and used to evaluate aspects of reliability. Second is the value of 2.43 for what is called person separation.

Separation is the signal-to-noise ratio in the data. Specifically, the separation coefficient gives us the square root value of the ratio between the true person variance and the error variance in the data (Linacre, 2012). Separation can range from 0 to infinity; thus, there is no ceiling to this index. For purposes of an introductory analysis, a higher value is "better" than a lower value. We most often use the separation values as those values that we consult if we are attempting to experiment with different analysis of the data. For example, if we think that it might be useful to combine attitudinal categories for an analysis, we will conduct separate analyses and then note the changes that occur in person separation and item separation.

We now point out that the term "reliability" is likely the most familiar term to researchers because of the commonly computed Cronbach's alpha and KR-20 for many analyses. However, in Rasch measurement we have two reliability indices (person reliability, item reliability) from 0 to 1 that we can report. Moreover, one has two separation indices (person separation, item separation) to report. Person separation and item separation indices represent an added and very important addition to an evaluation of a measurement instrument's function.

There exists a great amount of additional information about reliability within the "Person" parts of Winsteps Table 3.1. For example, readers should find the line containing the words "CRONBACH ALPHA (KR-20)." This line presents the alpha and KR-20 values, depending upon the type of data being evaluated. These values vary from 0.00 to 1.00. Why are these reliability indices reported in this table? Our experience is that many reviewers who know little of Rasch will demand reporting of a KR-20 or Cronbach's alpha. If we provide that value, we often attempt to explain in our articles or talks why the alpha or KR-20 really is not very useful.

From our own work and knowledge of Rasch theory, we understand that a Cronbach's alpha or KR-20 that is calculated from raw data is corrupted due to the nonlinearity of the raw data (raw data are used for the computation of the KR-20). Linacre (1997) has summarized this fact:

> "... **KR-20** is an index of the **repeatability of raw scores**, misinterpreted as linear measures" (p. 580).... "KR-20 (Cronbach Alpha) always exceeds the maximum reliability possible for the measures underlying these simulated data. This misleads the test-user into believing a test has better *measurement* characteristics than it actually has. Yet KR-20 has met its design criteria, because estimated *raw-score* "true" S.D.s in Figure 1 match their predicted values. It reports the reliability of raw scores accurately, but these are local, test-dependent rankings. KR-20 overstates the reliability of the test-independent, generalizable measures the test is intended to imply. For inference beyond the test, Rasch reliability is more conservative and less misleading." (p. 581)

Readers will note that in addition to the Person part of Table 3.1, there is also an Item portion of the table (Fig. 10.1). Perhaps not surprisingly, this part of the table provides information regarding the reliability of items. This information is an additional type of reliability that is almost never reported in research literature (prior to development of Rasch techniques). What aspect germane to our research does this part of the table address? Our answer is the reliability of the items.

Review of the third or item part of this table reveals a line that begins with the word "REAL" and is followed by the terms "SEPARATION" 9.95 and "RELIABILITY" .99. As was the case with the data for persons, the item separation index varies from 0 to infinity, and the reliability index varies from 0 to 1.00. The issue of immediate importance is that these two values give item reliability information. The ability to monitor both person reliability and item reliability of instruments and respondents represents an important additional tool to aid the development and use of measurement instruments in many fields.

Is there a good or accepted value for the person separation, person reliability, item separation, and item reliability that should be consulted? Honestly, it all depends upon what your measurement goals are. However, we provide a range of tips and criteria that are provided by Mike Linacre in the Winsteps manual. These tips are provided at the end of the chapter in the "Quick Tip Guidelines."

As You Experiment with Different Sets of Items, Keep Track of How the Indices Change

There is, however, an additional point that needs to be made regarding how the values of item and person (reliability and separation) might be used in an analysis. We suggest that indices such as "person separation" and "person reliability" not only be compared against some value (e.g., a project may aim for achieving an item reliability of .90) but rather these indices might be compared as steps are taken to improve the measurement precision of the instrument. We will discuss this suggestion shortly.

Formative Assessment Checkpoint #2

Question: Is there only one type of reliability?

Answer: No. There are a number of types of reliability. For instance, there are test–retest reliability and alternate form reliability. And, readers will be able to understand that within Rasch there are a number of reliability indices, such as item reliability and person reliability. Just as there is not one "validity," there is not one "reliability."

How, then, are these values used in an analysis? Certainly one can report person and item separation and reliability values to document the function of an assessment instrument. However, we suggest a use of these indices that has been at best underemphasized and at most overlooked by researchers in many fields. If one reviews definitions of reliability, one would typically find at least three definitions. All involve the idea of being able to depend upon the function of an instrument (like our police officer's radar unit analogy). The three commonly cited reliabilities are test–retest, alternate form, and internal consistency. Test–retest reliability is established by administering the same instrument twice to the same people. A major problem with test–retest reliability is that the first assessment experience can influence the responses to the second administration of the instrument (Nunnally, 1967). Alternate forms reliability is established by administering alternate forms of an instrument. Internal consistency reliability is based on the average correlation among the items of an instrument. Coefficient alpha is an index of internal consistency reliability.

Above we detail how reliability indices can be used to track and monitor changes in reliability that can occur in an effort to maximize or at least verify the reliability of an "as is" instrument. Figure 10.2 presents an evaluation of reliability as well as "fit" for the STEBI from a recent study by Boone, Townsend, and Staver (2011).

Scale	Person separation reliability	Person fit	Item separation reliability	Item fit
1 (1,2,3,4,5,6)	2.33	1.04	10.72	1.01
2 (1,1,3,4,6,6)	1.62	1.03	7.06	1.03
3 (1,2,2,5,5,6)	2.02	1.04	10.21	1.06
4 (1,1,3&4,6,6)	1.7	1.02	7.2	1.03

Fig. 10.2 Person separation and item separation as a function of different rating scale codings for one data set. A rating scale of *Strongly Agree* (6), *Agree* (5), *Barely Agree* (4), *Barely Disagree* (3), *Disagree* (2), *and Strongly Disagree* (1) was used for the initial coding of data

In particular, how might the reliability of the STEBI be influenced by potential recoding of the rating scale? Readers will recall that the chapter that considered the "most probable" response plots provided by Winsteps suggested that the two middle rating categories, BA (3) and BD (4), were rarely the most probable response observed for combinations of persons and items. In an effort to explore the impact that combining the categories might have upon measurement properties of the STEBI, the authors of that study recoded rating categories in 4 different ways. Scale 1 presents reliability information for the original 6-step scale. Scale 2 presents a recoding in which SA and A were combined into one category, SD and D were combined into a separate category, and the individual categories of BA and BD were retained. Scale 3 retained individual rating scale steps for SA and SD, but combined A and BA, and also combined D and BD. Scale 4 utilizes three combinations, SA and A, SD and D, and BA and BD. The values of reliability chosen for reporting were the separation values, which have no ceiling effect (this is because all separation values vary from a minimum of 0.00 to + infinity). The researchers' conclusion from this analysis suggested that no recoding of data was needed to improve reliability. There might be other issues associated with the instrument that could be addressed, but recoding of response selections did not improve reliability.

Formative Assessment Checkpoint #3

Question: If altering the mix of items increases the reliability of a test from .92 to .94, does that change in reliability possess the same meaning as an improvement in reliability from .85 to .87?

Answer: No. There exists a ceiling effect when using a reliability value that ranges from 0 to 1.00. We suggest using the person separation index and the item separation index when exploring the impact of different items, for instance, upon the reliability of a test or survey.

Many types of reliability can be considered as a measurement instrument is developed and data are evaluated. Rasch techniques provide both a person reliability index as well as an item reliability index. Furthermore, item separation and person separation indices, which have no ceiling effect, are available via a Rasch analysis. When reliability is evaluated, an analyst should experiment with techniques that can be used to maximize the reliability of an instrument. For instance, how do person reliability and item reliability increase when rating categories of a survey are combined? For tests and surveys, how might person reliability and item reliability be affected by removal of one or more items or removal of persons? If an instrument is administered pre and post, how does the reliability of persons and items compare over time?

Formative Assessment Checkpoint #4

Question: Common types of "reliability" (Cronbach's alpha, test–retest, alternate form) are often discussed in the education, medical, and psychology literature. What have been the problems with how these types of reliability have been evaluated?

Answer: By this part of the book, readers should now understand that "counting" is not measuring. When these three techniques of reliability have been used in the past, one of the major flaws has been the use of raw data as if the "counts" are measures. It is not a flaw to evaluate the reliability of alternate forms or to evaluate the reliability of test–retest, but those assessments of reliability should be expressed using the Rasch indices that we introduced in this chapter. Thus, if one were to evaluate the test–retest reliability of an instrument, we would suggest creating a spreadsheet in which the rows are the different types of reliability discussed in this chapter, the 1st column of the spreadsheet presents the values from the first administration of the instrument, and the second column of the spreadsheet presents the data from the second administration of the instrument.

In this chapter we have tried to explain the basics of many of the reliability terms for persons and items which are provided to the analyst when conducting a Rasch analysis. Much of our energy had been to consider the monitoring of increasing or decreasing reliability values. But what might the numbers mean beyond this issue? Below we provide some guidance which was provided to us, by Mike Linacre, and we suggest that this guidance and greater details provided in the Winsteps manual be used as researchers further explore and utilize person separation and item separation to evaluate the functioning of an instrument.

A Brief Discussion of Person Separation and Item Separation of the Authors with Mike Linacre (March 31, 2012)

Reply: Usually person and item separation have different applications and implications.

Person separation is used to classify people. Low person separation with a relevant person sample implies that the instrument may not be sensitive enough to distinguish between high and low performers. More items may be needed.

Item separation is used to verify the item hierarchy. Low item separation (<3 = high, medium, low item difficulties) implies that the person sample is not large enough to confirm the item difficulty hierarchy (= construct validity) of the instrument.

We close this chapter with a few summary comments. First, we hope that readers note the many aspects of reliability that can be evaluated with Rasch measurement. Moreover, these aspects range far beyond the common step of reporting a Cronbach's alpha in a paper or report. When Rasch measurement is employed, it is possible to report indices of reliability both for items and persons. These indices can be those that range from 0 to 1.0, but, in our view, other indices (person separation, item separation) are a superior way of assessing instrument function, in that there is no ceiling effect for the indices (the value has a minimum of 0 and has no maximum). Other strengths of using Rasch for reliability analyses as well as other aspects of instrument function include an understanding that including persons who are minimum in measure or maximum in measure may not provide a good estimate of how an instrument functions. When using Rasch measurement, it is easy to evaluate aspects of reliability with and without extreme respondents. In our work we usually use the indices that do not include the extreme measures, and we use the "Real" as opposed to "Model" values. We do so in that those values are more conservative.

Final tips that we find helpful are experimenting with an instrument (e.g., removal of misfitting items, removal of misfitting person, removal of selected responses of misfitting persons, combining of rating scale categories) and then creating a table in which the rows of a spreadsheet list the values of different indices introduced in this chapter, and the columns represent a particular Rasch analysis (column 1 of a spreadsheet might include 4 reliability indices for an analysis of all 13 STEBI items, column 2 of the spreadsheet might include the same indices but for an analysis that included only 12 of the STEBI Items, and so on).

Ending this chapter, we point out that, although the term "reliability" will probably be the most familiar term to researchers because of the commonly computed Cronbach's alpha and KR-20 of most analyses, in Rasch measurement we have two reliability indices from 0 to 1 that we can report. And, one has the separation indices that can be reported. Finally, person separation and item separation represent an added and very important addition to the way in which one can evaluate the function of a measurement instrument.

Formative Assessment Checkpoint #5

Questions: How does Rasch measurement help you evaluate reliability? What does item separation and person separation have to do with reliability?

Answer: When Rasch measurement is used, one is able to evaluate the reliability of both the person measures as well as the items of the instrument. This is a great advance over what has been done in the past with the computations of alpha or KR-20. We view item separation and person separation as additional techniques by which one can assess (1) how well a set of items is able to differentiate different respondents (in the case of the STEBI, how many groups of respondents are the set of items able to differentiate) and (2) how well the set of items is able to be differentiated by the group of respondents.

Isabelle and Ted: Two Colleagues Conversing

Isabelle: Ted, I am very impressed with what you have been picking up as we work through your data analysis. So I am going to quiz you. Can you tell me how you might address "reliability" using Rasch?

Ted: I love your questions Isabelle. Okay, with Rasch one of the major advantages over a standard analysis is that reliability indices are computed not based upon raw data. Remember how there are some real problems with the use of raw rating scale data? Well, I think it should make sense that any reliability indices, such as alpha, are influenced by the problem of using raw data.

Isabelle: So, are you telling me that using raw data and then computing an alpha are not the right way to conduct an analysis?

Ted: Yes! That is exactly what I am telling you!

Isabelle: Well that then begs the question, Ted. What is one to do?

Ted: One technique is to use Rasch reliability indices. But, notice I have said indices as opposed to index! With Rasch, we are able to compute a person reliability AND an item reliability. That means we have two ways of looking at the reliability of an instrument.

Isabelle: When we talked in the hallway, you mentioned a different way of presenting reliability. What did you mean?

Ted: In addition to using a Rasch reliability, which is not potentially influenced by the non-linearity of raw scores, Rasch provides what is called a Separation index. The cool thing that I never thought about was that a traditional reliability (from 0 to 1.00) can top out. Thus, even the reliability value that is computed in Winsteps and corrected for nonlinearity has a maximum of 1.00. Rasch provides a separation index that has a low value of 0 but no maximum. If I wish to experiment with the impact upon measurement of any sort of change in an analysis, it makes a lot of sense to use the separation values as a way to monitor attempts to improve a measurement instrument.

Keywords and Phrases

Alternate form reliability
Congeneric measures

Cronbach's alpha
Dimensionality
Internal consistency
Item reliability
Item separation
Person reliability
Person separation
Reliability
Standard error of measure
Strata
Test–retest reliability
Top out/bottom out
Uncorrelated error

Sample Article Text

Following data entry but prior to statistical data analysis of "final" person measures, a number of Rasch analyses were conducted to evaluate the reliability of a measurement instrument. A component of that analysis involved an assessment of person (respondent) reliability and item reliability using different coding of rating scale steps. A common misconception is that more rating scale steps are always better, always providing more certainty as to the views of respondents. Too many rating scale steps can confuse respondents, wear them out, and degrade the quality of data collected. Analyses were therefore conducted to investigate the reliability of the instrument as a function of rating scale categories used for an analysis. Furthermore, the separation index was employed to provide added details as to the impact of different rating scale categories for a final computation of person measures for statistical analysis. This separation index has the important advantage of being unbounded at the upper end of the scale. The separation index ranges from a minimum of 0 to positive infinity. It does not top out at 1.00 as do typical reliability coefficients.

Figure 10.2 presents an overview of four (4) analyses that were conducted to evaluate the impact of four different rating scales for the STEBI. The reason why combining categories needs to be explored in an analysis is because one cannot assume that more categories provide better measurement. Scale 1 is the original scale as presented to respondents (SA, A, BA, BD, D, SD). Scale 2 data were recoded so that SA and A were combined into one category, and SD and D were combined into one category. Scale 3 was created by combining A and BA and combining D and BD. Scale 4 was created by combining SA and A, combining SD and D, and combining BA and BD. There are many issues in the preparation of data for a final analysis and the computation of person measures that will be used for a statistical analysis. The steps taken to evaluate how reliability might change as a function of differences in category coding suggest that, from at least a person separation and item separation perspective, the original coding for this STEBI data set should be maintained.

Figure 10.2 Repeated here for readers' convenience

Scale	Person separation reliability	Person fit	Item separation reliability	Item fit
1 (1,2,3,4,5,6)	2.33	1.04	10.72	1.01
2 (1,1,3,4,6,6)	1.62	1.03	7.06	1.03
3 (1,2,2,5,5,6)	2.02	1.04	10.21	1.06
4 (1,1,3&4,6,6)	1.7	1.02	7.2	1.03

Quick Tip Guidelines

Person (sample, test) reliability depends chiefly on:

1. Sample ability variance. Wider ability range = higher person reliability.
2. Length of test (and rating scale length). Longer test = higher person reliability.
3. Number of categories per item. More categories = higher person reliability.
4. Sample-item targeting. Better targeting = higher person reliability.

It is independent of sample size. It is largely uninfluenced by model fit.

Item reliability depends chiefly on:

1. Item difficulty variance. Wide difficulty range = high item reliability.
2. Person sample size. Large sample = high item reliability.

It is independent of test length. It is largely uninfluenced by model fit.

Tentative Guidelines

Person reliability: Does your test discriminate the sample into enough levels for your purpose?

0.9 = 3 or 4 levels
0.8 = 2 or 3 levels
0.5 = 1 or 2 levels

Item reliability: Low reliability means that your sample is not big enough to precisely locate the items on the latent variable. (Linacre 2012, p. 644)

Guideline for person separation index discussed by Wright and Masters (1982) and Fisher (1992) as found in Duncan, Bode, Lai, and Perera (2003):

> *Person Separation*: A person separation index of 1.50 represents an acceptable level of separation, an index of 2.00 represents a good level of separation, and index of 3.00 represents an excellent level of separation. (p. 953)

Above as appearing in Duncan, Bode, Lai, and Perera (2003). In the work of Duncan et al. (2003) the authors utilized Wright and Masters (1982) and Fisher (1992).

Guideline for item separation index from Tennant and Conghan (2007):

> *Item Separation*: An item separation index value of 1.5 is required for analyzing at the individual level and 2.5 is required for analysis of groups.

Data Sets: (go to http://extras.springer.com)

cf 25 GCKA
cf 25 items 1st 34 People GCKA
cf first 13 GCKA items
cf naz oe 2007
cf naz wo mid rating oe 2007

Activities

Activity #1

Using the control file (cf 25 GCKA) from our colleague Kathy Trundle at the Ohio State University, run an analysis and find the person reliability, person separation, item reliability, and item reliability which is computed. Create a table in which you will enter these results as well as some additional analyses.

Take the control file that we provide and create a control file in which only the first 34 people are evaluated. An easy way to do this is to make a copy of the file and then remove the 35th–75th persons in the control file, by removing their responses. (Tip: You can also use the command in Winsteps PDFILE, look it up!) After you have created the file, run it, find the same reliability indices as we used earlier, and enter those values in your table. We provide the file for the run of the 34 people, so you can check your work (the file is named cf 25 items 1st 34 People GCKA).

Finally, take your original control file and complete an analysis with only the first 13 items, but all 75 people. We have provided that control file for you (cf 13 GCKA items). If you wish, you can look at the file and find a code called IDFILE. That code is one cool way to remove items from an analysis. Don't forget to enter these reliability results in your table.

	PS	PR	IS	IR
cf 25 GCKA	.67	.31	3.66	.93
cf 25 items 1st 34 People GCKA	.83	.41	2.45	.86
cf first 13 GCKA items	.55	.23	4.20	.95

Activity #2

Examining the results from activity 1, what might be some of the effects of removing items or persons upon the reliability of an instrument and the resulting measures?

Answer: Generally, the more people who complete an item, the better the reliability of items. Moreover, the more items a person completes, the better the reliability of persons.

Activity #3

Find ten science education articles (or articles of your own discipline) that report on, or use, a test and/or rating scale instrument and are published in peer-reviewed journals. Make a table and list the reliability information that is provided for each article. Take note of how the word "reliability" is used in each article.

Answer: Of course this activity will be partially dependent upon which articles you select. Generally you will find that more often than not, the only "reliability" that is reported is a Cronbach's alpha, if raw counts have been used.

Activity #4

Our Miami colleague in science education (Naz Bautista) has provided us with a subset of data from her collection of outcome-expectancy data from preservice science teachers. We provide two control files for you: One (cf naz oe 2007) is with a set of students, all outcome-expectancy items, and the original rating scale of *Strongly Agree, Agree, Uncertain, Disagree, and Strongly Disagree*. A second control file (cf naz wo mid rating oe 2007) has a very small change in it; the "codes" line in the control file does not contain the number "3." This means that all ratings of "3" will not be used in an analysis.

Compare the reliability statistics of an analysis with all rating categories and with the exclusion of a middle category. Does the reliability increase or decrease when the middle rating category is removed?

Answer: Below we provided a summary of the statistics for the two analyses.

	PS	PR	IS	IR
5 cats	1.45	.68	2.60	.87
4 cats	.75	.36	2.04	.81

Activity #5

Question: Do you think it is possible to have a measurement instrument in which the person reliability is high and the item reliability is low? Could it ever be the case that a measurement instrument could have a low person reliability estimate and a high item reliability? Please explain your thinking.

Answer: A part of the Quick Tips is provided below with respect to reliability. If one reviews the factors which impact reliability, certainly one can have a high reliability with respect to, for example, items but also have a low (or lower) reliability with respect to persons. For example, when multimatrix design of test booklets is used, a large number of students can be compared on the same metric, but colleagues of ours have noted very high item reliability values, but lower person reliability values. Key factors impacting this are that many students in a project may attempt all items on a test or survey, but there may be a limited number of items attempted by the respondent in question.

Person (sample, test) reliability depends chiefly on:

1. Sample ability variance. Wider ability range = higher person reliability.
2. Length of test (and rating scale length). Longer test = higher person reliability.
3. Number of categories per item. More categories = higher person reliability.
4. Sample-item targeting. Better targeting = higher person reliability.

It is independent of sample size. It is largely uninfluenced by model fit.

Item reliability depends chiefly on:

1. Item difficulty variance. Wide difficulty range = high item reliability.
2. Person sample size. Large sample = high item reliability.

It is independent of test length. It is largely uninfluenced by model fit. (Linacre 2012, p. 644)

Activity #6

Question: Do you think there are any limitations as to when you can use Rasch item reliability and Rasch person reliability?

Answer: If one is evaluating an instrument in which the items are viewed as marking the different parts of a single trait, then you can use these values to evaluate aspects of the reliability of the instrument. The limitations are really not whether there are situations in which Rasch reliability can or cannot be evaluated, but simply do your data lend themselves to a Rasch analysis?

References

AERA, APA, & NCME. (1999). *Standards for educational and psychological testing*. Washington, DC: Author.

Boone, W., Townsend, S., & Staver, J. R. (2011). Using Rasch theory to guide the practice of survey development and survey data analysis in science education and to inform science reform efforts: An exemplar utilizing STEBI self efficiency data. *Science Education, 95*(2), 258–280.

Duncan, P. W., Bode, R., Lai, S. M., & Perera, S. (2003). Rasch analysis of a new stroke-specific outcome scale: The stroke impact scale. *Archives of Physical Medicine and Rehabilitation, 84*(7), 953.

Fisher, W. P. (1992). Reliability statistics. *Rasch Measurement Transactions, 6*(3), 238.

Linacre, J. M. (1997). KR-20 or Rasch reliability: Which tells the "truth"? *Rasch Measurement Transactions, 11*(3), 580–581.

Linacre, J. M. (2012). A user's guide to Winsteps Ministeps Rasch-model computer programs [version 3.74.0]. Retrieved from http://www.winsteps.com/index.htm

Meyer, J. P. (2010). *Reliability*. New York: Oxford University Press.

Nunnally, J. (1967). *Psychometric theory*. New York: McGraw-Hill.

Tennant, A., & Conghan, P. (2007). The Rasch measurement model in rheumatology: What is it and why use it? When should it be applied, and what should one look for in a Rasch paper? *Arthritis Care and Research, 5*(8), 1358–1362.

Wright, B. D., & Masters, O. N. (1982). *Rating scale analysis (Rasch measurement)*. Chicago: MESA Press.

Additional Readings

Fisher, W. (1992). Reliability statistics. *Rasch Measurement Transactions, 6*(3), 238.

Linacre, J. M. (1996). True-score reliability or Rasch statistical validity? *Rasch Measurement Transactions, 9*(4), 455.

Schumaker, R., & Smith, E. V. (2007). Reliability: A Rasch perspective. *Educational and Psychological Measurement, 67*(3), 394–409.

Wright, B. D., & Masters, G. N. (2002). Number of person or item strata. *Rasch Measurement Transactions, 16*(3), 888.

Chapter 11
What Is an Ogive? How Do I Use It?

Isabelle and Ted: Two Colleagues Conversing

Isabelle: *One thing that I think is important for you to master is how to view the Rasch model in action – what it looks like with a real-world data set.*

Ted: *Yes, as I learn how to use Rasch with my data sets, a lot seems to be going on. But as I read it does seem as if it all works together, and I see better how it all fits together. When I attended a few Rasch talks, people often began their talks with brief comments on a general perspective of Rasch, then gave their talk, and finally returned to their opening theme. And, it did seem to fit together.*

Isabelle: *I have my own ideas, but what do you think is a meaningful way to start thinking about how real data can be used to show how things fit together, Rasch-wise that is?*

Ted: *Well, one way, I think, is to show an ogive in action and a score table. That helped me a lot.*

Isabelle: *What do you mean by "ogive"?*

Ted: *Well in Rasch measurement, we often talk of the ogive. I found that understanding the ogive helped me to understand Rasch measurement, and Rasch measurement theory also helped me understand the ogive. And, in the end I found the ogive is a good way to explain some aspects of Rasch measurement to those who are just learning about good measurement. For example, it's pretty easy to use an ogive to show that raw scores are not linear.*

Introduction

Throughout the chapters of this book, we have attempted to help readers think about Rasch analysis in a variety of ways so they might design new measurement instruments, revise existing instruments, and evaluate the quality of a data set (e.g., look at an instrument's items and/or at the responses of respondents).

As we help readers construct a more coherent understanding of Rasch, we have found it helpful to consider the curve presented in Table 20.1 of the Winsteps and Ministeps output. This curve has helped us and our students "put the pieces together,"

W.J. Boone et al., *Rasch Analysis in the Human Sciences*,
DOI 10.1007/978-94-007-6857-4_11,

in that the curve provides an excellent way of explaining the problem with raw scores and also provides fuel for explaining a little of the mathematics of Rasch.

The Ogive

To begin work with the ogive, we start with Winsteps Table 20.1, which is presented immediately below as our Fig. 11.1. The first part of the table is a raw score to measure conversion table. Figure 11.1 presents the results from the Rasch analysis of 13 self-efficacy items that were administered to a sample of 143 respondents. In this analysis UMEAN and USCALE were used to create a measurement scale from 0 to 1,000. Figure 11.1 (Winsteps Table 20.1) presents all possible raw scores that could be earned by respondents (from a low of 13 to high of 78). The table also

```
TABLE 20.1 SCIENCE TEACHER EFFICACY BELIEFS      ZOU397WS.TXT Dec 19  8:52 2010
INPUT: 143 Person  23 Item  MEASURED: 143 Person  13 Item  6 CATS          3.69.1.9
```

TABLE OF MEASURES ON COMPLETE TEST

SCORE	MEASURE	S.E.	SCORE	MEASURE	S.E.	SCORE	MEASURE	S.E.
13	.01E	120.16	35	423.33	22.36	57	568.52	21.54
14	87.22	70.52	36	431.08	22.01	58	575.99	22.03
15	145.07	53.17	37	438.60	21.68	59	583.83	22.59
16	182.55	45.00	38	445.90	21.36	60	592.10	23.24
17	210.68	39.79	39	452.98	21.06	61	600.88	23.99
18	233.26	36.10	40	459.87	20.77	62	610.26	24.85
19	252.20	33.36	41	466.58	20.51	63	620.37	25.82
20	268.61	31.28	42	473.13	20.28	64	631.32	26.91
21	283.23	29.70	43	479.55	20.08	65	643.24	28.11
22	296.54	28.47	44	485.84	19.92	66	656.25	29.37
23	308.87	27.52	45	492.05	19.78	67	670.44	30.64
24	320.46	26.76	46	498.18	19.69	68	685.82	31.84
25	331.47	26.14	47	504.27	19.64	69	702.35	32.93
26	342.02	25.63	48	510.34	19.62	70	719.97	33.92
27	352.19	25.19	49	516.41	19.65	71	738.62	34.91
28	362.03	24.80	50	522.52	19.72	72	758.43	36.06
29	371.57	24.43	51	528.68	19.83	73	779.76	37.61
30	380.84	24.08	52	534.92	19.99	74	803.35	39.93
31	389.85	23.74	53	541.27	20.19	75	830.69	43.71
32	398.60	23.39	54	547.77	20.45	76	865.22	50.63
33	407.09	23.05	55	554.46	20.75	77	917.74	67.55
34	415.34	22.70	56	561.36	21.11	78	1000.00E	118.13

```
CURRENT VALUES, UMEAN=496.360 USCALE=63.480
TO SET MEASURE RANGE AS 0-100, UMEAN=49.636 USCALE=6.348
TO SET MEASURE RANGE TO MATCH RAW SCORE RANGE, UMEAN=45.263 USCALE=4.126
Predicting Score from Measure: Score = Measure * .095 + -14.491
Predicting Measure from Score: Measure = Score * 9.931 + 173.673
```

Fig. 11.1 (Winsteps Table 20.1): The raw score to measure conversion table which resulted from a Rasch Winsteps analysis of data collected with the 13-item SE scale. A person who has a raw score of 26 is someone who could have answered a "2" to each of the 13 SE items ($2\times13=26$). This person has a "measure" of 342 on a scale that extends from a minimum of 0 to a maximum of 1,000. In this Rasch analysis, first an analysis was conducted, and then a rescaling using UMEAN and USCALE was conducted. The letter "E" that one sees for the highest and lowest measure indicates that the measure is an extreme measure (at the end of the scale)

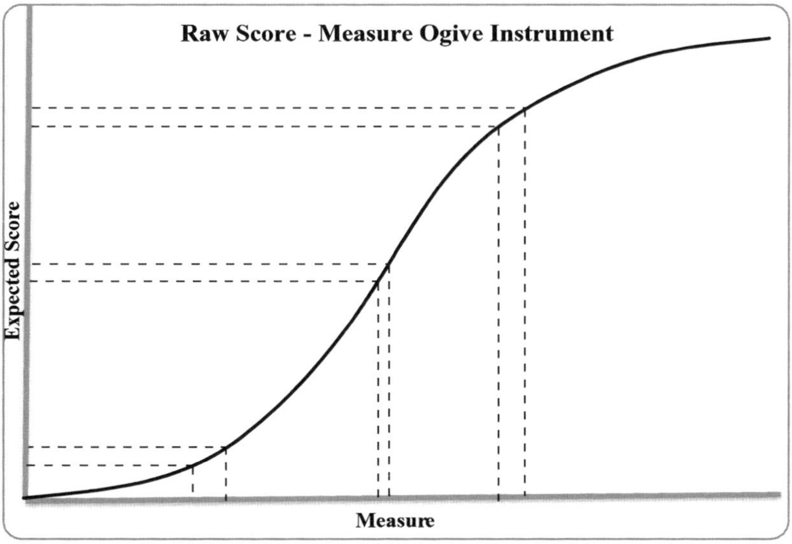

Fig. 11.2 Plot of Winsteps Table 20.1

presents all possible measures for respondents. Remember, only the measures can be used for parametric statistical comparisons of respondents because only the measures have been corrected for the nonlinear character of the raw data. (But, as we have stated in previous chapters, one needs to make sure the data fit the Rasch model requirements.)

The Score Measure Table

Many Rasch model users may view the conversion of raw data to linear measures as a concept not to be plumbed further. Basically they accept the work of previous researchers and move onto other Rasch concepts. We suggest that, even for beginning Rasch users, a very reasonable next step (to advance your facility with Rasch) is to pause and further consider both the table and the plot (Figs. 11.1 and 11.2).

Formative Assessment Checkpoint #1

Question: Is it okay to use raw data from an existing instrument because the instrument has been published?

Answer: We say no. Many instruments have been published, and the developers used raw scores. This is why, and for many added reasons, many medical certification tests use Rasch measurement and PISA uses Rasch.

The plot of Table 20.1 (Fig. 11.2) is created by plotting each measure for each potential raw score, and the curve is called the logistic ogive (the logistic ogive is a cumulative probability path/curve/distribution). When you see such a curve for test data or survey data (one axis is a measure, and one axis is a raw score), you should suspect that the Rasch model may have been applied to the analysis of a data set.

In our analysis that was used to create Fig. 11.2, we of course know that we used Rasch to conduct the analysis. But if we are trying to explain Rasch to someone, how could we make use of this curve to show the implications of the Rasch transformation of raw scores to measures? To begin, it is helpful to first identify the vertical and horizontal axes in Fig. 11.2. The vertical axis presents all possible raw scores for the 13 STEBI self-efficacy items using a 6-step scale with *Strongly Disagree* (SD) at the low end and *Strongly Agree* (SA) at the high end. Since SD was coded with a "1," the lowest possible raw score of respondents is 13 (13 items × 1 (for SD) = 13). The highest possible raw score is a 78 (13 × 6 (for SA) = 78). The horizontal axis represents the range of Rasch scale score measures from 0 to 1,000. Readers should remember that a scale of 0–1,000 was used in this analysis to facilitate communication with readers of a research article. The values from 0 to 1,000 represent a linear transformation or rescaling of initial Rasch logit person measures.

Formative Assessment Checkpoint #2

Question: What variables are plotted on the horizontal (X) and vertical (Y) axes, respectively?

Answer: Raw scores are plotted on the y-axis, and Rasch person measures are plotted on the x-axis.

The curve represents the relationship between any possible raw score (13–78) and each scale score (0–1,000). To see this relationship, pick any expected raw score on the vertical scale and draw a horizontal line to the curve. It should be clear to readers that if we plotted all 66 possible raw scores and measures and connected all the dots, we would end up with the solid black curve that we see in our Fig. 11.2 (plot of Winsteps Table 20.1).

We have shown how plotting the raw scores (y-axis) and the scale score measures (x-axis) results in the ogive. Where does the ogive come from? What's the big deal about the ogive? There are many properties of the ogive that are important, and following the last activity of this chapter, we provide a brief summary of observations which has been made by researchers such as Linacre (2006). The main point of our summary is that when the ogive, the Rasch model, and the work of various statisticians are reviewed, one sees that much of past thinking can be tied together through the ogive.

A second important point for readers as we discuss the ogive is where the ogive comes from. Numerous books (e.g., Wright & Stone, 1979) have been written in which the details of the derivation of the Rasch model are presented and discussed. Presenting such derivations is beyond the scope of our book. What is important for readers to appreciate is that the Rasch logistic ogive, which this chapter considers, is expressing the mathematics of the Rasch model. The "big deal" about the Rasch logistic ogive is that the Rasch model is the model that allows us (when the data fits the model), finally, in research with surveys and tests to confidently (1) compute linear measures, (2) make comparisons of persons and items, and (3) link forms. Of course in the previous chapters, and those to come, readers will see that there are many nuances to using the Rasch model beyond the comment just made. But to appreciate the Rasch model and the ogive, just think "measurement" and what is needed to "conduct measurement." When one reviews, for example, a plot of person raw scores for a survey such as the STEBI and person measures, one will see the Rasch logistic ogive (this is the line which is plotted in Fig. 11.2).

The Ogive Helps You See that Raw Scores Are Not Linear

The books that present the details of the derivation of the ogive are very important for understanding Rasch measurement. To help readers develop a feel for what is being shown by the ogive, we share an activity that we have used with classes. The activity makes use of many of the comments Ben Wright made to one of the authors when the author was a student at the University of Chicago. In Fig. 11.3 we present six pairs of raw scores/measures to show the consequences of using and not using the Rasch model. The measures are for 6 individuals who responded to the 13 self-efficacy STEBI items. The logit person measures have been rescaled as the result of using UMEAN and USCALE. The raw scores and measures can be found in Fig. 11.1.

To show the impact of using the Rasch transformation from raw data to measures, we begin by computing the difference in raw score and scale score measures of the six fictitious students. Each student is conveniently represented in Fig. 11.2 by a pair of the plotted points (a raw score and a measure) on the ogive. The quantitative comparisons are presented in Fig. 11.4.

This computation of the difference of one raw score point shows that a difference of one raw score point between pairs of respondents does not have the same meaning along the measure of the self-efficacy trait. So here we can see that one raw score point difference does not mean that same difference along the variable for all raw score/measure pairs.

We have found that altering this example a little bit can also help our students and colleagues grasp the impact of applying the Rasch model and also understand the ogive. To alter our example, we often talk of subgroup averages and subgroup measures. In Fig. 11.5 we provide the means for two comparison groups (a classroom "A" and a classroom "B) at two time points (pre, post). Using raw scores, classroom "B" appears to gain more than classroom "A." Using measures we can see that the reverse is the case.

	Raw Score	Scale Score Measure
Bill	14	87
Bob	15	145
Joe	45	492
Sam	46	498
John	68	685
Mel	69	702

Fig. 11.3 Raw score and measures for six fictitious students

Difference Between Bob & Bill	
1 raw score unit	15 – 14 = 1
58 Scale Score Measure	145 – 87 = 58

Difference Between Sam & Joe	
1 raw score unit	46 – 45 = 1
6 Scale Score Measure	498 – 492 = 6

Difference Between Mel & John	
1 raw score unit	69 – 68 = 1
17 Scale Score Measure	702 – 685 = 17

Fig. 11.4 Three quantitative comparisons of the difference between one raw score point at different points along the ogive in terms of the corresponding measure

	Mean Pre Raw Score	Mean Post Raw Score	Post-Pre
Classroom A (treatment)	23	35	35 – 23 = 12
Classroom B (control)	39	53	53 – 39 = 14

	Mean Pre Scale Score	Mean Post Scale Score	Post-Pre
Classroom A (treatment)	308	423	423 – 308 = 115
Classroom B (control)	452	541	541 – 452 = 89

Fig. 11.5 Comparisons between two groups with raw scores and scale score measures

The take-home message of these data is to emphasize the problem of comparing classrooms and time points using raw score data. Clearly, the meaning of increasing one raw score unit changes depending upon the position (low, medium, or high) of the raw score along the trait. This means looking at differences even for one classroom from time point pre to time point post is impacted greatly by the nonlinearity of raw scores. Moreover and on top of this issue, comparing the amount of change from pre to post for the two classrooms is also, perhaps not surprisingly, problematic for the meaning of the change depends upon what part of the scale is used. Only by using the scale score measure data for classrooms and time points can a researcher make useful comparisons and reach confident conclusions. This is because the scale score measures are linear data and therefore not affected by the specific part of the scale used to make the comparisons (be it a classroom where most students disagreed with the 13 items or a classroom where most of the students agreed with the 13 items).

Formative Assessment Checkpoint #3

Question: Whereas the ogive presents the relationship between raw scores and Rasch measures, are there other transformations one could use to take raw scores and compute measures?

Answer: No. The Rasch model is a definition of measurement. This model is the only model that meets the requirements of fundamental measurement. This is the model that must be used for the computation of measures.

Many important ramifications exist regarding the shape of the ogive. One nuance is that one segment (the middle of the graph) of the ogive is linear. This means there is a linear relationship between the raw scores and the linear Rasch metric in the middle of the graph. If it were the case (it is rarely so) that all persons of a sample fall within the linear portion of the ogive, then the raw scores would not be corrupted as they are at other portions (lower and upper curved sections) of the ogive. In this example, if one were to conduct a *t*-test comparing the performance of boys and girls whose raw scores and scale scores lie at or very near the middle, linear section of the ogive, then the parametric statistical test would yield an accurate comparison of the two groups. Rarely, however, will all persons fall within the linear portion of the ogive. Moreover, one cannot predict where additional respondents will be located along the ogive.

Consideration of where respondents may fall with regard to the ogive (linear portion or nonlinear portion) is also a very important issue with regard to much of the research that takes place in many settings, in that often some of the most carefully tracked individuals are those who are performing at a low level along a trait. Examples of such groups might be students who are performing far below grade level in science or teachers who have extremely low confidence in teaching science. Clearly if you want to evaluate high performers, low performers, high-performing schools, or low-performing schools, then you must set those raw scores aside and use Rasch measurement to compute "measures."

Changing the Starting Point of the Ogive Does Not Change the Meaning of the Ogive

We mentioned above that the scale used to express the Rasch measures does not make any difference for statistical computations. For instance, if the attitudes of males and females are compared through a *t*-test, it makes no difference if the original

TABLE 20.1 SCIENCE TEACHER EFFICACY BELIEFS ZOU864WS.TXT Dec 20 11:34 2010
INPUT: 143 Person 23 Item MEASURED: 143 Person 13 Item 6 CATS WINSTEPS 3.70.6

TABLE OF MEASURES ON TEST OF 13 Item

SCORE	MEASURE	S.E.	SCORE	MEASURE	S.E.	SCORE	MEASURE	S.E.
13	-7.82E	1.89	35	-1.15	.35	57	1.14	.34
14	-6.45	1.11	36	-1.03	.35	58	1.25	.35
15	-5.53	.84	37	-.91	.34	59	1.38	.36
16	-4.94	.71	38	-.79	.34	60	1.51	.37
17	-4.50	.63	39	-.68	.33	61	1.65	.38
18	-4.14	.57	40	-.57	.33	62	1.79	.39
19	-3.85	.53	41	-.47	.32	63	1.95	.41
20	-3.59	.49	42	-.37	.32	64	2.13	.42
21	-3.36	.47	43	-.26	.32	65	2.31	.44
22	-3.15	.45	44	-.17	.31	66	2.52	.46
23	-2.95	.43	45	-.07	.31	67	2.74	.48
24	-2.77	.42	46	.03	.31	68	2.98	.50
25	-2.60	.41	47	.12	.31	69	3.25	.52
26	-2.43	.40	48	.22	.31	70	3.52	.53
27	-2.27	.40	49	.32	.31	71	3.82	.55
28	-2.12	.39	50	.41	.31	72	4.13	.57
29	-1.97	.38	51	.51	.31	73	4.46	.59
30	-1.82	.38	52	.61	.31	74	4.84	.63
31	-1.68	.37	53	.71	.32	75	5.27	.69
32	-1.54	.37	54	.81	.32	76	5.81	.80
33	-1.41	.36	55	.92	.33	77	6.64	1.06
34	-1.28	.36	56	1.02	.33	78	7.93E	1.86

CURRENT VALUES, UMEAN=.0000 USCALE=1.0000
TO SET MEASURE RANGE AS 0-100, UMEAN=49.6357 USCALE=6.3481
TO SET MEASURE RANGE TO MATCH RAW SCORE RANGE, UMEAN=45.2632 USCALE=4.1262
Predicting Score from Measure: Score = Measure * 6.0090 + 32.4941
Predicting Measure from Score: Measure = Score * .1564 + -5.0833

Fig. 11.6 The relationship between the raw score data and the logit measures

logit data or rescaled logit data are used. As we discussed in Chap. 6, perhaps the best analogy is the collection and analysis of temperature data. It does not make any difference if temperature data are collected in Fahrenheit, Celsius, or Kelvin, and it does not make any difference if an analysis is conducted with Fahrenheit, Celsius, or Kelvin temperature data. And of course, data could be collected in any of the three scales (e.g., Celsius), then transformed into either of the other two scales (e.g., Fahrenheit), and subsequently evaluated.

To further aid readers' understanding of the ogive and concurrently ease your reservations of person measures that range from negative values to positive values, we present an analysis of the same data set that was used to produce the raw score to scale score conversion table (Fig. 11.1) as well as the ogive (Fig. 11.2). The only difference is that the two lines (UMEAN and USCALE) have not been added to the control file. This means that a logit scale will be presented in which the mean item logit value is 0.00.

Our Fig. 11.6 (Winsteps Table 20.1) presents the relationship between the raw score data and the logit measures. A raw score of 13 (SD to all 13 items) results in a person measure of –7.82. Following the conversion table is the ogive (Fig. 11.7) that presents the coordinates of each set of two points presented in the table (e.g., Raw

```
        RAW SCORE-MEASURE OGIVE FOR COMPLETE TEST
      -+------+------+------+------+------+------+------+------+-
    78 +                                                  *    E+
    76 +                                             * *        +
    74 +                                           **           +
    72 +                                        **              +
    70 +                                      **                +
    68 +                                     *                  +
    66 +                                   **                   +
    64 +                                  *                     +
    62 +                                 *                      +
  59.5 +                                *                       +
E 57.5 +                               *                        +
X 55.5 +                              **                        +
P 53.5 +                             **                         +
E 51.5 +                             *                          +
C 49.5 +                            *                           +
T 47.5 +                           **                           +
E 45.5 +                           *                            +
D 43.5 +                          *                             +
  41.5 +                         **                             +
S 39.5 +                         *                              +
C 37.5 +                        *                               +
O 35.5 +                       *                                +
R 33.5 +                      **                                +
E 31.5 +                     **                                 +
  29.5 +                    **                                  +
    27 +                   **                                   +
    25 +                  *                                     +
    23 +                 *                                      +
    21 +               **                                       +
    19 +              *                                         +
    17 +           **                                           +
    15 +        * *                                             +
    13 + E   *                                                  +
      -+------+------+------+------+------+------+------+------+-
      -8     -6     -4     -2      0      2      4      6      8
                 MEASURE

                                  1 11 21 1
Person                          140616700947981132 2 1         1
                               T   S    M    S   T
%TILE                           0 10 30 60 80 90    99

Item                     1  1  1 1 1231 11
                          T    S   M   S    T
%TILE                    0 10 20 30 70 90 99
```

Fig. 11.7 Raw score – measure ogive for Fig. 11.6

Score = 13, Logit Measure = –7.82). The vertical axis scale is identical to that presented in Fig. 11.2. The horizontal scale looks different, but is the same, in that each part of the logit scale from –7.82 to +7.93 (respondent answered SA to all 13 items) can be mapped onto the scale from 0 to 1,000.

	Raw Score	Scale Score
Bill	13	- 7.82
Bob	14	- 6.45
Joe	45	- .07
Sam	46	.03
John	68	2.98
Melissa	69	3.25

Bob-Bill Raw Score	= 1 pt	(14 – 13 = 1)
Bob-Bill Scale Score	= 1.37 logits	(-6.45 – (-7.82) = 1.37)
Sam-Joe Raw Score	= 1 pt	(46 – 45 = 1)
Sam-Joe Scale Score	= .10 logits	(.03 – (-.07) = .10)
Melissa-John Raw Score	= 1 pt	(69 – 68 = 1)
Melissa-John	= .27 logits	(3.25 – 2.98 = .27)

	Mean Pre Raw Score	Mean Post Raw Score	Post-Pre
Classroom A (treatment)	23	35	35 – 23 = 12
Classroom B (control)	39	53	51 – 39 = 14
	Mean Pre Scale Score	Mean Post Scale Score	Post– Pre
Classroom A (treatment)	-2.95	-1.15	= 1.80
Classroom B (control)	-.68	.71	= 1.39

Fig. 11.8 Comparisons between two groups with raw scores and measures in which no rescaling has been used

In Fig. 11.8 we also present some easy to follow, we hope, details of the relationship between the raw scores and the measures. One example concerns the misleading aspect of using raw scores to compare three groups of students who differ in raw score by 1 raw score point. A second example is presented in which the gain of a classroom A and classroom B is presented. In our example, comparison of the "gain" from pre to post (using raw scores) suggests that classroom B may have gained more than classroom A, but use of the measures suggests that the opposite is true.

We conclude this chapter with an overview of how we came to appreciate and understand the ogive. First, experts in measurement such as Ben Wright and others have shown in articles (many technical) that the Rasch model is the only model that successfully addresses the requirements of measurement laid out by many individuals such as Thorndike, Campbell, and Guttmann. When that model is applied to raw data, an ogive results when one graphs the relationship between raw scores and measures. This distinctive curve, the ogive, represents the mathematical function expressed by the Rasch model. Second, although not immediately clear to researchers who may routinely look at curves, the fact that the ogive is not linear shows that the relationship between any two raw scores depends upon which raw scores one selects. Of particular and great importance is the pronounced impact of nonlinearity of raw scores toward the low and high ends of the distribution of raw scores. For researchers interested in helping low-performing individuals "grow," the nonlinearity of the ogive should show that there are profound implications in the use of raw data. The implications are similarly profound for researchers who study high-performing students. Our view is that once researchers understand the nonlinear character of raw scores and grasp the impact of using raw scores, they will embrace the position that using raw scores of tests or surveys for statistical analysis of data is unacceptable.

Isabelle and Ted: Two Colleagues Conversing

Isabelle: Ted, can you tell me why this ogive is the right transformation to use to correct for this whole problem with nonlinear rating scale data and in fact the problems with raw data from tests too?

Ted: That is something that I have been working on for quite a while. First, for those who are interested in lots of details, books such as Best Test Design and Rating Scale Analysis present nice overviews of arguments that are presented here. Also, I found some introductory articles that outline the reasons for the model. Some of these books and articles are philosophical at times and highly mathematical. I think for most researchers such as me, there might be a little interest in the math, but since I have so many things to do, I really need to rely on experts such as Ben Wright to trust the model. A truly wonderful article "Measurement for social science and education: A history of social science measurement" by Ben Wright (1997) summarizes how experts came to understand that the Rasch model is the one that should be used.

Isabelle: Is there anything else you would like to tell me?

Ted: Actually, there is more to say. The ogive is really interesting and I have been using the ogive to explain to people the problem with raw scores. What I do is draw an ogive, and I put raw scores on the vertical axis, and then usually I will have a scale that looks similar to PISA on the horizontal scale. That means I might have a minimum of 200 and a maximum of 800. I pretend that we have two scores for two people, one score at the start of the year, and one score at the end of the year. I make sure to pick one person who is a low performer at the start of the year, and my second person is in middle of the pack. Then I show how the growth in the same number of raw score points from pre to post is quite different when measures are used instead of raw scores. Usually people are flabbergasted and ask me if this may have impacted data they evaluated in the past. When I tell them it probably has, they usually are concerned, but I tell them the main thing is just to take the time to carefully prepare their nonlinear data for statistical tests by using Rasch measurement techniques. Sometimes they will return to something I talked to them about when I first met them, and they will ask how do the experts know that the Rasch model is the one to use. I then usually explain that researchers have shown that to "do" measurement, the Rasch model is the only model that addresses a number of requirements of measurement.

Keywords and Phrases

Ogive
Raw scores
Measures

Logits
Linear
Nonlinear
Linear transformation

The relationship between raw scores and measures is nonlinear; therefore, raw scores must be converted to measures and then those measures used for parametric statistical tests.

Two respondents who differ by one raw score point on a test do not necessarily represent the same difference in ability as two other respondents who also differ by one raw score point at another point on the test. The meaning of a difference of one raw score point will depend on where respondents are along the metric.

Converting Rasch measures that might range from –3.0 to 3.0 logits to measures that might have a mean of 500 and a SD of 100 is not that different from converting Fahrenheit temperatures to Celsius temperatures.

Potential Article Text

Data were collected from a sample of 900 patients who exhibited a range of symptoms associated with autism. The scale of Timler (2011) was utilized for the collection of data. The scale contains 119 rating scale items; each item has 6 rating scale steps (coded as 1, 2, 3, 4, 5, 6). Data were evaluated utilizing the Rasch model (Rasch, 1960) because rating scale data are not linear and therefore must be converted to linear measures prior to analysis with parametric statistical tests.

Figure 11.9 presents the relationship between raw data and the linear Rasch measures. The vertical axis presents all possible raw scores for respondents (low of 117 and high of 702). The horizontal axis provides the range of measures for respondents. The scale ranges from a low of around –6 (equivalent to a raw score of 117) to a high of almost 7 (equivalent to a raw score of 702). Of particular importance is the nonlinear relationship of raw scores and measures. Many of the respondents who have numerous symptoms of autism are low on the scale. Thus, readers should be able to "see" that to evaluate such patients would be particularly problematic should raw scores be utilized.

Quick Tips

Winsteps Table 20 presents every possible raw score for an instrument. Also provided is every possible Rasch measure. In an analysis you might not observe all potential raw scores (and thus see all possible person measures reported in a person measure table). This table provides all potential raw scores and all potential measures.

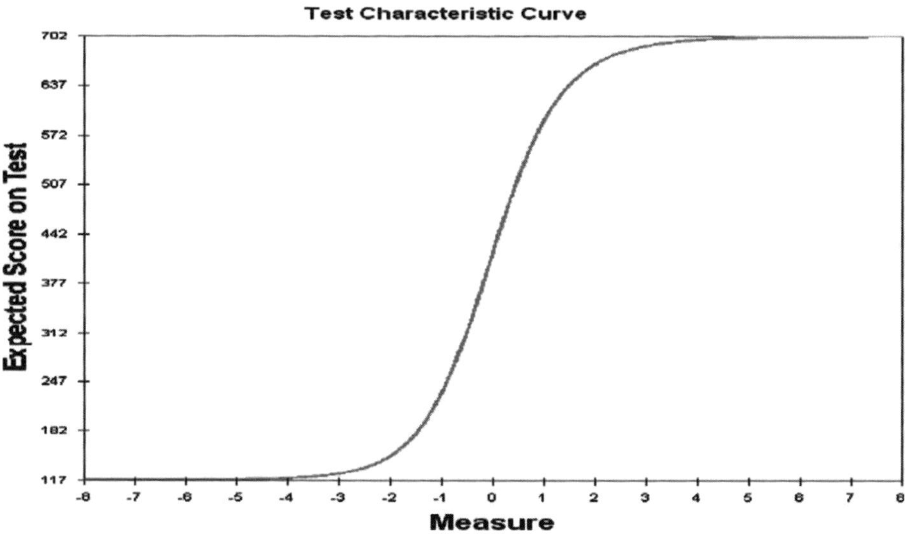

Fig. 11.9 The relationship between raw scores and Rasch measures for an instrument measuring aspects of autism

Table 20 also provides a plot of each raw score and each measure. The plot that results is called the logistic ogive. The portions of the plot that are curved are portions where there is a strong nonlinear relationship between the linear Rasch measures and the nonlinear raw scores.

You can show that raw scores are nonlinear: First plot two persons who differ by a set number of total raw score points (e.g., 1 pt), and pick a very low raw score total. Now compute their measure by using the graph (just draw a horizontal line from the person measure on the horizontal axis to the ogive; then at the point at which your line intersects the ogive, draw a vertical line that intersects the measure axis).

Then, plot two persons who differ by the same raw score amount for your first two people. But, pick people who earn a raw score about in the middle of what is possible for the instrument. Plot those two people and compute the measure of these two people. This plot will allow you to see visually and mathematically that the meaning of the same raw score difference does not mean the same difference in measure throughout the range of possible raw scores.

Data Sets: (go to http://extras.springer.com)

cf 13 GCKA Items
n75 Fall 2011Excel Jordan Data for activity o-give chp
cfjordan0givechp

Activities

Activity #1

We provide a ready to run control file named "cf 13 GCKA Items" for this activity. Please run the control file and generate a raw score-measure table as well as an ogive. Compute the difference in raw score and measure between Joe who earned a 1/13 and Bob who earned a 2/13 on the test. Also compute the difference in raw score and measure for Mark who earned a 6/13 on the test and Rich who earned a 7/13 on the test.

Answer: Below we provide the ogive and the score-measure table (Fig. 11.10).

	Raw score	Measure
Joe	1/13	−3.10
Bob	2/13	−2.20
Mark	6/13	− .15
Rich	7/13	.26

Joe and Bob's raw score difference is 1 and measure difference is .90 logits.
Mark and Rich's raw score difference is 1 and measure difference is .41 logits.

Activity #2

Write a brief paragraph in which you present the results of your computation in Activity 1 and explain why the results are important for research.

Answer: A 13-item multiple-choice test was administered to 75 respondents and Rasch analysis was performed, in part, to compute person measures. It is critical that person measures, not raw scores, be used for any subsequent parametric statistical analysis of data. The nonlinear nature of raw scores can be seen by comparing the raw scores of pairs of respondents with the Rasch linear measures of the same pairs of respondents. For example, the difference between two low performers on the test (1 item correct, 2 items correct) is expressed by a difference of .90 logits. Comparing two other students who also differ in number of items correct (6 items correct, 7 items correct) reveals a difference of .41 logits. The fact that the same raw score difference (a difference of 1 raw score point) is expressed by different measure differences demonstrates that the raw score scale is nonlinear. The one difference (.90) is more than twice the difference (.40) between the two individuals.

Activity #3

Our colleague Saed Sabah provided us with an Excel data set (n75 Fall 2011Excel Jordan Data for activity o-give chp). The data are from 75 respondents who answered a rating scale survey. The coding of items was 1 = almost never, 2 = seldom,

```
TABLE 20.1 GEKA Content only w/o true false item ZOU863WS.TXT  Dec 22 14:59 2011
INPUT: 75 PERSON  25 ITEM  REPORTED: 75 PERSON  13 ITEM  2 CATS    WINSTEPS 3.73
--------------------------------------------------------------------------------

                        TABLE OF MEASURES ON TEST OF 13 ITEM
----------------------------------------------------------------------------
| SCORE  MEASURE    S.E. | SCORE  MEASURE    S.E. | SCORE  MEASURE    S.E. |
|------------------------+------------------------+------------------------|
|     0   -4.43E   1.86  |     5     -.58     .66  |    10     1.57     .71 |
|     1   -3.10    1.09  |     6     -.15     .64  |    11     2.14     .80 |
|     2   -2.20     .84  |     7      .26     .64  |    12     2.98    1.06 |
|     3   -1.57     .75  |     8      .67     .64  |    13     4.27E   1.84 |
|     4   -1.04     .69  |     9     1.10     .66  |                        |
----------------------------------------------------------------------------
CURRENT VALUES, UMEAN=.0000 USCALE=1.0000
TO SET MEASURE RANGE AS 0-100, UMEAN=50.9248 USCALE=11.4952
TO SET MEASURE RANGE TO MATCH RAW SCORE RANGE, UMEAN=6.6202 USCALE=1.4944
Predicting Score from Measure: Score = Measure * 1.7324 + 6.5095
Predicting Measure from Score: Measure = Score * .5623 + -3.6602

        RAW SCORE-MEASURE OGIVE FOR COMPLETE TEST
       -+-----+-----+-----+-----+-----+-----+-----+-----+-----+-----+-
E    13 +                                                        E    +
X    12 +                                                *            +
P    11 +                                           *                 +
E    10 +                                       *                     +
C     9 +                                     *                       +
T     8 +                                  *                          +
E     7 +                                *                            +
D     6 +                             *                               +
      5 +                           *                                 +
S     4 +                        *                                    +
C     3 +                     *                                       +
O     2 +                 *                                           +
R     1 +           *                                                 +
E     0 +   E                                                         +
       -+-----+-----+-----+-----+-----+-----+-----+-----+-----+-----+-
       -5    -4    -3    -2    -1     0     1     2     3     4     5
                  MEASURE
```

Fig. 11.10 Raw score-measure table and ogive plot for Activity 1

3 = sometimes, 4 = often, and 5 = almost always. For the purposes of this activity, assume that all items define the same trait. Construct a control file and review the score-measure table and the ogive.

Answer: We provide the control file that one can make, but try to do so on your own. The name of the file that we made is cfjordan0givechp. Part of that control file is provided below, with some comment lines edited out.

```
&INST
Title= "n 75 Fall 2011 Excel Jordan Data for Activity
; ogive chp.xls"
ITEM1 = 1 ; Starting column of item responses
NI = 8 ; Number of items
NAME1 = 10 ; Starting column for person label in data
; record
NAMLEN = 7 ; Length of person label
XWIDE = 1 ; Matches the widest data value observed
```

```
CODES = "12345 " ; matches the data
TOTALSCORE = Yes ; Include extreme responses in reported
; scores
@gender = 1E1 ; $C10W1
@gpa = 3E6 ; $C12W4
&END ; Item labels follow: columns in label
C1 ; Item 1 : 1-1
C2 ; Item 2 : 2-2
C3 ; Item 3 : 3-3
C4 ; Item 4 : 4-4
D1 ; Item 5 : 5-5
D2 ; Item 6 : 6-6
D3 ; Item 7 : 7-7
D4 ; Item 8 : 8-8
END NAMES
55554455 2  2.8
```

Activity #4

Using the score-measure table and the ogive for Activity #3, conduct a comparison of respondents to show that those who provided middle-of-the-road ratings to most items should not be confidently compared to those who provided very high ratings for most items.

Answer: Below we provide the score-measure table as well as the ogive from the analysis of the data. The lowest potential raw score for a respondent who has answered all items is a raw score of "8" (8 items, lowest rating possible is a 1, $8\times1=8$). The highest potential raw score for a respondent who answered all items is a raw score of "40" (8 items, highest rating possible is a 5, $8\times5=40$). If one selects two respondents who have a middle-of-the-road raw score (e.g., 20 and 23), we see that the raw score difference is 3 raw score points ($23-20=3$) and that the difference in measures is .62 [$-.37-(-.99)=.62$]. Now let's select two respondents who exhibit high ratings (e.g., 40 and 37). These two respondents have a raw score difference of 3 ($40-37=3$) and a difference in measure of 2.58 ($5.92-3.34=2.58$). The difference in measure of the two high respondents is over 4 times ($2.58/.62=4.16$) the difference in measure of the two middle-of-the-road respondents! Clearly a comparison in which respondents would be high on the raw scale cannot be compared to those in the middle of the scale. Another way to look at the problem with raw scores in this activity is to look at the difference of 2.58 in the middle of the scale. The difference of 2.58 logits between the two high rating respondents is about the same magnitude of logits as the difference between someone who had a total rating of 25 and a person who had a rating of 13 [$.07-(-2.47)=2.54$]! The difference in these two respondents is 12 raw score points. Thus 12 raw score points represents the same difference along the trait as a difference of 3 raw score points! (Fig. 11.11)

```
TABLE 20.1 n 75 Fall 2011 Excel Jordan Data for  ZOU874WS.TXTv Dec 23  9:42 2011
INPUT: 75 PERSON  8 ITEM  REPORTED: 74 PERSON  8 ITEM  5 CATS       WINSTEPS 3.73
```

TABLE OF MEASURES ON TEST OF 8 ITEM

SCORE	MEASURE	S.E.	SCORE	MEASURE	S.E.	SCORE	MEASURE	S.E.
8	-5.46E	1.83	19	-1.19	.44	30	1.26	.50
9	-4.25	1.01	20	-.99	.45	31	1.50	.50
10	-3.53	.73	21	-.79	.45	32	1.76	.51
11	-3.08	.61	22	-.59	.46	33	2.02	.52
12	-2.75	.55	23	-.37	.46	34	2.30	.54
13	-2.47	.51	24	-.16	.47	35	2.60	.56
14	-2.22	.48	25	.07	.48	36	2.94	.60
15	-2.00	.47	26	.30	.48	37	3.34	.67
16	-1.79	.45	27	.53	.49	38	3.85	.78
17	-1.58	.45	28	.77	.49	39	4.65	1.05
18	-1.39	.44	29	1.01	.49	40	5.92E	1.85

```
CURRENT VALUES, UMEAN=.0000 USCALE=1.0000
TO SET MEASURE RANGE AS 0-100, UMEAN=47.9816 USCALE=8.7800
TO SET MEASURE RANGE TO MATCH RAW SCORE RANGE, UMEAN=23.3541 USCALE=2.8096
Predicting Score from Measure: Score = Measure * 3.6264 + 15.9754
Predicting Measure from Score: Measure = Score * .2668 + -4.2614
```

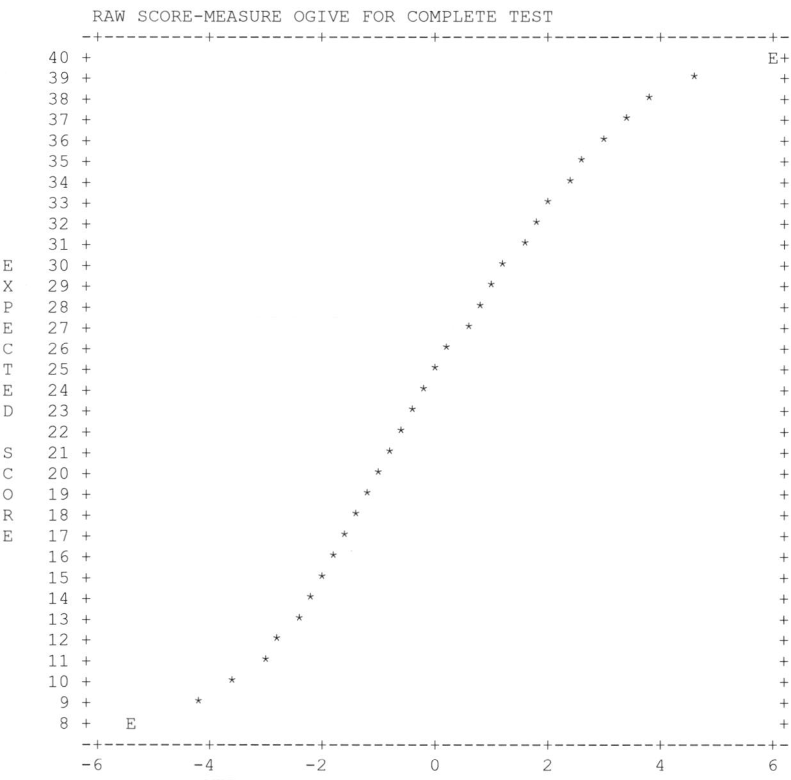

Fig. 11.11 Raw score-measure table and ogive plot for Activity 4

Activity #5

An excellent and thorough discussion of Rasch measurement is provided in University of Chicago Measurement, Evaluation and Statistical Analysis (MESA) Memo 62, which can be retrieved at http://www.rasch.org/memo62.htm. A component of this memo employs an ogive to explain why raw scores are not linear measures. Read this memo and find other aspects that help you to further understand Rasch measurement, in particular for this chapter the importance of the ogive.

Activity #6

The US state of Ohio utilizes Rasch and Winsteps for the analysis of high-stakes test data. Data from the analysis of October 2011 reading test data are supplied by the state on the Ohio Department of Education website (Office of Assessment, Ohio Department of Education. October 2011 Administration of the Ohio Achievement Assessment, Grade 3 Reading Test Statistical Summary). Create a plot of the raw score data against the scale scores. What do you predict you will see? The left-hand column presents the possible raw scores on the test and the possible measures. If a student gets a raw score of 1, they have a measure of 270.

0	251
1	270
2	290
3	302
4	311
5	319
6	325
7	331
8	336
9	340
10	344
11	348
12	352
13	355
14	359
15	362
16	365
17	367
18	370
19	373
20	376
21	378
22	381
23	383
24	386
25	388
26	390

(continued)

(continued)

27	393
28	395
29	398
30	400
31	402
32	405
33	407
34	410
35	413
36	415
37	418
38	421
39	424
40	428
41	432
42	435
43	440
44	445
45	451
46	459
47	470
48	488
49	506

Answer: Plotting that data will provide an ogive.

References

Linacre, J. M. (2006). Bernoulli trials, Fisher information, Shannon information and Rasch. *Rasch Measurement Transactions, 20*(3), 1062–1063.

Rasch, G. (1960). *Probabilistic models for some intelligence and attainment tests*. Copenhagen: Danish Institute for Educational Research.

Wright, B. D. (1997). *Measurement for social science and education. A history of social science measurement*. University of Chicago MESA (Measurement, Evaluation and Statistical Analysis) Research Memo 62.

Wright, B. D., & Stone, M. H. (1979). *Best test design*. Chicago: MESA Press.

Additional Readings

A very good comparison of the Rasch model and the 2P and 3P IRT models.

Wright, B. D. (1992a). IRT in the 1990s: Which models work best? *Rasch Measurement Transactions, 6*(1), 196–200.

Wright, B. D. (1992b, April). *Opening remarks in an invited debate with Ron Hambleton*. Presented at the annual meeting of the American Educational Research Association, San Francisco, CA. Retrieved April 22, 2013, from http://www.rasch.org/rmt/rmt61a.htm

Chapter 12
Some Wright Map Nuances: How to Set the Probability of Success at 65 % (or Whichever Percentage You Wish to Choose)

Isabelle and Ted: Two Colleagues Conversing

Ted: I know now that when an item measure and a person measure exhibit the same value, there is a 50–50 (50 %) chance that the person answers the item correctly. That makes sense to me, because it's like a teeter-totter. If a person is standing on a teeter-totter exactly at its fulcrum, then the teeter-totter will not tip either way. The same is true if a person measure and an item measure exhibit identical values. If the item is a right/wrong item, there is a 50–50 chance that such a person will answer the item correctly. But I do not understand two things. First, why might there be different sets of people above the hardest item? Second, I know that in many assessments the probability is set at around 65 % (PISA does 62 %), not 50–50, when a person and an item have the same measure value. Why do researchers do this?

Isabelle: Ted those are great points, and I really like your analogy with the teeter-totter! When I look at basic Wright Maps, I sometimes see that pattern of multiple groups of persons with measures above the hardest item measure. I ask myself: Shouldn't all the people who are above the hardest item get all the items right? But looking at the Wright Map, I saw a number of groups of respondents above the hardest item. It took me a while to understand what I was seeing and not seeing in the Wright Map. Here is how it might look in a Wright Map (Fig. 12.1).

Ted: What about this 65 % chance (or thereabouts) of correctly answering an item that is used in some international assessments? How do you alter the Wright Map, what changes can you make in the control file, and why do they make this change?

Isabelle: We will get to that in this chapter. The change in the control file is not difficult. Regarding why research groups might change the percentage of chance of correct, we will consider that as well. The main thing to tell you just now is that when you raise the percentage of chance of correctly answering an item, then of course you are able to say that you predict with more certainty that a respondent did or did not correctly answer the item. Actually, when we change the percentage from, say 50 % (50–50) to 65 %, some of the patterns that take some time to explain in the Wright Map will disappear. For example, one might have only one group of respondents above the hardest item.

Isabelle: When you are comparing a Wright Map set at 50/50 and say 65 %, you can think of this change as one in which the difficulty of items is increased. Visually, this looks like keeping the respondents in the same spot but sliding all the items on the right side of the Wright Map upward by the same amount. On the other hand, you can visualize making a change to 65 % as keeping the items in the same spot but moving persons down on the Wright Map (Fig. 12.2).

W.J. Boone et al., *Rasch Analysis in the Human Sciences*,
DOI 10.1007/978-94-007-6857-4_12, © Springer Science+Business Media Dordrecht 2014

```
XXX  |
     |
XX   |
     |
XXXX |
     |
     |Item 9 (Most Difficult Item)
```

Fig. 12.1 The top portion of a Wright Map, in which item #9 is the hardest item, but three separate groups of respondents are above the hardest item

```
XXX  |
     |Item 9
XX   |
     |
XXXX |
     |
     |
```

Fig. 12.2 A Wright Map using the same data as in Fig. 12.1, changing the probability to 65 %. The spacing of respondents is maintained

Introduction

In a number of chapters, we have presented Wright Maps, item measures, and person measures and discussed the fact that when an item measure and a person measure are identical, that person has a 50 % (50–50) chance of correctly answering that item. This relationship is a fundamental part of the Rasch model, and considering the interaction of persons and items in this manner is the easiest way to start learning and using Rasch measurement techniques. In this chapter, we focus on a slight change that is implemented in the relationship between items and persons. We will introduce the change and focus on how it is made for multiple-choice tests. Sometimes one may want to be more confident than 50/50 as to a person's response. For instance, a group administering a very important certification test for physicians may want to be 80 % certain that a "passing" physician is able to answer an item.

Why Make This Change?

First, why might an analyst investigating a multiple-choice data set decide not to use the 50–50 relationship between items and persons that Winsteps normally computes? In some testing situations, researchers are interested in being more certain than 50–50 that a student will correctly answer the question. One reason, from our vantage point, focuses on political considerations. If one is presenting test results to stakeholders, such as policy makers and politicians, who may know little about psychometrics, concern may be voiced about only knowing at a 50–50 level if a person has correctly answered an item. Another reason why one might choose a "probability of success" above 50/50 focuses on the very reasonable view that one simply wants to be more certain of a correct response. We can imagine of many situations in which one might want to select a higher probability of predicting a success.

Formative Assessment Checkpoint #1

Question: Is it difficult to change what it means for a person and an item to be at the same logit measure level?

Answer: No. It only requires the insertion of four lines of control file code that we provide.

How to Make This Change?

How, then, does one change all the item and person measures so that when an item and person exhibit the same measure value (i.e., are located at the same measure on a Wright Map), there is, say, a 65 % chance or even an 80 % chance that the student will correctly answer the item? Setting the specific meaning for a person and item to be at the same logit value (e.g., deciding that when a person and an item are at the same logit value, it means there is a 70 % chance of the person correctly answering the item) requires two simple steps. Step 1 makes use of the Rasch equation that was introduced previously herein, and step 2 adds a few simple lines in a Winsteps control file.

Step 1: Recall that we use probability in Rasch measurement. For each item and each person responding to an item, there are two probabilities to keep in mind. First, a probability that the respondent will correctly answer the item is expressed as a decimal or percentage. Second, there also exists a probability that the respondent will incorrectly answer the item, which is also expressed as a decimal or percentage. These two decimals or percentages add up to 1 or 100 %, respectively. So, if a person has a 50–50 chance of answering an item correctly, then one simply expresses the relationship of decimal chance correct and decimal chance wrong as .5 and .5, which add up to 1. Of course, the sum of the percentage chance correct (50 %) and the percentage chance incorrect (50 %) is 100 %. If a respondent completes an item and a researcher computes that the respondent has a 65 % chance of correctly answering the item, then this relationship between chance correct and chance incorrect would be expressed as .65 and .35, respectively (in the case of using percentages, it would be 65 and 35 %, respectively). Again, these two values will always add up to 1 or to 100 %.

To enable Winsteps to change the percentage chance of success on an item (from 50 % to another percentage) when the person measure and item measure are identical, the analyst must first choose the percentage of success needed on an item. Our example will use .65. Other values might be .70 and .80. Once a value is selected, the analyst must use an equation to compute a number that will be typed into a control file before an analysis is conducted with Winsteps. That equation is presented below for the selection of a 65 % chance of success on an item when a person has the same measure as an item:

$$\ln\left((100-65)/65\right) = -0.6190392084.$$

This result will be added to the control file to tell Winsteps how to analyze a correct response. A handheld calculator can be used to compute the value of

$$\ln\left((100-65)/65\right) = \ln\left(35/65\right) = -.619.$$

Step 2: The second step is taking the computed value (−0.619) and entering it into a Winsteps control file. In this case, only four lines are added using the command line "SAFILE=." These lines are provided in Fig. 12.3 for our example of changing the chance of correct from 50–50 to 65–35 (where if a person has the same measure as an item, it means there is a 65 % chance of her or him answering the item correctly). Briefly, these 4 command lines allow the researcher to examine the probability of answering items correctly at various levels. If readers are curious about the details of the lines, then read the "SAFILE" section of the Winsteps manual. The important point for Rasch beginners is that when one decides to set the probability values to different levels, then these four lines need to be entered into the control file. Moreover, the lines will always be identical except for one part of one of the four lines. This is where readers will have decided on another set of chances of correctly and incorrectly answering the item; thus, the resulting value to be subtracted from 1 will not be −0.619 (Fig. 12.3).

In Figs. 12.4 and 12.5, we provide two Wright Maps that were constructed using identical geoscience test data from our colleague Kathy Trundle. Looking at these Wright Maps, can you identify which map displays the data using a 50 % (50–50) chance of correctly answering an item? A 65 % chance? Figure 12.4 presents the data from an analysis with the 65–35 criterion. To figure this out, readers should

```
UASCALE=1 ; this tells the program the anchoring is in logits
SAFILE=*
0 0
1 - 0.619 ; ln ((100-65)/65) which is ln (35/65) which is -0.619
*
```

Fig. 12.3 Four lines to modify a control file to change the chances of correctly and incorrectly answering an item from 50–50 to 65–35

```
TABLE 12.2 GEKA Content only w/o true false item ZOU215WS.TXTr Jun 17 15:04 2011
INPUT: 413 PERSON  48 ITEM  MEASURED: 412 PERSON  42 ITEM  2 CATS WINSTEPS 3.70.6
--------------------------------------------------------------------------------

           PERSON - MAP - ITEM
               <more>|<rare>
    2                +
                     |
                     |
                     |  Q79,
                     |
                     |
                     |  Q23,        Q39,        Q65,
                     |  Q55,
                     |
                     |  Q25,        Q35,
    1               +S  Q3,
                     |  Q19,
                     |
                  .  |  Q37,        Q67,
                  .  |  Q31,
                  . T|  Q45,        Q87,        Q91,
                ##   |  Q21,        Q53,
               ###   |  Q43,        Q83,        Q85,
           #######   |  Q93,
            .#####   |
    0           ## S+M  Q27,        Q59,        Q77,
        .#########   |  Q7,
       .##########   |  Q13,        Q29,
         .#######    |
       .##########   |  Q71,
         .########   |  Q81,
     ############# M|  Q15,        Q57,        Q89,
            .#####   |  Q11,        Q69,        Q75,
           #######   |  Q51,        Q73,
      .###########   |
   -1    .########  +S
             ##### S|
              ####   |  Q47,
              ####   |
              .###   |  Q33,
                ##   |
               ###   |  Q41,
                .# T|  Q63,
                     |
                  .  |
   -2             #  +
                     |T
                     |
                     |
                  .  |
                     |
                     |

EACH "#" IS 3. EACH "." IS 1 TO 2
```

Fig. 12.4 The 21 (7×3) students located at the ability level noted by the horizontal arrow have a 65 % chance of correctly answering the items Q51 and Q73. Note that there are nine test items located below the mean (*M*) person measure

note that in Fig. 12.4, for most respondents, fewer items are predicted as having been correctly answered by each respondent than is the case in Fig. 12.5. For example, observe that the number of items above the highest performing student is less for Fig. 12.4 compared to Fig. 12.5. There is one additional nuance to point out,

```
TABLE 12.2 GEKA Content only w/o true false item ZOU377WS.TXTr Jun 17 15:15 2011
INPUT: 413 PERSON  48 ITEM  MEASURED: 412 PERSON  42 ITEM  2 CATS WINSTEPS 3.70.6
--------------------------------------------------------------------------------

           PERSON - MAP - ITEM
<more>|<rare>
    2                +
                     |
                     |
                     |  Q79,
                     |
                     |
                  .  |  Q23,       Q39,       Q65,
                  .  |  Q55,
                  . T|
                 ##  |  Q25,       Q35,
    1           ###  +S Q3,
             .#####  |  Q19,
                .##  |
            .######  |  Q37,       Q67,
         .######### S|  Q31,
          #########  |  Q45,       Q87,       Q91,
          .########  |  Q21,       Q53,
           .#######  |  Q43,       Q83,       Q85,
       .###########  |  Q93,
         ########## M|
    0      .#######  +M Q27,       Q59,       Q77,
           ########  |  Q7,
       .###########  |  Q13,       Q29,
             .#####  |
           .####### S|  Q71,
              .####  |  Q81,
               ####  |  Q15,       Q57,       Q89,
                .##  |  Q11,       Q69,       Q75,
                .##  |  Q51,       Q73,
                ###  |
   -1             # T+S
                  #  |
                  .  |  Q47,
                     |
                  #  |  Q33,
                     |
                     |  Q41,
                  .  |  Q63,
                     |
                     |
   -2                +
                     |T
                     |
                     |
<less>|<frequ>
 EACH "#" IS 3. EACH "." IS 1 TO 2
```

Fig. 12.5 The 12 (4×3) students located at the ability level noted by the horizontal arrow have a 50/50 chance of correctly answering the items Q15, Q57, and Q89. Also note that there are more items below the mean (*M*) person measure than when the plot with 65 % chance of success and the mean (*M*) person performance are reviewed. This makes sense in that by specifying a 65 % chance of success, one is requiring more certainty in terms of probability. For this plot, there are 20 test items below the mean person measure

something that has tricked colleagues in the past. Although the pattern of respondents looks a little different for Figs. 12.4 and 12.5, this difference is due only to the plotting of respondents in the Wright Map. If an analyst needs to prove to a colleague that the difference in the pattern of respondents is simply due to a graphing

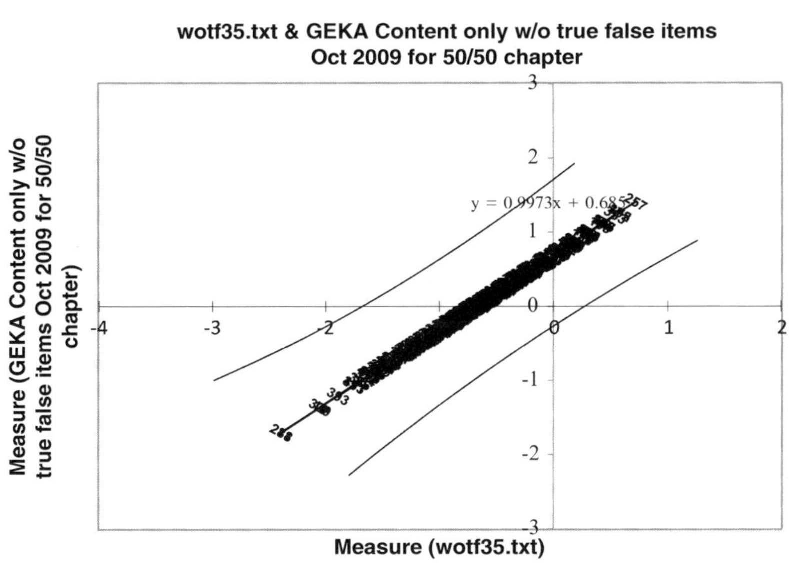

Fig. 12.6 A plot of the person measures computed for Figs. 12.4 and 12.5. The straight line shows that the change in the distribution of persons in the Wright Map is simply the manner in which the data are graphed due to the limited space available for graphing

protocol, then he or she needs only to plot the measures of respondents computed for both analysis techniques (50–50 and 65–35). When cross plotting the measures of the two techniques, a straight line will result. The straight line shows that the few differences in the person distributions are due only to graphing. Figure 12.6 presents a graph of the person measures presented in Figs. 12.4 and 12.5. The important point to note is when one alters the probability of success, it is the relative location of the persons with regard to items which changes. A comparison of the performance of males and females using the .50 chance of success could be computed and that p value will be the same as that which is computed using the performance of males and females using the .65 chance of success.

Formative Assessment Checkpoint #2

Situation: I have completed an analysis with a control file with the probability of success set at .50. I have also completed an analysis with the identical data, with the probability of success set at .65. But, when I put the item Wright Maps from the two analyses next to each other, the pattern of items seems to be a little different, and also the pattern of persons seems to be a little different.

Question: Is something wrong?

Answer: No. How items and persons are plotted can present the appearance of a change in a pattern. However, if one plots the person measures from the two analyses against each other, one will see only that all items have been shifted by the same amount (either harder or easier depending upon the change one has made in the control file). If you want you can also think of the change as being a shift of the same amount (also known as a linear transformation) in all person measures.

We have shown readers a technique for altering the chance of success of a person answering an item from 50–50 (0.5 or 50 %) to other values. Why would a researcher want to make such a change? How does making such a change alter, and not alter, what we have presented thus far? In many contexts it is important to be able to predict with greater surety (more than 50–50 or 50 % sure) that a student will correctly answer an item. This is why many assessments use values higher than 50/50. For example, PISA uses 62 %.

It is important to point out that using a different confidence level of a correct answer to item (e.g., 50/50 chance, 65/35 chance, 62/38 chance) does not change the techniques we have presented thus far in the book. For instance, if a researcher has interpreted the ordering and spacing of items on a Wright Map (and the researcher has only spent time thinking about the items), the interpretation of those items will not be affected by utilizing a different confidence level, say 65 % certainty of correctly answering an item, in contrast to the default level of 50/50. The ordering of items and the gaps between items remain unchanged. If the ordering and gaps of items appear different, such appearances result from a graphing issue as we have demonstrated in Fig. 12.6.

A second point needs to be made about altering the probability of success of a correct answer: The statistical calculations carried out with the item data set will also be unaffected. For instance, if a statistical comparison of the mean difficulty of items with pictures and items without pictures has been made, then the same p value will result from this comparison no matter if one used the default value of 50/50 or altered the value to that which we have been utilizing in this chapter (65/35).

A final point for readers focuses on what does and what does not change with an alteration in the change in the probability of success on items. A change in the probability level does change what items will be predicted to have been correctly answered by respondents and groups of respondents. However, when one is investigating only differences in person measures (and not thinking about item measures), there will be no changes in the conclusions, for example, if male measures and female measures from the Fig. 12.4 analysis have been compared and the results suggested a statistically different mean measure as a function of gender of .50 logits (SD = .321 logits, $p = .01567$). The exact same values would result if a statistical comparison were made of male and female students for the analysis of data presented in Fig. 12.5. The only difference is how these differences would be interpreted in light of items that one group or another would be predicted to answer correctly. This is the result of being more certain of a correct response in one scenario in contrast to another scenario.

Formative Assessment Checkpoint #3

Question: I have conducted an analysis using the default value of 50/50 on a data set. Will all of my calculations need to be redone if I change the confidence "of a respondent correctly answering an item" I want to use?

Answer: No. The calculations comparing items will be the same. The comparisons of the same persons will be the same. The only change will be any comment about the types of items that would have been typically correctly answered by a person or group of persons.

Throughout this book we have used the rating scale data from instruments to help readers master introductory Rasch analyses. The topics of this chapter are equally useful for rating scales, although we have rarely seen individuals consider probabilities other than 50/50 when evaluating rating scales. At the end of this chapter's Quick Tip section, we provide a step-by-step procedure by which readers can change the probability of success for a rating scale.

An End-of-Chapter Thought

In recent years, researchers in education have used Rasch analysis techniques to examine large-scale assessments of students. Those in charge of PISA have selected 62 % probability correct for those students and items that exhibit identical Rasch logit values. Thus, if Jack has a measure of 2.0 logits and items 20 and 11 of a PISA test have values of 2.0 logits, Jack has a 62 % probability of success when he attempts items 20 and 11. Certainly other values can be selected for equal probability of success, but it may make the most sense for researchers in education to make use of the decisions that were made for PISA when data are evaluated. Certainly there were deep, extensive, and wide-ranging discussions with respect to the selection of a PISA probability value. In our view, it makes little sense to reinvent the wheel. Also, since most educators are simply attempting to use Rasch to help them explore an issue of interest to them (e.g., using Rasch to develop an instrument to measure competency in biology), it makes sense to rely upon the decisions made by assessments such as PISA. Regarding our purpose herein, we are presenting, we hope, a helpful introduction to Rasch measurement and analysis; therefore, there are many extensions of topics we have not presented. We do, however, wish to mention that, for the analysis of survey data, it is possible to set the probability of success at 62 % instead of 50 %. Another advantage that we have found is that, when we have designed professional development workshops for teachers, we have learned that presenting Wright Maps using the 62 % threshold makes more sense to teachers, in that the interaction of students and items appears to be more realistic with respect to their observations and experiences in classrooms.

Ted: Two Colleagues Conversing

Isabelle: Ted, do you think you understand the procedures to change Wright Maps so that the probability of success on items is not 50/50?

Ted: I think so. The first thing, which was the easiest for me, was the sequence of mechanical steps that I must take to change the probability of success from 50/50 to, for example, 65/35. That made sense. Now, I could set any value that I wanted, but I suspect that I will almost always use 50/50 or 65/35.

Isabelle: What was the harder part?

Ted: The harder part for me was to make the mental change from 50/50 to something else. As I have learned Rasch, it really helps me to think of 50/50 because 50/50 is the result of flipping a coin. That is something that many people, myself included, are used to thinking about. I think another reason for my confusion at times had to do with how I taught myself to think about Wright Maps. For instance, if a person has the same measure as an item measure, then there is a 50/50 chance of that person getting that item correct. Item measures below the person measure are items I would predict the person would correctly answer, and item measures above the person measure are items I would predict that person would not correctly answer. To help me understand this use of 65 % probability of correctly answering an item, I first realized that all the techniques that I had used in the past to understand the Wright Map still worked. If I find a person and an item at the same measure, then I know there is a 65 % chance of correctly answering the item. And, I still know that the items below that person are those items the person is more likely to have answered correctly. Now we say for those items the person has over a 65 % chance of correctly answering the items. And, just as was the case with all the work with 50/50, the farther an item was below a person, the higher the likelihood of the person answering the item correctly.

Isabelle: Well explained. What about the use of 65 %, how are you doing with that?

Ted: I am okay with that. First, it makes sense to me that in some cases one might want to be a little more assured that a person would have gotten an item correct. I also looked over some of the PISA technical reports, and how they explained their use of 62 % made sense to me.

Isabelle: What about surveys and tests that have partial credit? Tell me what you think about that!

Ted: That is a tough one; I struggled over that one for quite a while. What I finally realized is that, yes, I could use this change in confidence level for both surveys and tests with partial credit items. Let me give you an example. If I had a physics item that could be worth 0, 1, 2, and 3 points, I would just talk about how we might want to be 65 % confident that a student had at least achieved 2 points on the item. I might even start an explanation off with items that are just right/wrong, and then I would move to partial credit items.

Ted: Also there were some other things I thought of too. In medicine, for instance, there might be a particular probability of success that one might want for a patient who is being assessed using a scale. So it makes sense that this issue of increasing the probability value would be very useful. Also in market research I can imagine that there are situations in which a high probability of success would be called for. Maybe before a company spends a large amount of money on an advertising campaign, they wish to be able to look at a Wright Map and compare items and persons at an 80 % confidence level?

Keywords and Phrases

Probability of a respondent successfully answering an item correctly

Potential Article Text

Data were collected from a sample of 1,000 8th grade students using public release items from past TIMSS and PISA tests. Data were evaluated using the Rasch model (Wright & Stone, 1979) and the Winsteps program (Linacre, 2011). Wright Maps were created from the computed person and item measures. Analysis conformed to a rule that a respondent's measure equal to an item's measure represented a 62 % chance of success for that respondent on that item. This threshold is utilized by PISA.

Quick Tips

Below is the code that would be inserted into a control file to set the probability of success to 65 %.

```
SAFILE=*
0 0
1-0.619 ; ln (100-65/65) which is ln (35/65) which is -0.619
*
```

A Quick Tip Addition

How to Set 62 % Confidence with a Rating Scale

1. Use the command SFILE= to create a file that will provide you with the "step calibrations" that will be used to anchor the rating scale steps in such a way as to change the expected success on items from approximately 50–62 %. This requires us to raise the item difficulty estimates. Since the item difficulties are centered on zero, raising the item difficulty estimates actually lowers the reported person ability estimates!

   We conducted an analysis on a rating scale data set in which we wanted to set the expected success to 62 %. This is what our SFILE looked like with the default expected success of around 50 %:

   ```
   ; STRUCTURE MEASURE ANCHOR FILE  Jul 14  8:02 2012
   ; CATEGORY  Rasch-Andrich threshold
     1      .00
     2     -.98
     3     -.25
     4     1.22
   ```

2. The next step is we needed to use an equation supplied to us by Mike Linacre to compute an "expected score."

We used the following equation to compute an "expected" score corresponding to 62 % success on the item:

$$1+(4-1)\times(.62)=1+(3)\times(.62)=1+1.86=2.86$$

3. Now we needed a table that would allow us to determine the "measure" for the "expected" score of 2.86. You can generate a table to do so, using the command GRFILE. The Winsteps manual says the following about what one gets by using GRFILE:

   …a file is output which contains a list of measures (x-axis coordinates) and corresponding expected scores and category probabilities (y-axis coordinates)

   When we used GRFILE= in our control file, a file was created that provides a large number of expected score and measures that corresponded to each expected score. Below we provide part of the file we created. From this table we can see that at a "measure" of .00 logits relative to the item difficulty, the expected score on the item is 2.55, which is about halfway up the rating scale. For our expected score of 2.86, we would have a "measure" of .44 (.44 is halfway between .40 and .48):

```
; PROBABILITY CURVES FOR SET AT 65% Ursprung_1909.sav
Jul 14 8:10 2012
ITEM    MEAS    SCOR    INFO      0       1       2       3
      ....
      1     .00    2.55     .73     .12     .33     .43     .12
      .....
      1     .32    2.78     .70     .07     .27     .47     .19
      1     .40    2.83     .68     .06     .25     .48     .21
      1     .48    2.89     .67     .05     .23     .48     .23
      1     .56    2.94     .65     .05     .22     .49     .25
```

4. Now in order to make the step calibrations .44 logits easier, we now go back to the data we computed in step 1, and we *subtract* .44 logits from each of the step 1 numbers, making the item appear to be .44 logits easier:

   $-.98-.44=-1.42$
   $-.25-.44=-.69$
   $1.22-.44=.78$

5. Now all we have to do is insert the following in our control file:

```
SAFILE=*
1   0
2   -1.42
3   -.69
4   .78
```

6. We can see what we have done by comparing Table 1 from the two analyses.

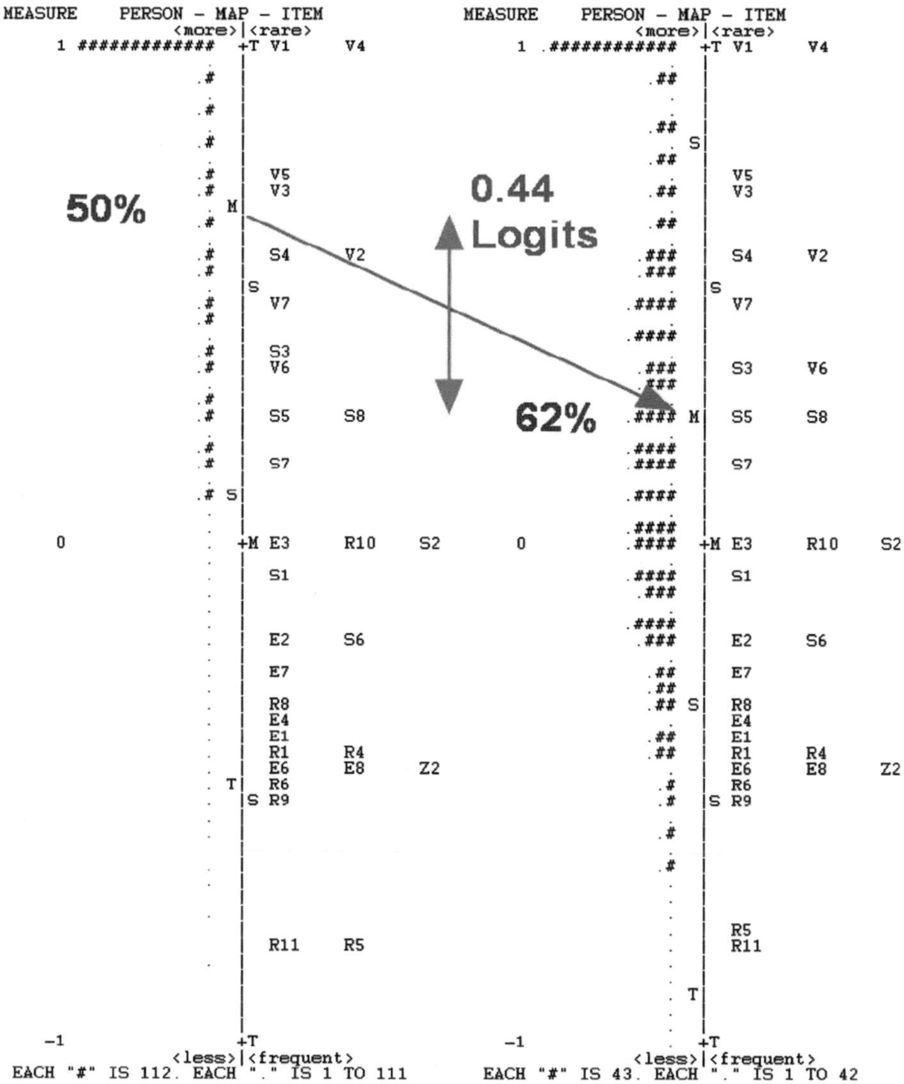

MEASURE PERSON - MAP - ITEM
<more>|<rare>
50%
0.44 Logits
62%
<less>|<frequent>
EACH "#" IS 112. EACH "." IS 1 TO 111
EACH "#" IS 43. EACH "." IS 1 TO 42

Data Sets: (go to http://extras.springer.com)

cf Chih-CheTai Chem Educ

Activities

Activity #1

Take a Wright Map that you have created for a test that you have developed (if you do not have your own data set, then use the Geology Earth Science Test data evaluated in this chapter). For the Wright Map, write out how you would explain that an item at the same logit value as a person would have a 50/50 chance of being correctly answered.

Answer: Pretend that a test taker Karen has the same logit measure as a test item 21 of the GCKA. A potential text might be the following: A respondent (Karen, measure 1.23 logits) has the same measure as item 21 of the GCKA. When a person has an identical measure as an item, this item is a good descriptor of the respondent's ability with respect to the latent trait as defined by the test items. This person has a 50/50 chance of correctly answering the item correctly. Think of this as akin to a diver attempting a dive that is exactly at his/her ability level. With such a dive, it really is a 50/50 shot that the diver will be able to complete the dive.

Activity #2

Pretend that you are teaching a class on using Rasch analysis. Write an explanation that you could provide to those attending your class as to why someone might want to change the threshold from 50/50 to a value of 62/38.

Answer: Class, we have talked quite a bit about how to read Wright Maps, and I think all of you can now pretty quickly read the maps. For multiple-choice test data, you now know that when an item measure is plotted below a person measure, the person has a greater than 50 % chance of correctly answering the item. Thus, we would predict the person will correctly answer the item. You also will remember that items plotted above a person on a Wright Map will be those items we would predict would not be correctly answered by the person. Finally, when items are at the same measure as the person, then we are really unsure if the person will get the item right or wrong; it is truly 50/50.

Okay, here is the deal; there are cases in which we want to be more sure than 50/50. For example, in medical certification, perhaps we want to be 80 % sure that candidates can correctly answer particular items. I have found an example from education that provides a good example of decision makers using a value higher than 50/50. Staffers at the Program for International Student Assessment (PISA) have decided to use 62 % as a threshold. Now if you think of it, and this is the way I thought about it, it makes sense that one might want to be a little more sure above 50/50. Below is drawn part of a potential Wright Map for the analysis of data.

```
      Q7

Brian Q20

      Q6
```

In this scenario, if the Wright Map was constructed using the 62 % threshold, then there exists a 62 % certainty that Brian correctly answered item 20. Our certainty about Brian and Q7 is less than 62 %. Finally, we can be higher than 62 % certain that Brian correctly answered item Q6.

Activity #3

We provide a control file using chemistry test data provided by our chemistry education colleagues Chih-CheTai and Keith Sheppard. The file is named cf Chih-CheTai Chem Educ. Create three copies of this control file and author additional lines of control file code to create three Wright Maps, one for a 62 % certainty, one for a 65 % certainty, and one for a 85 % certainty. Run the four control files (50, 62, 65, 85) and print out a Wright Map for each analysis. Place the 4 maps side by side. Can you see the changes in the location of persons in relation to items? From just reading each map, what seems to be the test item that is closest to the mean person of the data set (i.e., what item best describes the ability level of the mean respondent)?

Answer: Below parts of the Wright Maps to demonstrate such analysis Fig. 12.7. Recall that the "M" on the left side of the map marks the location of the mean person measure. The important aspects of these plots to note are that the mean person can be viewed as "moving down" to easier and easier item difficulties as the confidence in a correct answer is increased from 50–62 to 65–85. Note that the ordering of items remains the same. This means that the concept of what it means to increase along the trait does not change with changing the confidence of a correct answer by respondents (that should make sense!). The interpretation of what it means to have a particular ability level will change, as can be seen in these plots.

Activity #4

Using one of the chemistry education control files, change the confidence level to 90 and 95 %. Place the Wright Maps for 90 and 95 % side by side with the Wright Map for 65 %. What differences do you see?

Answer: Persons will appear to move downward in relation to items on the Wright Maps with increasing confidence levels.

```
50% Probability

  Higher Performers Harder Items

    2 XXXXXXXXXXX   +
                    |
                    |
          XXXXXX    |
                    |
              XX    |  q10
                    |  q18a
           XXXXX    |  q18b
            XXXX  M|S
           XXXXX    |  q3
    1         XXX   +
              XXX   |
         XXXXXXX    |  q11
               X    |  q7
           XXXXX    |
                    |
             XXX  S|   q15b

  Lower Performers  Easier Items

62% Probability

Higher Performers   Harder Items

                  S|
      XXXXXXXXXXX   |  q10
                    |  q18a
                    |S q18b
          XXXXXX    |
                    |  q3
    1          XX   +
           XXXXX    |
                  M|   q11
            XXXX    |  q7
        XXXXXXXX    |
                    |
             XXX    |  q15b

Lower Performers    Easier Items

65% Probability
Higher Performers   Harder Items
                  S|   q10
      XXXXXXXXXXX   |  q18a
                    |S q18b
                    |
                    |  q3
    1     XXXXXX    +
              XX    |
                    |  q11
           XXXXX    |  q7
            XXXX  M|
           XXXXX    |
             XXX    |  q15b
             XXX    |  q15a
         XXXXXXX    |  q2

Lower Performers    Easier Items

85% Probability
Higher Performers   Harder Items

                  S|S q18b
      XXXXXXXXXXX   |
```

Fig. 12.7 Segments of three Wright Maps modified to 62, 65, and 85 % probability of success

References

Linacre, J. M. (2012). Winsteps (Version 3.74) [Software]. Available from http://www.winsteps.com/index.html

Additional Readings

In this brief article Ben Wright shows some hows and whys for dichotomous test data.

Wright, B. D. (1992). Rasch model from ratio-scale counts. *Rasch Measurement Transactions, 6*(2), 219.

A key Rasch article in which Ben Wright lays out the reasons for the Rasch model.

Wright, B. D. (1977). Solving measurement problems with the Rasch model. *Journal of Educational Measurement, 14*(2), 97–116.

Chapter 13
Differential Item Functioning

Isabelle and Ted: Two Colleagues Conversing

Isabelle: *Ted I hear you chanting, it sounds like three letters. What's going on?*

Ted: *Easy does it Isabelle....DIF.....DIF.....DIF......that is short for Differential Item Functioning.*

Isabelle: *I've heard of that....why are you working with DIF?*

Ted: *Well, I have this 50-item test and this 25-item survey, and I want to make sure the test and the survey are not biased when I compare males and females. I have heard that if I learn how to think about DIF and also learn how to conduct a DIF analysis, then I can at least take one step toward insuring my instrument is not biased (in a measurement way). I want some assurance that I am being fair in my comparisons of males and females.*

Isabelle: *I could not have said it better. Have you found it difficult to understand DIF?*

Ted: *Well, after having analyzed some data sets, I definitely have a handle on how to consider DIF. It really is an issue to consider when developing and using measurement instruments!*

Introduction

Differential Item Functioning (DIF) is a key technique for analyzing survey and test data. Moreover, DIF should be conducted via Rasch measurement. Whereas the concept of DIF contains many complicated aspects, beginning Rasch measurement users can easily conduct an initial DIF analysis in a thoughtful manner.

It is perhaps easiest to begin by considering differences in item responses due to gender, a well-known issue in the education literature. To pose a question: Are the items of an instrument (e.g., test, survey) perhaps biased in some manner with respect to gender? Even without reviewing the literature, we are aware of a number of individual issues within the broad landscape of gender bias. For example, are the individuals named in a set of test or survey items all male or all female? Do all pronouns refer to one gender? A different type of bias can be present within the

W.J. Boone et al., *Rasch Analysis in the Human Sciences*,
DOI 10.1007/978-94-007-6857-4_13, © Springer Science+Business Media Dordrecht 2014

context of a survey or test item. For instance, an extreme type of bias is a test or survey item that emphasizes a particular sport with which males are more familiar. There is, however, a type of bias that is often rarely explored by researchers. This is a measurement bias in which a set of test or survey items defines a different measurement scale as a function of an attribute such as gender. This is the case in which the pattern (order and spacing) of items along the trait as a function of difficulty is different for a comparison of one group to another (e.g., males, females). This measurement bias can often be detected by DIF analysis. In this chapter, we will present a foundation that allows readers to reflect on measurement bias and document its presence or absence in an instrument. A wide range of potential biases can be explored. Some are well represented in the education literature, such as gender and racial bias. More recently, bias as a function of language has become important. Test bias that misrepresents the achievement levels of English language learners on high-stakes assessments is being scrutinized. Another source of bias that researchers should carefully consider is potential bias when comparing respondents pre and post. This issue is rarely addressed in many studies. It is very important to point out that DIF does not mean that one is searching for items that are somehow unfair to one group or another (e.g., items that involve a topic that is more familiar to boys as opposed to girls). Rather, in considering DIF, one is interested in evaluating if the manner in which items define a measurement scale does so in the same way for different groups of respondents. In the real world, when we measure the height of boys and girls in a class, although boys may be measured as being taller than girls, we know the meter stick does not change how it is measuring from one student to another. This is what we are attempting to evaluate with DIF; does our "item-defined meterstick" operate in the same way for different groups of respondents?

Meaning of DIF

From a measurement perspective, what do we mean when we state that an item functions in a different manner across subgroups? In Fig. 13.1, we present two Wright-like maps for a 10-item test. Think of the first map as the product of a Winsteps analysis of only female respondents. Think of the second map as the product of a Winsteps analysis of only male respondents. Now review the ordering and spacing of the test items from easy to difficult for each group of respondents. Such a review reveals two key differences in the pattern of items as a function of gender. First, nearly all of the items on the male map are shifted a similar distance "down" the measurement scale in comparison to the females. Second, one item (Q4) is in quite a different location relative to the other items as a function of gender.

Fig. 13.1 Two Wright Maps presenting the ordering and spacing of 10 test items. One map is based only upon an analysis of the females who took the test. The other map is based upon the analysis of only the males who took the test. It is the shift of Q4 that may suggest DIF

```
Females                  Males
|Q1 Q4                   |
|Q3 Q6 Q7 Q9 Q10         |
|Q2                      |
|                        |
|Q5 Q8                   |
|                        |Q1
|                        |Q3 Q6 Q7 Q9 Q10
|                        |Q2
|                        |Q4
|                        |Q5 Q8
```

DIF and Construct Validity

Are either or both of these observations examples of DIF? To interpret these plots properly, readers must recall our earlier discussion of construct validity. Within Rasch analysis, construct validity can be reviewed by comparing the actual ordering and spacing of items from easy to hard along a single trait compared to the predicted ordering of items. The prediction might be from experts, from the results of past studies, or a mixture of these two sources, but it must result from using a theory to predict how the items should be ordered from easy to hard for a test or from easiest to agree with to hardest to agree with for a Likert scale survey. When considering construct validity, researchers must think not only of the order of items from easy to hard (test) or easiest to agree with to hardest to agree with (survey) but also the spacing of items. Spacing of items indicates how much harder or easier the items are relative to each other. Items immediately above (harder) or below (easier) a given item will have less space than items far above (harder) or below (easier) a given item.

Formative Assessment Checkpoint #1

Question: If an item exhibits DIF, does this mean that the item is "unfair"?

Answer: Yes and No. If an item exhibits DIF, it means that the item defines a trait in a different manner when its performance is compared across two or more groups of respondents. An item that exhibits DIF is not necessarily unfair to different subgroups of respondents (e.g., vocabulary/terminology of the item is somehow more easily understood by one group as opposed to another). From a measurement perspective, DIF simply means that an item measures a trait in a different way for the two or more compared groups.

There exists yet another construct validity issue that is germane to our discussion of DIF. The invariance of the ordering and spacing of items must also be examined.

When an instrument has high construct validity, its items should not shift in order and spacing as a function of subgroup. We have found that a ruler or tape measure analogy works well to explain this issue. If a scientist were to measure the height of a random sample of 50 males and 50 females (all of whom are 20-years-old) using a 10 m long tape measure, all would agree that the markings on the tape measure (e.g., 100 cm, 300 cm, 451 cm) do not move as measurements of the student heights are made. We know that this observation of ours may sound a little odd, but this concept is crucial to understand a requirement of measurement. If some markings on the tape measure moved along the metal tape when the scientist walked from the group of female subjects to the male subjects, then all would agree that the scientist might not be able to make valid conclusions about the relative height of 20-year-old males and females. Fortunately, the markings on a tape measure do not move or change in any way, and confident conclusions can be stated about the height of 20-year-old males and 20-year-old females and how these two groups were compared.

However, researchers must carefully monitor the items on any test or survey used for a measurement scale for movement as a function of subgroup; one needs to make sure that the items marking the metric do not shift as a function of subgroup. Recall the centimeter marks on the metal tape measure do not move. We need to make sure the items that mark our measurement instrument of humans do not move about! When DIF is explored with Rasch measurement, the researcher carefully evaluates possible movement of test or survey items along the constructed tape measure. If moving items are identified, all is not lost. As we shall see, however, some important steps must be taken before a person measure is computed and a statistical analysis of data is completed. DIF takes a small amount of time to conduct, but there is no excuse for not carefully making sure an instrument works in the same manner for important subgroups of respondents.

Construct Validity and a Misconception

We continue our discussion of DIF with an important misconception regarding construct validity. Moreover, we believe that clarifying this misconception will promote deeper understanding of DIF. Immediately below in Figs. 13.2 and 13.3, we present two Wright Maps that display the ordering of items from easy to hard for a test or survey. We then plot the location of the mean on each Wright Map for males and females. Figure 13.2 is for a test in which items can be answered right or wrong. Figure 13.3 has been created from an analysis of survey data in which students could SA, A, D, or SD with items. The frequent confusion and misconception results from researchers' thoughts that each of these maps shows DIF because mean student performance as a function of gender appears to be different, and attitude as a function of gender also appears to be different. It does *not* mean that one can conclude test or survey bias (from a measurement perspective).

Such inferences are incorrect and reflect a misconception (a misunderstanding) as to the meaning of construct validity. The locations of the males and females on

```
                            |
                            |
                            |
                            |T
  2                         +
                            |  Q9
                            |
                            |
                            |
                            |  Q8
                            |
                            | Q10        Q11        Q6
  1                         +
                            |
                            |
                            |
                            |  Q3        Q4         Q7         Q7         Q12
                      F->   |  Q11
                            |  Q12
                            |
  0                         +
                            |  Q2
                      M->   |
                            |  Q5
                            |  Q3        Q6
                            |
                            |
                            |
 -1                         + Q1          Q4         Q5
                            |
                            |
                            |
                            |  Q10       Q8
                            |
```

Fig. 13.2 A Wright Map presenting test item difficulty and the average ability level of male (M) and female (F) test takers. Of importance is that although there is an apparent difference between males and females, this does not mean DIF is present. Furthermore, it would be possible for the males and females to have similar mean measures, and DIF could be present in terms of how the test items are measuring the males and females

the two maps may be accurate. The observed differences in locations of males and females on the maps may simply reflect an accurate difference in the males and females, just as the measurement of the height of 100 students by the scientist might reveal a true difference in average height as a function of gender. It may be that females outperform males on this test. And it may be that females are more agreeable toward the survey items than the males. Observing a difference in the location of two comparison groups (in this case males and females) does not necessarily mean DIF and is not necessarily evidence of a problem with the construct validity defined by a set of items.

Fig. 13.3 A Wright Map presenting survey item difficulty and the average measure of male (M) and female (F) survey respondents

```
                           |
2                          |
                           |
                           |
                           |
                           |
                           |
                           |
                           |  Q11
1                     F->  +  Q12      Q7
                           |  Q4
                           |
                           |  Q10
                           |
                           |
                      M->  |
                           |  Q14
                           |
                           |
0                          +  Q2
                           |
                           |  Q6
                           |  Q3       Q8
                           |  Q1       Q15
                           |
                           |
```

If, indeed, DIF means that there might be bias (mis-measurement of some sort) for a subgroup of respondents, then how might one see potential DIF? And how does DIF relate to construct validity? Now let's go back to the two Wright Maps of Fig. 13.1. After a careful review of these two pictures, what do you see? Hopefully, you see that item 4 is in a different location along the trait compared to other items, when one looks at the male and female plots. Item 4 might exhibit DIF because the manner in which the item defines the test trait appears to be quite different for males and females. Think of item 4 as being a particularly important mark on a meterstick that is used to compare the height of men and women, a mark that is used to find the height of numerous tall women and average-height men. From a measurement prospective, if this item is indeed as odd as it appears to be in these two side-by-side plots, then it will not only be important to spot the item but also confidently verify actual DIF using some set of rules for potential DIF, and then take a course of action so that measures of respondents to an instrument are not influenced. Ignoring the fact that the item defines a different portion of the latent trait as a function of subgroup (e.g., gender) influences the computed measures of all respondents and may influence the validity of conclusions drawn from an analysis of data. Fortunately, techniques exist that can address DIF. Later in this chapter, we will outline the steps one can take. For instance, a simple step is to drop the item from the analysis. If, however, the number of items on a test or survey is limited, researchers may retain the item but treat it as a different item for males and females. The text of the item is identical, but from a measurement perspective, the item is viewed as a different item.

Formative Assessment Checkpoint #2

True or False: DIF can exist only for test items that are scored right/wrong.

Answer: False. Survey items and partial credit items (e.g., an item on a test which can be worth up to a maximum of 3 points) can also exhibit DIF.

The Mechanics of Reviewing for DIF

Rasch measurement is based on the requirement that an instrument, as a set of items, must focus on a single trait. In earlier chapters, we discussed at length the meaning of a unidimensional trait. An analysis of DIF can be conducted using Winsteps by adding a command line to a control file. When the steps outlined in Chap. 3 are used to create a control file, the command line introduced here will be automatically inserted in the control file. Readers must remind themselves that by inserting this command line and completing the subsequent steps to be described, one in essence can put two or more Wright Maps next to one another and conduct statistical analyses that detect the amount of relative movement of items along the line of a trait. This is identical to what readers and authors of this book did when the ordering and spacing of items were weighed in our brains when the Wright Map for males only and the Wright Map for females only were placed side by side and we scanned for differences in item ordering and spacing. A final note before we get to the command line: readers should note our assertion above that one can in essence put two or more Wright Maps next to one another. This is indeed the case, which means that it is fairly easy to investigate DIF when comparing multiple groups, for instance, race (e.g., White, African American, Hispanic, Asian).

Figure 13.4 presents a portion of the control file we used to demonstrate an analysis of DIF. The Excel data set entitled "Raw Data for DIF Chp.xls" was used to create the control file. This data set includes responses from 75 persons to the 10 outcome-expectancy items of the STEBI (the 13 self-efficacy items are not included). In order to practice the techniques we present here, we have added fictitious gender data (denoted by letters M and F) and fictitious race information (White, W; Hispanic, H; African American, AA; Asian, AS). If you carefully review the steps for making a control file, which we outlined in Chap. 3, you will be able to easily construct this control file. This control file is also provided to readers and is named "cf for DIF chp OE Data From Excel." The lines in the control file that are germane to the evaluation of DIF are the line beginning with the phrase "@Gender" and the line that begins with the phrase "@Race." This first line tells Winsteps which column of data in the control file will be used to conduct a DIF analysis for gender. Specifically, this line tells Winsteps that the 12th column of data contains the

```
&INST
Title= "Raw Data for DIF Chp.xls"
ITEM1 = 1 ; Starting column of item responses
NI = 10 ; Number of items
NAME1 = 12 ; Starting column for person label in data
NAMLEN = 5 ; Length of person label
XWIDE = 1 ; Matches the widest data value observed
CODES = 12345 ; matches the data
TOTALSCORE = Yes ; Include extreme responses
@Gender = 1E1   ; $C12W1
@Race = 3E4 ; $C14W2
&END ; Item labels follow: columns in label
```

Fig. 13.4 A control file to be modified for an analysis of DIF

variable to be used in the DIF analysis for gender. The control line that begins with the phrase "@Race" tells the program that the race data in the control file are located in the 14th column of the control file data.

To investigate and document DIF in a data set, the first step is to run a Winsteps analysis (this means taking a completed control file and running Winsteps). When the analysis is complete (readers will see the iterations on the screen), select the output Table 30 (Item: DIF, between/within). Winsteps will request the analyst to "Please select grouping for this table." If the program has a number of command lines starting with the symbol "@", then choosing this option will present the different DIF analyses that could be run (this will be the number of lines which one has in the control file that begins with the term "@"). Once Table 30 is selected, a gray box will appear asking the analyst which variable will be used for the DIF comparison (you will see the phrase "DIF=" and then you will see the white box with a drop-down menu option). In our sample control file, two terms appear ("@GENDER" and "@RACE"); thus, a variable selection is needed. For the time being, let us pretend that we select "Gender" for our comparison. Our next step is to make sure only tables for the output are selected (keep the "plots" option unchecked for now) and then click on the button labeled "OK."

Formative Assessment Checkpoint #3

Question: Is it difficult to conduct a DIF analysis?

Answer: No. When a control file is constructed using Winsteps/Ministeps, researchers are asked to indicate which pieces of data are "person label variables." By indicating in the construction of the control file which variables are person variables, it is easy to conduct a DIF analysis.

TABLE 30.1 Raw Data for DIF Chp.xls ZOU098WS.TXT Dec 18 17:25 2011
INPUT: 74 PERSON 10 ITEM REPORTED: 74 PERSON 10 ITEM 5 CATS WINSTEPS 3.73

DIF class specification is: DIF=@GENDER

PERSON CLASS	DIF MEASURE	DIF S.E.	PERSON CLASS	DIF MEASURE	DIF S.E.	DIF CONTRAST	JOINT S.E.	Welch t	d.f.	Prob.	Mantel-Haenszel Chi-squ	Prob.	Size CUMLOR	ITEM Number	Name
F	-.23	.26	M	.34	.27	-.58	.37	-1.55	70	.1268	.4434	.5055	-.40	1	Q1OE
F	-.04	.25	M	.27	.27	-.31	.37	-.84	69	.4065	1.5946	.2067	-.75	2	Q4OE
F	.49	.24	M	.63	.27	-.14	.35	-.40	69	.6901	.2833	.5945	-.31	3	Q7OE
F	-1.21	.29	M	-1.65	.34	.44	.44	.99	68	.3280	1.4212	.2332	.75	4	Q9OE
F	1.12	.23	M	.91	.26	.21	.35	.61	68	.5453	.1451	.7032	.20	5	Q10OE-RC
F	.09	.25	M	-.20	.29	.28	.38	.75	68	.4585	.0427	.8363	.13	6	Q11OE
F	.38	.24	M	.12	.28	.26	.36	.71	68	.4808	1.3486	.2455	.64	7	Q13OE-RC
F	-.04	.25	M	.12	.28	-.16	.37	-.42	69	.6725	1.8002	.1797	-.96	8	Q14OE
F	-.37	.26	M	-.04	.28	-.33	.38	-.86	69	.3904	.0270	.8695	-.09	9	Q15OE
F	-.17	.26	M	-.63	.30	.46	.39	1.17	68	.2472	.2455	.6203	.29	10	Q16OE
M	.34	.27	F	-.23	.26	.58	.37	1.55	70	.1268	.4434	.5055	.40	1	Q1OE
M	.27	.27	F	-.04	.25	.31	.37	.84	69	.4065	1.5946	.2067	.75	2	Q4OE
M	.63	.27	F	.49	.24	.14	.35	.40	69	.6901	.2833	.5945	.31	3	Q7OE
M	-1.65	.34	F	-1.21	.29	-.44	.44	-.99	68	.3280	1.4212	.2332	-.75	4	Q9OE
M	.91	.26	F	1.12	.23	-.21	.35	-.61	68	.5453	.1451	.7032	-.20	5	Q10OE-RC
M	-.20	.29	F	.09	.25	-.28	.38	-.75	68	.4585	.0427	.8363	-.13	6	Q11OE
M	.12	.28	F	.38	.24	-.26	.36	-.71	68	.4808	1.3486	.2455	-.64	7	Q13OE-RC
M	.12	.28	F	-.04	.25	.16	.37	.42	69	.6725	1.8002	.1797	.96	8	Q14OE
M	-.04	.28	F	-.37	.26	.33	.38	.86	69	.3904	.0270	.8695	.09	9	Q15OE
M	-.63	.30	F	-.17	.26	-.46	.39	-1.17	68	.2472	.2455	.6203	-.29	10	Q16OE

Size of Mantel-Haenszel slice: MHSLICE = .010 logits

Fig. 13.5 (Winsteps Table 30.1): A DIF analysis of 75 persons' responses to 10 STEBI outcome-expectancy items

The first important observation is that identical information is presented in the top half and bottom half of Fig. 13.5 (Winsteps Table 30.1). The top half of the table is a comparison of how the 10 items of the outcome-expectancy scale define outcome-expectancy as a function of females and males. Stated another way, we are exploring how the 10 items define the construct of outcome-expectancy for males, how the same items define the construct of outcome-expectancy for females, and a possible difference in the manner in which the items define the construct of outcome-expectancy for the two groups. The data for females are presented first in each row for the top half of the table. For example, item Q1OE (short for Item #1 of the outcome-expectancy scale) had a logit measure of –.23 and an error of .26 logits. Think of this as the value one would get if one were to evaluate only the females in the control file which we have constructed. The 4th column of this table presents the symbol that we used in our data set to denote males ("M"); following that column, the measure of item Q10E is presented (.34 logits) as well as the error of the item (.27 logits). Again, as was the case for the female data, it is important to think of the logit measure reported for this item for males (.34) as what would have been computed had we evaluated this data set only for the male respondents. Now for the time being, we ask readers to focus on the right side of the table and find the first column containing the word "Prob.".

The next step in conducting a DIF analysis is to scan the column headed by the term "Prob." (the 11th column). Each value reported in this column gives an assessment of the magnitude of the difference in the location of the item along the construct for the two groups. As in the case of a statistical analysis, .05 is a typical threshold to use. Using this cutoff for "Prob." values to scan the "Prob." column suggests that DIF did not occur as a function of gender. This should not be surprising since we randomly inserted gender data into the data file. Can you find the item that exhibits the most potential DIF, even though it was not statistically significant? The answer is the first item presented at the top of the "Prob." column, which exhibits a probability of .1268. We ask readers now to look at Fig. 13.5. The identification of a value in this column at or below .05 suggests that the relative location of an item is different between males and females. This is the column of data that begins to help us identify items with potential bias. The next step the analyst should take is to evaluate, in essence, "the effect size" of the potential DIF. Why look at effect size? Recall that effect size in essence helps an analyst evaluate how meaningful the difference is. A DIF comparison may reveal a statistically significant p value, but examination of its effect size is needed to determine whether or not the difference is meaningful. Mike Linacre, the author of Winsteps, presents a table (2012, p. 548) which suggests using a DIF contrast of >.64 to flag items that exhibit not only statistically significant DIF but also meaningful effect size (>.64 DIF contrast indicates moderate to large DIF). This value will be the absolute value of any of the numbers presented in the "DIF contrast" column. For this data set, no DIF contrast greater than .64 is observed, as we might predict since we did not identify any items with a p value <.05. It is exceedingly important to remember that even if a potential DIF item were identified (Prob. <.05 and DIF contrast >.64), this does not mean there is a significant difference in the mean outcome-expectancy measures of females and males. Identification of this item means only that this item probably should not be treated as the same item in terms of where the item defines the construct for the two groups of respondents.

When DIF has been determined, what are a researcher's potential subsequent steps? The first step is a qualitative, conceptual application of Rasch theory: Review the text of the item and ask why the item may generate DIF. Equally important, review the other items and ask if any items surprisingly did not generate DIF. This step exemplifies that high-quality measurement work involves a mixture of applying Rasch theory to numbers and applying Rasch theory in a qualitative manner. This is naturally what is needed in many fields such as psychology, medicine, education, and market research.

Upon the detection of potential DIF, review of one's predictions of an item's behavior is necessary to inform subsequent action. In our example of outcome-expectancy items, if data (p value, DIF contrast) *and* subsequent reflection further suggest potential DIF, then one choice of action is to remove the item from the analysis. Recall that it is easy to delete an item without editing a data set and then rerun an analysis by adding the line (IDFILE=) to the control file.

Sometimes, however, items are at a premium because respondents have completed a limited number of items. In such situations, steps are available to retain an item. For the sake of explanation, let us pretend for the rest of this chapter that item Q1OE was an item that exhibited DIF, but we wished to somehow keep item Q1. However, we are cognizant that it is important to not let a difference in the manner in which Q1 defines the construct as a function of gender influence an analysis. This

```
&INST
Title= "Raw Data for DIF Chp.xls"
ITEM1 = 1 ; Starting column of item responses
NI = 10 ; Number of items
NAME1 = 12 ; Starting column for person label in data
NAMLEN = 5 ; Length of person label
XWIDE = 1 ; Matches the widest data value observed
CODES = 12345 ; matches the data
TOTALSCORE = Yes ; Include extreme responses
@Gender = 1E1 ; $C12W1
@Race = 3E4 ; $C14W2
SFILE=9OEItemsSteps; A file with the name 9OE...with
;                    steps will be created
; The 3 lines below remove item 1 from the analysis
IDFILE=*
1
*
&END ; Item labels follow: columns in label
```

Fig. 13.6 The control file in Fig. 13.4 now modified for a DIF analysis

last sentence is so very important that we will repeat the idea. In some cases we may identify an item with DIF, but we wish to retain the item in a study, but at the same time, we want to make sure that retaining the item does not impact our measures of respondents. Now what do we do?

One technique begins with evaluating the data set and the 9 OE items that did not exhibit DIF. By conducting such an analysis, we can determine the calibration (the measure) of the 9 non-DIF items and also determine the steps of the rating scale (think of these steps as the spacing between each of the attitude categories). To generate the "step" file, all we need to do is to insert a command line that begins with the phrase SFILE=. To analyze the data without item Q1OE, we can make use of the command lines that allow removal of an item before an analysis (the item is still in the data at the base of the control file, but the item is not used for an analysis). Above we provide the edits (Fig. 13.6) that we would make to the command file we presented in Fig. 13.4.

The next step in our goal of making use of item Q1 but simultaneously considering the potential DIF of the item as a function of gender involves the notation of the SFILE and also writing down the item calibrations of those 9 items that were calibrated using all respondents (males and females). Below we provide the SFILE that was generated from running the control file presented in Fig. 13.7 and the item entry table that resulted from the analysis. Note that one can see that item Q1OE was not used in the analysis.

Now that we have the item calibrations, what comes next? Our next steps are fairly simple and will lead to two goals: (1) computing female person measures using all 10 OE items and (2) computing male person measures using all 10 OE items. First, we make two copies of the initial control file (the file without the command lines SFILE and IDFILE). One file will be used to conduct a male-only analysis; the other file will be used to conduct a female-only analysis.

```
TABLE 14.1 Raw Data for DIF Chp.xls                 ZOU334WS.TXT  Dec 19  9:46 2011
INPUT: 74 PERSON  10 ITEM  REPORTED: 74 PERSON  9 ITEM  5 CATS     WINSTEPS 3.73
--------------------------------------------------------------------------------
PERSON: REAL SEP.: 1.39  REL.: .66 ... ITEM: REAL SEP.: 3.15  REL.: .91

         ITEM STATISTICS:  ENTRY ORDER

-------------------------------------------------------------------------------------------------
|ENTRY   TOTAL  TOTAL           MODEL|   INFIT  |  OUTFIT  |PT-MEASURE |EXACT MATCH|           |
|NUMBER  SCORE  COUNT  MEASURE  S.E. |MNSQ  ZSTD|MNSQ  ZSTD|CORR.  EXP.| OBS%  EXP%| ITEM      |
|------------------------------------+----------+----------+-----------+-----------+-----------|
|     1    DELETED                    |          |          |           |           | Q1OE      |
|     2    260     74      .12     .19| .95   -.3|1.05    .4|  .55   .54| 64.9  58.7| Q4OE      |
|     3    246     74      .59     .18| .68  -2.4| .70  -2.1|  .53   .55| 64.9  53.8| Q7OE      |
|     4    297     74    -1.51     .23| .99    .0| .95   -.2|  .56   .48| 73.0  72.6| Q9OE      |
|     5    230     74     1.10     .18|1.07    .5|1.08    .6|  .56   .56| 44.6  52.2| Q10OE-RC  |
|     6    264     74     -.03     .19| .85   -.9| .87   -.7|  .55   .53| 66.2  59.4| Q11OE     |
|     7    255     74      .29     .19|1.71   3.8|1.86   4.2|  .42   .54| 48.6  55.7| Q13OE-RC  |
|     8    262     74      .04     .19| .71  -1.9| .66  -2.1|  .61   .53| 67.6  59.0| Q14OE     |
|     9    269     74     -.22     .20| .83  -1.0| .87   -.7|  .55   .52| 66.2  61.8| Q15OE     |
|    10    273     74     -.38     .20|1.30   1.6|1.30   1.5|  .53   .52| 62.2  64.5| Q16OE     |
|------------------------------------+----------+----------+-----------+-----------+-----------|
1      .00
2    -4.64
3     -.76
4      .67
5     4.73
```

Fig. 13.7 (Winsteps Table 14.1): The item entry table from the analysis of the data set with the removal of item Q1 OE. The output of the command line SAFILE= which provides the step calibrations which are provided below the item entry table

Fig. 13.8 Form of command lines that will be inserted into the control file

```
; Below we are anchoring items to item calibrations
IAFILE=*
2 .12
3 .59
4 -1.51
5 1.10
6 -.03
7 .29
8 .04
9 -.22
10 -.38
*
; Below we are setting the "Steps" of the analysis
SAFILE=*
1      .00
2    -4.64
3     -.76
4      .67
5     4.73
*
```

Second, we insert the item calibrations that were computed for the 9 OE items as item anchors in both control files and insert the step values into both control files. In Fig. 13.8, we provide the form of the command lines to be inserted into the control file.

Formative Assessment Checkpoint #4

Question: Must an item be removed from an analysis if that item exhibits potential DIF?

Answer: No. Steps can be taken to retain the item, but the item may not be viewed as defining the same portion of a trait for compared groups (e.g., males vs. females).

One final command line, which is very helpful for a DIF analysis and for many other analyses, must be inserted into the two control files. This command line allows only a particular type of data in a control file to be evaluated. In our specific case, the command line can be used to evaluate only female respondents and only male respondents. At least two techniques exist to create a control file for an analysis of a subset of data. First, an analyst can use a command line "PDFILE=" to indicate the lines of data to be removed from an analysis. For example, if an analyst wanted to remove only persons 2, 7, 13, and 25 from an analysis, then the analyst would add the following to the control file:

```
PDFILE=*
2
7
13
25
*
```

An easier way to remove particular individuals from an analysis is to use the command line "PSELECT=." This command line allows only one particular type of respondent to be selected for an analysis. In our data set, this control line will be added to evaluate only males: "PSELECT=M*." Looking carefully at the data set, readers should note that the gender datum for each person is contained in the 1st column of the name portion of the data (NAME1=12 in the control file indicates that the start of the name information in the data at the base of the control file begins in the 12th column of data). The command line to be inserted into the control file to evaluate only females is "PSELECT=F*." Figure 13.9 presents the command lines for the female-only control file; the male-only file is different only in that an "M" is presented in the PSELECT line. We have also included the first line of data.

Having presented the control files for computation of the male and female respondents, let's now review what we have done before we move forward to "next steps" in this analysis. From a measurement perspective, we have conducted an analysis that produces male measures and female measures that are anchored with 9 of the 10 outcome-expectancy items that did not exhibit DIF. Each analysis retained item 1 (Q1OE), but from a measurement perspective, this item was not anchored to the same part of the OE trait for males and females.

```
&INST
Title= "Raw Data for DIF Chp.xls"
ITEM1 = 1 ; Starting column of item responses
NI = 10 ; Number of items
NAME1 = 12 ; Starting column for person label in data record
NAMLEN = 5 ; Length of person label
XWIDE = 1 ; Matches the widest data value observed
; GROUPS = 0 ; Partial Credit model: in case items have different rating scales
CODES = 12345 ; matches the data
TOTALSCORE = Yes ; Include extreme responses in reported scores
; Person Label variables: columns in label: columns in line
@Gender = 1E1 ; $C12W1
@Race = 3E4 ; $C14W2
;
PSELECT=F*
;
; Below we are anchoring items to item calibrations
IAFILE=*
2 .12
3 .59
4 -1.51
5 1.10
6 -.03
7 .29
8 .04
9 -.22
10 -.38
*
; Below we are setting the "Steps" of the analysis
SAFILE=*
1 .00
2 -4.64
3 -.76
4 .67
5 4.73
*
&END ; Item labels follow: columns in label
Q1OE ; Item 1 : 1-1
Q4OE ; Item 2 : 2-2
Q7OE ; Item 3 : 3-3
Q9OE ; Item 4 : 4-4
Q10OE-RC ; Item 5 : 5-5
Q11OE ; Item 6 : 6-6
Q13OE-RC ; Item 7 : 7-7
Q14OE ; Item 8 : 8-8
Q15OE ; Item 9 : 9-9
Q16OE ; Item 10 : 10-10
END NAMES
4444344444 F w
```

Fig. 13.9 The command lines and one line of data for the analysis of all females. The command "SAFLE" anchors the steps of this analysis to that determined from an analysis using all respondents (both male and female) and the items that did not show potential DIF. The command line "IAFILE" anchors the item calibrations for the analysis to the values computed for just items Q2OE-Q9OE. The line "PSELECT" ensures that only the female respondents are evaluated in this analysis. It is very important to note that this control file does not delete any items; this means that item Q1OE is retained for the analysis. But, item Q1OE is allowed to be calibrated to a value that results from an analysis of only the female respondents

This allows us to retain item Q1 yet ensures that its presence does not invalidate our measures of respondents, since evidence exists that item Q1 may exhibit DIF as a function of gender.

Maintaining Quality Controls in Our Work

We close this chapter with some "checks" in our analysis that we routinely conduct to ensure that our analysis has done what we want it to do (e.g., anchoring of steps, anchoring of selected item calibrations, computation of only female measures, computation of only male measures). Then we will present a point of confusion that often is exhibited by workshop participants.

When we conduct a check following an analysis with the control files in Fig. 13.9, we need to ensure that the item calibrations are set to our desired values. Also, we must make sure that the step calibrations are set to the values that we have set for the Rasch analysis. Furthermore, we make sure that, indeed, only the measures of males are computed in our male-only analysis. Of course, we also check to ensure that only the measures of females are compared in our female-only analysis. Finally, we need to ensure that items not to be anchored are not anchored. In Fig. 13.10, we provide three tables from the analysis of the male data that were anchored to the 9 OE items and anchored to the steps we have discussed. Each of these tables provides a "check" of whether or not what we intended to accomplish has indeed taken place. For Winsteps Tables 14.1 and 3.2, the key information to review is the presence of the letter "A" (Measure column for Table 14.1, Threshold column for Table 3.2). This letter indicates that the item in question (for Winsteps Table 14.1) and the step in question (for Winsteps Table 3.2) were indeed anchored. If you do not see an "A," it means that the item (or step) was not anchored. Of course, it is also important to make sure that the value used as an anchor is indeed anchored to the intended value. The third and final check (Winsteps Table 18.1) is to verify whether or not the subset of persons that we wished to evaluate were indeed the respondents whom we wished to evaluate. In Winsteps Table 18.1, one can see that indeed males were evaluated (e.g., the first respondent is a female respondent, and this respondent was deleted from the analysis; the third respondent is a male respondent and this respondent was not deleted).

In our presentation we have provided an overview of DIF and some key tables one can consult. There are other techniques that have been used by researchers that involve in essence the plotting of item measures against each other as a function of two analyses (e.g., male, female) and the computation of "control lines." Items outside the control lines are items which may exhibit DIF. For now, we will emphasize the points we have made throughout this chapter, but readers can contact us for details of these other techniques.

```
TABLE 14.1 Raw Data for DIF Chp.xls               ZOU948WS.TXT  Dec 19 14:15 2011
INPUT: 74 PERSON  10 ITEM  REPORTED: 32 PERSON  10 ITEM  5 CATS    WINSTEPS 3.73
--------------------------------------------------------------------------------
PERSON: REAL SEP.: 1.29  REL.: .62 ... ITEM: REAL SEP.: 1.71  REL.: .74

         ITEM STATISTICS:  ENTRY ORDER
```

ENTRY NUMBER	TOTAL SCORE	TOTAL COUNT	MEASURE	MODEL S.E.	INFIT MNSQ	INFIT ZSTD	OUTFIT MNSQ	OUTFIT ZSTD	PT-MEASURE CORR.	PT-MEASURE EXP.	EXACT MATCH OBS%	EXACT MATCH EXP%	DISPLACE	ITEM
1	108	32	.41	.28	1.97	3.3	2.69	4.8	.21	.56	50.0	55.1	.00	Q1OE
2	109	32	.12A	.29	1.32	1.3	1.45	1.6	.54	.54	59.4	58.8	.21	Q4OE
3	104	32	.59A	.28	.46	-2.9	.50	-2.4	.69	.57	71.9	54.2	.13	Q7OE
4	130	32	-1.51A	.35	1.17	.6	1.09	.4	.56	.48	71.9	71.3	-.22	Q9OE
5	100	32	1.10A	.27	1.03	.2	1.15	.7	.60	.58	59.4	51.6	-.09	Q10OE-RC
6	115	32	-.03A	.29	.77	-.9	.74	-.9	.58	.53	65.6	61.4	-.13	Q11OE
7	111	32	.29A	.28	1.67	2.4	1.94	3.0	.46	.55	40.6	57.8	-.11	Q13OE-RC
8	111	32	.04A	.29	1.03	.2	.98	.0	.58	.54	62.5	60.9	.13	Q14OE
9	113	32	-.22A	.30	.78	-.9	.84	-.5	.58	.52	71.9	62.2	.22	Q15OE
10	120	32	-.38A	.30	1.58	2.0	1.65	2.0	.49	.52	50.0	63.4	-.23	Q16OE

Fig. 13.10 These tables represent a full item entry table, part of the table that includes the steps of the rating scale, and part of the person entry table. The "A" letters following the item calibrations show that all items but item Q1OE are anchored. A comparison of the calibration of each item matches the values determined in earlier stages of the analysis. This shows that no errors were made in entering item measures. The same is true of the "step" values; the letter "A" appears after each step value, and those values match the intended values from earlier analyses. The table from the person entry table provides the first seven respondents for the data set when PSELECT was used to evaluate only male respondents. The person label column shows that indeed the females were removed from the analysis

```
TABLE 3.2 Raw Data for DIF Chp.xls                ZOU948WS.TXT  Dec 19 14:15 2011
INPUT: 74 PERSON  10 ITEM  REPORTED: 32 PERSON  10 ITEM  5 CATS    WINSTEPS 3.73
--------------------------------------------------------------------------------


SUMMARY OF CATEGORY STRUCTURE.  Model="R"
-------------------------------------------------------------------
|CATEGORY   OBSERVED|OBSVD SAMPLE|INFIT OUTFIT|| ANDRICH |CATEGORY|
|LABEL SCORE COUNT %|AVRGE EXPECT|  MNSQ  MNSQ||THRESHOLD| MEASURE|
|-------------------+------------+------------++---------+--------|
|  1   1       1   0|  3.23  -.29|  3.10  6.32||  NONE A |( -5.75)| 1
|  2   2      39  12|   .19* -.08|  1.30  1.51||   -4.64A|  -2.73 | 2
|  3   3     100  31|   .52   .67|   .80   .76||    -.76A|   -.04 | 3
|  4   4     158  49|  1.69  1.81|  1.00  1.00||     .67A|   2.73 | 4
|  5   5      22   7|  3.12  2.22|   .92  1.04||    4.73A|(  5.84)| 5
-------------------------------------------------------------------

TABLE 18.1 Raw Data for DIF Chp.xls               ZOU948WS.TXT  Dec 19 14:15 2011
INPUT: 74 PERSON  10 ITEM  REPORTED: 32 PERSON  10 ITEM  5 CATS    WINSTEPS 3.73
--------------------------------------------------------------------------------
PERSON: REAL SEP.: 1.29  REL.: .62 ... ITEM: REAL SEP.: 1.71  REL.: .74

         "M" PERSON STATISTICS:  ENTRY ORDER

------------------------------------------------------------------------------------------------
|ENTRY   TOTAL  TOTAL           MODEL|   INFIT    |  OUTFIT   |PT-MEASURE |EXACT MATCH|      |
|NUMBER  SCORE  COUNT  MEASURE  S.E. |MNSQ  ZSTD|MNSQ  ZSTD|CORR.  EXP.| OBS%  EXP%| PERSON|
|------------------------------------+----------+----------+-----------+-----------+-------|
|     1     DELETED                  |          |          |           |           | F w   |
|     2     DELETED                  |          |          |           |           | F w   |
|     3     32       10      .44  .46|1.24    .7|1.41   1.1|  .59   .38| 50.0  50.0| M w   |
|     4     29       10     -.20  .46| .40  -2.0| .43  -1.8|  .58   .41| 70.0  50.8| M h   |
|     5     35       10     1.13  .50| .89   -.1|1.04    .2|  .90   .35| 60.0  54.1| M h   |
|     6     28       10     -.42  .47| .33  -2.3| .35  -2.2|  .82   .41| 80.0  49.3| M h   |
|     7     DELETED                  |          |          |           |           | F h   |
```

Fig. 13.10 Continued

A Final Comment

One issue remains, an issue that is raised frequently in classes and in workshops: Where are the measures that I then use for any statistics involving respondents? In this data set, following the running of the male-only data with file "cf DIF m only 1 to 9 anch", a researcher would generate output files that contain each male's person measure and then incorporate those measures into a data set, such as an SPSS data set. The researcher would then make sure to follow the running of the female-only data set with file "cf DIF f only 1 to 9 anch." The researcher would generate output files that contain each female person measure and then incorporate those measures into the same data set that contains the male person measures. We have included a copy of the Excel file with male measures (file: male measures final) and the Excel file with the female measures (file: female measures final). These files can be created by Winsteps. When using these files, a researcher should note that the value of "0" reported for a "measure" does not represent a person measure to be used (those persons with a measure of 0 in the Excel sheet are those respondents who were deleted from the analysis. For example, in the male measure final file, all respondents with a value of "0" for measure are the females, who were not evaluated in the analysis using only males).

Isabelle and Ted: Two Colleagues Conversing

Ted: *Whew! DIF is different, but I think I get it.*

Isabelle: *Okay smarty! Tell me what you think you know!*

Ted: *I think the first thing that took me a while to understand is that when people talk about item bias for tests, there is an immediate assumption that an item is bad and must be thrown out. What I now appreciate is that bias is not always bad, and that a different word might be better to use when one considers the issue of bias from a measurement perspective. An item might measure in a different manner for two (or more) groups of students, but the item might not be bad. The item simply does not act in the same manner for both groups, and the net effect is that blindly including the item in an analysis can influence the measures we are trying to compute for our respondents.*

Isabelle: *That makes sense to me. Now, tell me about construct validity and what it has to do with DIF.*

Ted: *In Rasch measurement, we are really careful about what sets of items we use to compute a person measure. We want to make sure that the all items involve only one trait. Then and only then can I use a set of items to measure a respondent and compare that person to another respondent.*

Isabelle: *What does that have to do with DIF and Rasch?*

Ted: *Well, I think of DIF as an extension of our careful concern regarding a single trait. When we consider DIF, we not only implicitly address a single trait, but we also check to see if the items define the trait in the same manner for different subgroups, for instance, males and females. It does not necessarily mean that there is DIF if boys do better on test items than girls. However, there may be DIF if the order and spacing of hardest to easiest items are different for boys and girls.*

Isabelle: *Can you draw me a picture? That might help me better understand the issue. Can we try an attitude survey?*

Ted: *Sure. Here, I am drawing a line to represent the trait of outcome-expectancy. The left side of the line is low outcome-expectancy, and the right side of the line is high outcome-expectancy. Items on the left side are easier for respondents to agree with. They might not "agree" with the items per se (they might be circling "disagree" on the survey), but they find it easier to agree in some manner with these items compared to the outcome-expectancy items on the right-hand side of the line (maybe they are circling strongly disagree for these items).*

1 3 4 2 5

Low OE *High OE*

The whole point of DIF is that when measuring respondents, the general picture of items (from easier to agree with to harder to agree with) should be very similar for the comparison groups. If the picture of how the items define the trait looks different, then I must definitely think and perhaps react. If I don't at least think (and potentially react by taking some simple steps), then my measures of people may not be accurate.

Isabelle: *If we are comparing males and females, and there is no DIF, does it mean that males answer in the same way as females?*

Ted: No! No! No! This was tough for me to get, but now I get it. It could be that the intensity of responses selected differs from males to females. Maybe males tend to select SA and A on the OE scale, and females select D and SD. That would NOT mean DIF. It could be that there IS a difference in the response pattern of females and males. But I want to ensure that the manner in which items define the construct, in this case from low OE to high OE, does not differ across the groups.

Keywords and Phrases

Differential Item Functioning
Contrast

The presence of DIF does not mean that an item is definitely "biased"; it simply means that the manner in which the item defines the trait differs from the way the item defines the trait for a comparison group.

Potential Article Text

Data were collected from a sample of 143 males and females in the first week of a preservice science teacher methods course in an effort to evaluate potential differences in self-efficacy. The goal of the data analysis was to fine-tune the course curriculum in the event that differences were present at the onset of the course as a function of gender.

The PST STEBI (Enochs and Riggs, 1990) was used for the collection of data. Only the 13 items of the self-efficacy subscale were utilized for this analysis. The Rasch model and Winsteps (Linacre, 2011) were employed to analyze the data. In order to conduct an accurate comparison of males and females, a Differential Item Functioning (DIF) analysis was performed to explore if the STEBI SE items defined the same construct in the same way for males and females. Evaluating the stability of a construct is similar to steps that are taken to link the measurement of time to a known standard. For purposes that require an exact knowledge of time, it is common to align all timing devices to Greenwich Mean Time (GMT).

Only one item of the SE scale, item 2, potentially exhibited DIF as a function of gender ($p < .05$). In addition, a conceptual review of the difference in the pattern and spacing of items along the construct of self-efficacy as a function of gender was conducted. Finally, an evaluation of effect size was conducted using a suggested threshold of .64 for DIF contrast. This value is suggested by Linacre as the result of work conducted by Rebecca Zwick (Zwick, Thayer, & Lewis, 1999) while at the Educational Testing Service. Item 2 exhibited a contrast greater than .64.

One potential analysis technique is to simply remove item 2 from an analysis and compute the measures of all respondents in a single run of the data. However, given the small number of items presented to respondents with respect to SE, a decision was made to retain item 2 but to not view item 2 as defining the construct in the same way for males and females. This step allowed the computation of person measures not impacted by having a survey item potentially measure the construct in a different manner as a function of gender.

Initially, a Rasch analysis was completed utilizing the 12 SE items that did not exhibit DIF. This analysis allowed the computation of item measures for each of the 12 items. These measures can be viewed as marking 12 parts of the SE construct. Then an analysis was conducted on only the female respondents. For this analysis of females, all items except item 2 were anchored to the values of the item measures. Item 2 was included in the analysis, but the item was not anchored. The female person measures were then placed in the project master data set.

An analysis of the male data was also completed. That analysis also made use of the 12 item anchors, and the item that was not anchored, item 2, was incorporated into the analysis. The completed analysis resulted in person measures for each male. These measures were also not influenced by the potential of item 2 defining a different portion of the SE trait for males and females. The computed male measures were also placed into the project data set (Zwick et al., 1999).

Quick Tips

In an analysis, if you compare the test performance (or attitude measure) of males and females on the instrument and discover a statistically significant difference, that does not mean you have DIF. Likewise you could be comparing two groups (e.g., again males and females) and you might find no statistical difference between males and females, but that does not mean you do not have DIF.

Table 30 of Winsteps allows you to explore DIF.

Use a DIF contrast of greater than .64 as a way of identifying moderate to large DIF.

In our control file, these lines were used to enable Winsteps to look at DIF.

```
@Gender=1E1 ; $C12W1
@Race=3E4 ; $C14W2
```

The first line tells the program where the gender data are located in the data set. The second line tells the program where the race data are located in the data set.

Remember if you are making numerous comparisons, you will need to make a Bonferroni correction.

Data Sets: (go to http://extras.springer.com)

cf created for activity #2 DIF chapter
For DIF activity #4 from Sibel.xls
cf for DIF chp OE Data From Excel

Raw Data for DIF Chp.xls
cf DIF m only 1 to 9 anch
cf DIF f only 1 to 9 anch
Male measures final
Female measures final
Excel data SE STEBI used for DIF exercise

Activities

Activity #1

```
White                                         Hispanic                                   Asian
    |
    |  Q19                                         Q19                                        Q19
    |  Q6
    |S Q5                                          Q5                                         Q5
    +
    |  Q20       Q23                               Q20     Q23                                Q20     Q23
    |  Q17                                         Q17                                        Q17
    |  Q12                                         Q12                                        Q12
    |  Q18       Q21        Q3                     Q18     Q21     Q3                         Q18     Q21
  S+M
    |
    |                                              Q6                                         Q6
    |
    |
    +  Q8                                          Q8                                         Q8
   T|S
    |
    |                                                                                         Q3
    |  Q22       Q2                                Q22     Q2                                 Q22     Q2
```

Throughout most of this book, we have utilized the 13-item SE component of the STEBI. Using horizontal lines to represent the trait of self-efficacy, show potential DIF for the 13 SE items as a function of three racial subgroups (African American, White, Asian). We have created this plot to demonstrate DIF, and it does not represent a real data set, but the patterns are those that one might observe when considering DIF in many data sets.

Answer: There are as many answers to this item as there are leaves in a forest! Below we present some observations regarding ways in which 13 SE items define the trait for the three subgroups.

Readers should note that the relative spacing and ordering of almost all of the items are the same for the three comparison groups. There are two items that appear in quite different relative locations as a function of subgroup. Item 6 is in quite a different relative location for the White subgroup in comparison to the other subgroups. Item #3 is also in quite a different relative location. Item #3 is in a similar relative location for White and Hispanic. These are two potential items that may exhibit DIF.

Activity #2

We provide an Excel data set that includes the answers of 75 respondents to the 13 self-efficacy (SE) items of the STEBI (excel data SE STEBI used for DIF exercise). Create a control file that will allow you to examine possible DIF for the two school types, School A and School B, that students attended. Also, when you create your control file, make sure you can conduct a DIF comparison as a function of gender.

Answer: We provide as an electronic file the control file that we created (cf created for activity #2 DIF chapter). The file is below, with comments edited out and also only the first line of data provided.

```
&INST
Title="SE STEBI data used for DIF Exercise.xlsx"
ITEM1=1 ; Starting column of item responses
NI=13 ; Number of items
NAME1=15 ; Starting column for person label in data
; record
NAMLEN=73 ; Length of person label
XWIDE=1 ; Matches the widest data value observed
CODES=123456x ; matches the data
TOTALSCORE=Yes ; Include extreme responses in reported
; scores
@ID=1E68 ; $C15W68
@Gender=70E70 ; $C84W1
@Schl-A-or-B=72E72 ; $C86W1
&END ; Item labels follow: columns in label
Q2 ; Item 1 : 1-1
Q3 ; Item 2 : 2-2
Q5 ; Item 3 : 3-3
Q6 ; Item 4 : 4-4
Q8 ; Item 5 : 5-5
Q12 ; Item 6 : 6-6
Q17 ; Item 7 : 7-7
Q18 ; Item 8 : 8-8
Q19 ; Item 9 : 9-9
Q20 ; Item 10 : 10-10
Q21 ; Item 11 : 11-11
Q22 ; Item 12 : 12-12
Q23 ; Item 13 : 13-13
END NAMES
6526525545555 21141    PR 4655265555425445554 5555    ;
spring 2008 PRE      M A
```

Activity #3

Using the control file that you just created, evaluate potential DIF of the self-efficacy scale as a function of gender and as a function of school type.

Answer: Winsteps Table 30.1 immediately below allows one to generate the table for a comparison of gender and the table for a comparison of school. Table 30.1 (immediately below) for gender suggests that item Q5 and item Q19 may exhibit potential DIF. The two items have a probability below .05, so there is a significant difference. But, the absolute value of the DIF contrast is not greater than .64, which suggests the statistical difference is not meaningful.

```
TABLE 30.1 SE STEBI data used for DIF Exercise.x ZOU450WS.TXT  Dec 19 22:03 2011
INPUT: 75 PERSON  13 ITEM  REPORTED: 75 PERSON  13 ITEM  6 CATS  MINISTEP 3.72.3
--------------------------------------------------------------------------------

DIF class specification is: DIF=@GENDER
```

PERSON CLASS	DIF MEASURE	DIF S.E.	PERSON CLASS	DIF MEASURE	DIF S.E.	DIF CONTRAST	JOINT S.E.	Welch t	Welch d.f.	Welch Prob.	Mantel-Haenszel Chi-squ	Mantel-Haenszel Prob.	Size CUMLOR	ITEM Number	ITEM Name
F	-2.46	.30	M	-2.54	.33	.08	.45	.19	71	.8536	2.7293	.0985	1.50	1	Q2
F	.19	.18	M	.26	.20	-.07	.27	-.26	71	.7951	.0935	.7598	.21	2	Q3
F	.88	.17	M	1.44	.18	-.56	.25	-2.27	71	.0260	2.4511	.1174	-1.15	3	Q5
F	-.34	.20	M	-.67	.24	.34	.31	1.08	70	.2832	.0001	.9927	-.01	4	Q6
F	-.91	.23	M	-.91	.26	.00	.34	.00	71	1.000	.0256	.8729	-.11	5	Q8
F	.38	.18	M	.29	.19	.08	.26	.31	71	.7564	.9723	.3241	-.73	6	Q12
F	.53	.18	M	.47	.20	.06	.27	.23	65	.8208	.2653	.6065	-.37	7	Q17
F	.06	.19	M	.18	.21	-.12	.28	-.43	65	.6677	1.2911	.2559	-.76	8	Q18
F	1.37	.18	M	1.95	.19	-.59	.27	-2.21	64	.0304	3.6860	.0549	-1.60	9	Q19
F	.85	.18	M	.66	.19	.19	.26	.71	65	.4802	2.7034	.1001	1.37	10	Q20
F	.30	.18	M	.00	.22	.30	.28	1.07	64	.2902	.9100	.3401	.77	11	Q21
F	-1.61	.28	M	-2.13	.33	.51	.43	1.20	64	.2344	.2109	.6460	.39	12	Q22
F	.94	.18	M	.62	.19	.32	.26	1.21	65	.2312	4.5562	.0328	1.61	13	Q23

There does not appear to be potential DIF as a function of school type. No items have a p value of <.05.

```
TABLE 30.1 SE STEBI data used for DIF Exercise.x ZOU450WS.TXT  Dec 19 22:03 2011
INPUT: 75 PERSON  13 ITEM  REPORTED: 75 PERSON  13 ITEM  6 CATS  MINISTEP 3.72.3
--------------------------------------------------------------------------------

DIF class specification is: DIF=@SCHL-A-OR
```

PERSON CLASS	DIF MEASURE	DIF S.E.	PERSON CLASS	DIF MEASURE	DIF S.E.	DIF CONTRAST	JOINT S.E.	Welch t	Welch d.f.	Welch Prob.	Mantel-Haenszel Chi-squ	Mantel-Haenszel Prob.	Size CUMLOR	ITEM Number	ITEM Name
A	-2.66	.33	B	-2.36	.30	-.30	.45	-.66	70	.5113	.0047	.9454	-.04	1	Q2
A	.26	.19	B	.18	.18	.09	.27	.33	71	.7420	.5959	.4402	.44	2	Q3
A	1.16	.19	B	1.16	.17	.00	.25	.00	70	1.000	.5041	.4777	-.41	3	Q5
A	-.38	.21	B	-.58	.22	.20	.31	.65	71	.5197	.3187	.5724	.39	4	Q6
A	-.67	.23	B	-1.20	.25	.53	.34	1.56	71	.1233	1.2199	.2694	.93	5	Q8
A	.26	.19	B	.40	.18	-.14	.26	-.52	70	.6028	.6577	.4174	-.43	6	Q12
A	.34	.21	B	.61	.17	-.27	.27	-.99	60	.3267	2.8645	.0906	-1.04	7	Q17
A	.16	.22	B	.07	.19	.08	.29	.30	61	.7677	.2742	.6005	-.31	8	Q18
A	1.78	.21	B	1.57	.17	.21	.27	.76	60	.4495	1.0162	.3134	.65	9	Q19
A	.56	.21	B	.89	.17	-.33	.27	-1.25	60	.2156	.5210	.4704	-.41	10	Q20
A	.21	.21	B	.14	.18	.06	.28	.22	61	.8267	.9364	.3332	.61	11	Q21
A	-1.92	.33	B	-1.77	.28	-.15	.43	-.36	61	.7214	.0612	.8046	.16	12	Q22
A	.79	.20	B	.79	.17	.00	.26	.00	61	1.000	.0005	.9822	-.01	13	Q23

Activity #4

We provide readers with an Excel sheet (file: For DIF activity #4 from Sibel.xls) that contains a nonrandom sample of data collected by our colleague Sibel Telli using the The Test of Science Related Attitudes (TOSRA) (see Fraser, 1981).

For the purposes of this activity, let us treat all the TOSRA items as defining a single trait. First create a control file including the following variables: gender, teacher, and class. Then conduct a DIF analysis for "class" (there are three types of classes: class 1, class 2, and class 3).

Answer: Below we provide Winsteps Table 30.1 for the DIF item analysis of the variable "class." In this evaluation we provide the entire first table as produced by

Winsteps/Ministeps. This is because in the case of the class variable there are three comparisons for each item. For example, for the first item of the data set (item t2) there is a comparison of class 1 and class 2 for item t2, there is a comparison of class 1 and class 3 for item t2, and there is a comparison of class 2 and class 3 for item t2. We have underlined all the comparisons for which there is a *p* value <.05, but not all items are above the threshold of .64 DIF contrast. So, some of the comparisons may be statistically significant, but may not be meaningful. It is also important for readers to note that each comparison appears twice. For example, for t19 there is a comparison for class 1 and class 2 in the first third of the table. Then in the middle third of the table there is a comparison for item t19 for class 2 and class 1.

```
TABLE 30.1 For DIF Activity #4 from Sibel.xls        ZOU118WS.TXT  Dec 20  8:44 2011
INPUT: 75 PERSON  10 ITEM  REPORTED: 75 PERSON  10 ITEM  5 CATS     WINSTEPS 3.73
------------------------------------------------------------------------------------

DIF class specification is: DIF=@CLASS
```

PERSON CLASS	DIF MEASURE	DIF S.E.	PERSON CLASS	DIF MEASURE	DIF S.E.	DIF CONTRAST	JOINT S.E.	Welch t	Welch d.f.	Welch Prob.	Mantel-Haenszel Chi-squ	Mantel-Haenszel Prob.	Size CUMLOR	ITEM Number	Name
1	1.58	.27	2	1.35	.22	.23	.35	.67	41	.5083	2.9930	.0836	1.58	1	t2
1	1.58	.27	3	.78	.18	.79	.32	2.45	40	.0189	6.7189	.0095	1.90	1	t2
1	-1.16	.30	2	-1.04	.25	-.12	.39	-.31	41	.7614	.9263	.3358	-.82	2	t6
1	-1.16	.30	3	-.35	.18	-.81	.35	-2.30	39	.0268	6.8948	.0086	-2.33	2	t6
1	.38	.20	2	.43	.18	-.05	.27	-.20	42	.8396	.1821	.6696	.25	3	t10
1	.38	.20	3	.53	.17	-.15	.26	-.58	43	.5669	.2063	.6497	.26	3	t10
1	-.99	.28	2	-.92	.24	-.07	.37	-.20	42	.8448	.6001	.4385	-.61	4	t13
1	-.99	.28	3	-.32	.18	-.68	.33	-2.02	40	.0500	6.6934	.0097		4	t13
1	.99	.22	2	1.13	.20	-.14	.30	-.45	42	.6545	.0759	.7829	.23	5	t17
1	.99	.22	3	.68	.17	.31	.28	1.12	42	.2686	3.1361	.0766	1.23	5	t17
1	-1.26	.31	2	-2.56	.51	1.31	.60	2.19	43	.0338	.7186	.3966	.53	6	t19
1	-1.26	.31	3	-1.08	.24	-.18	.39	-.45	42	.6523	.0922	.7614	-.24	6	t19
1	1.20	.24	2	1.44	.23	-.24	.33	-.75	42	.4569	.1846	.6675	-.27	7	t23
1	1.20	.24	3	.71	.17	.49	.29	1.69	41	.0995	2.4623	.1166	.97	7	t23
1	-.42	.23	2	-.71	.22	.28	.32	.89	42	.3762	.4947	.4818	.48	8	t26
1	-.42	.23	3	-.53	.19	.10	.30	.35	43	.7288	.0073	.9319	-.06	8	t26
1	.42	.20	2	.76	.18	-.34	.27	-1.23	42	.2254	1.8526	.1735	-.84	9	t29
1	.42	.20	3	.37	.17	.05	.27	.19	42	.8525	.0288	.8653	.11	9	t29
1	-.71	.25	2	-.86	.23	.15	.34	.44	42	.6600	.0188	.8909	.08	10	t31
1	-.71	.25	3	-.46	.19	-.26	.32	-.81	41	.4246	.4547	.5001	-.46	10	t31
2	1.35	.22	1	1.58	.27	-.23	.35	-.67	41	.5083	2.9930	.0836	-1.58	1	t2
2	1.35	.22	3	.78	.18	.56	.28	1.99	51	.0517	.1868	.6656	.28	1	t2
2	-1.04	.25	1	-1.16	.30	.12	.39	.31	41	.7614	.9263	.3358	.82	2	t6
2	-1.04	.25	3	-.35	.18	-.69	.31	-2.21	50	.0316	3.3932	.0655	-1.19	2	t6
2	.43	.18	1	.38	.20	.05	.27	.20	42	.8396	.1821	.6696	-.25	3	t10
2	.43	.18	3	.53	.17	-.10	.24	-.40	52	.6919	.3578	.5498	-.39	3	t10
2	-.92	.24	1	-.99	.28	.07	.37	.20	42	.8448	.6001	.4385	.61	4	t13
2	-.92	.24	3	-.32	.18	-.60	.30	-2.00	51	.0507	4.2137	.0401	-1.45	4	t13
2	1.13	.20	1	.99	.22	.14	.30	.45	42	.6545	.0759	.7829	-.23	5	t17
2	1.13	.20	3	.68	.17	.45	.27	1.69	52	.0965	6.5453	.0105	2.01	5	t17
2	-2.56	.51	1	-1.26	.31	-1.31	.60	-2.19	43	.0338	.7186	.3966	-.53	6	t19
2	-2.56	.51	3	-1.08	.24	-1.48	.56	-2.65	45	.0111	3.0662	.0799	-1.41	6	t19
2	1.44	.23	1	1.20	.24	.24	.33	.75	42	.4569	.1846	.6675	.27	7	t23
2	1.44	.23	3	.71	.17	.74	.28	2.59	51	.0124	4.7617	.0291	1.66	7	t23
2	-.71	.22	1	-.42	.23	-.28	.32	-.89	42	.3762	.4947	.4818	-.48	8	t26
2	-.71	.22	3	-.53	.19	-.18	.29	-.61	52	.5433	.1557	.6931	-.25	8	t26
2	.76	.18	1	.42	.20	.34	.27	1.23	42	.2254	1.8526	.1735	.84	9	t29
2	.76	.18	3	.37	.17	.39	.25	1.53	50	.1320	2.8499	.0914	1.02	9	t29
2	-.86	.23	1	-.71	.25	-.15	.34	-.44	42	.6600	.0188	.8909	-.08	10	t31
2	-.86	.23	3	-.46	.19	-.41	.30	-1.34	51	.1847	1.4550	.2277	-.91	10	t31
3	.78	.18	1	1.58	.27	-.79	.32	-2.45	40	.0189	6.7189	.0095	-1.90	1	t2
3	.78	.18	2	1.35	.22	-.56	.28	-1.99	51	.0517	.1868	.6656	-.28	1	t2
3	-.35	.18	1	-1.16	.30	.81	.35	2.30	39	.0268	6.8948	.0086	2.33	2	t6
3	-.35	.18	2	-1.04	.25	.69	.31	2.21	50	.0316	3.3932	.0655	1.19	2	t6
3	.53	.17	1	.38	.20	.15	.26	.58	43	.5669	.2063	.6497	-.26	3	t10
3	.53	.17	2	.43	.18	.10	.24	.40	52	.6919	.3578	.5498	.39	3	t10
3	-.32	.18	1	-.99	.28	.68	.33	2.02	40	.0500	6.6934	.0097		4	t13
3	-.32	.18	2	-.92	.24	.60	.30	2.00	51	.0507	4.2137	.0401	1.45	4	t13
3	.68	.17	1	.99	.22	-.31	.28	-1.12	42	.2686	3.1361	.0766	-1.23	5	t17
3	.68	.17	2	1.13	.20	-.45	.27	-1.69	52	.0965	6.5453	.0105	-2.01	5	t17
3	-1.08	.24	1	-1.26	.31	.18	.39	.45	42	.6523	.0922	.7614	.24	6	t19
3	-1.08	.24	2	-2.56	.51	1.48	.56	2.65	45	.0111	3.0662	.0799	1.41	6	t19
3	.71	.17	1	1.20	.24	-.49	.29	-1.69	41	.0995	2.4623	.1166	-.97	7	t23
3	.71	.17	2	1.44	.23	-.74	.28	-2.59	51	.0124	4.7617	.0291	-1.66	7	t23
3	-.53	.19	1	-.42	.23	-.10	.30	-.35	43	.7288	.0073	.9319	.06	8	t26
3	-.53	.19	2	-.71	.22	.18	.29	.61	52	.5433	.1557	.6931	.25	8	t26
3	.37	.17	1	.42	.20	-.05	.27	-.19	42	.8525	.0288	.8653	-.11	9	t29
3	.37	.17	2	.76	.18	-.39	.25	-1.53	50	.1320	2.8499	.0914	-1.02	9	t29
3	-.46	.19	1	-.71	.25	.26	.32	.81	41	.4246	.4547	.5001	.46	10	t31
3	-.46	.19	2	-.86	.23	.41	.30	1.34	51	.1847	1.4550	.2277	.91	10	t31

References

Fraser, B. J. (1981). *Test of science related attitudes handbook (TOSRA)*. Melbourne, Australia: Australian Council for Educational Research.

Additional Readings

Two generally nontechnical discussions of DIF.

Luppescu, S. (1991). Graphical diagnosis. *Rasch Measurement Transactions, 5*(1), 136.
Luppescu, S. (1993). DIF detection examined. *Rasch Measurement Transactions, 7*(2), 285–286.
Zwick, R., Thayer, D. T., & Lewis, C. (1999). An empirical Bayes approach to Mantel-Haenszel DIF analysis. *Journal of Educational Measurement, 36*(1), 1–28.

Chapter 14
Linking Surveys and Tests

Isabelle and Ted: Two Colleagues Conversing

Ted: *I have a survey that I previously administered to students, and I want to change the survey a little bit. When I use it next time, can I still look at all of the data I collected?*

Isabelle: *Are you asking if you can give two surveys that are a little different and still compare students?*

Ted: *Exactly! I do not want to give the identical survey to both sets of students. When I gave the survey the first time – in September – it was clear that the survey was a little long. I would like to shorten the survey for the second time I collect the data in January from a different group of students. But, I want to be able to compare the students from both the initial survey and the second survey. I know I need to make sure the students are expressed on the same metric. I think Rasch will help me with this, what do you think?*

Isabelle: *It will – as long as you are carful and thoughtful. Let me explain…..*

Introduction

In previous chapters, we introduced and discussed a wide range of issues that not only help one think about measurement but also will help one use measurement on a test or a survey. Thus far, we have generally presented an entire test or survey to explain a concept, strategy, or process. Here we will introduce readers to Rasch techniques and thinking that allow one to present different versions of an instrument – be it a survey or a test – to respondents, but in a way that still allows a researcher to measure respondents with the same metric. This aspect of Rasch measurement is powerful for researchers. Tests and surveys can be altered over time. New items may be added and older items removed in order to improve a test or survey. Items may also be edited to improve wording. But, when the instruments involve the same trait, a researcher can employ Rasch techniques to express all respondents on the same metric. This then allows data to be collected over time and with a number of slightly different versions of instruments, yet the data can still be evaluated.

W.J. Boone et al., *Rasch Analysis in the Human Sciences*,
DOI 10.1007/978-94-007-6857-4_14, © Springer Science+Business Media Dordrecht 2014

No longer must a researcher discard "old" data collected with a different version of an instrument.

Rasch measurement provides an ability to collect data over time and, with different versions of an instrument, yet still evaluate data as if all respondents completed an identical instrument. Additionally, since Rasch measurement allows different forms of an instrument to be linked, researchers can fine-tune instruments over time, thereby improving the measure of respondents yet retaining the ability to compare respondents who completed earlier forms. Those interested in books and articles which will support your understanding of the material we present here are suggested to review portions of *Best Test Design* written by Wright and Stone (1979). Another useful article which you may want to review when you complete this chapter is "Equating and Item Banking with the Rasch Model" (Wolfe, 2000).

Formative Assessment Checkpoint #1

Question: If items are added to a survey or test, then can data collected with the "old" instrument and the "new" instrument be compared?

Answer: Yes. If the same metric is used with the old and the new instrument, then it *can* be possible to express all respondents on the same measurement scale. For beginners, think of making sure that some identical items are used for both versions of an instrument.

The Linking Process

To introduce the technique used to "link" one form of an instrument to another, we present a "real-world" example of "linking." Figure 14.1 presents a sketch of scaffolding built alongside a building. The scaffolding may look a little different than scaffolding one might see alongside a real building. Close inspection, reveals that some ladders serve as walkways from one part of the scaffold to another, but the whole length of the scaffold is not traversed. These ladders could still be viewed as not a complete waste, in that the boards might allow a worker to inspect part of a building's façade!

This drawing provides a visual overview of what takes place when common items are used to link a test to another test or a survey to another survey. Using Rasch measurement, a researcher can express respondents on the same metric – even if they have taken different versions of a survey or test – as long as both instruments contain some common items. Most importantly, linking different forms of a survey or test only makes sense if both instruments measure the same trait!

There are numerous considerations when linking instruments. Here we present the basics so that readers can confidently begin the process of linking appropriate

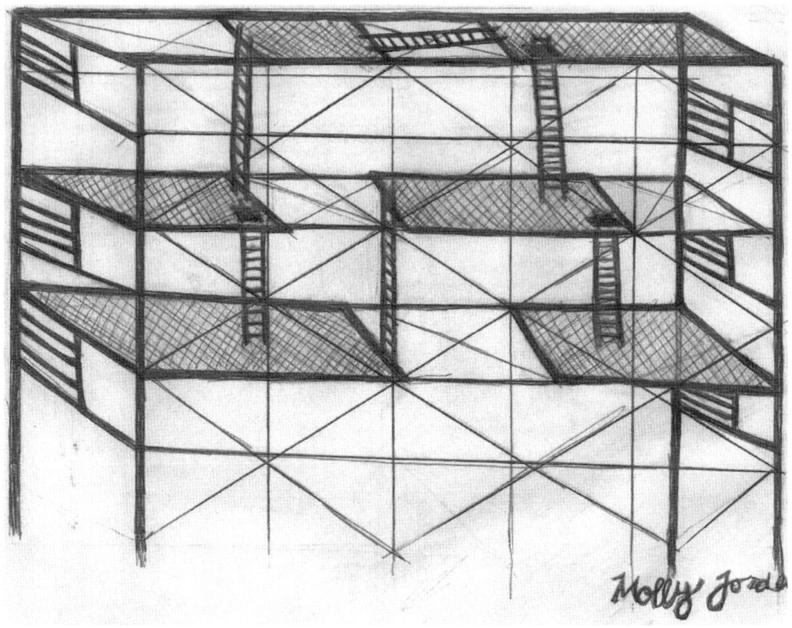

Fig. 14.1 Scaffolding of a building under construction (Figure created by Molly Jorden)

instruments. Throughout this book, we have used a data set from the self-efficacy subscale of Enochs and Riggs' well-known instrument. Let us pretend that we wish to compare the self-efficacy of a group of 50 teachers at the end of a year-long intervention to the self-efficacy of another group of 25 teachers at the end of a similar year-long intervention. The 25 teachers completed their intervention a year after the 50 teachers completed their intervention. Let us also pretend that all 13 self-efficacy STEBI items were given to the group of 50 teachers, but we wished to collect data using a shortened version of the 13-item STEBI when data were collected from the group of 25 teachers. So, we are just juggling so much data collection, and we want to use the self-efficacy STEBI, but we hope we might be able to present fewer items to respondents. Yes, we know that shortening the survey from 13 items to a smaller number of items does not save that much time for respondents, but this example will allow readers to learn how to link when they might need to cut down the number of items from a far larger survey or test.

To explain the set of steps that might be used to link an original form of an instrument to a revised form of the same instrument, we used the data set with the 50 respondents we mentioned immediately above. The control file (cf50SEitemlinkingchp) that we created with this data set is provided in Fig. 14.2; however, we provide only the first line of data and the last line of data. Readers will also note that a new line, which we have not seen before (SFILE=SFILE50), appears in the control file. This is a key line that will, in part, allow us to link the full 13-item

```
&INST
TITLE = '50 Teachers SCIENCE TEACHER EFFICACY BELIEFS'
; This is the title that gets printed on pages of output
NAME1 = 1
;This is telling the program that the first column of data
; is the start of the survey taker ID
NAMELENGTH = 10
; This is telling the program that the length of the respondent
; ID is 10 columns wide
ITEM1 = 11
; This is telling the program that that the 1st piece of data is
; in the 11th column of data
NI = 23
; This is telling the program that there are 23 items in the data set
CODES = "123456"
; This is telling the program that the codes for data are 123456
FORMAT=(10A1,1X,23(A1))
; This is an old fashioned way of reading BIGSTEPS and Winsteps data
; All STEBI items read in, but we only want to look at 13 SE items,
; these 7 lines remove the OE STEBI items so that we only are evaluating the SE items
IDFILE=*
1
4
7
9-11
13-16
*
; All STEBI items read in, but we only want to look at 13 SE items
; these 7 lines remove the OE STEBI items so that we only are evaluating the SE items
SFILE=sfile50
; This is the line that generates the information we need to link! Yea!
;
&END
Q1oe
Q2se
Q3se-rc
Q4oe
Q5se
Q6se-rc
Q7oe
Q8se-rc
Q9oe
Q10oe-rc
Q11oe
Q12se
Q13oe-rc
Q14oe
Q15oe
Q16oe
Q17se-rc
Q18se
Q19se-rc
Q20se-rc
Q21se-rc
Q22se
Q23se-rc
END LABELS
21141    PR 46552655554254455545555
.
.
.ALL DATA NOT DISPLAYED FOR THIS FIGURE
.
.
.
68028    PR 56555546645525555636366
```

Fig. 14.2 The control file used for the analysis of 50 respondents. Only the 13 self-efficacy items of the STEBI are evaluated. Only the first line of data and the last line of data are presented. The key line in the control file that will help in future linking is the line that begins with the word SFILE. We have underlined that line for readers

```
; STRUCTURE MEASURE ANCHOR FILE FOR 50 Teachers SCIENCE TEACHER EFFICACY BELIEFS Dec 11 12:28
2011
; CATEGORY  Rasch-Andrich threshold
  1      .00
  2    -3.05
  3     -.34
  4     -.32
  5      .39
  6     3.32
```

Fig. 14.3 A Winsteps SFILE created by running the 50 respondents' self-efficacy data. Of importance for readers to note is that the Rasch-Andrich thresholds (taus, steps) from 3 (a –.34) to a 4 (a –.32) to a 5 (a .39) are not the same

self-efficacy STEBI survey to an abbreviated self-efficacy STEBI survey (let's say only 7 SE items) which is given to another group of respondents.

The command line "SFILE=SFILE50" tells the Winsteps program to compute the steps from one rating scale step to another. Think of this as the length of the gap from SA to A, the length of the gap from A to BA, the length of the gap from BA to BD, and so on. Readers should recall our earlier admonition that one cannot assume the gaps are uniform from SA to A to D to SD. This control file line "SFILE=SFILE50" instructs Winsteps to create a file that includes each of these steps, and the name of this file with the steps is named sfile50. You can give this file any name you wish. This information regarding the steps is critical for later procedures. Figure 14.3 shows the SFILE that was created by running the Winsteps control file cf50SEitem-linkingchp using the 50 respondents.

For our purposes herein, readers need only to be aware that uneven gaps can exist regarding the manner in which a rating scale operates, and it is important to know what the gaps are. If researchers are linking forms of an instrument, they must use their knowledge of these gaps in runs of the data, and the command line SFILE allows one to quickly and confidently take note of the size of gaps in a rating scale which can later be used to ensure the same gaps are used in later analyses of data (e.g., in our example, the later 25 teachers who completed a shortened STEBI).

Following the computation of the steps, the analyst must take note of the value of each item in logits that will be used to set the location of the item along the metric when that item is used in a later data collection. Why is it important to note (to write down and later use) the logit value of the items from the analysis of the SE STEBI data? Readers now need to think back to the rulers that we talk about in almost every chapter. When we conduct measurement, we need to build good rulers. When researchers think about good measurement, they should *always* strive to measure a single trait that will help them advance in their quest to learn and add to human knowledge. When administering a survey, such as the SE portion of the STEBI, the SE items serve to mark the locations of different parts of the trait (some items are easier to agree with than others). If we are attempting to link forms, we need to make sure our rulers (e.g., a ruler from the collection of SE data from the 50 respondents and a ruler from the collection of data from the 25 respondents) line up. One step in lining up the rulers is to make sure that common items mark the same spot on the line of the trait. One way to do so is to simply "set" the location of items

```
&INST
TITLE = '50 Teachers SCIENCE TEACHER EFFICACY BELIEFS'
; This is the title that gets printed on pages of output
NAME1 = 1
;This is telling the program that the first column of data
; is the start of the survey taker ID
NAMELENGTH = 10
; This is telling the program that the length of the respondent
; ID is 10 columns wide
ITEM1 = 11
; This is telling the program that that the 1st piece of data is
; in the 11th column of data
NI = 23
; This is telling the program that there are 23 items in the data set
CODES = "123456"
; This is telling the program that the codes for data are 123456
FORMAT=(10A1,1X,23(A1))
; This is an old fashioned way of reading BIGSTEPS and Winsteps data
; All STEBI items read in, but we only want to look at 13 SE items, these 7 lines remove the
; OE STEBI items so that we only are evaluating the SE items
IDFILE=*
1
4
7
9-11
13-16
*
; All STEBI items read in, but we only want to look at 13 SE items
; these 7 lines remove the OE STEBI items so that we only are evaluating the SE items
SFILE=sfile50
; This is the line that generates the information we need to link! Yea!
;
IFILE=ifile50
; This is the command line which creates a file of item calibrations.
; This file then can be used for linking
&END
```

Fig. 14.4 A copy of the control file cf50SEitemlinkingchp with an added line for the command line IFILE. That line is underlined for readers

using the logit measures of items that have been determined through analysis of the data. We will walk readers through this step, so just sit tight.

In order to set the logit value of an item, one must know the logit values of the item. When we ran Winsteps to compute these step values of Figure 14.3, we also computed the logit measures values for each item. These values can be found in tables such as the item measure table and the item entry table. It is also possible to add a command line to the Winsteps control file so that an output file is created that lists each item and the logit value of that item. That command line is named "IFILE". Below we provide a copy of the control file "cf50SEitemlinkingchp", but readers will see we have added a line for the command line IFILE (Fig. 14.4).

Readers will be able to see in our sample control file "cf50SEitemlinkingchp" that such a line is present. Figure 14.5 presents the file that this command generates.

We need only the entry number of the item to be used to link and the measure of that item; this information appears in the first two columns of data in the table (if in the instrument to be anchored the entry number of an item has changed, then make sure to use the new entry number when anchoring and the item measure which has been computed in the initial analysis). To stay organized through the linking process, readers should know the name of each item. For instance, knowing only that an item

```
; ITEM   50 Teachers SCIENCE TEACHER EFFICACY BELIEFS  Dec 11 13:02 2011
;ENTRY MEASURE ST COUNT SCORE  ERROR IN.MSQ IN.ZST OUT.MS OUT.ZS  DISPL PTMEAS WEIGHT OBSMA EXPMA DISCRM LOWER UPPER PVALU PME-E  RMSR G M R NAME
     2  -2.50  1 50.0  275.0   .28   1.05    .29   1.07    .34    .00    .32   1.00  57.1  65.9    .89   .00  4.95  5.50   .43    .53 1 R . Q2se
     3    .15  1 50.0  216.0   .17   1.50   2.19   1.45   1.90    .00    .61   1.00  40.8  46.1    .49   .00  5.00  4.32   .61   1.05 1 R . Q3se-rc
     5   1.19  1 50.0  175.0   .16   1.27   1.35   1.27   1.23    .00    .58   1.00  42.9  42.6    .75   .00  5.00  3.50   .70   1.03 1 R . Q5se
     6   -.44  1 50.0  235.0   .19   1.13    .63   1.03    .20    .00    .54   1.00  46.9  51.5    .97   .00  5.00  4.70   .56    .82 1 R . Q6se-rc
     8   -.85  1 50.0  246.0   .20   1.10    .50   1.01    .10    .00    .59   1.00  59.2  56.8   1.11   .00  5.00  4.92   .53    .74 1 R . Q8se-rc
    12    .34  1 50.0  209.0   .16    .84   -.77    .89   -.45    .00    .61   1.00  55.1  46.1   1.19   .00  5.00  4.18   .63    .80 1 R . Q12se
    17    .49  1 44.0  180.0   .17    .73  -1.36    .79   -.91    .00    .72   1.00  39.5  44.2   1.21   .00  5.00  4.09   .63    .76 1 R .  Q17se-rc
    18    .09  1 44.0  193.0   .18    .72  -1.31    .68  -1.47    .00    .64   1.00  44.2  47.4   1.30   .00  5.00  4.39   .60    .71 1 R . Q18se
    19   1.69  1 44.0  137.0   .17   1.22   1.05   1.32   1.38    .00    .68   1.00  44.2  43.0    .68   .00  5.00  3.11   .71    .99 1 R . Q19se-rc
    20    .89  1 44.0  166.0   .17   1.23   1.12   1.28   1.25    .00    .65   1.00  41.9  42.7    .83   .00  5.00  3.77   .66   1.02 1 R . Q20se-rc
    21    .12  1 44.0  192.0   .18    .79   -.93    .87   -.49    .00    .64   1.00  46.5  47.3   1.21   .00  5.00  4.36   .60    .75 1 R . Q21se-rc
    22  -1.92  1 44.0  235.0   .27    .61  -1.70    .63  -1.57    .00    .57   1.00  79.1  63.3   1.40   .00  5.00  5.34   .46    .44 1 R . Q22se
    23    .78  1 44.0  170.0   .17    .80   -.99    .87   -.51    .00    .71   1.00  46.5  43.0   1.12   .00  5.00  3.86   .65    .81 1 R . Q23se-rc
```

Fig. 14.5 A copy of the file generated by the command IFILE. For clarity, we have only provided the data for the self-efficacy items, ignoring Winsteps notations regarding the deleted outcome-expectancy items (see the PDFILE command in the control file). For purposes of linking, the important information is solely the entry number of the item and the logit calibration. For example, the 22nd item (Q22se) that appeared in the survey exhibits a calibration of −1.92 logits

is the 17th item in the survey will cause some confusion. It is also important to know the measures of each item and if it was reverse coded or not.

The information in the table created by the command "IFILE" can be used to master the art of linking. In our work, however, we often take a "shortcut," in that we just skip using IFILE. Instead, we copy and paste the Entry Item Measure Table of Winsteps. This table contains all the pertinent information available in the IFILE table, and it is a table that we and our readers are by now quite familiar with in terms of its organization. When linking is carried out, the SAFILE previously discussed is used, and a simple file is created through use of the item statistics table (or alternatively the table produced through use of the IFILE command).

To link (in addition to using the step calibrations), one must list the entry number of the item and the item measure calibration on a line for each item to be used for linking from one form of an instrument to another. Figure 14.6 shows how a control file would appear if items 2, 3, 5, 6, 8, 12, and 17 of the STEBI were used as the items to link one form of the STEBI to another. So when you wish to anchor, just mimic what is written for the line IAFILE and then, of course, put the ENTRY number of the item being used to anchor as well as the calibration to which the item will be set. Also, one must add the values of the steps that have been determined from the analysis of the $n=50$ data set. To simplify our presentation of the steps needed to link one form to another, we have pretended that only the first 17 items of a survey were presented in the shortened version of the survey to respondents. To see this, readers should review the data at the base of the control file. Readers should see a series of X's starting in the 18th column of the data. For instance, the first person in the $n=25$ data set had the following data:

94827 *PR* 56545445434564 44*XXXXXXX*

Following the insertion of the "step anchors" (Rasch-Thurstone Thresholds) in the control file, as we might call them, and insertion of the "item anchors" into the control file, readers will find it important (after they have run their control file which includes the step anchors and the item anchors) to look at two tables from Winsteps to make sure that the anchoring of the steps and the items did indeed take place after the analysis has been run.

The control file of Fig. 14.6 is used to anchor the data collected with only 17 STEBI items to data collected with all 23 STEBI items. Since outcome-expectancy items are not used for either analysis (OE is a different metric), those items are removed from both analyses. Note the control file contains information for anchoring steps and anchoring items. Also note the presence of two key lines, SAFILE and IAFILE. To save space, only the first two respondents' answers are presented.

```
&INST
TITLE = 'SCIENCE TEACHER EFFICACY BELIEFS'
NAME1 = 1
NAMELENGTH = 10
ITEM1 = 11
NI = 23
CODES = "123456"
; Remember for the linking we are using data in which only the first 17 STEBI items are
; presented to respondents. That is why we are reading only the first 17 items in.
FORMAT=(10A1,1X,17(A1))
IDFILE=*
1 oe
4
7
9-11
13-16
*
; The lines below anchor the steps. Notice the command line is SAFILE, not SFILE (there is a
;letter "A" after the "S")
; The SFILE from the other control file generates the "step" information. The SAFILE command
;will ;"set" the gapsbetween
; the rating scale categories.
;
SAFILE=*
  1     .00
  2   -3.05
  3    -.34
  4    -.32
  5     .39
  6    3.32
*
; The lines below anchor the items for this run of data to the values in logits determined
;from ;the run of the data
; using the 50 teachers. If you look at the analysis of the 50 people and look up any of the
;items listed below
; you will see the noted logit value. That is where these values came from!
;
IAFILE=*
2 -2.50
3 .15
5 1.19
6 -.44
8 -.85
12 .34
17 .49
*
&END
Q1oe
Q2se
Q3se-rc
Q4oe
Q5se
Q6se-rc
Q7oe
Q8se-rc
Q9oe
Q10oe-rc
Q11oe
Q12se
Q13oe-rc
Q14oe
Q15oe
Q16oe
Q17se-rc
END LABELS
94827    PR 56545445434564444XXXXXXX
36206    PR 55562545556255555XXXXXXX
```

Fig. 14.6 Control file would appear if items 2, 3, 5, 6, 8, 12, and 17 of the STEBI were used as the items to link one form of the STEBI to another. We have underlined and bolded the key lines which facilitate anchoring

Double-Checking Your Linking

Winsteps provides a simple way to conduct this important check of assurance that the steps and the items were indeed anchored. Below find our Fig. 14.7 (Winsteps Table 3.2), which lists the step values used for the analysis of the $n=25$ data set. Notice that the second to last column of the table does indeed present the numbers that were presented as the step anchors. Also, those numbers are followed by the letter "A," which confirms that anchoring of steps occurred. In this table the word "NONE" is used for the number "0." Of course, it is important to make sure that the step anchors are correctly entered in the control file used for the step anchoring. The program will not know if the correct number is entered into the control file to accomplish anchoring. An advantage in using the command line "SFILE" in the $n=50$ run of the data to create the step anchor file is the diminished danger of placing incorrect numbers in the $n=25$ control file for the step anchors (if you were to hand type). Remember again, if one wants to link forms of a survey, then the first of two key steps is to make sure the step anchors used to evaluate the two sets of data are identical.

```
TABLE 3.2 SCIENCE TEACHER EFFICACY BELIEFS          ZOU086WS.TXT  Dec 11 13:59 2011
INPUT: 25 PERSON  23 ITEM  REPORTED: 25 PERSON  6 ITEM  6 CATS       WINSTEPS 3.73
--------------------------------------------------------------------------------


SUMMARY OF CATEGORY STRUCTURE.  Model="R"
-------------------------------------------------------------------
|CATEGORY   OBSERVED|OBSVD SAMPLE|INFIT OUTFIT|| ANDRICH |CATEGORY|
|LABEL SCORE COUNT %|AVRGE EXPECT|  MNSQ  MNSQ||THRESHOLD| MEASURE|
|-------------------+------------+------------++---------+--------|
|  1   1       3  2|  -.60  -.28|  1.08    .98||  NONE A |( -4.20)| 1
|  2   2      14  9|  -.35  -.37|   .97    .96||  -3.05A|  -1.89 | 2
|  3   3      12  8|   .07   .11|   .96    .85||   -.34A|   -.55 | 3
|  4   4      28 19|   .63   .79|  1.32  1.25||   -.32A|    .35 | 4
|  5   5      77 51|  1.53  1.32|   .96    .95||    .39A|   1.96 | 5
|  6   6      16 11|  3.05  3.73|   .91    .87||   3.32A|(  4.46)| 6
-------------------------------------------------------------------
OBSERVED AVERAGE is mean of measures in category. It is not a parameter estimate.


8 Lines from control file used to anchor data:

SAFILE=*
  1     .00
  2   -3.05
  3    -.34
  4    -.32
  5     .39
  6    3.32
*
```

Fig. 14.7 (Winsteps Table 3.2): Output of the analysis with the control file that included the step anchors. Below the table we provide the portion of the cf used to anchor the Rasch-Andrich thresholds

---

Formative Assessment Checkpoint #2

Question (True/False): The entry numbers are just the names of items in a test; for example, the 3rd entry number is always entry number 3.

Answer: False. The entry numbers cannot be assumed to always be the item number of the test. For instance, if we were evaluating the outcome-expectancy scale of the STEBI and we wanted to use the control file of this chapter, we would make sure we removed all SE items from our analysis. We might do this with IDFILE as we did in the sample control file. In an analysis of the OE items only, the first OE item would have an entry number of 4 since it was the fourth item presented in the original survey. Just make sure to look at one of the item tables (e.g., item entry); there you will be able to find the entry number.

The second step is to make sure that the items common to the two forms were indeed anchored. Figure 14.8 (Winsteps Table 14.1) presents the items statistics from the analysis of the group of 25 teachers. Carefully review the MEASURE column and note that the measure values contain the letter "A" for items with an entry number of 2, 3, 5, 6, 8, 12, and 17. This confirms that these items were indeed anchored to the values entered into the $n=25$ control file. Although it takes a little more editing to edit the output of the "IFILE" command in the $n=50$ table to create the values for the item anchoring, we strongly suggest there is less room for error in terms of correctly entering the item anchor values than if one were to enter by hand the values for anchoring, as could be done when reading the appropriate item calibration values presented in the item entry table of $n=50$. This means that when constructing the control file for the analysis of the 25 respondents, one could just edit the item entry table to create the lines which are needed for IAFILE=*.

Formative Assessment Checkpoint #3

Question: Is IAFILE the same as IFILE?

Answer: No. IFILE is a command that creates a file with (among other things) the entry number of items and the logit calibration of items. IAFILE is the command that is used to anchor items to a particular logit value.

Question: Is SAFLE is the same as SFILE?

Answer: No. SFILE is the command line that provides the details of the step calibrations; SAFILE is the file that is used to anchor step calibrations.

TABLE 14.1 SCIENCE TEACHER EFFICACY BELIEFS ZOU0?6WS.TXT Dec 11 13:59 2011
INPUT: 25 PERSON 23 ITEM REPORTED: 25 PERSON 6 ITEM 6 CATS WINSTEPS 3.73

PERSON: REAL SEP.: .71 REL.: .33 ... ITEM: REAL SEP.: 4.07 REL.: .94

ITEM STATISTICS: ENTRY ORDER

ENTRY NUMBER	TOTAL SCORE	TOTAL COUNT	MEASURE	MODEL S.E.	INFIT MNSQ	INFIT ZSTD	OUTFIT MNSQ	OUTFIT ZSTD	PT-MEASURE CORR.	PT-MEASURE EXP.	EXACT MATCH OBS%	EXACT MATCH EXP%	DISPLACE	ITEM
1	DELETED													Q1oe
2	135	25	-2.50A	.37	.95	-.1	.94	-.1	.39	.34	64.0	60.4	.17	Q2se
3	101	25	.15A	.22	1.57	1.9	1.55	1.7	.37	.49	24.0	43.4	.18	Q3se-rc
4	DELETED													Q4oe
5	83	25	1.19A	.21	1.02	.2	1.02	.2	.64	.58	32.0	35.3	-.09	Q5se
6	117	25	-.44A	.26	.69	-1.0	.67	-1.0	.28	.43	44.0	50.5	-.08	Q6se-rc
7	DELETED													Q7oe
8	123	25	-.85A	.29	.60	-1.2	.63	-1.1	.47	.40	64.0	58.5	-.13	Q8se-rc
9	DELETED													Q9oe
10	DELETED													Q10oe-rc
11	DELETED													Q11oe
12	101	25	.34A	.21	1.08	.4	.91	-.2	.56	.51	40.0	40.0	.00	Q12se
13	DELETED													Q13oe-rc
14	DELETED													Q14oe
15	DELETED													Q15oe
16	DELETED													Q16oe
17	DROPPED													Q17se-rc
18	DROPPED													I0018
19	DROPPED													I0019
20	DROPPED													I0020
21	DROPPED													I0021
22	DROPPED													I0022
23	DROPPED													I0023
MEAN	110.0	25.0	-.35	.26	.99	.0	.95	-.1			44.7	48.0		
S.D.	17.0	.0	1.15	.06	.31	1.0	.30	.9			15.0	9.3		

Fig. 14.8 (Winsteps Table 14.1): The item statistics table results from the analysis of the 25 respondents. The analysis is the result of anchoring rating steps and items to that determined from an analysis of 50 respondents. The anchoring of steps and items facilitates the computation of 25 person measures on the same scale as that used to express the self-efficacy of the 50 respondents who completed the entire 13-item self-efficacy instrument. The letter "A" that is presented in the table indicates that the item calibration was set to a value prior to an analysis of the 25 respondents. The Winsteps manual provides the following information when detailing specifics of Table 14.1 concerning the column DISPLACEMENT (Linacre, 2012):

DISPLACE is the displacement of the reported MEASURE from its data-derived value. This should only be shown with anchored measures. The displacement values can be seen in IFILE= and PFILE= output files. The displacement is an estimate of the amount to add to the MEASURE to make it conform with the data.
+1 logit displacement for a person ability indicates that the observed person score is higher than the expected person score based on the reported measure (anchor value).
+1 logit displacement for an item difficulty indicates that the observed item score is lower than the expected item score based on the reported measure (anchor value).
Unanchored measures: If small displacements are being shown, try tightening the convergence criteria, LCONV=.
Anchored measures: We expect half the displacements to be negative and half to be positive and for them to be normally distributed according to the standard errors of the measures. (p. 418)

Let's reflect briefly on what we have accomplished through the steps outlined in this chapter. We have measured students with two similar forms of the SE scale of the STEBI, and we have expressed students' scores on the same metric, even though they did not complete the same mix of items. In our example, this was done through:

1. The collection of data from a group of fifty ($n=50$) teachers using all STEBI items. This allows item calibrations of all items to be computed. This also allows step calibrations to be computed. Knowledge and later use of item calibrations and step calibrations open the door to anchoring similar surveys measuring the same trait.
2. The collection of data from a group of twenty-five ($n=25$) teachers using a shortened form of the STEBI. Analysis of the $n=25$ data was conducted, but that analysis used anchored steps and anchored items. This was to express the 25 students on the same metric as that used for the 50 students.
3. Researchers can double-check their anchoring procedures by reviewing a number of Winsteps tables in which the letter "A" should appear.

What is the point of all these steps? Anchoring of items and steps permits us to administer similar but not identical forms of a survey to different groups of respondents, yet still express the respondents' measures on the same scale. Only by taking this issue into consideration can one then compare the two data sets. Below we present two score-to-measure conversion tables. These tables show how the raw scores of the $n=50$ data collection are related to the Rasch measures. The second table shows how the $n=25$ data raw scores are related to the Rasch measures. Take a moment and look over these two tables. Notice that each table ranges from a different minimum raw score to a different maximum raw score. This is because each survey ($n=50$ and $n=25$) presented a different number of items (and a different mix of item difficulties) to respondents! If a researcher had no knowledge of Rasch, she or he would think there would not be any method of comparing the performance of a respondent to the full SE STEBI items to a respondent who completed the shorter version of the SE STEBI. It is through the anchoring of steps and items that the measurement scales (SE from 13 items, SE from 6 items) can be expressed on the same metric. Rasch measurement is the technique that will allow you to do this and conduct real measurement (Fig. 14.9).

Examination of these two tables shows that a ($n=50$) respondent with a measure of .58 (raw score of 52) would be predicted to have a raw score of between 25 and 26 if this same respondent had been given the survey form completed by the 25 respondents. Knowledge of the raw score earned by a respondent completing the shorter survey allows a measure (in logits) to be computed. But also, examination of the score-to-measure table for the $n=50$ sample allows a researcher to compute how many raw score points that respondent would have been predicted to have earned had he or she completed the shorter survey. For example, a respondent completing the 13-item survey and earning 65 points (2.15 logit measure) would have been predicted to earn between 30 and 31 points on the shorter survey.

Using Rasch to anchor steps and to anchor items opens the door for researchers to accomplish something amazing: namely, different forms of an instrument can be

TABLE 20.1 SCIENCE TEACHER EFFICACY BELIEFS ZOU086WS.TXT Dec 11 13:59 2011
INPUT: 25 PERSON 23 ITEM REPORTED: 25 PERSON 6 ITEM 6 CATS WINSTEPS 3.73

TABLE OF MEASURES ON TEST OF 6 ITEM

SCORE	MEASURE	S.E.	SCORE	MEASURE	S.E.	SCORE	MEASURE	S.E.
6	-7.01E	1.94	17	-1.20	.48	28	1.17	.53
7	-5.52	1.18	18	-.97	.47	29	1.46	.56
8	-4.49	.89	19	-.76	.46	30	1.81	.61
9	-3.84	.74	20	-.56	.45	31	2.23	.67
10	-3.36	.65	21	-.36	.44	32	2.72	.73
11	-2.96	.60	22	-.16	.44	33	3.30	.80
12	-2.62	.58	23	.04	.45	34	4.01	.89
13	-2.29	.56	24	.24	.45	35	4.99	1.13
14	-1.99	.54	25	.45	.46	36	6.39E	1.90
15	-1.70	.52	26	.67	.48			
16	-1.44	.50	27	.91	.50			

TABLE 20.1 SCIENCE TEACHER EFFICACY BELIEFS ZOU367WS.TXT Jan 31 14:41 2011
INPUT: 43 PERSON 23 ITEM MEASURED: 43 PERSON 7 ITEM 6 CATS WINSTEPS 3.70.6

TABLE 20.1 50 Teachers SCIENCE TEACHER EFFICACY ZOU206WS.TXT Dec 11 15:35 2011
INPUT: 50 PERSON 23 ITEM REPORTED: 50 PERSON 13 ITEM 6 CATS WINSTEPS 3.73

TABLE OF MEASURES ON TEST OF 13 ITEM

SCORE	MEASURE	S.E.	SCORE	MEASURE	S.E.	SCORE	MEASURE	S.E.
13	-7.53E	1.89	35	-1.03	.34	57	1.08	.33
14	-6.17	1.10	36	-.92	.33	58	1.18	.33
15	-5.28	.83	37	-.81	.33	59	1.30	.34
16	-4.70	.70	38	-.70	.32	60	1.42	.35
17	-4.26	.62	39	-.60	.32	61	1.55	.36
18	-3.91	.56	40	-.50	.31	62	1.68	.37
19	-3.62	.52	41	-.41	.31	63	1.83	.39
20	-3.37	.49	42	-.31	.31	64	1.98	.40
21	-3.15	.46	43	-.22	.30	65	2.15	.42
22	-2.95	.44	44	-.13	.30	66	2.34	.44
23	-2.76	.42	45	-.04	.30	67	2.54	.46
24	-2.59	.41	46	.05	.30	68	2.75	.48
25	-2.42	.40	47	.14	.30	69	2.99	.49
26	-2.26	.40	48	.23	.30	70	3.24	.51
27	-2.10	.39	49	.31	.30	71	3.51	.53
28	-1.95	.38	50	.40	.30	72	3.80	.55
29	-1.81	.38	51	.49	.30	73	4.12	.58
30	-1.67	.37	52	.58	.30	74	4.48	.62
31	-1.53	.37	53	.68	.31	75	4.89	.68
32	-1.40	.36	54	.77	.31	76	5.43	.79
33	-1.27	.35	55	.87	.31	77	6.24	1.06
34	-1.15	.35	56	.97	.32	78	7.53E	1.86

Fig. 14.9 (Winsteps Table 20.1): Two score-to-measure conversion tables. A student with a raw score of 28 on the 6-item survey is the equivalent of a student with a raw score of 58 on the 13-item survey. The raw score of 28 is 1.17 logits and the raw score of 58 is 1.18 logits

administered, yet respondents can be expressed on the same scale. This, then and only then, allows comparisons to be made. As one thinks to research fields there are many many instances in which it is very useful to be able to link forms. For instance, perhaps as a patient improves in her or his condition from an illness she or he is administered a different form of an instrument that is target at her or his perception of her or his health level. This would allow the presentation of items that are targeted to the patient, as opposed to being too easy or too difficult. This in the end results in the better measurement of the patient and thus (one hopes) better medical diagnosis and care.

A Few Closing Observations

If we were explaining some of a project's results to readers of a journal, we might simply present the table from the $n=50$ data and explain that all measures of respondents for both the $n=50$ and $n=25$ are expressed on the scale presented in the table up to 78 raw score points. We typically would state that different versions of the instrument were administered to the two groups ($n=50$ and $n=25$) of respondents, but through Rasch analysis, we are able to compute person measures that are expressed on a metric as if ALL the respondents had taken the full STEBI. We have found it easy to present reviewers and readers of articles with a table such as the $n=50$ table, in which measures for all possible raw scores are presented. We then explain, in simple terms, that we were able to administer different forms of a survey, but we were still able to express all measures the same scale.

In this chapter we have presented the procedure to anchor steps and items using the scenario of initial completion of a survey with 13 items and then the later completion of a 6-item survey which measures the same trait by using 6 of the 13 items presented to respondents using the long survey form. A common question that participants often pose in our workshops is whether or not it is possible to alter the form of a survey by adding items. Attendees might ask: "I have collected some data using a survey with 10 items to measure the trait of XYZ. A year into my project I realize that I could really improve my measurement of respondents by adding items to the survey. Is it possible to add 8 items to a survey but still express all respondents, past and present, on the same scale? This is important because we want to evaluate respondents on the XYZ trait over time." The answer to this question is yes, one can add items to a survey. The steps the analyst would take would be almost identical to the steps presented in this chapter. First the data with the 10 items would be evaluated (maybe in the fall of 2030), and a report of some sort authored using person measures anchored to a 10 scale using the rating scale steps from the analysis. Then perhaps data were collected using the original 10 items, and items 11–18 were added, with that data being collected in the fall of 2031. The analyst would first compute the steps for anchoring, most likely computing the steps, by just rerunning the fall 2030 data with a SAFILE line added to the control file. The analyst would need the calibrations of items 1–10 for anchoring of the fall 2031 data. More than likely, the analyst would examine the rerun of the fall 2030 data, write down the entry number of each item, and note the calibration of each item. Then the analyst would create a control file for the fall 2031 data (a data set with 8 more items). That control file would have a number of lines using the command line SAFILE, and also there would be 10 items anchored (using the command IAFILE) utilizing the values computed from the fall 2030 analysis. When the analyst completes the analysis of the fall 2031 data, she or he would pull up the item entry table and see that there would be the letter "A" for items 1–10, but for items 11–18 there would not be an "A." After a few moments of reflection, the analyst would understand that this was what she or he should see. This is because the fall 2030 scale and the fall 2031 scale were anchored using "steps" and only items 1–10.

Formative Assessment Checkpoint #4

Question: Why were items 11–18 not anchored?

Answer: There could not be any anchors of those items because those items did not appear in the fall 2031 data collection.

A second question sometimes posed to us during workshops is as follows: Would these techniques of step anchoring and item anchoring be useful to researchers if there were changes in the wording of items between pre and post data collections? The answer, of course, is "yes." In the case of a change in wording, a researcher should consider any modified items to be new items. So, let's pretend data were collected with a 12-item survey in the spring of 2011 from German physics students in Kiel, Germany. The researchers discover, through no fault of their own, that item 9 could be improved. When data are collected in the spring of 2012, the improved item 9 is presented to students. To analyze these data, the researcher would have probably already evaluated the spring 2011 data, and as a result, the researcher would already have the step calibrations as well as the logit values for items 1–12. Protecting against possible mistakes, the researcher might rerun the spring 2011 data control file to get output with the step values and the item calibrations. Then she or he would author a control file for the spring 2012 data. That file would look almost identical to the spring 2011 file, but there would be an additional control line with SAFILE information as well as the item anchors for the items that were not changed from spring 2011 to spring 2012. This means the item anchoring might look like the following (we are just making up the logit values for the items).

```
IAFILE=*
1 -1.23
2 .48
3 .89
4 -2.12
5  -.32
6  1.73
7  3.30
8  .94
10 .21
11  -.06
12 -.33
*
```

A Quick Review of Linking: What, How, and Why?

To complete this chapter, let's just review what the main points are for readers. First, with Rasch measurement, it is possible to link different forms of instruments. This means that a set of respondents might complete different forms of an instrument, but by having common items which "link" the forms, it is possible to express *all* respondents on a scale as if *all* respondents completed the same instrument. This ability to link means that different forms of a survey or test can be constructed, and the forms confidently linked. Tests of all types (right/wrong, partial credit) can be linked and surveys can be linked. Creating different forms of a test or survey is just one scenario in which linking is of massive help to researchers. Another amazing gain with being able to link is that surveys/tests can be shortened and also improved with new items. This means that quality control can take place, but respondents can still be compared with data collected at an earlier time point. Researchers collecting data for a project might discover that the length of a survey collected in January was far too long for respondents. When the survey is administered in May, it is possible to present respondents with a shorter survey, but still express all respondents on the same scale.

Now one grand finale. Anchoring allows test forms (and survey forms) targeted to respondents to be administered and linked. For example, students in 5th grade can be administered a 20-item math test. In 6th grade, the same students can be administered a test measuring the same variable, but by using selected items which appeared on the 5th grade test (these are the links), the 6th graders can be administered a test targeted to their ability level. This linking allows the *growth* of respondents to be confidently measured on one metric!

Isabelle and Ted: Two Colleagues Conversing

Isabelle: So Ted, do you get this idea of anchoring of items and steps?

Ted: Yes, it's easy, at least in this simple case. What I do is run my survey data as I usually do. But, I make sure to take note of the item calibrations and the step calibrations. I can get the item calibrations and the step calibrations from Winsteps tables, but to be less paranoid, I usually use IFILE and SFILE in order to get Winsteps to dump the item calibrations to a file and to dump the step calibrations to a file.

Isabelle: Okay, you have these item calibrations and step calibrations; now, what is the next step?

Ted: Okay, pretend at a later time I need to give the survey again. I usually have to do some sort of report or paper right after the data collection, so I cannot just sit around and wait for the second data collection…so anyway time has passed, and maybe I want to shorten the survey. If I do, then I need to "link" the metric of the first survey to the metric of the second survey. But, I can do this only by making sure that the items and the rating scale steps of the second survey are "aligned" with the items and rating scale steps of the initial survey. If you have not thought about measurement, this may seem a little odd, but if you do not do this linking, you can really goof up your results.

Now it works this way, pretend you gave a survey to 100 students at the end of a class, and then a year later you gave the same survey to 75 new students, but you removed 2 questions from the original survey. Maybe you had completed a Rasch analysis of the 100 students, and you wanted to do more analysis with the added 75 students, but you did not want to throw your previous work out…well it would be through anchoring that you could express everything on the same scale and extend your work to include the old and new data sets. More data sometimes provide more statistical power, so being able to improve the survey and also being able to use the old data are a BIG plus.

Keywords and Phrases

IAFILE
IFILE
SFILE
SAFILE
Entry number
Item anchoring
Rasch-Andrich Thresholds
Step anchoring
To anchor an item
To anchor a rating scale

Potential Article Text

In 2010, XYZ University received a multi-year grant to address documented deficiencies in preservice science teachers' self-efficacy that have been discussed in the literature. The goal of the project was to increase preservice teachers' confidence in teaching science. One hundred (100) preservice teachers participated in year 1. At the end of year 1, participants completed the STEBI (Enochs and Riggs, 1990). Only the 13-item self-efficacy subscale of the STEBI was used.

The respondents' self-efficacy measures were computed using the Rasch analysis Winsteps program (Linacre, 2011). The Rasch Rating Scale model was used to evaluate these data. An analysis of the year 1 data was completed following the completion of the year 1 curriculum. Person measures were computed and expressed on a linear logit scale.

Prior to the year 2 intervention, a decision was made to limit the amount of data collection requested of respondents. Data collection was limited by shortening the self-efficacy STEBI survey administered to respondents. At the end of year 2, only 7 self-efficacy items were completed by this group of year two respondents.

Although year 1 attendees and year 2 participants completed a different mix of self-efficacy items, Rasch techniques were applied in order to express all respondents (year 1 and year 2) on the same linear metric. So-called step anchoring and item anchoring were used with Winsteps in order to express the responses of the year 1 and year 2 participants on the same scale. Use of Rasch analysis allowed the year 2 survey to be shortened while simultaneously permitting the self-efficacy measures of the year 1 and year 2 participants to be expressed on the same scale (Fig. 14.10).

```
Year 1                                                  Year 2


|   Q19se-rc
|
|S  Q5se-------------------------------------------Q5se
+
|   Q23se-rc
|   Q17se-rc  Q20se-rc------------------------------Q17se-rc
|   Q12se-------------------------------------------Q12se
|   Q18se     Q21se-rc  Q3se-rc---------------------Q3se-rc
S+M
|
|
|   Q6se-rc-----------------------------------------Q6se-rc
|
+   Q8se-rc-----------------------------------------Q8se-rc
|S
|
|
|   Q22se
+
|T
|
|   Q2se--------------------------------------------Q2ase
```

Fig. 14.10 A visual of the linking of two groups using items 2, 3, 5, 6, 8, 12, and 17. Six items are not used as anchors

Quick Tips

To successfully link two steps you must use are to note the item measures of a survey and to use those item measures to "anchor" the same items when the items appear in another data set. The other step is to note the "steps" between rating categories when anchoring. The key codes for your control file will be IAFILE, IFILE, SFILE, and SAFILE.

Data Sets: (go to http://extras.springer.com)

cf50SEItemLinkingChp
Sabah8ItemsFall2010
Sabah6ItemsFall2011
cf n 40 Jordan activity

Activities

Activity #1

Situation: Our colleague Dr. Saed Sabah of the Hashemite University (Jordan) has kindly provided us with a sample of data that he collected from students in Jordan as part of a study of students' perceptions of inquiry experiences in science laboratories. Dr. Sabah's specialty areas within the field of science education are assessment and technology integration. The data were collected using the scale of Campbell, Abu-Hamid, and Chapman (2010) in which respondents could answer using a frequency scale (1 = almost never, 2 = seldom, 3 = sometimes, 4 = often, 5 = almost always). The scale included two items that needed to be reverse coded. The data in the spreadsheets have already been corrected for the reverse item wording of the two items. Below we provide eight rating scale items that, for purposes of our activity, we will consider all eight items to be part of one construct, meaning one metric. The data for these items are provided in two Excel sheets. One sheet is labeled Sabah8ItemsFall2010. You should view these data as collected in the fall of 2010. A second Excel sheet, named Sabah6ItemsFall2011, is provided, and you should pretend these data were collected in the fall of 2011. In our activity we are pretending items D3 and D4 were not administered to respondents in the fall of 2011 in order to shorten the survey.

C. Conducting investigations: in the science classroom	
C1	I conduct the procedures for my investigation
C2	The investigation is conducted by my teacher in front of the class
C3	I am actively participating in investigations as they are conducted
C4	I have a role as investigations are conducted
D. Collecting data: in the science classroom	
D1	I determine which data to collect
D2	I take detailed notes during each investigation along with other data I collect
D3	I understand why the data I am collecting is important
D4	I decide when data should be collected in an investigation

Task: Create a control file for the fall 2011 data.

Answer: Use the procedures we have detailed earlier herein to construct a control file. Even though we are not using the data for an analysis of the subgroups, make sure to read in the GPA data and gender data as person label variables. Remember there are some items that were flipped and entered as flipped data in the spreadsheet, so you do not have to recode any data with the control file. However, if there is a flipped item, then you need to make sure the item has an "item description" that reflects the flip! Below is the control file we made using the fall 2011 data. We removed the comments (with the semicolons) provided by Winsteps. Also, note the one item that was flipped has the word "NOT" added to the text. The data were flipped when they were entered, but we need to remember to change the wording for

```
&INST
Title= "n 40 Fall 2010 Excel Jordan Data for Activity.xls"
ITEM1 = 1 ; Starting column of item responses
NI = 8 ; Number of items
NAME1 = 10 ; Starting column for person label in data record
NAMLEN = 10 ; Length of person label
XWIDE = 1 ; Matches the widest data value observed
CODES = "12345 " ; matches the data
TOTALSCORE = Yes ; Include extreme responses in reported scores
@id = 1E2 ; $C10W2
@gender = 4E4 ; $C13W1
@gpa = 6E9 ; $C15W4
&END ; Item labels follow: columns in label
C1 I conduct the procedures for my investigation.
C2 The investigation is NOT conducted by my teacher in front of the class.
C3 I am actively participating in investigations as they are conducted
C4 I have a role as investigations are conducted.
D1 I determine which data to collect.
D2 I take detailed notes during each investigation along with other data I collect.
D3 I understand why the data I am collecting is important.
D4 I decide when data should be collected in an investigation
END NAMES
55554455  1 2  2.8
55555454  2 2 2.45
.
.
.
.
45454545 40 1 2.35
```

Fig. 14.11 Control file for Activity #1

the item! To save space, the first two lines of data are provided as well as the last line of data. We also provide the control file (cf n 40 Jordan activity) we created, but push yourself to make your own file! (Fig. 14.11)

Activity #2

Task: Having created the control file, run the data for the 40 students who answered the survey in the fall of 2010. Find a table that provides the step calibrations, and then find a table that provides the item calibrations.

Answer: Winsteps Table 3.2 provides the step calibrations. The item calibrations are provided in a number of tables, and we find the item entry table to be the one that keeps us most organized, as the items are presented in the order in which they were entered into the spreadsheet, which means item C1 is presented first and item D4 is presented last.

Activity #3

Situation: In the fall of 2010, you collected and analyzed survey data (you had a report that you needed to finish). Time passed, and you then collected the fall 2011 data with a shortened version of the survey. The fall 2011 survey does not have

```
; STRUCTURE MEASURE ANCHOR FILE FOR n 40 Fall 2010 Excel Jordan Data
; for Activity.xls Dec 12 11:04 2011
; CATEGORY  Rasch-Andrich threshold
  1      .00
  2    -1.60
  3    -1.09
  4      .37
  5     2.31
```

Fig. 14.12 Item calibration file for fall 2011 data for Activity #3

```
TABLE 14.1 n 40 Fall 2011 Excel Jordan Data for  ZOU485WS.TXT  Dec 12 11:04 2011
INPUT: 40 PERSON  8 ITEM  REPORTED: 40 PERSON  8 ITEM  5 CATS      WINSTEPS 3.73
--------------------------------------------------------------------------------
PERSON: REAL SEP.: 1.63  REL.: .73 ... ITEM: REAL SEP.: 1.71  REL.: .75

         ITEM STATISTICS:  ENTRY ORDER

-------------------------------------------------------------------------------------------------
|ENTRY   TOTAL  TOTAL           MODEL|   INFIT  |  OUTFIT  |PT-MEASURE |EXACT MATCH|
|
|NUMBER  SCORE  COUNT  MEASURE  S.E. |MNSQ  ZSTD|MNSQ  ZSTD|CORR.  EXP.| OBS%  EXP%| ITEM
|
|------------------------------------+----------+----------+-----------+-----------+------------
|     1    157     40    -.07    .22|1.12    .6|1.05    .3|  .55   .61| 55.0  53.3| C1 I conduct
|     2    163     40    -.38    .23|1.75   2.7|1.57   2.2|  .47   .59| 45.0  55.4| C2 The inve
|     3    166     40    -.55    .24| .76  -1.0| .72  -1.2|  .69   .58| 65.0  55.5| C3 I am act
|     4    167     40    -.61    .24| .98    .0| .90   -.4|  .58   .57| 62.5  55.5| C4 I have a
|     5    149     40     .31    .21| .75  -1.2| .76  -1.1|  .61   .63| 57.5  50.3| D1 I dete
|     6    139     40     .74    .20| .85   -.6| .86   -.6|  .71   .65| 45.0  47.3| D2 I take d
|     7    151     39     .03    .22| .87   -.5| .84   -.7|  .71   .62| 56.4  51.9| D3 I underst
|     8    144     40     .53    .21| .99    .0|1.02    .2|  .60   .64| 37.5  49.2| D4 I decide
```

Fig. 14.13 List of the item calibrations of the survey items

items D3 and D4. You know that you will need to compute person measures for the fall 2011 data, but you want to make sure the data are expressed on the same metric as that used for the fall 2010 data. You know that before you can start evaluating the fall 2011 data, you will need to have the step anchors and item anchors close at hand from the fall 2010 data.

Task: Edit the fall 2010 control file so that you have the anchors. Also provide the anchor values for the steps and the anchor values for the items you are able to compute!

Answer: You will need to add two lines to your control file for the fall 2010 data. These two lines are "SFILE=" and "IFILE=". Following the equal sign (=) will be the name of the file. The SFILE that was computed when we made this change is provided below, and we also provide the item calibrations that will be used to anchor the fall 2011 data (Fig. 14.12).

The item calibrations are provided in the IFILE and also in the item entry table. Below are the results of editing the item entry table to create a list of the item calibrations of the survey items. We present first the item table (with the item names shortened), then the results of editing the item table (Figs. 14.13 and 14.14).

```
1        -.07
2        -.38
3        -.55
4        -.61
5         .31
6         .74
7         .03
8         .53
```

Fig. 14.14 The item entry table resulting from the analysis of the fall 2010 data

This is the file that can be then used to confidently anchor items. Editing the entry table helps minimize the chance of inserting an incorrect item calibration value. Recall that if not all items are used in later data collections, then only those items that repeat will be used as anchors.

Activity #4

Question: Now that you have created a control file for the fall 2010 data and also created the step anchor file and the item calibration file, what is the first step that you must take to evaluate the fall 2011 data, which has 40 students and utilized items C1, C2, C3, C4, D1, and D2?

Answer: In order to evaluate the fall 2011 data and ensure that the data are expressed on the same metric as the fall 2010 data, you need first to create a control file for the fall 2011 data. This control file will not be anchored in terms of steps or items.

Activity #5

Create the control file, run it, and make sure your data have been read correctly. Also make sure to change the wording of any items that were flipped. Then add the information for SAFILE and IAFILE. This can be done in a number of ways, but if you repeat the techniques we used in the chapter, you will insert the following lines for SAFILE:

```
SAFILE=*
1 .00
2 -1.60
3 -1.09
4 .37
5 2.31
*
```

For the line IAFILE, you should enter the following into your control file. Notice that you do not see an anchor value for items 7 and 8. The reason for this is that item 7 (D3) and 8 (D4) were not presented in the survey.

```
IAFILE=*
1 -.07
2 -.38
3 -.55
4 -.61
5 .31
6 .74
*
```

Activity #6

Task: Create the control file for the fall 2011 data that will allow you to express that data on the same metric that was used to express the fall 2010 data.

Answer: Below we provide the fall 2011 control file that facilitates anchoring to the fall 2010 data. Note the insertion of lines for the step anchoring and item anchoring. Again, we present only part of the person responses (Fig. 14.15).

Activity #7

Question: How could you verify that you have successfully anchored the steps and the items?

Answer: Look at Winsteps Table 3.2 below and also look at one of the tables that list the item calibrations, such as the item entry table. If you see the letter "A" in the tables following the step anchor values and the item anchor values, then you have anchored. In this particular set of exercises, you have succeeded in expressing the measures of the fall 2011 respondents on the same scale as that used to express the measures of the fall 2010 respondents, even though a different mix of items was used at the two data collection time points (Fig. 14.16).

Activity #8

Question: A student who completed the fall 2011 survey has a raw score of 24. What is this student's measure on the fall 2010 scale?

Answer: Run the fall 2011 data with the step and item anchors.

Go to the Score Table and find the raw score of 24. The logit measure (1.36) reported in the fall 2011 table is the measure of the student on the fall 2011 survey, and due

```
&INST
Title= "n 40 Fall 2011 Excel Jordan Data for Activity.xls"
ITEM1 = 1 ; Starting column of item responses
NI = 6 ; Number of items
NAME1 = 8 ; Starting column for person label in data record
NAMLEN = 10 ; Length of person label
XWIDE = 1 ; Matches the widest data value observed
CODES = "12345 " ; matches the data
TOTALSCORE = Yes ; Include extreme responses in reported scores
@id = 1E2 ; $C8W2
@gender = 4E4 ; $C11W1
@gpa = 6E9 ; $C13W4
SAFILE=*
1 .00
2 -1.60
3 -1.09
4 .37
5 2.31
*
IAFILE=*
1 -.07
2 -.38
3 -.55
4 -.61
5 .31
6 .74
*
&END ; Item labels follow: columns in label
C1 I conduct the procedures for my investigation.
C2 The investigation is NOT conducted by my teacher in front of the class.
C3 I am actively participating in investigations as they are conducted
C4 I have a role as investigations are conducted.
D1 I determine which data to collect.
D2 I take detailed notes during each investigation along with other data I collect.
END NAMES
233234 41 2 3.22
334545 42 1  2.8
.
.
.
.
333322 75 2 2.94
```

Fig. 14.15 Control file for anchoring to the fall 2010 data

to anchoring, this is the measure of the student on the fall 2010 metric. This is because we have anchored the items and the steps.

Activity #9

Question: For a student who exhibits a 1.36 logit measure on the fall 2011 instrument, what would have been the student's raw score had she or he been administered the fall 2010 instrument?

Answer: Below is the Score Table for the fall 2010 analysis. To find out what the person taking the fall 2011 instrument (measured at 1.36 logits) would have earned in terms of raw score on the fall 2010 instrument, find 1.36 logits in the table above. The measure of 1.36 falls between 31 and 32. This means that if this same person had completed the fall 2011 version of the instrument, she or he would have been predicted to receive a raw score of 31 or 32 (Fig. 14.17).

```
TABLE 3.2 n 40 Fall 2011 Excel Jordan Data for A ZOU194WS.TXT  Dec 12 12:25 2011
INPUT: 35 PERSON  6 ITEM  REPORTED: 34 PERSON  6 ITEM  5 CATS       WINSTEPS 3.73
--------------------------------------------------------------------------------


SUMMARY OF CATEGORY STRUCTURE.  Model="R"
-------------------------------------------------------------------
|CATEGORY   OBSERVED|OBSVD SAMPLE|INFIT OUTFIT|| ANDRICH |CATEGORY|
|LABEL SCORE COUNT %|AVRGE EXPECT|  MNSQ  MNSQ||THRESHOLD| MEASURE|
|-------------------+------------+------------++---------+--------|
|  1   1       2   1| -1.91 -2.60|   .21   .25||  NONE A |( -3.01)| 1
|  2   2      12   6|  -.41  -.40|   .85   .86||  -1.60A|  -1.49 | 2
|  3   3      70  34|   .39   .29|   .90   .91||  -1.09A|   -.23 | 3
|  4   4      63  31|  1.35  1.60|   .95   .91||    .37A|   1.42 | 4
|  5   5      57  28|  2.52  2.36|   .89   .93||   2.31A|(  3.51)| 5
```

```
TABLE 14.1 n 40 Fall 2011 Excel Jordan Data for  ZOU194WS.TXT  Dec 12 12:25 2011
INPUT: 35 PERSON  6 ITEM  REPORTED: 34 PERSON  6 ITEM  5 CATS       WINSTEPS 3.73
--------------------------------------------------------------------------------
PERSON: REAL SEP.: 1.70  REL.: .74 ... ITEM: REAL SEP.: 1.75  REL.: .75

         ITEM STATISTICS:  ENTRY ORDER

-------------------------------------------------------------------------------------------------
|ENTRY   TOTAL  TOTAL            MODEL|   INFIT  |  OUTFIT  |PT-MEASURE |EXACT MATCH|         |
|NUMBER  SCORE  COUNT  MEASURE  S.E. |MNSQ  ZSTD|MNSQ  ZSTD|CORR.  EXP.| OBS%  EXP%|DISPLACE|
ITEM                                                                                |
|-----------------------------------+----------+----------+-----------+-----------+--------+----
|      1    121     34    -.07A    .24| .89   -.3|1.08    .4|  .77    .66| 48.5  53.9|     .42| C1
|      2    137     34    -.38A    .25| .69  -1.3| .69  -1.3|  .74    .64| 66.7  55.3|    -.18| C2
|      3    144     34    -.55A    .25| .80   -.7| .74  -1.0|  .63    .62| 60.6  55.4|    -.50| C3
|      4    131     34    -.61A    .26| .87   -.4| .97    .0|  .72    .62| 63.6  55.5|     .42| C4
|      5    123     34     .31A    .23| .76  -1.0| .78   -.9|  .75    .68| 54.5  49.6|    -.06| D1
|      6    117     34     .74A    .22| .82   -.7| .87   -.5|  .64    .71| 42.4  44.9|    -.17| D2
|-----------------------------------+----------+----------+-----------+-----------+--------+----
```

Fig. 14.16 Tables for activity 7 (*To facilitate presentation of this table, only the letter name (e.g., C2) of the 6 items is presented)

```
TABLE 20.1 n 40 Fall 2010 Excel Jordan Data for  ZOU662WS.TXT  Dec 12 12:35 2011
INPUT: 40 PERSON  8 ITEM  REPORTED: 40 PERSON  8 ITEM  5 CATS       WINSTEPS 3.73
--------------------------------------------------------------------------------

                TABLE OF MEASURES ON TEST OF 8 ITEM
----------------------------------------------------------------------------
| SCORE  MEASURE     S.E. | SCORE  MEASURE     S.E. | SCORE  MEASURE     S.E. |
|------------------------+------------------------+------------------------|
|     8   -4.99E   1.82 |    19    -1.02     .40 |    30      .97     .47 |
|     9   -3.80    1.00 |    20     -.86     .40 |    31     1.20     .49 |
|    10   -3.11     .71 |    21     -.70     .40 |    32     1.45     .50 |
|    11   -2.69     .59 |    22     -.53     .40 |    33     1.71     .52 |
|    12   -2.38     .53 |    23     -.37     .41 |    34     1.99     .54 |
|    13   -2.12     .48 |    24     -.20     .41 |    35     2.30     .57 |
|    14   -1.90     .46 |    25     -.03     .42 |    36     2.65     .61 |
|    15   -1.70     .44 |    26      .16     .43 |    37     3.06     .67 |
|    16   -1.52     .42 |    27      .34     .44 |    38     3.58     .79 |
|    17   -1.35     .41 |    28      .54     .45 |    39     4.39    1.05 |
|    18   -1.18     .41 |    29      .75     .46 |    40     5.67E   1.86 |
```

Fig. 14.17 Score Table for the fall 2010 data analysis

Activity #10

Question: Can you explain in words why anchoring of steps and items is needed if you intend to express respondents of two different surveys on the same metric? Are there any requirements for the linking?

Answer: If two versions of a survey are administered, then it is often possible to express all respondents (regardless of survey form completed) on the same metric. One caveat is that the surveys must involve the same variable. One must anchor the steps from one administration to another administration so that the functioning of the rating scale is maintained from one data set to another. The same is true with items. Those items that appear on both surveys must be set to the same values. Think of this as insuring that the items mark the variable in the same way for both surveys.

Activity #11

Question: Can you explain in words why being able to link surveys is useful in research?

Answer: Being able to link surveys means that if researchers want to administer a large number of items (perhaps far more than the respondents have to answer), they can create two forms (or more) of a survey. As long as there are common items for anchoring, using two forms of the survey will allow the number of items that will be presented to be decreased. This saves time for respondents and often improves the quality of collected data. Being able to link surveys also means that a survey can be altered over time (e.g., items improved, items added), but "old" data collected with a previous form of the survey can be expressed on the same metric as that defined with the new instrument. Perhaps most importantly for both tests and surveys, forms can be linked to measure growth.

References

Wolfe, E. W. (2000). Equating and item banking with the Rasch model. *Journal of Applied Measurement, 1*(4), 409–434.

Wright, B. D., & Stone, M. H. (1979). *Best test design*. Chicago: MESA Press.

Additional Readings

A clearly written discussion of Rasch measurement.

Callingham, R., & Bond, T. (2006). Research in mathematics education and Rasch measurement. *Mathematics Education Research Journal, 18*(2), 1–10.

A good discussion of Rasch as well as selected details concerning linking and anchoring techniques.

Albano, A. D., Rodriguez, M. C., McConnell, S., Bradfield, T., & Wackerle-Hollman, A. (2011, April). *Scaling measures of early literacy*. Paper presented at the annual meeting of the National Council on Measurement in Education, New Orleans, LA.

Luo, G., Seow, A., & Chin, C. (2001). *Linking and anchoring techniques in test equating using the Rasch model*. http://hdl.handle.net/2134/1817

Yu, C. H., & Osborn Popp, S. E. (2005). Test equating by common items and common subjects: Concepts and applications. *Practical Assessment Research & Evaluation*, *10*(4), 1–19. Retrieved April 22, 2013, from http://pareonline.net/getvn.asp?v=10&n=4

Chapter 15
Setting Pass–Fail Points and Competency Levels

Isabelle and Ted: Two Colleagues Conversing

Ted: *Isabelle dear, I have read the chapter on competency levels and procedures one can follow to compute the boundary between levels on a Wright Map. That chapter describes a very good technique when I am evaluating a data set where the competency levels have already been determined by researchers. But, what do I do if I am starting a project and I want to be the one to set the competency levels?*

Isabelle: *Ted that is a great question. There is a fairly simple way of setting competency levels. The technique involves a Wright Map and the item calibrations from a Winsteps analysis. The technique also involves using theory and thinking about the meaning of a number, something similar to what we have been doing all along.*

Ted: *I mentioned competency levels, but could this technique also be used for figuring out a pass–fail point for a test? It seems to me that figuring out the location of a pass–fail point is a special case of computing competency levels.*

Isabelle: *Yes, we can use this technique for both pass–fail levels and competency levels.*

Introduction

In education numerous research projects are being conducted that make use of summarizing student performance with respect to competency levels. For instance, in the USA, schools in the state of Ohio have been classified as Academic Emergency, Academic Watch, Continuous Improvement, Effective, Excellent, or Excellent with Distinction (Ohio State Report Card 2009–2010). Such classification systems are both advantageous and disadvantageous. One obvious disadvantage is that a school is indeed much more than a set of words being used to summarize academic performance. That said, such classification systems are popular with the public and policy makers, many who have limited backgrounds in technical data analysis. The authors of this book wish that such classifications were not used, but wishing alone will not change the use of such classifications. Our purpose in this chapter is to present techniques that help researchers within and beyond education deeply and confidently

W.J. Boone et al., *Rasch Analysis in the Human Sciences*,
DOI 10.1007/978-94-007-6857-4_15, © Springer Science+Business Media Dordrecht 2014

understand how they might think about their definitions of competence levels. The techniques we present also apply to pass–fail decisions. Pass–fail decisions are most commonly observed in credentialing tests in which one wishes to document one's competency in a field (e.g., teaching, medicine, law, accounting). As readers will be able to note in this chapter we have chosen a right/wrong test to be the core of our discussion. It is possible to expand what we introduce here to survey instruments such as the STEBI, but we find that those first starting with Rasch can better understand the points we are making by considering a test.

A Historical Approach to Pass–Fail Decisions

To think about pass–fail points, we have found it helpful to consider how such determinations were made prior to the advent of Rasch analysis (this is similar to the steps we have taken elsewhere in this book when we consider how other measurement issues were poorly tackled in the past). Let's pretend that a 100-item multiple-choice chemistry test has been developed to evaluate the competency of potential secondary school chemistry teachers. More than likely, the test developers employed some sort of guiding document to create a test blueprint as they developed the instrument. Current guiding documents usually are standards of a geographical region and/or an organization (e.g., Schleswig Holstein (Germany), Indiana (USA), New South Wales (Australia), Japan, American Association of Physics Teachers).

Following instrument construction, the test is administered, and then a "cut" point is determined. Often, the cut point is an outcome of many issues (e.g., instructional goals, public perception, political points). We suspect that in the past, typically a "number" was suggested, based, in part, upon one's own school experience. For instance, a common grade scale in the USA is A = 100–90 %; B = 89–80 %; C = 79–70 %; D = 69–60 %; and F = 59 % and below. A score of 70 % correct on a test is often viewed as the minimum acceptable score; any lower score is deemed not acceptable. This percentage (70 % correct) for the pass–fail point is a "cut point" that decision makers have experienced, for right or wrong, throughout their lives. There are many reasons why this technique is flawed – one key issue is there is no meaning to selecting 70 %, for it is dependent upon the difficulty of test items.

Formative Assessment Checkpoint #1

Question: Is using past experiences to select a percentage of items correctly answered an acceptable manner of setting a pass–fail point on a test? For example, if students correctly answer 90 % of items on a test, they should be considered as exhibiting a "high pass" on a test.

Answer: No. Decisions regarding students passing a test or being classified in two or more categories should not be based on the use of past experiences to choose the percentage of test items that should be correctly answered. This is because the difficulty or ease of items greatly impacts the meaning of student knowledge supposedly expressed with the use of percentages.

A typical next step is to determine how many students will pass or fail if a particular value is used as a cut point. In this case, if 70 % (70 correctly answered items out of 100 items) is used as a cut point, how many students will pass or fail? If too high a percentage of students fail, this pass–fail point will often be adjusted. There may be political pressures (if too many respondents fail, then someone is going to look bad). There may also be financial issues (if too many students fail in a school system and are forced to repeat a grade, then would there be enough free seats in a classroom to accommodate the students?).

A number of problems exist with the procedure outlined above. The political issues we will not address; however, we will address the measurement flaws in the procedure. By applying Rasch measurement theory and the data available from a Rasch analysis, we will describe how a thoughtful, theory-driven criterion point can be determined. A number of advanced Rasch techniques are used by many groups to set cut points, but herein we will present the basics. At the least, if you think about the basics, you will be well ahead of many groups in your sophistication of thinking.

What Does That Number Mean?

The central error in the procedure described above is the value of 70 % (70 out of 100). This value is meaningless because 70 % correct on the chemistry test does not tell a teacher, researcher, or policy maker what a learner who scores a 70 % correct knows and does not know. Furthermore, a 69 % correct, the highest "fail score," does not tell a teacher, researcher, or policy maker in what way this score is "less" in chemistry knowledge than the person who scores a 70 % correct and passes. This issue, the meaning of a number, has been discussed in different contexts throughout this book. For us, the most important aspect of learning and applying Rasch measurement is our central focus on thinking about the substantive, qualitative meaning of a number. So often in our field, we observe cut values that are determined by policy makers' past experiences as students (90 % correct is an "A", 85 % correct is a *very good*, 70 of 100 items correct is a *pass*), with little thought devoted to considering whether or not these values accurately represent the desired level of competence (e.g., pass or fail, excellent or only very good).

In our classes and workshops, we have developed a number of hands-on, inquiry-based measurement activities. One of these activities is described and applied below, thanks to a unique data set from our colleagues Mark Haugan and Lynn Bryan in Physics and Physics Education at Purdue University.

The data set we use is a nonrandom sample of 75 student responses to the well-known Force Concept Inventory (FCI) (Hestenes, Wells, & Swackhamer, 1992). Our goal is to demonstrate how one can thoughtfully apply Rasch measurement theory and analysis results in the computation of a pass–fail point informed by sound research. Step 1 is to convene a group of experts who are intimately familiar with the trait measured by the measurement instrument. These experts should be able to predict which items are the easiest items of the trait, which items are more difficult, and so on. If experts cannot confidently predict the manner in which individual test items define a trait being measured by a set of test items, that may mean the set of items should not be used to define a single trait. Perhaps there is not a single trait?

For our example, let's pretend that the experts were able to reach consensus in the manner in which they thought the FCI items defined a single trait. Step 2, then, is to present a group of test items, such as those 30 FCI items, to a group of experts and ask the group of experts to determine through group consensus the ordering and spacing of items from easy to hard. In essence, although we do not tell them so, we ask them to predict item difficulty on a logit scale!

Step 3 is to ask the experts to discuss the location of a cut point based upon the ordering and spacing of the FCI items. We direct them to draw a line between any two adjacent items such that the line will mark the boundary between the group of candidates who have at least exhibited the minimum level of performance to be considered a "pass" and the candidates who have not exhibited at least a minimum level of performance. Think of the Wright Maps we have discussed throughout this book. In essence, we are asking the experts to create their own Wright Map from theory, and then we are asking them to select a location that marks the point where there is over a 50–50 chance that the lowest passing candidate will answer correctly the required pass items of a test.

In Fig. 15.1 we present a schematic of a portion of the predicted spacing and ordering which could be made by experts evaluating the FCI. Since there are so many FCI items, we display only those items near the group consensus boundary.

Step 3, while simple, is also an epiphany for almost all of our experts. We ask the experts to count the number of items below the pass–fail point. In our example there are 12 FCI items below the pass–fail line (item 14, item 3, item 27, item 29, etc.). This means that the minimum number of items that must be answered correctly to pass the FCI is 12+1. In almost all counts, the percentage correct required for the lowest pass is not some nice, familiar, round percentage such as 70 % or 80 %. In this example the percentage of items needed to pass is 43 % (13/30). This is because the pass–fail boundary has been determined based upon the content of the items, not on a raw score. More specifically, the pass–fail point has been determined based upon its conceptual meaning in terms of the trait measured by the FCI to have passed or failed the FCI.

Taking Step 4, we present the experts with a duplicate set of items, in this case FCI items. Each item has a logit value written on it. These logit values are the item difficulties computed from the Winsteps analysis of the FCI data. These items are then organized along a scale that is side-by-side with the ordering and spacing of items as predicted by the experts. We pose a number of questions to the experts,

Fig. 15.1 A schematic of predicted spacing and ordering of items near the 50–50 point

More Difficult Items
19
|
|
|
|
22 28

|
14
|
3
|
|
27
|
|
29
Less Difficult Items

such as: Are there differences in their prediction scale and the data scale? If large differences exist, we suggest, when possible, that an item be dropped if a disconnect exists between the experts' prediction and the data. When it is politically not possible to drop an item, we guide discussion toward the experts' potential acceptance, based upon their theory, of the item ordering and spacing as presented in the Wright Map. If they cannot agree with the ordering, then we suggest that they not use the test to make a pass–fail decision. Needless to say, this is often not a popular suggestion.

If the experts agree that the Wright Map does match their theory (for one since no theory is 100 % correct), then we ask the expert group to review the Wright Map and reach a group consensus for the boundary between a pass and a fail using the meaning of each item as follows: Items above a pass–fail line are items that might not be answered by those passing the test; items below a pass–fail point will be the items that are, probabilistically, answered correctly by those passing the test, no matter how low the pass is. Needless to say they will make use of their predicted pass–fail work.

Once the experts reach a consensus as a group, then one possible next step is simply counting the number of items that fall below the pass–fail point. Then, add one to that number. So if the line is drawn so that 12 items are below the pass–fail line then if students correctly answer 13 items, they will have demonstrated the minimum level of competence that is needed to "pass," where "pass" is based on the group consensus of the experts who determined the pass–fail point. The pass–fail point is based upon a review of the trait, not upon an arbitrary, raw score.

The example delineated above is a new aspect of Rasch measurement that has and has not been presented previously in this book. It has not, in that we have not described the procedures to determine a cut point. This information has been presented previously in discussions about not confusing counting with measuring. Counting in social science research is almost always misleading in that the counts are treated as measures and are then evaluated as such.

Formative Assessment Checkpoint #2

Question: Are counting items and measuring items the same process?

Answer: No. Counting items correctly answered is not measuring. As a result counts must not be used as if they were measures.

Repeatedly in his career at the University of Chicago, Ben Wright stressed that "counts are not measures" (see Wright, 1999). The problem with counting should be abundantly clear in this example with respect to cut points.

There are a number of experts in the field of Rasch measurement such as Greg Stone who have conducted a great amount of work with respect to what is called standard setting. What we have presented is an introduction, and we encourage readers to review Stone's work as well as others involved in standard setting.

Isabelle and Ted: Two Colleagues Conversing

Isabelle: *Pass–Fail levels and Rasch. Proficiency levels and Rasch. Tell me about them Ted.*

Ted: *Really Isabelle, I cannot believe how interesting this is. If one remembers that counting mindlessly is meaningless, it makes so much sense that a traditional number used for a pass–fail point is equally meaningless. What one must be able to do is organize test items along a construct, and then decide what is the location of the pass based upon the construct. Only then does one count the items on the test to determine the pass–fail point.*

Isabelle: *Wait a minute. So you're telling me that if a set of students completes a 100-item physics test, then we cannot say that 90/100 or higher is the highest of passes?*

Ted: *Exactly. It all depends upon the items. I know you would not expect it, but what if those 100 test items were far too easy for the students. Wouldn't you agree that 90/100 might not be indicative of a high pass on such an easy test?*

Isabelle: *Well yes, of course.*

Ted: *There are lots of detailed and complicated but correct techniques used in Rasch measurement to determine a pass–fail point. But, this simple example is an excellent beginning.*

Keywords and Phrases

Standard setting
Pass–fail point
Group consensus regarding a pass–fail point
A pass-fail point of 70 % on a test tells one nothing. 70 % of what?

Potential Article Text

As part of an initiative to improve the instruction of students completing an introductory university physics class, FCI data were collected from students at the start and end of a semester-length physics class. In total, 100 students completed the FCI.

In order to communicate the performance level of students in terms of easily understood categories, a pass–fail boundary was determined. Detailing a pass–fail boundary allowed all students to be classified into one of these two categories. Although there are certainly some disadvantages in classifying the students into one of two categories, for communication with stakeholders, it was important to be able to express student performance in terms of these two classifications.

Data were evaluated using the Rasch analysis program Winsteps. A Wright Map was constructed. Then a group of four expert judges determined the pass–fail boundary. This boundary was then expressed in terms of the minimal number of FCI items that must be answered to be classified as a "pass."

Using a Wright Map and expert judges is far superior to using only a percentage of items correctly answered. The fundamental flaw in using only a percentage is that the qualitative, conceptual meaning of the construct is not taken into consideration nor are the ranges of item ease or item difficulty.

Quick Tips

When setting a pass–fail level, first decide what it will mean to "pass." For the person who barely "passes," what would that person be expected to exhibit?

When setting a competency level, what would a person who represents the lowest performance level of that competency level be able to exhibit? What would the person who represents the highest performance level of that competency level be able to exhibit (and not exhibit)?

Once you have decided what it means to pass a test, or what it means to be at the lowest level and highest level of a competency level, then review a Wright Map and find the items which mark these parts of the trait.

Do not be tempted to define a competency level or a pass–fail level using a percent correct (e.g., 70 % correct on a test is a pass). Counts are not measures. Also percent correct is meaningless unless one knows what this percentage represents in terms of what one can do and what one cannot do.

Base the pass–fail upon the meaning of items that you see in your Wright Map. Ultimately draw a line to separate the two items which define what it means to pass or fail. Once you have drawn that line, on that test, the number of items for a pass will be the number of items below the line +1.

If you later plan to administer a test that is linked to the test you have just used, make use of the location (in logits) of the pass–fail line you have just drawn. When you evaluate the data from a new test, which measures the same variable and is linked to your first test, you can still draw your pass fail line by marking a line at the logit value of the line from your first test! It may very well be that with the new test, the percentage of items that need to be correctly answered is different than with the initial test. This is due to differing test item difficulty on the new test.

Data Sets: (go to http://extras.springer.com)

None

Activities

Activity #1

Task: Author a paragraph or two in which you explain why setting a pass–fail point at 70 % for a 100-item test (due to the common use of 70 % to mean the lowest of "C" grades) is almost a worthless setting of a pass–fail point.

Answer: A unique answer from each reader, of course. The key points to mention might be that 70 items correct does not give one any idea of what the person could or could not do. Perhaps the person does not know key material.

Activity #2

Setting: Imagine that you are meeting with a group of 10 doctors who specialize in emergency room medicine. They have been selected to set the pass–fail point for a certification test. The test has 250 multiple-choice test items.

Task: Write a set of directions for the 10 doctors to help guide them determining a pass–fail point.

Answer: You will want to author directions that are clear, accurate, and absent of jargon. You will want to guide them in terms of what they are doing, how they are doing it, and why they are doing it. Groups of experts setting standards in their field are very interested in their discipline, but they also enjoy learning new material. Explain the basics of measurement in your directions. This will provide you with added credibility. And the experts will enjoy learning something new.

Activity #3

Task: Take a multiple-choice test that you are very familiar with. It would be of added benefit if the test is used to determine pass–fail performance. Print out the test, and cut each item with a pair of scissors. Then order and space test items in terms of difficulty. Then determine a pass–fail line, based upon the context of the items passed. Then compute the raw number of items needed for a "pass."

Answer: It is most useful and enjoyable to use a test with which you have familiarity. To complete the activity, just mimic the steps we have outlined in the text.

Activity #4

Task: We present some text we have authored. Review the text, and write a response that first makes use of the topic of this chapter; second, make use of one topic of your choosing that has already been presented in previous chapters.

In short, we frequently observe educated people setting policy with little or no attention to research that is relevant to the policy being made. This brings us to a question: What role should research play in setting educational policy? Research possesses limitations and simultaneously makes contributions to policy. Regarding its limitations, research cannot determine goals or standards, which are primarily a reflection of values. Research alone cannot establish what is best nor can it prescribe a curriculum or pedagogical approach for all students at all times. Regarding contributions, research can inform decisions based on probabilities that a specific outcome will result. Research can prevent mistakes, and it can identify what is possible and what holds promise.

References

Hestenes, D., Wells, M., & Swackhamer, G. (1992). Force concept inventory. *The Physics Teacher, 30*, 141–158.

Wright, B. D. (1999). Common sense for measurement. *Rasch Measurement Transactions, 13*(3), 704.

Additional Readings

Excellent extensions to the topic we have introduced in this chapter.

Grosse, M. E., & Wright, B. D. (1986). Setting, evaluating, and maintaining certification standards with the Rasch model. *Evaluation and the Health Professions, 9*, 267–285.

Wang, N. (2003). Use of the Rasch IRT model in standard setting: An item mapping method. *Journal of Educational Measurement, 40*(3), 231–253.

Wright, B. D., & Grosse, M. (1993). How to set standards. *Rasch Measurement Transactions, 7*(3), 315.

Chapter 16
Expressing Competency Levels

Isabelle and Ted: Two Colleagues Conversing

Ted: Isabelle, I know sometimes researchers have already defined competency levels, or they are working with competency levels that have already been defined. How would I go about marking the boundary from one competency level to another on a Wright Map?

Isabelle: Can you give me a little more detail on what you want to do?

Ted: Okay, suppose I have a data set in which persons have been evaluated with respect to 5 different competency levels (I, II, III, IV, V). I am interested in showing on a Wright Map where the boundaries are located so that I can very quickly show where a sample of respondents is in one quick picture. Also then someone would be able to see what items separate persons in one competency level from persons in an adjoining category.

Introduction

In many studies, the researcher collects a wide variety of data, evaluates the data set, and then attempts to summarize results in a succinct manner. We now present an important Rasch technique that can be used to bring added meaning to Wright Maps beyond the introductory techniques presented in Chap. 6. The specific technique presented here focuses on how to compute accurately and add competency bands to a Wright Map. What we call "competency bands" may go by many different names, but the idea is similar. Consider a state or a country (say Indiana, USA, or Nord Rhein-Westfalen, Germany). The state may collect test data from 6th graders. Rasch measures are computed for students from 1 to 1,000, but a technique must be used to classify students quickly and accurately for stakeholders such as policy makers, teachers, and parents. Such a classification can be used to sort students into as many different groups as one might wish, but all classifications are attempts to summarize the performance of students. In states such as Indiana or Nordrhein-Westfalen, classifications such as Below Basic, Basic, Intermediate, and Advanced might be used. The point in using such classifications of competency (e.g., PISA, State of

DOI 10.1007/978-94-007-6857-4_16,

Ohio, USA, Achievement Tests) is an attempt to be able to quickly characterize the performance of a student.

In Germany, our colleagues Andrea Moeller at the University of Trier, Jürgen Mayer of the University of Kassel and others have assembled a large data set to explore the growth of student competence with regard to life science. In Chap. 19 we will use these data to help explain a technique for conceptualizing and correctly analyzing a complex rating scale. In this chapter we use a nonrandom subset of data to guide readers through the steps for computing and visually displaying competency bands. We have simplified our analysis for this book. Our simplification involves the use of only a subset of items, for it is much easier to explain what one does with an edited data set. Readers will, we trust, be able to apply the steps we outline to many of their own data sets.

The first step in our journey is to recall that the data set was collected using a number of partial credit test items. Students could receive raw score ratings ranging from 0 to 5 (0, 1, 2, 3, 4, 5) for each item they were administered. The numerical values a student received on an item were predetermined from theory. Also, and of great importance for readers of this chapter, students received a rating that represented their "Level" of competency. For example, if Tina completed "Item A3A-Formulating Questions Category" and she received a partial credit score of "3," this would mean that she had exhibited a competency level of "3" for the item. That level for that item category is briefly summarized by the text "pose scientific question(s) based on biological concept knowledge".

The theory (Fig. 16.1) envisions that competency levels can be exhibited using one of four item types (Formulating Questions, Generating Hypotheses, Planning an Investigation, Interpreting Data). A student receiving a partial credit score of "4" for her or his answer to a Generating Hypothesis item is classified at Level IV with regard to scientific competency. This student would have been able to "Generate Hypotheses that are Generalizable and/or Quantifiable." She or he would not have discussed alternative hypotheses and therefore would not have made it to Level V. Needless to say, by having exhibited what was detailed for Level IV, this student would have been classified as being able to do what is described in Levels I, II, and III.

The items administered and scored for each respondent were entered into an SPSS data set, and Winsteps was utilized to create a control file. Then a person competency measure (how well she or he did on the test) was computed. A higher person measure represents higher competency level with respect to the field of life science, a lower measure means lower competency with regard to the field of life science as measured by the set of test items.

Use of the Rasch model for such data is required for the computation of person measures prior to statistical analysis. Throughout this book we have emphasized this point, for example, consider the figure in the ogive chapter (Chap. 11) in which one can observe that a difference of 1 point on a test does not have the same meaning when comparing two respondents; however, we revisit the point in order to emphasize the importance of clear, accurate communication of research findings. The computation of a Rasch measure allows standard parametric tests to be conducted.

Levels I–V as expressed by "Formulating Questions"

FQ -Level V Pose creative scientific question(s) which incorporate problem solving techniques.
FQ-Level IV Pose scientific question(s) that are generalizable and/or quantifiable.
FQ-Level III Pose scientific question(s) based on biological concept knowledge.
FQ-Level II Pose scientific question(s) in which two variables are correlated.
FQ- Level I Pose simple scientific question(s) regarding an observed phenomenon.

Levels I–V as expressed by "Generating Hypotheses"

GH-Level V Discuss alternative hypotheses.
GH- Level IV Generate hypotheses that are generalizable and/or quantifiable.
GH-Level III Generate hypotheses based on biological concept knowledge.
GH-Level II Generate hypotheses based on analogies from everyday life.
GH-Level I Generate simple testable hypotheses without explanation.

Levels I–V as expressed by "Planning an Investigation"

PI-Level V Reflect own experimental methodology (e.g. measurement accuracy etc.).
PI-Level IV Consider duration and repetition of the experiment.
PI-Level III Consider control variables.
PI-Level II Correlate independent and dependent variables.
PI-Level I Identify one variable.

Levels I–V as expressed by "Interpreting Data"

ID-Level V Consider alternative data or observation interpretations.
ID-Level IV Consider the generalizability of the observation or data interpretation.
ID- Level III Use biological concepts knowledge to interpret observation or data.
ID-Level II Interpret observations or data.
ID-Level I Report observations or data.

Fig. 16.1 Details of the meaning of five levels (I–V) for the topic of "science competency" as envisioned and developed by researchers. This work utilized the Giessen Competence Model of Scientific Inquiry (GCMSI) proposed by Mayer (2007), previously proposed techniques of defining inquiry competence skills (e.g., Bybee, 2002; Hammann, 2004; Schauble, Glaser, Duschl, Schulze, & John, 1995) and the synthesis and extension of these concepts (see Möller, Grube, Hartmann, & Mayer, 2009; Möller, Grube, & Mayer, 2008, for details). Figure developed from the presentation of Modeller, Mayer, and others

Parametric tests allow researchers to make clearer inferences, comments, and interpretations with regard to a variety of trends in a data set. This is why parametric statistical tests such as ANOVA are conducted in medical research and social science research. However, if raw data, not Rasch measures, are used, then spurious conclusions may result. A second major topic is germane to what we will do in this chapter. Communication of research findings is key to any endeavor, and Rasch techniques of communicating the meaning of results are particularly powerful. For example, in Chap. 6, we discussed a Wright Map in which the means of male and female performances were presented along the person measure side of the map. The advantage of presenting the male and female measures was that those measures could be viewed in relation to the items presented to the test takers.

The competency levels that are so well defined in Fig. 16.1 are very useful as stand-alone text. Student answers can be reviewed, and students classified for each item. However, the use of numerous items, using different categories of item type, and use of the multimatrix design can make determining the overall competency level of a respondent daunting. Below we will show readers how use of Winsteps to calculate person measures, an ability to locate the boundaries between level on the same scale as that used to express the overall measure of each respondent, and use of the Wright Map can bring clarity to expressing analysis results. Such Wright Maps (with bands, a person or persons, and a few examples items) are now commonly used visuals for presenting high-stakes test data for assessments.

Formative Assessment Checkpoint #1

Question (True or False): If one is going to classify students who completed a 40-item multiple-choice test in one of four groups, it is easy to classify the students. One need only divide the number of test items by the number of classification groups one wants to achieve the needed width of the classification groups. So this test will have groups 10 items wide (40/4 = 10). This will mean Level I students will earn between 0 and 9 points. Level II students will earn between 10 and 19 points. Level III students will earn between 20 and 29 points. Level IV students will earn between 30 and 40 points.

Answer: False. In this example more than one fatal error has been made. First, raw scores are treated as if they express linear measures, which they should not be assumed to express. Second, the analyst has fallen into a trap of thinking that all levels need to be the same width in terms of raw score. Third, the levels are not tied to items that define what it means to exhibit a particular "level." The levels as they stand are close to ad hoc and provide very little useful information.

The Control File

The question is: What are the steps to the computation of the boundaries between each of these levels when one uses a set of test items to define the trait and to compute a person measure along the trait?

As with any Rasch analysis, the first step is to create a control file that correctly runs the data set. Following is a part (we do not show all the person responses) of a control file for a subset of the data set. In this sample, the responses of 250+ persons are provided. A brief review of the control file (Fig. 16.2) is provided to aid readers' understanding.

The first five lines tell the program how to read the person and item information in the data file. The control file reads the first 6 items of the data set (NI = 6) and specifies the column that begins the presentation of the 6 items (Item1 = 1). A key aspect of the data file is that the codes to be viewed as valid and (potentially) useful for the analysis are the numbers 0, 1, 2, 3, 4, and 5. The control file indicates that

```
&INST
Title= "Competency Bands"
ITEM1 = 1 ;
NI = 6 ;
NAME1 = 26 ;
NAMLEN = 37 ;
XWIDE = 1 ;
CODES = 012345 ;
&END ;
Item A1A ;
Item A3A ;
Item A8A ;
Item A11A ;
Item A13A ;
Item A16A ;
END NAMES
....20.1...1..2......2..    1 1 ADJO05102 15 10 0 1 0 0 0 0 0
..1.1.1.2.......0.2.....    2 1 ADMI03 71 12  7 1 0 0 1 0 0 0
.0.2..10......0...0.....    3 1 AGUL06 52 10  5 0 0 0 0 1 1 0
.
. DATA EXCLUDED FROM FIGURE
.
..2.....2......4.223....  249 1 CHGÜ02102 15 10 0 0 0 1 0 0 0
.1.0..11......0...1.....  250 1 CHGÜ03 51 10  5 1 0 0 0 0 0 0
333332                   Fake  3
233333                   Fake  2
344444
433333
355555
533333
```

Fig. 16.2 Portions of the control file utilized for this chapter. Data are from a multimatrix data collection from students throughout Germany. Due to space limitations, data from the first three students are provided, data from the last two students are provided, as well as data person responses are included to ensure that all rating scale steps would be observed in the analysis

each student response is 1 column wide (XWIDE = 1). Finally, the control file shows that the name of the student (e.g., a real name, a student ID, or a student ID which includes all manner of demographic information) begins in column 26 (NAME1 = 26), and this information is 37 columns wide (NAMELENGTH = 37). The names and labels of the items to be evaluated are presented between the line "&END" and "END NAMES." These two lines of code are in all control files. "&END" tells the program the control variable instruction section of the file has ended. "END NAMES" tells the program the list of item labels has ended and typically is followed by the data.

There are two unique pieces of information to note in this control file. First, students completed different mixes of the six items being evaluated. Review the data in the control file to see the variation in the items that were answered by respondents. The administration (and confident analysis) of varied items of differing difficulty to students in a data set was impossible to evaluate prior to the use of Rasch measurement. By using Rasch measurement, a researcher takes great care to think about a latent trait; then it can be possible for respondents to complete different sets of items and still be measured on the same scale because the scale represents a single trait. Presenting a mix of items but remaining able to measure and compare respondents does not happen by chance. There must be a plan for deciding which respondents take which items. The plan is named a "multimatrix" design.

A second notable aspect of the control file is that a few fake student responses have been inserted at the end of the control file (we name these people "fake 1," "fake 2," etc.). We sometimes do this in an analysis to ensure that unobserved ratings in a data set are not dropped. And we add such people to help us make sure we understand the Wright Map and the meaning of going up or down in logits. For instance, perhaps no student received a rating of 5 for item A3A. In the event of more data and thus the possibility of a "5" being observed, it is then important for a number of reasons to retain the rating scale step that is one step "better" (at least in this example) than a value of "4." Readers should also note the inclusion of a fake person who received scores of "3" to all items. This person was included because analysis of all other data revealed that for this data set no one received a score of 3 for some items. Inclusion of this fictitious person ensures that all categories are observed for each item. We were not lazy, but instead of finding those items that no one marked "3," we simply entered a person and then typed in the number 3.

We now make a final comment before we detail the steps to delineate the competency bands (the levels) on the Wright Map and thus classify students based upon levels. For this example we used only "Formulating Questions" items. As a result, the location of respondents has a larger error than if we had used all completed items. We used only the "Formulating Questions" items simply because we have found it is easier to explain the steps readers would take to compute the bands with this subset of items from the data set.

When the data are evaluated, person measures can indeed be computed and devices such as Wright Maps constructed. In Fig. 16.3, we present a Wright-like map from the analysis of this sample data set. Notice that we present only the measures of respondents. The fictitious people have been removed from the plot.

```
TABLE 16.3 Chp X Competency Bands                  ZOU773WS.TXT  Dec 30 13:21 2011
INPUT: 256 PERSON  6 ITEM  REPORTED: 256 PERSON  6 ITEM  6 CATS    WINSTEPS 3.73
--------------------------------------------------------------------------------

       ITEM - MAP - PERSON
         <rare>|<more>
    4          +  **
               |  **
               |
               |
    3          +
               |  **
               |
               |
    2          +
               |
               |
               |
    1          +
               |
               |
               |
    0          +
               |  **
               |  *
               |
   -1          +
               |
               |
               |
   -2          +
               |  **
               |
               |  **************
   -3          +  *****
               |  ********
               |
               |
   -4          +
               |
               |
               |
   -5          +  **
               |  *
               |  *****
               |  **
   -6          +
               |  ************
               |  *********
               |  ****
   -7          +  *************
               |
               |  *
               |  ******
   -8          +  *****
               |
               |
               |
   -9          +  ****************************************
               |

--------------------------------------------------------------------------------
```

Fig. 16.3 (Winsteps Table 16.3): A Wright-like map with only the plots of person measures. Each "*" represents that location of one respondent. Lines marking the start and end of levels will be added to this figure and the performance of respondents related to the levels

Adding the Competency Level Lines

Having guided readers through the nuances of our control file, we move onto the core of this chapter: The Wright-like map above presents the measures of respondents expressed along the linear logit metric. But, how can one add the lines for the competency categories? As before, we will make sure to make use of our prior thinking. In this case, we used the thinking, the conceptualization that our colleagues have already done with respect to what it means for a student to be in a specific category. Therefore, a key aspect of our work to determine competency levels is to use the rating scale values that have already been so well defined (see Fig. 16.1).

To compute the location of the boundaries and classify respondents within one level or another, an analyst needs to take a few simple steps to map predetermined levels already defined. For instance, Fig. 16.1 can be used to determine that a student Phil is at Level III when his answer for an item is evaluated. Using the point scheme determined by the researchers, Phil receives 3 points for his answer to this item. This 3 points is used to note that he is at a Level III for this item.

To determine the location of boundaries between such levels, one should first run a Winsteps analysis (Step 1) and then click on the part of the toolbar that says "Output Files" (Step 2). Next, click on the option "ITEM-Structure File" (Step 3). The analyst is then prompted to indicate that this file can be temporary or permanent. Also, the analyst is asked to select a format (e.g., Excel, SPSS, STATA) for the data. For the example below, the temporary file will be created in Excel (Fig. 16.4).

The next step (Step 4) is the computation of the boundaries. When the "Item-Structure File" in Excel format is brought up, the columns needed for the boundary calculation are the column with the heading CAT and the column with label 50 % PRB. To compute the boundary between competency level 1 and competency level 2, simply find the column labeled 50 % PRB that is located between the column labeled "CAT 1" and the column labeled "CAT 2." In this case, six numbers can be seen; each number is information to compute the boundary between CAT 1 and CAT 2 for this data set which will then be used to plot the boundary on the Wright Map in Fig. 16.5. Readers should think of the six numbers (–7.43, –7.34,

F	G	H	I	J	K	L	M	N	O	P
CAT	STRUCTUR	MEASURE	ERROR	CAT-0.5	AT CAT	50%PRB	CAT	STRUCTUR	MEASURE	ERROR
1	1	-7.31	0.18	-7.6	-6.36	-7.43	2	2	-5.41	0.18
1	1	-7.21	0.18	-7.5	-6.26	-7.34	2	2	-5.31	0.18
1	1	-7.83	0.18	-8.12	-6.89	-7.96	2	2	-5.93	0.18
1	1	-8.3	0.18	-8.59	-7.35	-8.42	2	2	-6.4	0.18
1	1	-6.79	0.18	-7.08	-5.84	-6.91	2	2	-4.89	0.18
1	1	-8.22	0.18	-8.51	-7.28	-8.35	2	2	-6.32	0.18

Fig. 16.4 Part of the "Item-Structure File" provided by a Winsteps analysis of the student data set. Although numerous columns of data are provided, only particular columns are utilized for the computation of the location of boundaries between the levels expressed by the numerical coding of 0, 1, 2, 3, 4, and 5 used for the partial credit scoring of each item administered to each student

```
TABLE 16.3 Chp X Competency Bands                    ZOU773WS.TXT  Dec 30 13:21 2011
INPUT: 256 PERSON  6 ITEM  REPORTED: 256 PERSON  6 ITEM  6 CATS    WINSTEPS 3.73
--------------------------------------------------------------------------------

       ITEM - MAP - PERSON
         <rare>|<more>
    4          +  **
               |  **
               |
               |
    3          +
               |  **
               |
               |
    2          +
               |
               |
               |
    1          +
               |
               |
               |
    0          +
               |  **
               |  *                                               Level 4
               |  ________________________________________________________
               |                                                  Level 3
   -1          +
               |
               |
               |
   -2          +
               |  **
               |
               |  **************
   -3          +  *****
               |  ********
               |
               |
   -4          +
               |
               |
               |
   -5          +  **
               |  *                                                 Level 3
               |__________________________________________________________
               |  *****                                             Level 2
               |  **
   -6          +
               |  ************
               |  *********
               |  ****
   -7          +  *************
               |
               |  *                                                 Level 2
               |  ________________________________________________________
               |  ******                                            Level 1
   -8          +  *****
               |
               |
               |
   -9          +  ****************************************
               |
```

Fig. 16.5 The computation of the median value of the appropriate 50 % probability columns presented in Fig. 16.4. Since there was an even number of 50 % probability values, for a median, the average is computed of the two middle values of each set of six values

−7.96, −8.42, −6.91, −8.35) provided between the columns for CAT 1 and CAT 2 as estimates of the location of the boundary as determined from the data for each item. The median value of the categories is often used as the reference point in computing the boundary between the categories. In this example, the median value is −7.69 [(−7.96+−7.43)/2=−7.69]. The median value of −7.69 can then be plotted on the Wright Map to bring more meaning to the location of the respondents as a function of category.

Figure 16.5 presents the same Wright Map as presented in Fig. 16.3, but with the computed boundary between categories 1 and 2. Also plotted are the computed boundaries that help delineate the boundaries between the other partial credit categories. Note that we have removed the locations of items from the Wright-like map. This information can cause information overload when viewing the Wright Map with the boundaries as well as the partial credit items. We prefer to remove the items that have been used to define the boundaries and simply present the performance of the respondents (or a subgroup thereof) and the boundaries. Below are the boundaries between the other levels.

[(6.06+6.59)/2]=6.32 for the boundary between IV and V
[(−.75+−.22)/2]=−.48 for the boundary between III and IV
[(−5.81+−5.29)/2]=−5.55 for the boundary between category II and III
[(−7.96+−7.43)/2]=−7.69 for the boundary between category I and II

The steps outlined above facilitate computation of the boundaries from one competency level to another. Why is this sort of presentation – without items on the Wright Map – advantageous? One could certainly present the calibration of each test item; however, in many studies one is interested in succinctly summarizing the overall performance of students. By including only the person measures (right side of the Wright-like map we created) and the bands for competency levels, one is able to quickly and clearly show the location of each respondent with regard to the competency levels of the trait as defined by the numbers used to indicate competency levels. There are, to be sure, some aspects of this type of presentation that at first bothered us. For instance, when a student is classified as being at one particular "level" in terms of her or his overall performance, some generalization is made. For example, a student could be at the very low end of Level II, and another student could be at the high end of Level II, meaning that there might be a significant difference between these two students' competency levels, even though they are both in Level II. We have, however, come around to accepting, even embracing, this type of classification. We know that there will always be this type of problem in classifying student performance in this way; however, we now appreciate that it is much easier for stakeholders to visualize and comprehend general trends in the data. We suspect that is why many US states and PISA classify student performance for stakeholders (e.g., a newspaper story might report "23 % of 8th grade students in the German State of Bayern exhibited an 'Advanced' ability level with respect to Physics where the Advanced level is one of five potential levels for students

(Superior, Advanced, Basic, Below Basic, and Not Acceptable)"). When we are communicating results, we might classify students with categories. When we conduct our analyses, we use the actual person measures of students, therefore in essence taking into consideration that students may be at a range of locations within a competency band.

Formative Assessment Checkpoint #2

Question (True/False): It is more important to present analysis results in a manner that highlights what you know and what stakeholders do not know than it is to simplify your explanation of results.

Answer: False. We realize that some stakeholders may need to feel they have paid for an expert who knows more than they do, but it is important for experts to apply their knowledge to provide stakeholders with research-based information that will help inform their decisions. Using Rasch measurement theory to establish competency levels is one technique researchers can use to simplify data in a meaningful way, thereby aiding decision makers.

Close inspection of Fig. 16.5 reveals that all bands are not the same width. The different widths are, in part, the result of using an analysis technique that does not treat the ratings 0, 1, 2, 3, 4, and 5 as if the ratings were linear. What one observes in this plot, with the bands of differing widths, is the real width that was expressed along the single trait by each of the rating scale categories.

In this chapter we have presented the ins and outs of (when a set of categories for a trait has been defined [e.g., Below Basic, Basic, Advanced; Cat I, Cat II, Cat III] ahead of time), how one can determine where boundaries fall on a Wright Map. As readers will be able to imagine, there are many variations of the techniques which are yet to be discussed in this chapter. However, we feel that we have provided specific guidance which can allow other researchers to apply what we present (and expand) in a variety of settings.

A Closing Thought

In closing this chapter, we want to emphasize the need for strong, sound connections between research and policy and their effect on practice. We frequently observe educated people setting policy with little or no attention to research that is relevant to the policy being made. This brings us to a question: What role should research play in setting policy? Research possesses limitations and simultaneously makes contributions to policy. Regarding its limitations, research cannot determine goals or standards, which are primarily a reflection of values. Research alone cannot establish what is best nor can it prescribe a curriculum or pedagogical approach for all students at all times. Regarding contributions, research can inform decisions based on probabilities that a specific outcome will result. Research can prevent mistakes, and it can identify what is possible and what holds promise. This chapter represents an example of the value of making policy decisions based on sound research and in particular sound measurement techniques.

Isabelle and Ted: Two Colleagues Conversing

Isabelle: *I am going to summarize what I think I heard you say, Ted. Tell me if you agree, ok?*

Ted: *Sounds good.*

Isabelle: *In education and medicine sometimes we talk of levels. This is not a bad way of communicating some complex ideas, because people are familiar with the concept of levels. For instance, when we go to a department store, we are used to the idea that when we go up to higher and higher floors, we are farther "up" from the ground level. We could talk of "going up" in terms of meters or feet when we move up in the store, but that would be odd and cumbersome. The number of meters (or feet) from one level to the next might not be the exact same amount. Also, sometimes when you are in a store, there may be parts of a floor that have a few steps to take you to a special area of that floor, and that would be hard to measure. So, to make life easier, when we tell someone where we are going ("I am going to the 3rd floor to buy some clothing"), we just use the floor "level." It's not perfect in terms of describing how far up we are, but using that information is good enough to provide a quick communication to someone else.*

Ted: *I really like that analogy.*

Isabelle: *To figure out where to mark our levels, we need to use a file in Winsteps that provides us with the 50 % probability for each item for each adjoining level. Since this value will differ for each item, a useful procedure is to compute the median value for all the items between any two adjoining categories. That is the value we plot. Also, when we finally plot our boundaries, it will not be surprising if the width of a level is not the same for all levels. This has to do with this idea that we can get tricked with our coding. All we know, for example, is that level 2 means more than level 1, and that level 1 means more than level 0.*

Isabelle: *Also Ted, I can see this same issue that comes up in education could come up in many fields. For instance, perhaps in terms of "Physical Mobility" there has already been a five-step standard (I, II, III, IV, V) that has been developed in which step V means normal mobility and step I means very poor mobility. It makes sense to me that the same steps we follow here could be used in that situation as well!*

Keywords and Phrases

Item-structure file
Median
Competency levels

Using theory to define competency levels
Classifying students as a function of competency levels

Potential Article Text

There are a number of studies in the education literature that apply the idea of levels to classify students. Researchers in Germany have reported on the development of a definition of scientific inquiry that makes use of 6 levels (0, I, II, III, IV, V). Data were collected from a large sample of German students, and student answers were evaluated in light of the defined levels. Since students completed a unique subset of open-ended items, a multimatrix design was employed along with a Rasch analysis to facilitate equating the student measures on a unidimensional scale. Wright Maps were created using computed person measures. The boundaries between levels were computed by reviewing the 50 % probability thresholds between each pair of adjoining categories (0/1, 1/2, 2/3, 3/4, 4/5) and computing a median value for each boundary by using the set of 50 % thresholds that were calculated for each of the test items.

Quick Tips

To compute the boundaries between defined categories:

1. Run Winsteps.
2. Select "ITEM-Structure file ISFILE=" from Output files menu.
3. Find the Category column (e.g., column headed with the word CAT and the number "2" is presented below the word CAT) and 50 % PRB column in the Item-Structure file.
4. Calculate the median value of the 50 % PRB for each of the category levels listed under the CAT column (e.g., 1, 2, etc.).
5. Use each median value that you compute as the boundary line between the category levels. Plot those lines on the Wright Map.
6. Repeat steps 3–5 for all category levels.

   Note: Remember that the boundary lines are ordinal values.

Data Sets: (go to http://extras.springer.com)

None

Activities

Activity #1

Task: Prior to the development of Rasch measurement, "cut scores" were often selected based upon common grading systems people grew up with. In the USA, 70 % or above was often viewed as "Average or Above." Write a paragraph in which you explain to a stakeholder who knows nothing of measurement issues why the selection of 70 % as the minimum of average is potentially greatly flawed.

Answer: Frau Berg, thank you for inviting me to help your organization in the analysis of the recent test data that were collected. When you asked me to help with the development of cut scores, you mentioned that you had heard that it was not as simple as computing a percent correct and picking values that are commonly used in school for the "excellent" students, the "very good" students, and so on. You are right; it is not that simple, but if you and your team provide some guidance to me as to what an "excellent" student should be able to do, then I can quickly help you. I think the fastest way for you to understand that a 70 % correct cannot just be immediately used is to imagine that the test taken by students of Saxony was very easy. In that case, I think you would agree that the "cut point" might need to be much higher, maybe 87 % for a minimum "average." If the test had been much harder than expected for students, then a 70 % for a minimum acceptable "average" might be far too hard to achieve. Perhaps a fairer and more accurate value would be 53 % of the items correctly answered. So you should see that the percent correct value needed for a particular level of achievement must be determined based upon the difficulty of items. Also, the level of achievement must be based upon what skills you want students to demonstrate mastery of, and that means which items need to be correctly answered by the students. It is very important to later consider what the meaning of say 70 % is, but we will consider that later.

Activity #2

Task: Below is a table summarizing data provided by the State of Ohio to summarize the raw score range and measure range for one of Ohio's high-stakes tests. Please explain what topics of this chapter can be explained by this table.

		Raw Score	Measures
Limited	Below	24	Below 385
Basic	Cut Point	24	385
Proficient	Cut Point	30	400
Accelerated	Cut Point	36	415
Advanced	Cut Point	41	432

RS	M	RS	M
0	251	25	388
1	270	26	390
2	290	27	393
3	302	28	395
4	311	29	398
5	319	30	400
6	325	31	402
7	331	32	405
8	336	33	407
9	340	34	410
10	344	35	413
11	348	36	415
12	352	37	418
13	355	38	421
14	359	39	424
15	362	40	428
16	365	41	432
17	367	42	435
18	370	43	440
19	373	44	445
20	376	45	451
21	378	46	459
22	381	47	470
23	383	48	488
24	386	49	506

Data provided in: http://www.ode.state.oh.us/GD/Templates/Pages/ODE/ODEDetail.aspx?page=3&TopicRelationID=285&ContentID=9479&Content=11772

Answer: This table presents classifications of proficiency similar to those presented in this chapter. In this example, there are five proficiency levels: Advanced, Accelerated, Proficient, Basic, and Limited. These are similar to the "Levels" (I, II, III, IV, and V) that were defined for the chapter data set. The data table presents the range of raw scores and Rasch measures for each proficiency level. The Rasch measures are used for statistical calculations, but the raw scores are presented to communicate results in terms of values that will be familiar to teachers and administrators. Of particular importance are the unequal widths of individual levels in terms of scale score units. Of note, the raw score width (and measure width) of "levels" is not identical. Also UMEAN and USCALE must have been used at some point to rescale.

Activity #3

Task: Many topics have been discussed in earlier chapters. Please explain a topic that was presented earlier in this book by using any of the data presented in Activity #2.

Answer: One of the issues that has been emphasized throughout the book is the problem with raw scores not being measures. In looking at the raw score to measure table, we can see that the 1-point difference between a raw score of 1 point and 2 points is 20 on the linear scale score. And we can see that the 1-point difference between a 23 and a 24 is 3 on the linear scale score. If we plot the raw scores against the linear scale scores, we would see an ogive.

Activity #4

Task: Pick a topic that you are very familiar with that could be measured with a single trait. An example of such measurement is the testing of students in mathematics. Pick your topic, and then author one short paragraph that describes what would constitute the lowest level of performance (Limited), author a second paragraph for one step higher (Basic), author a third paragraph for one step higher (Proficient), author a fourth paragraph for one step higher (Accelerated), and author a fifth paragraph for the highest step (Advanced). These descriptors are the same ones that were selected by the US State of Ohio to classify test takers who complete an Ohio high school graduation test.

Answer: The topic that you pick will determine what the paragraphs say; however, no matter the topic you pick, you must understand the trait you wish to measure. And, you must also decide what it means to be classified at a particular level.

Activity #5

Task: You have been asked to convene a group of experts in the field of medicine to evaluate the competency level of pediatricians. The goal is to be able to ultimately administer a multiple-choice test to pediatricians and then to express their performance using one of four competency levels (I, II, III, IV). What would be one technique you could use? Please detail the steps.

Answer: As readers will appreciate, there are many ways to move from A to Z in a task. What we present is one potential set of steps:

1. The group of experts should define the variable to be tested. What are skills/competencies that pediatricians should have, ranging from low-level skills (that all pediatricians should possess) to attributes that only the highest-level pediatricians would be able to demonstrate? These skills should be placed on the variable line.
2. The experts could then decide to describe, with words, what it means to be classified at different points on the variable line. Also, the experts are asked to define competency levels I, II, III, and IV.
3. A multiple-choice test could be developed that uses steps 1 and 2 to guide the authoring of test items. There should be test items which range from easy to hard. Items are to be authored with the goal of being able to determine a competency level for each test taker.

4. The multiple-choice test is administered to a sample of pediatricians. A Wright Map is created.
5. The experts are presented with the Wright Map, but the map only has item measures. Not the performance of test takers. The experts are asked to review the Wright Map and the definition of each of the 4 competency levels. The experts are asked to reach a consensus as to the boundary between competency levels as defined by the test items. Experts are urged not to count items and not to worry if a competency band is wide or narrow.

References

Bybee, R. W. (2002). Scientific literacy – Mythos oder realität? In W. Gräber, P. Nentwig, T. Koballa, & R. Evans (Eds.), *Scientific literacy. Der beitrag der naturwissenschaften zur allgemeinen bildung*. Opladen, Germany: Leske+Budrich.

Hammann, M. (2004). Kompetenzentwicklungsmodelle: Merkmale und ihre Bedeutung – dargestellt anhand von Kompetenzen beim Experimentieren. *Merkmahle und ihre Bedeutung, 57*(4), 196–203.

Mayer, J. (2007). Erkenntnisgewinnung als wissenschaftliches problemlösen. In H. Vogt & D. Krüger (Eds.), *Handbuch der theorien in der biologiedidaktischen forschung*. Berlin, Germany/Heidelberg, Germany: Springer.

Möller, A., Grube, C., Hartmann, S., & Mayer, J. (2009, April). *Increase of inquiry competence: A longitudinal large-scale assessment of students' performance from grade 5 to 10*. Paper presented at the international conference of the National Association of Research in Science Teaching (NARST), Garden Grove, CA.

Möller, A., Grube, C., & Mayer, J. (2008, March–April). *Skills and levels of students' inquiry competence in lower secondary biology education (grade 5–10)*. Paper presented at the international conference of the National Association of Research in Science Teaching (NARST), Baltimore, MD.

Schauble, L., Glaser, R., Duschl, R. A., Schulze, S., & John, J. (1995). Students' understanding of the objectives and procedures of experimentation in the science classroom. *The Journal of the Learning Sciences, 4*, 131–166.

Additional Readings

An article that presents the details of using Rasch techniques for standard setting and proficiency classification.

Jiao, H., Lissitz, R., Macready, G., Wang, S., & Liang, S. (2011). Exploring levels of performance using the mixture Rasch model for standard setting. *Psychological Test and Assessment Modeling, 53*(4), 499–522.

Chapter 17
Quality of Measurement and Sample Size

Isabelle and Ted: Two Colleagues Conversing

Ted: Isabelle, I know in a lot of projects, where people are planning on collecting data, one always asks what should my sample size be. It seems to me that this is important for our work, in that ultimately in a study, we will conduct parametric statistical tests using, for example, person measures. But in thinking about sample size, I think Rasch measurement has helped me think about new issues pertaining to sample size.

Isabelle: Can you elaborate on this?

Ted: Well when I look at a Wright Map, person measure tables, and item measure tables, I think much of my selected sample size will be impacted by my measurement instrument, by the trait I want to measure, and by where respondents are on the trait. Also sample size seems to impact item error and person error. So I will need to think about that issue.

Introduction

A common question that we are often asked is: "What sample size do I need to conduct a Rasch analysis of data?" The answer is both simple and complex. Sample size impacts aspects of a Rasch analysis. In this chapter we present some basic examples to guide learning about how issues of sample size influence a Rasch analysis. But before we begin, we want to address a misconception which exists in some quarters that large numbers of respondents are required to conduct a Rasch analysis. We wish to emphasize that this notion is false. We hypothesize that one source of this misconception is the extensive general publicity associated with very large international evaluations (e.g., PISA) that utilize Rasch measurement. Less broadly publicized are many studies in medicine, where small sample sizes are analyzed with Rasch measurement.

W.J. Boone et al., *Rasch Analysis in the Human Sciences*,
DOI 10.1007/978-94-007-6857-4_17,

The Measurement Instrument as a Source of the Sample Size Problem

This chapter consists of two sections concerning sample size. The first part of the chapter helps readers think about sample size and consider the issues which impact sample size. The second part of this chapter provides guidance as to "rules of thumb" that may be used when considering sample size in a future Rasch analysis that you plan to conduct.

To begin our discussion of sample size, we return to the discussion of item measures of instruments that was presented earlier in our book. A crucial difference between high- and low-quality instruments was explained by discussing the construction of a ruler from a blank piece of wood. After marks were placed on the ruler (the test, the survey), measurements were made for numerous students. In this chapter, we return to our ruler (our meterstick). Figure 17.1 contains a list of fictitious student pairs and the number of items that separate the students in each pair along the trait continuum. Pretend that these items are self-efficacy items that one can either agree or not agree with (so dichotomous items).

Figure 17.2 presents a horizontal Wright Map of ruler 1 marks as given in Fig. 17.1. Inspection of the Wright Map in Fig. 17.2 reveals that the students (below the trait line continuum) are evenly spaced across the continuum from less (left) to more (right); however, the items (above the trait line continuum) are quite varied in their spacing. On the left side, we see two items clustered tightly together almost directly above Sam and three clustered items on the right side above Tom. The remaining three items are spaced farther, but not equally apart.

In contrast, inspection of Fig. 17.3, also a horizontal Wright Map, shows both items (above) and students (below) that are evenly spaced along the continuum of the trait. The location of the same 10 students along the same trait remains unchanged compared to Fig. 17.2; however, a different distribution of items is displayed.

Given earlier chapters of this book and Figs. 17.2 and 17.3, readers should now appreciate that the distribution of marks along any ruler will strongly influence the probability of that ruler's ability to differentiate between respondents.

Figure 17.3 presents the same group of students who are located at the same spots along the continuum; however, they are measured by a second ruler, but this one has equal spacing of marks. The location of each person is a person-specific measure, and a more precise differentiation of these respondents can be seen to be possible with the ruler presented in Fig. 17.3. This ruler is not perfect; no ruler is perfect – but the ruler shown in Fig. 17.3 facilitates better measurement and differentiation of this group of respondents. Thus, one factor that does influence the necessary sample size of respondents is the quality of the measurement instrument. For example, are some marks – items – along the meterstick wasted? Examples of potential wasted marks (marks that are redundant or close to redundant) are the 3 very-close-to-each-other marks above Tom in Fig. 17.2 and the 2 very-close-to-each-other marks between Joe and Sam, also in Fig. 17.2. There of course may be some good reasons (e.g., content of a course, specifics of an intervention, or a physical activity that should be observed

Student Pairs	Number of Items Separating Student Pair	
	Ruler 1	Ruler 2
Bob-Joe	0	1
Joe-Sam	2	0
Sam-Sue	1	1
Sue-Al	1	1
Al-Rick	0	1
Rick-Tom	0	1
Tom-Pam	3	1
Pam-Glo	0	1
Glo-Alex	1	1

Fig. 17.1 The number of items separating each student pair for two examples of measurement marks on a ruler

```
Item
            II      I     I                III               I
-------------------------------------------------------------------
 X     X       X      X      X      X       X      X      X       X
Bob   Joe    Sam    Sue     Al    Rick    Tom     Pam    Glo    Alex
Person
Less                                                            More
```

Fig. 17.2 A horizontal Wright Map for students and items using Ruler 1. The symbol "I" denotes the location of each of the 8 items along the trait. The symbol "X" denotes the location of each student along the trait. The student Tom is located very near the location of three items. One item is exactly at the attitude level of Tom, and two items a little more "agreeable" than Tom's attitude level

```
Item
       I        I       I      I        I       I      I        I
-------------------------------------------------------------------
 X     X       X      X      X      X       X      X      X       X
Bob   Joe    Sam    Sue     Al    Rick    Tom     Pam    Glo    Alex
Person
Less                                                           More
```

Fig. 17.3 A second horizontal Wright Map of the same students as in Fig. 17.2 using Ruler 2

in a patient) to keep an item, but it is important to at least be aware that, from a measurement perspective, some items may be redundant.

When we discuss this issue with colleagues who are planning a project, we find, almost uniformly, no consideration that the selected instruments and their quality exert an influence on the sample size that might need to be required to answer a research question. Often, there seems to be an assumption that because an instrument has been used previously with a sample, then one can just proceed with one's work. Whereas careful selection of an instrument takes time, and the potential addition or subtraction of items may require additional time, it is paramount to maximize the strength of measurement instruments in a research study. A likely consequence of using a low-quality measurement device is the need to collect data from a very large sample of respondents, and even then it may be problematic as to whether true differences (if they exist) are being detected in a project.

Formative Assessment Checkpoint #1

Question (True/False): When investigating a research question, the only psychometric/ statistical issue that needs to be considered is sample size because a large sample size trumps all other issues.

Answer: False. Suppose you wish to investigate 12th graders' knowledge of physics and you use an instrument that is very easy for almost all respondents. Most respondents will correctly answer a very high percentage of test items. In this case you might be able to differentiate two broad groups of respondents, but you might not be able to discern any differences between high performers (the majority of respondents).

Interplay of Persons and Items

When considering the size of the sample, one helpful component of Rasch measurement is an ability to consider the interplay of persons and items. This interplay is expressed on the same linear equal-interval metric of logits and begs the question: How might person measures influence the sample size needed for a research project? Whereas this issue was introduced in Figs. 17.2 and 17.3, it has not been explicitly discussed. Figure 17.3 exhibits an evenly distributed range of person self-efficacy, with Alex having the highest self-efficacy and Bob having the lowest self-efficacy. In Fig. 17.4, we now present a different group of respondents who were measured with the same metric as that presented in Fig. 17.3.

Figure 17.3 presents 8 items and 10 respondents. With this distribution of items, 7 of 10 respondents could be differentiated. Joe and Sam could not be differentiated because their person measures lie between the same two-item measures on the meterstick. Therefore, two respondents. Figure 17.4 presents the same meterstick with 8 items marking the same locations along the "self-efficacy" continuum. Ten (10) new respondents are located along the continuum. Whereas these 8 items performed well in differentiating the views of 10 respondents such as Glo and Alex, they do not perform as well with this new group of respondents, for example, Chi and Cha. Thus, both the distribution of respondents and the locations of items along a trait influence the necessary sample size in a study. This is where piloting of an instrument with a similar sample (as the group you have to study) can be very helpful. By piloting an instrument with a similar sample, one can evaluate the detail with which respondents can be differentiated.

```
Item
       I       I        I      I        I       I       I       I
  ------------------------------------------------------------------
X   X   X   X        X         X   X  X                  X    X
JJ DiDi  Si Jill     Blu       ZZ  TT GG                Chi  Cha
Person
Less                                                            More
```

Fig. 17.4 A Wright Map displaying the same trait presented in Figs. 17.2 and 17.3 but different students are plotted along the trait. The location of items in Fig. 17.3 is identical to Fig. 17.4

Several Factors Influence the Sample Size Needed

Several additional issues must be taken into account in properly answering the sample size question. These nuances can be explained by using the self-efficacy control files and data from earlier chapters.

Figures 17.5 and 17.6 are a summary of key Rasch statistics that were, in part, discussed in Chap. 13. Figure 17.5 presents the summary statistics for an analysis of 75 respondents who answered the 13-item self-efficacy scale of the STEBI. Immediately below Fig. 17.5 is Fig. 17.6, a summary table of 143 respondents who completed the STEBI. This set of 143 people includes the 75 respondents utilized for Fig. 17.5.

These tables provide a wide range of helpful information as we work through our consideration of sample size. Regarding the issue at hand, we call readers' attention to Figs. 17.5 and 17.6. Analysis of the entire data set of 143 respondents yielded an item separation of 10.05.

Think of this separation as an index of the number of different groups of items that can be discerned with our sample. In the Winsteps manual, Mike Linacre (2012) writes of item separation:

> Item separation is used to verify the item hierarchy. Low item separation (<3 = high, medium, low item difficulties, item reliability <0.9) implies that the person sample is not large enough to confirm the item difficulty hierarchy (= construct validity) of the instrument. (p. 644)

Analysis of the smaller data set of 75 respondents yielded an item separation of 7.00. Readers should note the decrease in the item separation, the precision with which groups of items can be differentiated. As readers can see, this decrease seems to be caused by the smaller sample size. Why is this important? Thinking back to Figs. 17.2 and 17.3, a researcher's level of confidence in differentiating respondents is directly related to his or her level of confidence in locating survey items along the continuum used to differentiate respondents. Another index, called item reliability, is provided in a Rasch analysis of a data set. Item reliability is also impacted by sample size, specifically a large sample size will create a high item reliability and a low reliability indicates an inadequate sample size for proper estimation of item location along the trait (Linacre 2012).

```
SUMMARY OF 75 MEASURED (EXTREME AND NON-EXTREME) Person
-------------------------------------------------------------------------------
|          TOTAL                         MODEL         INFIT          OUTFIT   |
|          SCORE      COUNT    MEASURE   ERROR      MNSQ    ZSTD   MNSQ   ZSTD |
|------------------------------------------------------------------------------|
| MEAN      53.0       12.4       1.05     .38                                 |
| S.D.      11.7        1.9       1.29     .20                                 |
| MAX.      78.0       13.0       7.53    1.85                                 |
| MIN.      21.0        6.0       -.84     .29       .19   -2.6    .19   -2.5 |
|------------------------------------------------------------------------------|
| REAL RMSE    .47 TRUE SD    1.20  SEPARATION  2.52  Person RELIABILITY  .86 |
|MODEL RMSE    .43 TRUE SD    1.21  SEPARATION  2.81  Person RELIABILITY  .89 |
| S.E. OF Person MEAN = .15                                                    |
-------------------------------------------------------------------------------
Person RAW SCORE-TO-MEASURE CORRELATION = .68 (approximate due to missing data)
CRONBACH ALPHA (KR-20) Person RAW SCORE "TEST" RELIABILITY = .96 (approximate due to missing
data)

ITEM RELIABILITY VALUES USING 75 RESPONDENTS
     SUMMARY OF 13 MEASURED (NON-EXTREME) Item
-------------------------------------------------------------------------------
|          TOTAL                         MODEL         INFIT          OUTFIT   |
|          SCORE      COUNT    MEASURE   ERROR      MNSQ    ZSTD   MNSQ   ZSTD |
|------------------------------------------------------------------------------|
| MEAN     306.0       71.7        .00     .15      1.00      .0   1.01     .1 |
| S.D.      53.9        3.1       1.12     .03       .26     1.5    .26    1.4 |
| MAX.     410.0       75.0       1.66     .22      1.69     3.5   1.64    3.1 |
| MIN.     206.0       68.0      -2.49     .12       .66    -2.1    .62   -2.3 |
|------------------------------------------------------------------------------|
| REAL RMSE    .16 TRUE SD    1.11  SEPARATION  7.00  Item   RELIABILITY  .98 |
|MODEL RMSE    .15 TRUE SD    1.11  SEPARATION  7.27  Item   RELIABILITY  .98 |
| S.E. OF Item MEAN = .32                                                      |
-------------------------------------------------------------------------------
                DELETED:     10 Item
UMEAN=.0000 USCALE=1.0000
Item RAW SCORE-TO-MEASURE CORRELATION = -.95 (approximate due to missing data)
919 DATA POINTS. LOG-LIKELIHOOD CHI-SQUARE: 2052.76 with 829 d.f. p=.0000
Global Root-Mean-Square Residual (excluding extreme scores): .8368
```

Fig. 17.5 Summary person statistics for analysis of 75 respondents and 13 items

> **Item reliability**: Low reliability means that your sample is not big enough to precisely locate the items on the latent variable. (p. 575 Winsteps Manual)

Although Linacre (2012) reports in the Winsteps manual that item reliability can also be impacted by item difficulty, our point is for researchers to use item reliability and item separation as one tool by which they can evaluate whether or not a sample may allow them to investigate what they wish to investigate. When you evaluate the item separation you compute for a particular sample, for a specific instrument, are you distinguishing item difficulty at the level which is needed for your study?

The impact of the sample size can also be seen in the person separation values; think of person separation as the number of groups of respondents that can be differentiated as the result of an analysis (as we provided for our item separation discussion, we provide useful text from the Winsteps manual regarding person separation) (Linacre 2012):

> Person separation is used to classify people. Low person separation (<2, person reliability <0.8) with a relevant person sample implies that the instrument may not be not sensitive enough to distinguish between high and low performers. More items may be needed. (p.644)

```
     SUMMARY OF 143 MEASURED (EXTREME AND NON-EXTREME) Person
-------------------------------------------------------------------------------
|          TOTAL                         MODEL         INFIT         OUTFIT    |
|          SCORE     COUNT     MEASURE   ERROR      MNSQ   ZSTD   MNSQ   ZSTD |
|------------------------------------------------------------------------------|
| MEAN      31.7       7.0       1.80      .56                                 |
| S.D.       4.4        .2       1.36      .20                                 |
| MAX.      42.0       7.0       7.44     1.89                                 |
| MIN.      20.0       6.0       -.86      .43       .13   -2.3    .15   -2.1 |
|------------------------------------------------------------------------------|
| REAL RMSE    .66 TRUE SD    1.19  SEPARATION  1.81  Person RELIABILITY   .77 |
|MODEL RMSE    .59 TRUE SD    1.22  SEPARATION  2.07  Person RELIABILITY   .81 |
| S.E. OF Person MEAN = .11                                                    |
-------------------------------------------------------------------------------
Person RAW SCORE-TO-MEASURE CORRELATION = .93
CRONBACH ALPHA (KR-20) Person RAW SCORE "TEST" RELIABILITY = .76

     SUMMARY OF 7 MEASURED (NON-EXTREME) Item
-------------------------------------------------------------------------------
|          TOTAL                         MODEL         INFIT         OUTFIT    |
|          SCORE     COUNT     MEASURE   ERROR      MNSQ   ZSTD   MNSQ   ZSTD |
|------------------------------------------------------------------------------|
| MEAN     647.4     142.1        .00      .12      1.03     .2   1.01     .0 |
| S.D.      88.3       2.1       1.27      .03       .20    1.4    .21    1.5 |
| MAX.     794.0     143.0       1.67      .17      1.49    3.5   1.46    3.1 |
| MIN.     509.0     137.0      -2.54      .10       .86   -1.2    .79   -1.7 |
|------------------------------------------------------------------------------|
| REAL RMSE    .13 TRUE SD    1.27  SEPARATION 10.05  Item   RELIABILITY   .99 |
|MODEL RMSE    .12 TRUE SD    1.27  SEPARATION 10.34  Item   RELIABILITY   .99 |
| S.E. OF Item MEAN = .52                                                      |
-------------------------------------------------------------------------------
          DELETED:     16 Item
UMEAN=.0000 USCALE=1.0000
Item RAW SCORE-TO-MEASURE CORRELATION = -.97
981 DATA POINTS. LOG-LIKELIHOOD CHI-SQUARE: 1970.95 with 830 d.f. p=.0000
Global Root-Mean-Square Residual (excluding extreme scores): .7592
```

Fig. 17.6 Summary person statistics for analysis of 143 respondents and 13 items

Person separation, at least for the authors of this book, is more difficult to understand and explain than item separation. Looking at the two analyses, readers should note that the person separation values for the smaller and the larger data sets are 2.52 and 1.81, respectively. What might that tell us about sample size? It seems that an enhanced ability exists to differentiate the smaller sample compared to the larger sample, when the same set of survey items was used. How can this be? As we looked at this data initially, to be honest we were confused; however, we realized that the observed increase in person separation with the smaller sample size had revealed yet another of the many wrinkles in sample size. In this case, the range of views of the 75 respondents will determine in part what we can or cannot say about the respondents. Linacre (2012) alerts us to this issue when discussing five issues that impact person reliability. Two of his points for us to note are "person (sample, test reliability depends chiefly on sample ability variance. Wider ability range = higher person reliability…It [person (sample, test) reliability] is independent of sample size" (p. 644).

Additional guidance regarding sample ability variance was also provided to the authors by Mike Linacre:

> If the additional person sample is more central than the original sample, then the person separation will reduce. Separation = person S.D./person S.E. (personal communication, October 20, 2011)

When possible herein, we offer concrete guidance, when such guidance exists (e.g., what values of MNSQ should be selected as cutoffs for further investigation of persons and/or items?). Therefore, we now present a range of guidance with concrete numbers offered by researchers in the field of Rasch measurement vis-à-vis the issue of sample size. We do not provide all available information, but rather what we have found to be most useful. The Additional Readings list at the end of this chapter includes readings for those who desire greater details.

Below we provide guidance from two key articles that we have used in much of our work. In several instances, we have added bold print for emphasis and included only those portions of the article we think will help beginning Rasch users. Wright and Tennant (1996) state:

> In most cases... data is required to estimate each item's "one parameter" of difficulty. With a **reasonably targeted sample of 50 persons**, there is 99 % confidence that the estimated item difficulty is within +/−1 logit of its stable value - this is close enough for most practical purposes, especially when persons take 10 or more items. With 200 persons, there is 99 % confidence the estimated value is within +/−10.5 logits (see RMT 7:4 p. 328). **But for pilot studies, 30 persons are enough to see what's happening** (see Best Test Design). Even if you plan to test 200, start the analysis as soon as the first data become available: **200 incorrect administrations are never as good as 50 correct ones**. (p. 468)

Linacre (1994) considers the issue of sample size when rating scales are used in the following:

The extra concern with polytomies is that you need at least 10 observations per category; see, for instance, Linacre J.M. (2002) Understanding Rasch measurement: Optimizing rating scale category effectiveness. Journal of Applied Measurement 3:1 85–106 or Linacre J.M. (1999) Investigating rating scale category utility. Journal of Outcome Measurement 3:2, 103–122.

For the Andrich Rating Scale Model (in which all items share the same rating scale), this requirement is almost always met.

For the Masters Partial Credit Model (in which each item defines its own rating scale), then 100 responses per item may be too few.

Otherwise the actual sample sizes could be smaller than with dichotomies because there is more information in each polytomous observation. (p.328)

A question about sample size can therefore be quite circular; the necessary sample size will depend upon the distribution of items, but knowing the distribution of items will depend, in part, on the number (and location) of respondents answering a question. We hope the examples presented in this chapter help readers understand that a necessary sample size for a study depends upon many issues. Examples of issues are: number of items; location of items along a trait; overlap of items along a trait; distribution of respondents along a trait; number of respondents; targeting of items to persons along a trait; and the goal of an instrument. Regarding the goal of an instrument, is the goal to determine the pass/fail status of respondents, or is the goal to measure a number of respondents whose measures are unknown? For data collected in the real world, all of these issues need to be considered. They may not be resolved, but at the least, they must be considered. In so doing, a researcher can better understand what simple steps might be taken to optimize a study. And, by taking such simple steps, a researcher might better understand the parametric analyses that make sense once measures are computed. In this chapter we have generally considered the issue of sample size as it relates to a dichotomous test/survey, but we have provided some details regarding some of the issues which impact sample size with a rating scale. For those interested in more detail, we suggest consulting more advanced texts on the issue.

A Closing Comment

So, what's the moral of this story? We implore our colleagues not only to think about the issue of sample size but also to understand that sample size is influenced by a number of factors (the distribution of items along a trait and the distribution of persons along a trait). However, even if one is not an expert in Rasch measurement or statistics, he or she can take some simple steps to greatly enhance an analysis. First, as we have done in this chapter, researchers may conduct sample analyses as they explore the measurement possible with different sample sizes (and also different numbers of items, and maybe even distributions of items along a trait). Second, researchers can greatly improve the quality of a study by simply thinking about sample size. Rasch measurement allows one to consider and think about items and persons. Just as thinking about a variable can be greatly enhanced by also considering where respondents fall along the continuum of a variable, the same type of thinking can help researchers think of sample size issues. There may be many instances in which successful measurement can be made with fewer respondents (that of course would save a market research company money), and of course there can be instances in which it is very important to realize what cannot be concluded from a sample that is too small to optimize measurement.

Isabelle and Ted: Two Colleagues Conversing

Isabelle: *Ted, test time, okay?*

Ted: *Okay…throw the questions at me, I am tough.*

Isabelle: *Some people think that Rasch can only be used on large data sets. And, they wonder what sort of sample size they need.*

Ted: *Isabelle, that wasn't a question. Try again.*

Isabelle: *You got me that time. OK, here's a question, actually two in one: Is the use of Rasch restricted to large data sets, and, if so, how large do sample sizes need to be?*

Ted: *At first I thought Rasch could be only used for large data sets, but as I learned more about Rasch, I came to understand that the sample size one "needs" depends upon so many things. For example, who is taking the instrument? What do you want to learn about the respondents? Do you want to know how the respondents differ? And/or, do you want to know if they pass some sort of threshold (did they pass, did they fail)? Also, so much depends upon the quality of the instrument. If there are a lot of items that overlap on a trait, basically you have to work with a smaller number of items than you thought.*

Isabelle: *You are throwing a lot at me, can we do some little pieces?*

Ted: *Okay. Well, some people might not understand this, but one of the things that helps me think about sample size is to look at what other people have done with Rasch. Have they all used huge data sets? What I noticed is that in medicine, where Rasch is used a lot, there are a number of studies that have used small sample sizes. Often these small sample sizes are due to the expense and the rarity of a medical issue that is being studied. So, at least in medicine, researchers have used small sample sizes and still have been able to advance medical thinking. That is what we want to do in education. I did a quick search and found a number of sample medical studies that used Rasch with small sample sizes. For example, Björkdahl, A., Nilsson, A. L., Grimby, G., and Stibrant Sunnerhagen, K. S.* (2006) *conducted a Rasch analysis of n = 58 patients recovering from a stroke. Another example that I found was a Rasch investigation of 51 patients after cataract surgery (Gothwal, Wright, Lamoureux, & Pesudovs,* 2010).

Isabelle: *Anything else?*

Ted: *What struck me most is the large number of factors that can influence the sample size. For items, it's not only the number of items, but also where they fall on the trait line. For people, it's where they fall on the trait line, too. Running the data set and looking at indices such as person separation and item separation, one can learn some of what is possible in the data set.*

Isabelle: *Anything else that you think was useful?*

Ted: *Well…yes. Remember last week when we talked about how Rasch helps us think about quality control, for instance, trying to make sure we have high-quality items and also high-quality responses from our survey takers (or test takers)?*

Isabelle: *You mean fit? What does this have to do with sample size?*

Ted: *The important issue is all of these concepts are interconnected. In Mike Linacre's article considering sample size, he said "****30 items administered to 30 persons (with reasonable targeting and fit) should produce statistically stable measures****." The part of this tip that helped me is not only the information on the 30 items and 30 persons but also the comment about "reasonable fit." You see, it is not enough to give a lot of items or to have many people take a survey or test. We also have to think of the quality of the data. If the items involve one trait (they do not misfit) and the respondents are not messing around (wildly answering to get done), then sometimes we can use a smaller sample than we might think.*

Keywords and Phrases

Sample size
Person separation
Item separation

Potential Article Text

The goal of the NSF funded XYZ project was to present meaningful workshops over a period of time to K-6 teachers of Boiler Town School District. Goal 1.1 of the project was to increase the self-efficacy of participating teachers. In order to confidently measure change over time, a review of existing self-efficacy instruments was conducted, and three potential instruments were identified. These instruments were used in five or more previous studies that were presented at peer-reviewed conferences such as AERA, NCME, or NARST and subsequently published in AERJ or JRST.

A number of psychometric issues were evaluated with respect to each instrument prior to final, large-scale data collection from the district's 1,500 K-6 teachers. Particular attention was paid to the amount of person measurement error in order to identify and select the instrument with the highest precision for assessing the self-efficacy of the district's K-6 teachers. This is a critical issue if respondents are to be tracked over time and compared. If a measurement instrument has a large amount of error, then trustworthy comparisons of respondents are more difficult to carry out. Additionally Rasch item reliability and Rasch person reliability were reviewed.

In early fall, the school district facilitated the collection of pilot data from a purposeful subsample of the districts' teachers. A total of 200 teachers completed each of the three potential instruments. Data were entered into an Excel spreadsheet and were analyzed using the Rasch program Winsteps. Person measures were computed for each respondent for each instrument. Also, item measures were computed for each item presented in the three instruments. The interplay of items and persons can be complex, and the purpose of the analysis is not to provide a measurement lesson but rather to describe briefly the measurement steps taken to prepare data for statistical analyses that can be used to document and improve teacher learning. Therefore, here we report the results of this analysis in terms of so-called person separation and item separation. Results of the analysis suggest the following: Instrument A-item separation 5.32, person separation 2.32; Instrument B-item separation 6.44, person separation 3.71; and Instrument C-item separation 3.77, person separation 1.23. The higher values of person separation and item separation for Instrument B provide two pieces of evidence that support the selection of this instrument for the project.

Quick Tips

As for other topics presented in this book, we suggest that when you experiment on the impact of sample size, you conduct experiments with a data set. Create a spreadsheet in which the columns are different indices that are impacted by your experiments. In the case of sample size, consider topics such as item error, person error, item separation, item reliability, person reliability, and person separation. The rows of your spreadsheet will be the different forms of your data set. One row might be your full data set, one row might be all the items of your data set, but only half of the respondents are included in an analysis and so on. Conducting such experiments on your data will allow you to see the impact of the topics of sample size on a particular data set:

> data is required to estimate each item's "one parameter" of difficulty. With a **reasonably targeted sample of 50 persons**, there is 99 % confidence that the estimated item difficulty is within +/−1 logit of its stable value. This is close enough for most practical purposes, especially when persons take 10 or more items. With 200 persons, there is 99 % confidence the estimated value is within +/−0.5 logits (see RMT 7:4 p. 328). **But for pilot studies, 30 persons are enough to see what's happening** (see Best Test Design). Even if you plan to test 200, start the analysis as soon as the first data become available: **200 incorrect administrations are never as good as 50 correct ones**.
>
> Wright and Tennant (1996).
>
> Note: For Dichotomous Data
>
> **Rasch is the same as any other statistical analysis with a small sample:**:
>
> 1. **Less precise estimates (bigger standard errors)**
> 2. **Less powerful fit analysis**
> 3. **Less robust estimates (more likely that accidents in the data will distort them)**
>
> **Polytomies – The extra concern with polytomies is that you need at least 10 observations per category**.
>
> **Person Measure Estimate Stability – 30 items administered to 30 persons (with reasonable targeting and fit) should produce statistically stable measures**.
>
> **As a rule of thumb, at least 8 correct responses and 8 incorrect responses are needed** for reasonable confidence that an item calibration is within 1 logit of a stable value.

Item Calibrations stable within	Confidence	Minimum sample size range (best to poor targeting)	Size for most purposes
± 1 logit	95%	16 -- 36	30 (minimum for dichotomies)
± 1 logit	99%	27 -- 61	50 (minimum for polytomies)
± ½ logit	95%	64 -- 144	100
± ½ logit	99%	108 -- 243	150
Definitive or High Stakes	99%+ (Items)	250 -- 20*test length	250
Adverse Circumstances	Robust	450 upwards	500

Linacre (1994)

Data Sets: (go to http://extras.springer.com)

cf naz for chp
cf Saed Sabah
cf asking questions

Activities

Activity #1

Task: Using the control file "cf naz for chp", compare the person and item separations for an analysis using only the first 35 respondents and using all the respondents. What do you see? Why? (Hint): You can create the control file with the 35 respondents by simply editing out the remaining respondents. Another easy way to edit out the respondents, without removing them, can be facilitated through the use of the command line "PDFILE". Look it up in the Winsteps manual! We provide the text in the control file.

Answer: In Fig. 17.7, we first present the analysis with the full data set of 75 respondents who completed the 10 items of the STEBI outcome-expectancy scale. In Fig. 17.8, we present the analysis of the first 35 respondents in the data set to the same set of items.

In this sample with these items, the person separation and the item separation are higher for the large sample size. Recall that we cannot always assume that more people mean being able to discern differences between groups of respondents. In this sample analysis, unlike our example in the text, the person separation reliability did increase for the larger sample size.

Activity #2

Question: Why does the analysis of 75 respondents report "5 cats," while the analysis of 35 respondents report "4 cats"?

Answer: If you carefully review the data, you will see that all 5 rating categories appear somewhere for at least one of the 10 items for at least one of the respondents. If you review the data for the 35 respondents, you will note that one of the rating categories was never used.

Activity #3

Setting: Our colleague Science Educator Dr. Saed Sabah at the Hashemite University (Jordan) has kindly provided us with a sample of data which he collected from students in Jordan as part of a study of students' perceptions of inquiry experiences in science laboratories. Dr. Sabah's specialty areas within the field of science education are assessment and technology integration. The data were collected using the scale of Campbell, Abu-Hamid, and Chapman (2010) in which respondents could answer using a frequency scale (1 = almost never, 2 = seldom, 3 = sometimes,

```
INPUT: 75 PERSON  10 ITEM  MEASURED: 75 PERSON  10 ITEM  5 CATS  WINSTEPS 3.70.6
--------------------------------------------------------------------------------

     SUMMARY OF 75 MEASURED PERSON
-------------------------------------------------------------------------------
|          TOTAL                         MODEL         INFIT        OUTFIT    |
|          SCORE     COUNT     MEASURE   ERROR      MNSQ   ZSTD   MNSQ   ZSTD |
|-----------------------------------------------------------------------------|
| MEAN       35.4      10.0       1.18     .52      1.05    -.1   1.05    -.1 |
| S.D.        3.8        .0        .99     .06       .85    1.6    .86    1.5 |
| MAX.       45.0      10.0       4.35     .62      5.31    4.8   5.31    4.8 |
| MIN.       26.0      10.0       -.92     .45       .09   -4.2    .10   -4.0 |
|-----------------------------------------------------------------------------|
| REAL RMSE     .60 TRUE SD     .78  SEPARATION  1.30  PERSON RELIABILITY  .63 |
|MODEL RMSE     .52 TRUE SD     .84  SEPARATION  1.62  PERSON RELIABILITY  .72 |
| S.E. OF PERSON MEAN = .11                                                   |
-------------------------------------------------------------------------------
PERSON RAW SCORE-TO-MEASURE CORRELATION = .99
CRONBACH ALPHA (KR-20) PERSON RAW SCORE "TEST" RELIABILITY = .70

     SUMMARY OF 10 MEASURED ITEM
-------------------------------------------------------------------------------
|          TOTAL                         MODEL         INFIT        OUTFIT    |
|          SCORE     COUNT     MEASURE   ERROR      MNSQ   ZSTD   MNSQ   ZSTD |
|-----------------------------------------------------------------------------|
| MEAN      265.8      75.0        .00     .19      1.00    -.1   1.05     .1 |
| S.D.       17.0        .0        .60     .01       .31    1.8    .38    2.0 |
| MAX.      302.0      75.0       1.05     .22      1.66    3.6   1.82    4.0 |
| MIN.      233.0      75.0      -1.41     .17       .65   -2.6    .65   -2.4 |
|-----------------------------------------------------------------------------|
| REAL RMSE     .20 TRUE SD     .57  SEPARATION  2.89  ITEM   RELIABILITY  .89 |
|MODEL RMSE     .19 TRUE SD     .58  SEPARATION  3.08  ITEM   RELIABILITY  .90 |
| S.E. OF ITEM MEAN = .20                                                     |
-------------------------------------------------------------------------------
UMEAN=.0000 USCALE=1.0000
ITEM RAW SCORE-TO-MEASURE CORRELATION = -.99
750 DATA POINTS. LOG-LIKELIHOOD CHI-SQUARE: 1365.86 with 663 d.f. p=.0000
Global Root-Mean-Square Residual (excluding extreme scores): .6259
```

Fig. 17.7 Results of an analysis of 75 respondents on 10 items of the STEBI outcome-expectancy scale

4 = often, 5 = almost always). The scale included two items that needed to be reverse coded; the data in the SPSS spreadsheet have already been corrected for the reverse item wording of the two items. The 20-item instrument of Campbell has 5 scales (4 items each). For this activity we provide readers with a control file that evaluates only the items for the Design Investigations scale.

The control file that enables readers to run the analysis of the 75 respondents who answered the Design Investigations scale is provided: "cf Saed Sabah". Addition of the three successive control lines immediately below will allow analysis of only the first 35 respondents:

```
PDFILE=*
36–75
*
```

Task: Conduct an analysis of the effect of sample size on item separation and person separation.

Answer: In Fig. 17.9, we provide the summary statistics for the analysis of the 75 respondents. In Fig. 17.10, we present the analysis of the 35 respondents.

```
TABLE 3.1 NAZ OE 07                                  ZOU078WS.TXT  Oct 24   9:03 2011
INPUT: 75 PERSON  10 ITEM  MEASURED: 35 PERSON  10 ITEM  4 CATS  WINSTEPS 3.70.6
--------------------------------------------------------------------------------

     SUMMARY OF 35 MEASURED PERSON
-------------------------------------------------------------------------------
|          TOTAL                         MODEL         INFIT        OUTFIT    |
|          SCORE     COUNT     MEASURE   ERROR      MNSQ   ZSTD   MNSQ   ZSTD |
|-----------------------------------------------------------------------------|
| MEAN      33.9      10.0        -.52     .50      1.03    -.2   1.03    -.2 |
| S.D.       3.5        .0         .86     .04       .73    1.6    .73    1.6 |
| MAX.      40.0      10.0        1.15     .59      3.70    3.2   3.61    3.1 |
| MIN.      26.0      10.0       -2.38     .46       .11   -3.7    .10   -3.4 |
|-----------------------------------------------------------------------------|
| REAL RMSE     .58 TRUE SD     .64  SEPARATION  1.11  PERSON RELIABILITY  .55 |
|MODEL RMSE     .50 TRUE SD     .70  SEPARATION  1.39  PERSON RELIABILITY  .66 |
| S.E. OF PERSON MEAN = .15                                                   |
-------------------------------------------------------------------------------
                DELETED:      40 PERSON
PERSON RAW SCORE-TO-MEASURE CORRELATION = 1.00
CRONBACH ALPHA (KR-20) PERSON RAW SCORE "TEST" RELIABILITY = .63

     SUMMARY OF 10 MEASURED ITEM
-------------------------------------------------------------------------------
|          TOTAL                         MODEL         INFIT        OUTFIT    |
|          SCORE     COUNT     MEASURE   ERROR      MNSQ   ZSTD   MNSQ   ZSTD |
|-----------------------------------------------------------------------------|
| MEAN     118.8      35.0         .00     .27      1.00     .0   1.03     .1 |
| S.D.       9.2        .0         .65     .01       .30    1.4    .34    1.5 |
| MAX.     135.0      35.0        1.40     .30      1.46    1.9   1.63    2.4 |
| MIN.      98.0      35.0       -1.23     .26       .48   -3.0    .46   -3.0 |
|-----------------------------------------------------------------------------|
| REAL RMSE     .28 TRUE SD     .58  SEPARATION  2.05  ITEM   RELIABILITY  .81 |
|MODEL RMSE     .27 TRUE SD     .59  SEPARATION  2.20  ITEM   RELIABILITY  .83 |
| S.E. OF ITEM MEAN = .22                                                     |
-------------------------------------------------------------------------------
UMEAN=.0000 USCALE=1.0000
ITEM RAW SCORE-TO-MEASURE CORRELATION = -1.00
350 DATA POINTS. LOG-LIKELIHOOD CHI-SQUARE: 651.07 with 304 d.f. p=.0000
Global Root-Mean-Square Residual (excluding extreme scores): .6358
```

Fig. 17.8 Results of an analysis of 35 respondents on 10 items of the STEBI outcome-expectancy scale

A comparison of the item separation and person separation as a function of sample size reveals that both item separation and the person separation are higher for the sample with 75 respondents than with 35 respondents.

Activity #4

Setting: Much of our work in this chapter involves using Table 3 (the Winsteps Summary Table), which provides the item separation and person separation information for an analysis. Readers will note that the titles of the tables are a little different for the analysis of the 35 respondents and the analysis of the 75 respondents.

Questions: What are those differences? Why do the differences exist? What could you do to test your theory? Is your theory correct?

Answer: The reason why there are differences in the titles of the tables emphasizes whether or not there are extreme persons (persons who answered using only one end of the scale) or extreme items (items which were answered by all respondents using only one extreme rating scale category).

```
   SUMMARY OF 75 MEASURED (EXTREME AND NON-EXTREME) PERSON
-------------------------------------------------------------------------------
|          TOTAL                         MODEL         INFIT        OUTFIT    |
|          SCORE      COUNT     MEASURE   ERROR     MNSQ   ZSTD   MNSQ   ZSTD |
|-----------------------------------------------------------------------------|
| MEAN      12.9        4.0        .23     .73                                 |
| S.D.       2.9         .1       1.38     .13                                 |
| MAX.      17.0        4.0       2.44    1.75                                 |
| MIN.       4.0        3.0      -4.69     .64      .00   -3.1    .01   -1.5 |
|-----------------------------------------------------------------------------|
| REAL RMSE    .86 TRUE SD    1.08  SEPARATION  1.27  PERSON RELIABILITY  .62 |
|MODEL RMSE    .74 TRUE SD    1.16  SEPARATION  1.56  PERSON RELIABILITY  .71 |
| S.E. OF PERSON MEAN = .16                                                    |
-------------------------------------------------------------------------------
PERSON RAW SCORE-TO-MEASURE CORRELATION = .98
CRONBACH ALPHA (KR-20) PERSON RAW SCORE "TEST" RELIABILITY = .63

     SUMMARY OF 4 MEASURED (NON-EXTREME) ITEM
-------------------------------------------------------------------------------
|          TOTAL                         MODEL         INFIT        OUTFIT    |
|          SCORE      COUNT     MEASURE   ERROR     MNSQ   ZSTD   MNSQ   ZSTD |
|-----------------------------------------------------------------------------|
| MEAN     242.3       74.8        .00     .17     1.01    -.2   1.49    -.1 |
| S.D.      82.3         .4       1.87     .02      .30    1.4   1.23    2.1 |
| MAX.     305.0       75.0       3.15     .20     1.52    2.1   3.62    3.5 |
| MIN.     103.0       74.0      -1.42     .15      .78   -1.3    .67   -1.8 |
|-----------------------------------------------------------------------------|
| REAL RMSE    .18 TRUE SD    1.86  SEPARATION 10.12  ITEM   RELIABILITY  .99 |
|MODEL RMSE    .17 TRUE SD    1.86  SEPARATION 10.98  ITEM   RELIABILITY  .99 |
| S.E. OF ITEM MEAN = 1.08                                                     |
-------------------------------------------------------------------------------
             DELETED:      16 ITEM
UMEAN=.0000 USCALE=1.0000
ITEM RAW SCORE-TO-MEASURE CORRELATION = -1.00
295 DATA POINTS. LOG-LIKELIHOOD CHI-SQUARE: 536.99 with 215 d.f. p=.0000
Global Root-Mean-Square Residual (excluding extreme scores): .6871
```

Fig. 17.9 Summary statistics of the analysis of the 75 respondents from Dr. Sabah's research

Task: To test this theory for persons, take the 35 persons and add a single person after the 35th person. This new person answers only 1s for all items. Then run the data (make sure to change your PDFILE code to remove persons; 37 and 76. We have added a person who chose all 1s; thus, you should focus on the responses of the first 36 respondents and ignore the responses of the 35 remaining respondents. Since you have added a line of data for the fake person after the 35th person, you should now delete the lines 37 and 76. These are your previous lines 36 and 75). Notice with this new analysis, we do indeed see new headings as the result of having a person who put all extreme answers.

Activity #5

Setting: It is possible to "see" a 4-item scale (items 5–8 of the survey), which is named "Designing Investigations: In the Science Classroom." For this activity (challenge), we provide a control file that allows readers to evaluate the scale from the same survey that is built using items 1–4. That scale is named "Asking questions/Asking Research Questions: In the Science Classroom," and the name of the control file is "cf asking questions."

```
TABLE 3.1 inquiry_dataset-1.sav                         ZOU178WS.TXT  Oct 24 10:16 2011
INPUT: 75 PERSON  20 ITEM  MEASURED: 35 PERSON  4 ITEM  5 CATS   WINSTEPS 3.70.6
--------------------------------------------------------------------------------

     SUMMARY OF 35 MEASURED PERSON
-------------------------------------------------------------------------------
|          TOTAL                         MODEL         INFIT         OUTFIT    |
|          SCORE     COUNT     MEASURE   ERROR      MNSQ   ZSTD   MNSQ   ZSTD |
|-----------------------------------------------------------------------------|
| MEAN      13.5       4.0         .48     .76       .91    -.2   1.13     .0 |
| S.D.       2.5        .0        1.29     .08      1.12    1.2   1.91    1.1 |
| MAX.      17.0       4.0        2.67     .88      4.70    2.8   9.90    3.7 |
| MIN.       6.0       4.0       -2.96     .63       .01   -2.8    .02   -1.2 |
|-----------------------------------------------------------------------------|
| REAL RMSE    .88 TRUE SD     .94  SEPARATION  1.06  PERSON RELIABILITY  .53 |
|MODEL RMSE    .77 TRUE SD    1.03  SEPARATION  1.34  PERSON RELIABILITY  .64 |
| S.E. OF PERSON MEAN = .22                                                   |
-------------------------------------------------------------------------------
          DELETED:      40 PERSON
PERSON RAW SCORE-TO-MEASURE CORRELATION = .99
CRONBACH ALPHA (KR-20) PERSON RAW SCORE "TEST" RELIABILITY = .56

     SUMMARY OF 4 MEASURED ITEM
-------------------------------------------------------------------------------
|          TOTAL                         MODEL         INFIT         OUTFIT    |
|          SCORE     COUNT     MEASURE   ERROR      MNSQ   ZSTD   MNSQ   ZSTD |
|-----------------------------------------------------------------------------|
| MEAN     118.5      35.0         .00     .26      1.03     .0   1.18    -.2 |
| S.D.      42.8        .0        2.09     .03       .26    1.1    .68    1.3 |
| MAX.     151.0      35.0        3.51     .31      1.29     .9   2.32    1.5 |
| MIN.      46.0      35.0       -1.64     .22       .60   -1.9    .55   -2.1 |
|-----------------------------------------------------------------------------|
| REAL RMSE    .28 TRUE SD    2.07  SEPARATION  7.30  ITEM   RELIABILITY  .98 |
|MODEL RMSE    .26 TRUE SD    2.07  SEPARATION  7.87  ITEM   RELIABILITY  .98 |
| S.E. OF ITEM MEAN = 1.20                                                    |
-------------------------------------------------------------------------------
          DELETED:      16 ITEM
UMEAN=.0000 USCALE=1.0000
ITEM RAW SCORE-TO-MEASURE CORRELATION = -1.00
140 DATA POINTS. LOG-LIKELIHOOD CHI-SQUARE: 234.05 with 99 d.f. p=.0000
Global Root-Mean-Square Residual (excluding extreme scores): .6534
```

Fig. 17.10 Summary statistics of the analysis of the 35 respondents from Dr. Sabah's research

Questions: Can you explain why the separation values for the two scales are not the same? Because the same people are used for the analysis, how can this be? Can you make a prediction as to the values for the person separation and item separation for the two analyses?

Answer: Remember that the power and usefulness of sample size is not only the number of people. Sample size also depends upon the way in which items define a trait. Below we provide the Wright Map for the analysis of the *Designing Investigations* control file. Also, we provide the Wright Map for the analysis of the Asking Questions control file. The same persons answered all 8 items, but remember the two scales are different traits, and most importantly, one set of items does a better job of defining a trait than a different set of items does at defining a different trait. What can be "seen" in the two Wright Maps can also be seen in a review of the summary statistics for these two scales. The biggest difference is that the *Designing Investigations* items do a better job of defining three parts of the trait (two items define a very similar portion of the trait). The Asking Questions items define only one part of the trait. Although the items are spread out a little, the same area of the trait is measured by the four items (Figs. 17.11 and 17.12).

```
TABLE 12.2 inquiry_dataset-1.sav                 ZOU270WS.TXT  Oct 24 11:16 2011
INPUT: 75 PERSON  20 ITEM  MEASURED: 75 PERSON  4 ITEM  5 CATS   WINSTEPS 3.70.6
--------------------------------------------------------------------------------

         PERSON - MAP - ITEM
             <more>|<rare>
    4              +
                   |
                X  |
                   |
                   |
                  T|
                   |
    3        XXXX  +
                   |
                   |
                   |
                   |
              XXX  |
                   |
    2              +
             XXXX  |
                  S|
                   |
               XX  |
                   |
                X  |
    1    XXXXXXXX  +
                   |
                   |T
        XXXXXXXXX  |
                   |  ITEM2
                   |S
         XXXXXXXX  |  ITEM1
    0             M+M
                   |
                   |S ITEM4
               XX  |  ITEM3
                   |
                   |T
        XXXXXXXXX  |
   -1              +
                   |
      XXXXXXXXXXX  |
                   |
                  S|
                   |
              XXX  |
   -2              +
                   |
                   |
              XXX  |
                   |
                   |
                   |
   -3              +
                X T|
                   |
                   |
                   |
                   |
                   |
   -4      XXXXXX  +
             <less>|<frequ>
```

Fig. 17.11 Asking Questions Wright Map

```
TABLE 12.2 inquiry_dataset-1.sav                  ZOU730WS.TXT  Oct 24 11:19 2011
INPUT: 75 PERSON  20 ITEM  MEASURED: 75 PERSON  4 ITEM  5 CATS   WINSTEPS 3.70.6
--------------------------------------------------------------------------------

           PERSON - MAP - ITEM
               <more>|<rare>
    4                +
                     |
                     |T
                     |
                     |
                     |  ITEM5
    3                +
                    T|
                     |
                 .#  |
                     |
                     |
    2                +
              .####  |S
                     |
                    S|
                     |
             ######  |
    1           .    +
                     |
           ########  |
                     |
                    M|
             .#####  |
    0                +M
                     |
                .##  |  ITEM8
                     |
                     |
                 ##  |
   -1               S+
                 ##  |
                     |  ITEM6   ITEM7
                     |
                 ##  |
                     |S
   -2           .    +
                    T|
                     |
                  .  |
                     |
                 .#  |
   -3                +
                     |
                     |
                     |
                     |T
                     |
   -4           .    +
               <less>|<frequ>
EACH "#" IS 2. EACH "." IS 1.
```

Fig. 17.12 Designing Investigations Wright Map

References

Björkdahl, A., Nilsson, A. L., Grimby, G., & Stibrant Sunnerhagen, K. S. (2006). Does a short period of rehabilitation in the home setting facilitate functioning after stroke? A randomized controlled trial. *Clinical Rehabilitation, 20*, 1038–1049.

Gothwal, V. K., Wright, T. A., Lamoureux, E. L., & Pesudovs, K. (2010). Measuring outcomes of cataract surgery using the visual function index-14. *Journal of Cataract & Refractive Surgery, 36*(7), 1181–1188.

Linacre, J. M. (1994). Sample size and item calibration stability. *Rasch Measurement Transactions, 7*(4), 328.

Linacre, J. M. (1999). Investigating rating scale category utility. *Journal of Outcome Measurement, 3*(2), 103–122.

Linacre, J. M. (2002). Understanding Rasch measurement: Optimizing rating scale category effectiveness. *Journal of Applied Measurement, 3*(1), 85–106.

Linacre, J. M. (2012). Winsteps (Version 3.74) [Software]. Available from http://www.winsteps.com/index.html

Wright, B. D., & Tennant, A. (1996). Sample size again. *Rasch Measurement Transactions, 9*(4), 468.

Additional Readings

Smith, E. V., Jr., & Smith, R. M. (Eds.). (2004). *Introduction to Rasch measurement: Theory, models, and applications*. Maple Grove, MN: JAM Press.

Smith, A., Rush, R., Fallowfield, L., Velikova, G., & Sharpe, M. (2008). Rasch fit statistics and sample size considerations for polytomous data. *BMC Medical Research Methodology, 8*, 33. doi:10.1186/1471-2288-8-33.

Stone, M., & Yumoto, F. (2004). The effect of sample size for estimating Rasch/IRT parameters with dichotomous items. *Journal of Applied Measurement, 5*(1), 48–61.

Chapter 18
Missing Data: What Should I Do?

Isabelle and Ted: Two Colleagues Conversing

Ted: Isabelle, I have been thinking about something. By using Rasch measurement, we have a great advantage in our data analyses, in that all students need not answer the same set of items, yet we can still compare them on the same scale.

Isabelle: That is of course correct. Go on.

Ted: Well…I have been thinking, this advantage might also provide other benefits in analyses of data that really can help researchers. If it is the case that not all items need to be answered, I wonder if that might help alleviate some common problems in many types of data (education, health fields, psychology). For example, maybe all students/patients take the same test, but maybe there are some participants who do not finish the test, or maybe they may not be given all the items, because the last page of the test booklet might have been misprinted for some of the respondents. I am sure that these same issues come up all the time in other fields as well.

Isabelle: Yes, tell me more.

Introduction

In previous chapters, we presented and discussed a great amount of theory and numerous applications, ranging from authoring control files to interpreting data tables. Readers should recall that we sometimes mentioned a point and then set it off to the side, saying that we would take up the point again in a later chapter. The ideas in this chapter are among those we perhaps alluded to but touched upon only briefly in earlier chapters. Specifically, we believe this chapter is especially important because its focus are the problems and implications of missing data and how to address such problems. Rasch measurement provides researchers within and beyond education with important and needed additional research tools for addressing problems of missing data.

W.J. Boone et al., *Rasch Analysis in the Human Sciences*,

DOI 10.1007/978-94-007-6857-4_18,

Two Problem Scenarios and How to Address Them

Below we provide two scenarios that occur frequently when data are collected:

Scenario #1

A researcher collected self-efficacy data from a group of preservice teachers using the STEBI. Unbeknownst to the researcher, when the surveys were copied, assembled, and stapled, the copier malfunctioned, but the person supervising the process did not catch the malfunction. This resulted in the absence of the last page of the STEBI for approximately half of the students. Consequently, about half of the students completed the full 23-item STEBI, but approximately half of the students completed only the first 13 items.

Scenario #2

A researcher collected data from 200 students using a 27-item multiple-choice test. The instrument was correctly duplicated, and students' tests contained all 27 items. The researcher was informed by the cooperating teacher who administered the test and supervised the students that some students did not finish the test. Upon reviewing the data, the researcher noticed that a number of the students reached a particular item and then did not answer any of the following items. Many of those who did not finish the test were female students. The researcher knows that "not-reached" test items are often viewed as "incorrect."

The two scenarios above describe common occurrences that can have profound impact upon the results of a study. Often these and similar issues are ignored. Steps are taken to correct for these problems, but the steps are not reported, and/or the implications of such steps are not discussed. In Scenario #1, with so many items not answered, a common technique to confront this lack of data is to simply throw out the students who did not complete the entire instrument. Throwing out missing data sometimes works, or is not a terrible solution of sorts, as long as a data set is large and as long as those students not receiving the full survey are randomly distributed in terms of issues (e.g., race, gender) that might be of interest to the researcher. Of course, a better solution would be to not have to discard any data.

When all items are not answered, a technique we have seen used is for a researcher to pretend that all items are equal and compute a raw mean answer for a set of items (e.g., if 40 of 50 students answered item Q21, then a mean response to Q21 is computed using just the 40 responses). Such a solution ignores the nonlinearity of raw data (the raw mean response of a sample of students to an item just gives a mean of a nonlinear scale). Of course, another problem with this solution is a stacking of errors upon errors in an attempt to correct for the missing data. For example the researcher is ignoring that all items are not equally easy.

Scenario #2 occurs more often than one might want to admit. For instance, if teachers in several classrooms are asked to collect data, the amount of time the teachers are able to allocate for data collection might vary. In some situations students may be rushed and may not finish no matter how hard they try to complete the test. Moreover, if a test is not part of a student's grade, then students may vary extensively in the degree of effort they devote in taking the test. Frequently, in piloting instruments there may be a misjudgment of the time it may take to answer a set of items.

Formative Assessment Checkpoint #1

Question (True/False): Missing data really do not make any difference in an analysis, as there is usually so little missing data.

Answer: False. There can often be a large amount of missing data whose absence does impact an analysis. Even when there is a small amount of missing data, this lack of data can impact conclusions made about subgroups of respondents.

Consequences in Results When Respondents Do Not Complete the Test

An important issue accompanies the problem of missing test data when the missing data are the result of students not finishing a test. If we collect "achievement data," then our goal is to get a fair and accurate estimate of students' abilities. It is quite plausible that unanswered test items might be items not reached as the result of students' reading speed and/or reading comprehension, particularly when such items are at the end of a test. Most researchers would concur that respondents' reading speed could indeed be one reason for not answering an item. In education, the issue of missing data (e.g., not-reached items or skipped items) is almost never discussed. In other fields we suspect it is a topic also not often discussed. We hypothesize that most researchers simply count unanswered test items as "incorrect" and may remove respondents who do not answer any items from the data set. Items skipped are also never discussed. In such a case, a respondent answers a number of items, skips an item, and then answers a subsequent item or items. For a 10-item multiple-choice test, this pattern of answers might appear in the following manner for a respondent: ACCD_ABACD.

We discussed some problems associated with corrections made in Scenario #1. Several flawed correction strategies are commonly used in education. First, researchers might pretend that all respondents answered all items. This means unanswered items are viewed as wrong. Most researchers would view this as a solution, because all respondents could then earn the same possible raw score total. A problem with this strategy is that respondents can be penalized for issues such as reading speed and/or the attention they give to a test when researchers count unanswered items as wrong. Consider the implications, if a subgroup of respondents reads more slowly, the subgroup's performance will be underestimated, and the type of assistance the subgroup needs with respect to a particular topic is then mis-targeted.

Let's take another tack; maybe a subgroup of respondents is more deliberate in answering items, perhaps taking more time to carefully read and answer the items. In this instance, counting not-answered items as wrong would mis-measure the subgroup and mis-target interventions inferred as needed from an analysis of test scores.

How Can Rasch Measurement Help Solve These Missing Data Problems?

Of course, what do these scenarios of missing data and traditional practices in addressing missing data have to do with Rasch measurement in fields such as education, psychology, medicine, and market research? Knowledge of Rasch measurement can help researchers confront missing data in two important ways. First, by using a deep understanding of what it means to measure, we can identify issues that potentially affect the computation of measures, which in turn can influence any and all subsequent analyses. Educating one's mind to identify and comprehend issues present in a data set with missing data helps one to reflect in depth about what it means to measure. This in turn facilitates the development of measurement instruments that help researchers advance the knowledge base and practices in all fields – be it science education, math education, psychology, medicine, or market research.

A second point is that Rasch offers a viable solution to the problem of missing data. By using Rasch, researchers possess a tool to evaluate the responses of students who do not complete all items of an instrument. Researchers can still compute these respondents' measures on the same scale used to express the performance of respondents who did complete all items. To reiterate, given the properties of Rasch measurement, as long as one is measuring a single trait, not all items need to be completed by all respondents (recall that in Chap. 14 we discussed this issue in terms of linking two tests). This means that if items are not reached or skipped, then the researcher has the option of not counting unanswered items as wrong. In the case of a survey, items not reached are not used in the calculation of a person measure.

Thinking About Missing Data

Prior to presenting the technical details of altering a control file to consider missing data, we need to consider how thinking about measurement can help researchers not only understand the issue of missing data but also learn how to use the same thinking techniques as detection devices for additional measurement problems as researchers of a multitude of fields design and use measurement instruments.

Rasch measurement helps us remember a number of points. First, if we want to compare respondents in any meaningful way, we need to use a set of items that measure respondents. This set of items should involve a single trait, at a level that is appropriate for the respondents.

In Chap. 5 – Item Measures – we discussed at length the importance of thinking about items as involving a single trait, but it is critical to think how items define a trait. Not all items are created equally. For a test, some items are more difficult than other items. For a survey in which respondents can agree or disagree with items, some items may be harder to agree with than other items. Thinking about items in this manner should help researchers spot the problem inherent with any sort of raw data computation in which all items are treated as equal.

Formative Assessment Checkpoint #2

Question: If one has missing data, is the one and only solution to drop the person or fill in most likely answers of the person for the missing data (if the person answered mostly *Agree* for items, then it is okay to insert an *Agree* answer for the missing data for that person)?

Answer: Use of the Rasch model allows the analysis of a person even if she or he does not complete all items on an instrument, as long as the instrument involves one trait (and the data fit the Rasch model).

Thinking about what it means to measure and what goes into a measure should help researchers think about the possible implications of issues such as missing data and the potential implications of any steps made to correct for missing data. Think of "thinking" as having your "antennae up," just as the way your "antennae are up" when you enter a school. For example, upon entering a school to administer the Force Concept Inventory (FCI) to a class of physics students, you note that a pep session has been scheduled in order to show support for the varsity basketball team's upcoming state tournament game. As a consequence, each class period has been reduced by 15 min. You decide to work with the physics teacher to reschedule administering the FCI on another day because you fear that the students will not have sufficient time to complete the instrument. However, if you knew how to use Rasch measurement, you would at least be aware that not all the FCI items would have to be completed. We cannot present all of the measurement issues that you will confront in the research you conduct in your chosen field, but Rasch can help raise your "antennae" and cause you to question, wonder, and in many cases resolve measurement issues that can influence an analysis.

The Nuts and Bolts of Exploring a Missing Data Measurement Issue

To include or exclude missing data, two key control lines are central. First consider the control line with the phrase "CODES=". This line tells Winsteps which codes will be used in an analysis of each item. If any symbols or blanks are used to indicate a response (or lack thereof in the case of a blank) and if those symbols are not in the information for the "CODES=" line, then that response is ignored for an analysis. The other line you will need to use is a line that will start off with the word KEY. This line will be used to tell the program the codes for the correct item answers. For example, a test with 10 items, coded with a "1" for correct, will have the following form: KEY1 = 1111111111. For our examples below, readers only have to now consider the CODES line, and just remember how the KEY line functions.

Suppose a student has completed a 10-item multiple-choice test. The student's answers were graded using a "1" to indicate a correct response and a "0" to indicate an incorrect response. An "X" indicates that the student skipped the item. A "9" indicates that the student selected two answers for a single item. This student's line of data might look like this "CODE=".

101091X0X0

If the researcher wanted to use the items scored only as correct or incorrect, then the following CODES = line would be used in the control file:

$$CODES = 10$$

In this case, the data for all persons and all items would be evaluated only if the person clearly answered the item and thus could be clearly scored as having correctly or incorrectly answered the item. This is the form of the codes statement to be used if the researcher did not intend to count not-reached items and skipped items as wrong.

If the researcher did intend to count unanswered items as wrong, then the codes statement would be of the following form:

$$CODES = 10\mathrm{X}$$

In this coding, items double marked or not clearly marked are omitted from a respondent's measure. A student's performance is based on correctly answering an item (coded as a "1"), incorrectly answering an item correct (coded as a "0"), or skipping an item (coded as an "X"). In this second case, both the wrong answer and the skipped item would be scored as "wrong."

Our final example is a codes statement to be used if the researcher wished to use all items for all respondents. If an item was correctly answered, incorrectly answered, skipped, or double coded, the following codes statement would be used:

$$CODES = 10\mathrm{X}9$$

When conducting a Rasch analysis, researchers have access to a number of techniques to explore and document the impact of how they evaluate missing data. To demonstrate how such an evaluation might proceed, we have enlisted the assistance of our colleague Dr. Kathy Trundle of the Ohio State University. Kathy is a specialist in earth science assessment and misconceptions in earth science (contact her if you need her help!). Kathy has kindly provided us with a nonrandom sample of data collected from OSU students who completed an earth science assessment Kathy developed. Our purpose here is to show the "ins and outs" of Rasch data analysis (emphasizing the topic of missing data). The data are nonrandom and, as a result, should not be used to reach any policy conclusions.

In Fig. 18.1, we provide a control file, which also includes the first line of data in the sample data set. We do not provide all the lines of code here, but the electronic copies of the files provided are ready to run with all control lines. The control file provides two "CODES=" lines. One CODES line would be used when all items (answered, not answered (k), unclearly marked and cannot be scored(d)) are used in computing the measure of a respondent (CODES = 12345kd). So all students would be viewed as having attempted all items. In our data set, we use the letter "k" to indicate an item not answered and we use a "d" for a double-marked item. You could also, for example, denote a not-answered item with a blank in the data set. In that case the CODES line would look like this: CODES = "123 45d" (to help readers see the blank, we have placed the blank between the 3 and 4; the blank can be put anywhere). The other CODES line (CODES = 12345) is one in which not-answered items and unclearly marked items are not included in computing the measure of a respondent. In such a case, if a student skipped one item of a 20-item test, the student's measure was computed based on 19 items. In our particular example, the items in which there were double markings of answers would also not be counted as answered. So, CODES = 12345 is the code one would use if only clearly answered items are to be used for the computation of person measures.

Readers will see additional control lines in the control file that were added to delete items and persons from the analysis. This was done so that readers will be able to use Ministeps to test out the control files. Remember that Ministeps (identical to Winsteps) allows only 75 respondents and 25 items to be answered.

As we have stressed herein, a number of ways exist to evaluate a data set using Rasch measurement. What we suggest here is only one set of several sets of steps that can be used. Our assessment of the possible implications of counting or not counting not-answered items as wrong begins by conducting a Rasch analysis of the data to generate an item measure table. Figures 18.2 and 18.3 (Winsteps Table 14.1) were produced by running the control files provided above. The control file that produced Fig. 18.2 contained the statement "CODES = 12345kd"; the control file that produced Fig. 18.3 contained the statement "CODES = 12345". Using only tables can be somewhat misleading, and subsequently herein we will use Wright Maps, but examining the item entry tables is a good place to begin the process of furthering one's understanding about missing data.

```
TITLE='Kathy Trundle Geology Test Subset of Test Items'
;
;A title which appears on each page
;
; Reading the data, a 11 col name ID is read then 48 items which are each separated with a comma
;
FORMAT=(11A1,26(1X,1x),22(1A1,1X))
;
; The answer key for the test
;
KEY1="1334141221342344313221"
; The first item answer is the 12th piece of information read in and used by Winsteps
;
ITEM1=12
; 22 content items
 NI=22
;The person ID starts in the 1st col of data
;
 NAME1=1
; The person ID is a total of 11 columns wide
 NAMLEN=11
; This is the line that is important for our work in this chapter
; This line is saying the numbers 1, 2, 3, and 4 and the letter K will be used to grade items
; Against the key
;
CODES="12345kd"
; Each answer is one column wide
 XWIDE=1
;
;
;
;
*
&END
Q27
Q28
.
.
.
Q46
Q47
Q48
END NAMES
1762354xxx,1,1,1,4,1,4,1,3,1,4,4,2,2,3,3,3,4,1,2,4,1,2,2,3,2,1,1,4,3,4,1,4,3,2,5,4,3,4,5,1,3,k,k,k,k,k,k,k,,k,k,k
```

Fig. 18.1 Control file for Kathy Trundle's Geology test with a subset of test items

---

Formative Assessment Checkpoint #3

Question: Since item measures are expressed in logits in both tables, can the item measures in the two tables be compared immediately?

Answer: No. The logit scale is centered to a mean item measure of 0.00 logits in each run of the data. If you wish to compare logit values of items and/or persons from different analyses of data, remember there must be a linkage of the analyses. Return to our chapters on the logit and also on item anchoring if you need a refresher.

Our examination focuses first on the information at the top of the two figures. This information lists the number of items and persons "INPUT" and the number of persons and items "REPORTED". The same information should be listed for "INPUT" in both figures. For any data set, it is important to study the numbers reported for "INPUT". For example, do the numbers look reasonable based upon the researcher's knowledge of the data set? This is a valuable technique for noticing all sorts of errors that can accidently occur in data management. Whereas the control files will evaluate the data a little differently due to the two different CODES statements, the values for the INPUT data should be the same. If you

```
TABLE 14.1 Kathy Trundle Geology Test Subset of  ZOU477WS.TXT  Nov 28  9:09 2011
INPUT: 64 PERSON  22 ITEM  REPORTED: 64 PERSON  22 ITEM  2 CATS  MINISTEP 3.72.3
--------------------------------------------------------------------------------
PERSON: REAL SEP.: 1.62  REL.: .72 ... ITEM: REAL SEP.: 2.71  REL.: .88

         ITEM STATISTICS:  ENTRY ORDER

-------------------------------------------------------------------------------------------
|ENTRY   TOTAL  TOTAL           MODEL|   INFIT  |  OUTFIT  |PT-MEASURE |EXACT MATCH|      |
|NUMBER  SCORE  COUNT  MEASURE  S.E. |MNSQ  ZSTD|MNSQ  ZSTD|CORR.  EXP.| OBS%  EXP%| ITEM |
|------------------------------------+----------+----------+-----------+-----------+------|
|     1     34     64     -.43    .27| .98   -.2| .94   -.4|  .42   .40| 64.5  65.9| Q27  |
|     2     18     64      .81    .30|1.25   1.9|1.62   2.5|  .06   .31| 66.1  72.4| Q28  |
|     3     40     64     -.90    .29|1.07    .7|1.19   1.2|  .35   .42| 71.0  70.4| Q29  |
|     4     30     64     -.13    .27| .99   -.1| .98   -.2|  .39   .38| 66.1  64.8| Q30  |
|     5     33     64     -.36    .27|1.04    .5|1.04    .3|  .36   .39| 64.5  65.5| Q31  |
|     6     54     64    -2.40    .40| .88   -.4| .65   -.9|  .59   .48| 88.7  87.2| Q32  |
|     7      9     64     1.77    .37|1.08    .4|1.19    .6|  .14   .23| 85.5  85.5| Q33  |
|     8     24     64      .32    .28| .95   -.5| .94   -.4|  .39   .35| 71.0  66.6| Q34  |
|     9     25     64      .24    .28|1.01    .2|1.06    .5|  .33   .35| 66.1  66.1| Q35  |
|    10     24     64      .32    .28| .92   -.8| .98   -.1|  .40   .35| 74.2  66.6| Q36  |
|    11     41     64     -.98    .29| .95   -.3|1.09    .6|  .44   .43| 79.0  71.4| Q37  |
|    12     39     64     -.82    .28|1.05    .5|1.02    .2|  .39   .42| 64.5  69.5| Q38  |
|    13     32     64     -.28    .27| .92   -.9| .87  -1.1|  .46   .39| 72.6  65.1| Q39  |
|    14     11     64     1.52    .35|1.18    .9|1.19    .7|  .11   .25| 80.6  82.4| Q40  |
|    15     35     64     -.51    .28|1.01    .2|1.00    .1|  .39   .40| 62.9  66.5| Q41  |
|    16     23     64      .40    .28| .92   -.8| .90   -.6|  .40   .34| 75.8  67.2| Q42  |
|    17     25     64      .24    .28|1.10   1.1|1.04    .3|  .29   .35| 59.7  66.1| Q43  |
|    18     19     64      .72    .29|1.05    .5|1.05    .3|  .27   .31| 71.0  71.2| Q44  |
|    19     35     64     -.51    .28| .87  -1.4| .82  -1.5|  .51   .40| 72.6  66.5| Q45  |
|    20     21     64      .56    .29| .96   -.3| .88   -.7|  .37   .33| 71.0  69.1| Q46  |
|    21     23     64      .40    .28| .81  -2.0| .73  -1.8|  .50   .34| 75.8  67.2| Q47  |
|    22     28     64      .02    .27| .97   -.3|1.08    .7|  .37   .37| 64.5  64.8| Q48  |
|------------------------------------+----------+----------+-----------+-----------+------|
| MEAN    28.3   64.0      .00    .29|1.00   -.1|1.01    .0|           | 71.3  69.9|      |
| S.D.    10.2     .0      .87    .03| .10    .8| .19    .9|           |  7.3   6.4|      |
-------------------------------------------------------------------------------------------
```

Fig. 18.2 (Winsteps Table 14.1). Item entry table with CODES=12345kd. Review of the TOTAL COUNT column shows that all items were evaluated using data from 64 respondents

intend to experiment and think about missing data, needless to say, it is important to make sure that the data you want to compare are read into an analysis as you hope they would be.

The other piece of information to review immediately is the "REPORTED" information at the top of the tables. The "REPORTED" values may differ when comparing the analyses of the two control files. If the number of items REPORTED differs from the number of items INPUTTED, does the analyst understand why? If one item is not answered by any student (not impossible for a very long test), and only answered items are used for an analysis, then there would be no answers for that item. As a result one would not be able to compute a measure of the item. In such a case, the item would be dropped by Winsteps in the analysis; therefore, the value given for REPORTED would be at least a number one less than the number reported for number of items INPUT.

Next, we focus on the columns TOTAL COUNT and TOTAL SCORE. Scanning the numbers in the TOTAL COUNT column of Fig. 18.2, the item entry table in which skipped items or double-marked items are counted as incorrect, readers will see identical numbers (64) as the TOTAL COUNT reported for each item. This should make sense, in that when not-answered items or double-marked items

TABLE 14.1 Kathy Trundle Geology Test Subset of ZOU442WS.TXT Nov 28 9:10 2011
INPUT: 64 PERSON 22 ITEM REPORTED: 62 PERSON 22 ITEM 2 CATS MINISTEP 3.72.3

PERSON: REAL SEP.: .80 REL.: .39 ... ITEM: REAL SEP.: 2.66 REL.: .88

ITEM STATISTICS: ENTRY ORDER

ENTRY NUMBER	TOTAL SCORE	TOTAL COUNT	MEASURE	MODEL S.E.	INFIT MNSQ	INFIT ZSTD	OUTFIT MNSQ	OUTFIT ZSTD	PT-MEASURE CORR.	PT-MEASURE EXP.	EXACT MATCH OBS%	EXACT MATCH EXP%	ITEM
1	34	61	-.35	.27	.98	-.2	.97	-.2	.34	.31	70.5	63.5	Q27
2	18	61	.87	.29	1.19	1.4	1.26	1.5	.02	.29	65.6	71.7	Q28
3	40	61	-.80	.28	.99	-.1	1.08	.6	.29	.30	73.8	68.5	Q29
4	30	61	-.05	.27	.99	-.1	1.01	.1	.32	.31	67.2	62.9	Q30
5	33	61	-.27	.27	.98	-.2	.96	-.3	.34	.31	65.6	63.1	Q31
6	54	61	-2.31	.41	.94	-.1	.71	-.7	.35	.21	88.5	88.5	Q32
7	9	61	1.83	.37	1.09	.4	1.28	.9	.06	.23	85.2	85.2	Q33
8	24	60	.37	.28	.91	-1.0	.86	-1.2	.45	.31	70.0	65.2	Q34
9	25	60	.29	.28	1.00	.0	.98	-.1	.32	.31	65.0	64.6	Q35
10	24	60	.37	.28	.91	-1.0	.90	-.8	.43	.31	73.3	65.2	Q36
11	41	58	-1.03	.30	1.00	.0	1.16	.9	.25	.29	75.9	72.2	Q37
12	39	59	-.79	.29	1.00	.0	.95	-.3	.32	.30	64.4	69.0	Q38
13	32	58	-.29	.28	.97	-.4	.94	-.6	.37	.31	69.0	63.5	Q39
14	11	58	1.53	.35	1.17	.9	1.25	.9	.01	.25	79.3	81.2	Q40
15	35	58	-.53	.28	1.07	.8	1.08	.7	.21	.31	58.6	65.6	Q41
16	23	57	.37	.28	.93	-.8	.94	-.5	.40	.31	73.7	65.0	Q42
17	25	58	.25	.28	1.13	1.5	1.11	1.0	.15	.31	56.9	63.9	Q43
18	19	57	.70	.29	1.07	.6	1.10	.7	.20	.30	70.2	69.0	Q44
19	35	56	-.62	.29	.93	-.7	.87	-1.0	.43	.31	69.6	66.8	Q45
20	21	53	.45	.30	1.02	.2	.98	-.1	.29	.31	64.2	65.0	Q46
21	23	52	.23	.29	.83	-2.0	.79	-2.0	.55	.31	73.1	63.5	Q47
22	28	52	-.21	.29	.93	-.8	.90	-.9	.41	.31	65.4	62.7	Q48
MEAN	28.3	58.3	.00	.30	1.00	-.1	1.00	-.1			70.2	68.4	
S.D.	10.2	2.8	.86	.03	.09	.8	.15	.8			7.4	7.2	

Fig. 18.3 (Winsteps Table 14.1). Item entry table with CODES=12345. Review of the TOTAL COUNT column shows that not all items were answered using codes of 1, 2 3, 4, or 5. Item 48 was in fact answered by only 52 respondents

are counted as wrong, there exists a complete set of data for each respondent. The only time when one would not see, in this example, a "64" reported as a TOTAL COUNT for each item would be the case in which some sort of unknown code is present in the data of respondents. Now let's look at Fig. 18.3, the item entry table for the analysis that did not count not-answered items as wrong, and try to understand what we see. First, the TOTAL COUNT is not "62" for all the items. Why is this? For the previous analysis, "answers" were viewed to be both true answers (selection of a, b, c, d or e) as well as nonanswers or unclear answers! However, only clear answers were used for the second analysis. This means that if two persons skipped an item (and all others answered the item), then a 62 would be reported for total count.

The TOTAL SCORE column can also be used to explore and understand the implications of one type of code statement in contrast to another type of code statement. As we have demonstrated, researchers must use Rasch measures for statistical analyses, and it is critical to use a Rasch perspective to design and validate instruments. However, at times quick appraisals can be used to begin.

For example, a researcher wants to show one of the many implications of counting and not counting not-answered items as wrong to someone who does not understand Rasch well. One could take the data for the last item on this test (Q48) and compute the percentage of respondents who correctly answered the item. In the case of using all respondents counting not-answered responses as wrong, the percentage correct for this item is 43.8 % ((28/64) × 100 %); however, when using only those persons who answered the item, the percentage correct ((28/52) × 100 %) is 51.9 %. At this introductory level of analysis, researchers can show clearly that one way of evaluating data suggests an item of greater difficulty compared to another technique of evaluating item difficulty. In our workshops we present such an analysis to try to help attendees to see that missing data is yet another issue which must be considered, because the issue affects not only the measure computed for a respondent, but the technique employed will also impact the relative difficulty computed for an item.

A next step in evaluating the role of missing data directly related to the computation of a simple percent correct is to review the order of items from easy to difficult. A rough overview can be conducted by means of Winsteps item measure tables, but these tables do not show the spacing and/or overlap of items. We therefore prefer to compare Wright Maps. In Figs. 18.4 and 18.5, we present two Wright Maps: one (18.4) for the analysis with CODES = 12345kd and one (18.5) for CODES = 12345.

The Wright Map in Fig. 18.4 was constructed from the control file used to compute the item entry table for CODES = 12345kd. Because not-answered items and double-marked items are used in the analysis, some students will have a lower person measure compared to their person measure when CODES = 12345 is used. Using CODES = 12345kd will also result in some items appearing more difficult compared to their item difficulty when CODES = 12345 is used.

The Wright Map in Fig. 18.5 was constructed using CODES = 12345. Since only clearly answered items were used for the analysis, some not-reached items will be displayed as relatively easier in comparison to other items than the same comparisons made from the Wright Map for CODES = 12345 kd. This is because if "not reached" are counted as wrong, the item will appear to be more difficult.

Formative Assessment Checkpoint #4

Question: In Fig. 18.5 (CODES = 12345), two people are above Q32. Does this mean these people are guaranteed to have correctly answered this item?

Answer: No. Nothing is guaranteed; we can only use probabilities. Our analysis suggests there is a better than 50/50 chance that these two people (the two Xs above the Q32 at about −1.8 logits) would correctly answer this item.

```
TABLE 12.2 Kathy Trundle Geology Test Subset of  ZOU439WS.TXT  Nov 28  9:16 2011
INPUT: 64 PERSON  22 ITEM  REPORTED: 64 PERSON  22 ITEM  2 CATS  MINISTEP 3.72.3
--------------------------------------------------------------------------------

       PERSON - MAP - ITEM
           <more>|<rare>
    2            +
                 |
                 |
              X  |T Q33
                 |
                 |
                T|  Q40
              X  |
                 |
                 |
                 |
                 |
    1            +
          XXXXX  |
                 |S Q28
                 |  Q44
             XX S|
                 |  Q46
                 |
      XXXXXXXXX  |  Q42    Q47
                 |  Q34    Q36
           XXXX  |  Q35    Q43
                 |
                 |
    0   XXXXXXX  +M Q48
                 |
       XXXXXXXX  |  Q30
                M|  Q39
                 |  Q31
          XXXXX  |  Q27
                 |  Q41    Q45
                 |
        XXXXXXX  |
                 |
              X  |S Q38
                 |  Q29
   -1            +  Q37
          XXXXX S|
                 |
                 |
                 |
              X  |
                 |
                 |
                 |
            XXX  |T
                 |
                T|
   -2            +
             XX  |
                 |
                 |
                 |
                 |  Q32
                 |
              X  |
                 |
                 |
                 |
                 |
   -3        XX  +
           <less>|<frequ>
```

Fig. 18.4 (Winsteps Table 12.2): A Wright Map using CODES=12345kd

```
TABLE 12.2 Kathy Trundle Geology Test Subset of  ZOU442WS.TXT  Nov 28  9:10 2011
INPUT: 64 PERSON  22 ITEM  REPORTED: 62 PERSON  22 ITEM  2 CATS  MINISTEP 3.72.3
--------------------------------------------------------------------------------

      PERSON - MAP - ITEM
          <more>|<rare>
    2           +
                |
                |  Q33
             X  |T
                |
                |
                |  Q40
             X  |
               T|
                |
                |
             X  |
    1           +
          XXXX  |
                |S Q28
             X  |
            XX  |  Q44
               S|
                |
      XXXXXXXX  |  Q46
                |  Q34     Q36     Q42
         XXXXX  |  Q35     Q43     Q47
                |
             X  |
    0  XXXXXXX  +M
               M|  Q30
      XXXXXXXX  |
                |  Q31     Q39     Q48
                |  Q27
      XXXXXXXX  |
             X  |  Q41
                |  Q45
          XXXX  |
               S|
            XX  |S Q29     Q38
                |
   -1        X  +  Q37
          XXXX  |
                |
             X  |
                |
                |
               T|
                |
            XX  |
                |T
                |
                |
   -2           +
                |
                |
                |
                |  Q32
                |
                |
                |
                |
                |
                |
   -3           +
          <less>|<frequ>
```

Fig. 18.5 (Winsteps Table 12.2): A Wright Map using Codes=12345

Before comparing these two maps, we must discuss one additional issue. When a Rasch analysis is run with a control file, the default analysis centers the logit scale on the mean item measure. Recall that in one case, Fig. 18.4 (CODES = 12345kd), the person measures and item measures are based upon viewing items not answered as wrong. In the other case, Fig. 18.5 (CODES = 12345), the person measures and item measures are based upon only those items answered. When we make a side-by-side comparison of the two Wright Maps, we should look for general patterns to better understand the implications of the two different techniques of coding. Remember since we have not done any linking, then the numerical values are not directly comparable across the two analyses.

What does the side-by-side comparison suggest? One key difference is the placement of person measures in relation to the placement of items. With CODES = 12345, all respondents have a greater than .50 probability of answering the easiest item of the test (item Q32). With CODES = 12345kd, that is not the case. There are three respondents below a .50 probability of answering item Q32.

Additional techniques can be used to understand the different item placements via the two coding techniques. One technique mentioned many times by Winsteps author Mike Linacre is to make simple plots of data. Just as Wright Maps provide a quick visual presentation of data, so do other types of plots. Regarding this comparison of coding techniques, two useful plots are to (1) plot the item measures computed using the two techniques against each other and (2) plot the person measures computed using the two techniques against each other. Figures 18.6 and 18.7 present these two types of plots using this technique. If readers hypothesize that the two techniques of analysis will result in very similar person measures and very similar item measures, then the plot of item measures should result in the placement of items lying along a diagonal, and the plot of person measures should result in the placement of persons lying along a diagonal. Items that are "off diagonal" and persons who are "off diagonal" are items that have moved and persons who have moved in terms of their relative location with regard to other items or people. In essence, these plots represent yet another way to conduct the same sort of visual analysis that was conducted using side-by-side Wright Maps. As in any analysis, it is very important to remember that the errors of person measures and item measures are critical to keep in mind. Since we have authored code and put together data sets that work with Ministeps, the errors of items and persons will be generally larger than would be found when analyses involve more items and persons. The point of providing examples herein is to show readers the types of techniques they can use. A useful example of cross plotting as part of an analysis is provided by Baghaei (2007) and Roorda et al. (2004).

Figures 18.6 and 18.7 show that, indeed, there are items and persons whose relative locations change with respect to one coding in comparison to another. For the time being, the most important technique for readers to grasp is that "cross plotting" provides a very quick, intuitive technique for detailed comparisons of the different relative locations of items or persons on two Wright Maps. Readers will recall that in our chapter considering DIF, we made sure to consider the error of items. That certainly is a very important added component of some cross plotting, and we are by

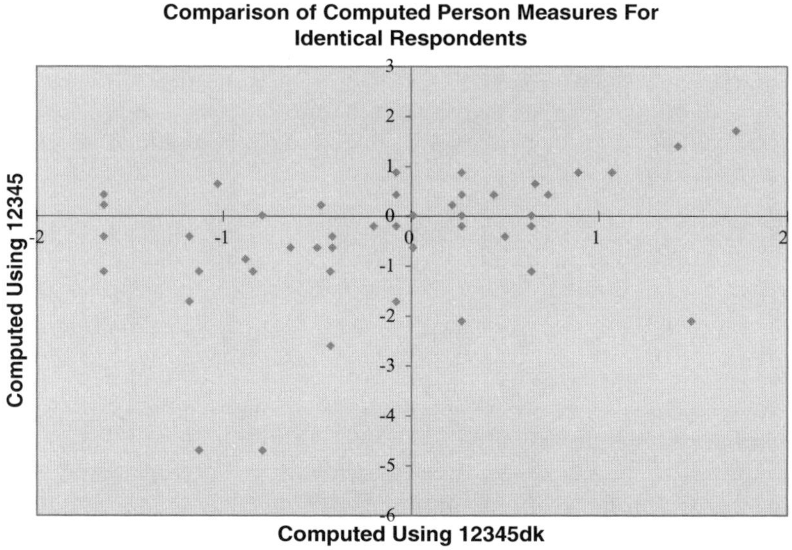

Fig. 18.6 A cross plot of the same respondents in single data set. One set of measures was computed in which missing or unclear data were not counted as a "wrong" answer (coding 12345). The other set of measures is computed based upon counting an unanswered item or an unclearly answered item a "wrong." Numerous respondents are "off diagonal"

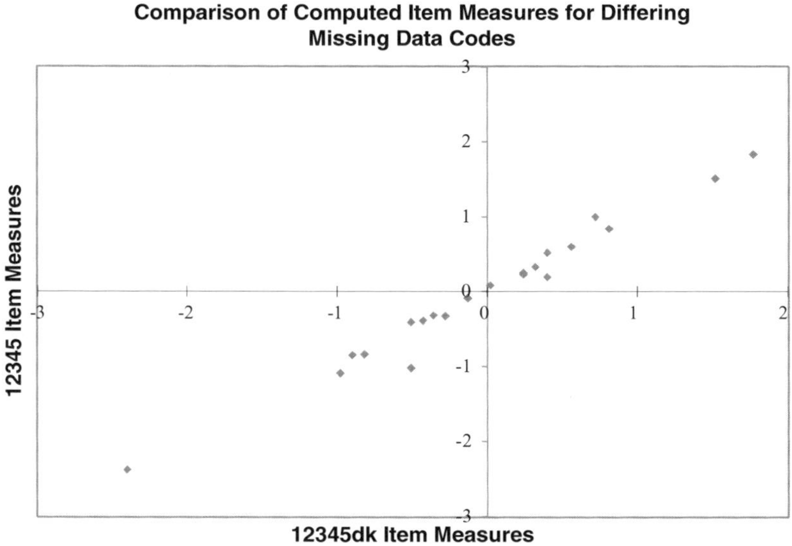

Fig. 18.7 A cross plot of item measures. One set of measures was computed in which missing or unclear data were not counted as a "wrong" answer (coding 12345). The other set of measures is computed based upon counting an unanswered item or an unclearly answered item a "wrong." Some items may be "off diagonal"

no means suggesting that consideration of error (be it of persons or items) is not important in cross plotting. But for the point that we wish to make in this chapter, we want to encourage readers as they experiment with the impact of different missing data codings that one important technique is the simple plotting of person measures as a function of coding scheme as well as the cross plotting of item measures as a function of coding scheme.

Formative Assessment Checkpoint #5

Question (True/False): A simple plot such as a cross plot is too simple; thus, it can't tell one very much.

Answer: False. Quite often, very simple analysis techniques provide as much, if not more, useful information regarding an analysis. Think back to the analyses that sought reasons for the Challenger disaster in America's space program. Professor Richard Feynman of the California Institute of Technology used a cold ice–water mixture in a glass to show that the Challenger's O-ring acted very differently in cold conditions.

Are there still other steps (to address missing data) that could be taken to investigate the potential impact of different coding techniques upon an analysis? Yes, yet another technique is to investigate differences in statistical results as a function of coding. Figure 18.8 presents the results of a simple *t*-test comparing one half of the respondents (students 1–32) to the other half of the respondents (students 33–64).

Coming to Closure on Missing Data

In this chapter we focused on the issue of missing data. Data can be missing for a myriad of reasons. For example, students may skip items, not reach items, or instrument pages might not have been duplicated. Deciding how to look for the impact of missing data and how to evaluate data are issues that can be fully considered because of the unique properties of the Rasch model. Namely, when a set of items involves a single trait, respondent measures can be computed based upon the items answered. No longer must researchers throw out data or crazily try to predict what a person might have selected for an item. The techniques presented herein are certainly, in part, Rasch techniques, such as thinking about what it means to measure, investigating the distribution of items along a trait, and digesting the implications of an item measuring a different part of a trait as a function of the type of coding of missing responses. Yet, there exist still other techniques, such as cross plotting. If we were to ask Ben Wright when he might have used cross plots first, he might respond: While conducting physics research with University of Chicago's Nobel Laureate

Coding CODES=12345

Respondents	Mean Person Measure	*t*-test *p* value
Students 1-32	.06	.09
Students 33-64	-.22	

Coding CODES=12345kd

Respondents	Mean Person Measure	*t*-test *p* value
Students 1-32	-.13	.11
Students 33-64	-.60	

Fig. 18.8 A comparison of the impact of coding upon statistical tests. Use of CODES=12345 (counting only clear answers for items) suggested a *p* value of .09 for the comparison of the first 32 respondents of the data set to the last 32 respondents of the data set. The same comparison of person measures for the values which were computed using CODES=12345kd suggests a different *p* value

Robert Mulliken. He would likely then say the whole point is that by cross plotting we are thinking about measurement, no matter the field.

How to confront missing data is an issue that seems to be ignored more often than not in research. We hope that we have shown how important it is to at least consider this issue, and also that, with Rasch measurement techniques, an analysis of only answered items is not difficult because a person's measure can be computed based upon items answered.

We assert that there are cases where a researcher may very well count not-answered items as wrong, for example, situations in which the political pressures of comparing students who do not complete the same number of items are so great that one has to use all items of an instrument and all skipped items are counted as wrong. However, when such pressure is low or absent, we suggest that it is better to use only the answers provided by students. We have discussed some of the potential reasons why a student may not answer an item (e.g., reading level, test taking technique), and it seems prudent in studies to think through what one wishes to measure and attempts to learn. For the monitoring of poorly performing student groups in education, this issue is particularly important. Those interested in equity must, in our view, take the issue of missing data seriously in their analysis. In medical studies we can well imagine that there are many situations in which missing data is an issue which can impact a study. For example, from some populations of respondents, it might be very difficult for an entire instrument to be completed.

The technique which we present here is one of many Rasch techniques which allow the researcher to evaluate the quality of data in a data set. In this chapter we showed how one might compare, for example, the distribution of respondents using each of the two coding schemes. There are added aspects of the two coding techniques which might also be evaluated, that being how "fit" changes for persons and items as a function of the CODES line. We hope that readers will see that there are many issues associated with missing data, but that with Rasch techniques there can be many solutions to the presence of missing data in a data set.

Isabelle and Ted: Two Colleagues Conversing

Isabelle: This entire idea of missing versus non-missing data is really interesting.

Ted: Yes it is. With Rasch techniques, it is very easy to compute a person's measure based upon just the person's answers to a set of items. I was truly amazed at the difference it made in a person's measure when we either counted or did not count not-answered items as wrong.

Isabelle: What I think is cool is that not only does Rasch software help us confront this issue, but also Rasch again forces us to think about what it means to measure. In this case, do we really want to count not-answered items as wrong? There may be students who are just taking their time to show us what they know. Maybe we are getting higher quality data from those students even though they answer fewer items?

Ted: Yes, but you know what? I think it is very interesting that there is no single right way to do this. I mean, just as there are multiple things to consider when we evaluate a set of items for a measurement scale, there is also no one right way to evaluate the role of missing data. And there may be situations in which one might count not-answered items as wrong.

Isabelle: I agree, but you know what? If there is an ability to report students having completed different numbers of items, my gut tells me that might be a technique that would help me better measure what I want to measure. So in the case of tests, I compute a measure of a student not based upon her or his speed but based upon what she or he did or did not know.

Keywords and Phrases

CODES=
Skipped items
Not reached items
Cross plot

Potential Article Text

A component of the analysis of the 22-item instrument included an assessment of the role of missing data, which were defined as items not answered by respondents. This analysis was conducted because the research goal was to monitor the performance of students over time and to compare selected subsamples of students as a function of time. It has been well documented that the test taking strategies exhibited by females and males are often quite different. Females often take a longer time than do males to answer items. It was deemed important, therefore, to explore the issues associated with missing data prior to the computation of respondent measures to be used for statistical tests.

Answering patterns of respondents and the implications of these answering patterns were reviewed through analyses of frequencies, of Wright Maps, cross plotting of item measures as a function of data coding, and cross plotting of person measures. These techniques suggested that not only did the computed relative performance measures of respondents differ as a function of coding scheme, but also the apparent difficulty of some test items was greatly affected by the coding scheme. Since the goal of the analysis project reported herein is to compare the performance of students, a decision was made not to count not-answered items as incorrect. This provided a measure of respondents not affected by test taking strategy and/or motivation.

Quick Tips

Missing data can impact an analysis. Conduct an initial analysis of your data. Then look at any of the item tables provided by Winsteps (e.g., item entry table). When you pull up the table, you will first see a table that summarizes information about each item (Table 14.1). Then below this initial table, you will see Table 14.3. This table will help you see how much data were missing for each item. This is a good place to start looking at your missing data.

Another fast technique of looking at missing data is to look at both an item table (such as 14.1) and a person table (such as Table 18). Look at the "COUNT" columns in both tables and you will be able to see for which items and persons data are missing.

Conduct an analysis with missing data included and missing data not included. Then cross plot the person measures from the two analyses. Do the different techniques make a difference in person measures?

Data Sets: (go to http://extras.springer.com)

cf exercise codes 12345dk
cf exercise 12345dk
cf exercise 12345

Activities

Activity #1

Problem: We supply a control file named cf exercise codes 12345dk.

Questions: How many items are evaluated in this analysis? What are the names of the items?

Are there data missing? Can you hypothesize why the data might be missing from the pattern of missing data?

Answers: (1) There are 22 items in the analysis. (2) Q1 to Q22. (3) Yes. (4) It looks like a sheet of questions was not presented to some students. There is more missing data for Q15–Q22 than Q1–Q14.

Activity #2

Situation: We supply two control files (cf exercise 12345dk, cf exercise 12345).

Task: Run a simple Rasch analysis. Look at the item entry table for each of the two analyses and compare the two tables. What do you see? Why?

Answer: The key difference can be seen in the TOTAL COUNT columns. For the CODES = 12345 analysis, there are a range of values present. This is because missing data and double coding of answers are not used for the computation of a person measure.

Codes=12345 analysis

```
TABLE 14.1 Kathy Trundle Geology Test Subset of  ZOU651WS.TXT  Nov 28 11:36 2011
INPUT: 64 PERSON  22 ITEM  REPORTED: 62 PERSON  22 ITEM  2 CATS  MINISTEP 3.72.3
--------------------------------------------------------------------------------
PERSON: REAL SEP.: .80  REL.: .39 ... ITEM: REAL SEP.: 2.66  REL.: .88

        ITEM STATISTICS:  ENTRY ORDER
```

ENTRY NUMBER	TOTAL SCORE	TOTAL COUNT	MEASURE	MODEL S.E.	INFIT MNSQ	INFIT ZSTD	OUTFIT MNSQ	OUTFIT ZSTD	PT-MEASURE CORR.	PT-MEASURE EXP.	EXACT MATCH OBS%	EXACT MATCH EXP%	ITEM
1	34	61	-.35	.27	.98	-.2	.97	-.2	.34	.31	70.5	63.5	Q27
2	18	61	.87	.29	1.19	1.4	1.26	1.5	.02	.29	65.6	71.7	Q28
3	40	61	-.80	.28	.99	-.1	1.08	.6	.29	.30	73.8	68.5	Q29
4	30	61	-.05	.27	.99	-.1	1.01	.1	.32	.31	67.2	62.9	Q30
5	33	61	-.27	.27	.98	-.2	.96	-.3	.34	.31	65.6	63.1	Q31
6	54	61	-2.31	.41	.94	-.1	.71	-.7	.35	.21	88.5	88.5	Q32
7	9	61	1.83	.37	1.09	.4	1.28	.9	.06	.23	85.2	85.2	Q33
8	24	60	.37	.28	.91	-1.0	.86	-1.2	.45	.31	70.0	65.2	Q34
9	25	60	.29	.28	1.00	.0	.98	-.1	.32	.31	65.0	64.6	Q35
10	24	60	.37	.28	.91	-1.0	.90	-.8	.43	.31	73.3	65.2	Q36
11	41	58	-1.03	.30	1.00	.0	1.16	.9	.25	.29	75.9	72.2	Q37
12	39	59	-.79	.29	1.00	.0	.95	-.3	.32	.30	64.4	69.0	Q38
13	32	58	-.29	.28	.97	-.4	.94	-.6	.37	.31	69.0	63.5	Q39
14	11	58	1.53	.35	1.17	.9	1.25	.9	.01	.25	79.3	81.2	Q40
15	35	58	-.53	.28	1.07	.8	1.08	.7	.21	.31	58.6	65.6	Q41
16	23	57	.37	.28	.93	-.8	.94	-.5	.40	.31	73.7	65.0	Q42
17	25	58	.25	.28	1.13	1.5	1.11	1.0	.15	.31	56.9	63.9	Q43
18	19	57	.70	.29	1.07	.6	1.10	.7	.20	.30	70.2	69.0	Q44
19	35	56	-.62	.29	.93	-.7	.87	-1.0	.43	.31	69.6	66.8	Q45
20	21	53	.45	.30	1.02	.2	.98	-.1	.29	.31	64.2	65.0	Q46
21	23	52	.23	.29	.83	-2.0	.79	-2.0	.55	.31	73.1	63.5	Q47
22	28	52	-.21	.29	.93	-.8	.90	-.9	.41	.31	65.4	62.7	Q48

Codes=12345kd analysis

```
TABLE 14.1 Kathy Trundle Geology Test Subset of  ZOU070WS.TXT  Nov 28 11:37 2011
INPUT: 64 PERSON  22 ITEM  REPORTED: 64 PERSON  22 ITEM  2 CATS  MINISTEP 3.72.3
--------------------------------------------------------------------------------
PERSON: REAL SEP.: 1.63  REL.: .73 ... ITEM: REAL SEP.: 3.21  REL.: .91

        ITEM STATISTICS:  ENTRY ORDER
```

ENTRY NUMBER	TOTAL SCORE	TOTAL COUNT	MEASURE	MODEL S.E.	INFIT MNSQ	INFIT ZSTD	OUTFIT MNSQ	OUTFIT ZSTD	PT-MEASURE CORR.	PT-MEASURE EXP.	EXACT MATCH OBS%	EXACT MATCH EXP%	ITEM
1	34	64	-.85	.28	1.11	1.1	1.09	.6	.36	.43	58.1	67.1	Q27
2	18	64	.45	.30	1.32	2.2	1.58	2.1	.10	.36	66.1	74.0	Q28
3	40	64	-1.33	.29	1.05	.4	1.29	1.5	.40	.45	67.7	71.4	Q29
4	30	64	-.54	.28	1.03	.3	1.01	.2	.40	.42	61.3	66.6	Q30
5	33	64	-.77	.28	1.23	2.2	1.24	1.6	.27	.43	56.5	66.7	Q31
6	54	64	-2.87	.40	.86	-.5	.75	-.4	.57	.48	88.7	87.6	Q32
7	9	64	1.46	.38	1.15	.7	1.59	1.3	.14	.28	85.5	85.8	Q33
8	24	64	-.07	.28	.95	-.4	.90	-.5	.43	.40	71.0	69.3	Q34
9	25	64	-.15	.28	1.10	.9	1.08	.5	.33	.40	66.1	68.7	Q35
10	24	64	-.07	.28	1.10	.9	1.08	.5	.33	.40	64.5	69.3	Q36
11	41	64	-1.42	.29	.89	-.8	1.13	.7	.50	.45	75.8	72.3	Q37
12	39	64	-1.25	.29	1.13	1.1	1.21	1.2	.35	.45	66.1	70.4	Q38
13	32	64	-.69	.28	1.01	.1	.98	-.1	.43	.43	64.5	66.3	Q39
14	11	64	1.19	.35	1.22	1.1	1.14	.5	.17	.31	79.0	83.0	Q40
15	23	64	.02	.29	.81	-1.8	.72	-1.6	.54	.39	79.0	69.9	Q41
16	12	64	1.06	.34	.82	-.9	.60	-1.2	.46	.32	80.6	81.5	Q42
17	14	64	.84	.33	.98	-.1	.81	-.6	.37	.33	77.4	78.8	Q43
18	9	64	1.46	.38	.94	-.1	.91	.0	.31	.28	88.7	85.8	Q44
19	18	64	.45	.30	.72	-2.2	.63	-1.7	.56	.36	88.7	74.0	Q45
20	11	64	1.19	.35	.93	-.3	.75	-.6	.37	.31	85.5	83.0	Q46
21	13	64	.95	.33	.72	-1.6	.54	-1.6	.54	.32	85.5	80.1	Q47
22	13	64	.95	.33	.78	-1.2	.58	-1.4	.50	.32	82.3	80.1	Q48
MEAN	24.0	64.0	.00	.32	.99	.0	.98	.0			74.5	75.1	
S.D.	12.2	.0	1.10	.04	.16	1.2	.29	1.1			10.3	7.0	

Activity #3

Task: Using the two control files, run an analysis and then compare the two Wright Maps (one from the coding 12345 and one for the coding 12345kd). What do you see? Why might you need to be careful with the comparison you make?

Answer: We will leave it up to readers to conduct the two analyses, print out the two Wright Maps, and conduct the comparison. Readers should look for differences in the pattern of items, look for differences in the pattern of person measures, and also look for differences in the pattern of persons with respect to items.

Activity #4

Task: Cross plot the person measures from the two analyses. What do you observe?

Activity #5

Task: Cross plot the item measures from the two analyses. What do you observe?

Activity #6

A 20-item physics test was administered in Germany. Students can answer using the letters A, B, C, or D. For one student, write out a potential item answering pattern as if the student does not reach the end of the test.

Answer:

```
ABCDDABDBDACC
```

Only 13 answers are presented, the last 7 items are not answered, and as a result the items are skipped.

Activity #7

Please write out a potential answering pattern of a student who skips some items on a 20-item physics test.

Answer: To keep track, we placed digits on a line to mark each column: the first "1" denotes the 1st column, the second appearance of a "1" denotes the 11th column, and so forth.

```
12345678901234567890
ABC DDBB ADCAABB BCB
```

This student skipped the 4th item, the 9th item, and the 17th item. It looks like the student completed the test, for she or he answered items at the end of the test.

Activity #8

Task: For the same administration of the physics test, write a potential answering pattern of a student who skips items and does not reach the end of the test.

Answer: To keep track we have placed digits on a line to mark each column: the first "1" denotes the 1st column, the second appearance of a "1" denotes the 11th column, and so forth.

```
12345678901234567890
BBCAA CCDA
```

A good guess is that this student skipped item 6, and the last item the student attempted was the 10th item of the test.

References

Baghaei, P. (2007). Applying the Rasch rating-scale model to set multiple cut-offs. *Rasch Measurement Transactions, 20*(4), 1075–1076.

Roorda, L. D., Jones, C. A., Waltz, M., Lankhorst, G. J., Bouter, L. M., van der Eijken, J. W., et al. (2004). Satisfactory cross cultural equivalence of the Dutch WOMAC in patients with hip osteoarthritis waiting for arthroplasty. *Annals of the Rheumatic Diseases, 63*(1), 36–42.

Additional Readings

A number of added missing data articles of use to researchers.

DeAyala, R. J. (2003). The effect of missing data on estimating a respondent's location using ratings data. *Journal of Applied Measurement, 4*, 1–9.

Ludlow, L., & O'Leary, M. (2000). What to do about missing data? *Rasch Measurement Transactions, 14*(2), 751.

Wang, S., Zhang, H., & Young, M. J. (2005). *The effect of missing data of rating design on parameter estimations using the many-facets Rasch model.* Paper presented at the annual meeting of the American Educational Research Association, Montreal, Canada. Retrieved April 2005.

Chapter 19
Combining Scales

Isabelle and Ted: Two Colleagues Conversing

Ted: *Isabelle, I have a problem. Actually, it's not a problem, but something I have been thinking about.*

Isabelle: *Okay, what is bothering you?*

Ted: *A colleague sent me data sets they would like me to look at. It is sort of serendipity, but the thinking that I have been doing of how to best tackle each data set really focuses on the same issue.*

Isabelle: *What is the issue?*

Ted: *Well, learning about Rasch has helped me think about measurement in education, medicine, and beyond. This "thinking" has helped me avoid some mistakes that have been made in the past in our field. Also my "thinking" has helped me understand new ideas of how some data analyses might be conducted. The one data set focuses on students' drawings of scientists. A 9-item scale was used to evaluate the level of sophistication of students' drawings. The 9 items contain a 5-point rating scale, so for each student, we have 9 ratings from 0 to 4, where 4 represents more sophisticated drawing.*

Isabelle (interrupting): *Wouldn't you just use the rating scale model that we have been using for the analysis of the STEBI data?*

Ted: *Actually, we need to do one other thing to correct for a nuance in the data and a nuance in the measurement scales of the project. We need to combine scales.*

W.J. Boone et al., *Rasch Analysis in the Human Sciences*,
DOI 10.1007/978-94-007-6857-4_19,

Introduction

In research of all fields, it is common to use numbers to record data collected from respondents. As we have emphasized throughout this book, when numbers are collected from respondents, and there is an interest in pooling responses to a set of items, be they test or survey items, to determine a student measure (e.g., an attitude measure with regard to "interest in science," a measure of physics content knowledge), then a number of critical steps must be taken. First, there must be a thoughtful review of what it means to define a trait. By doing so, then, one can validate whether it makes sense to try to pool a set of items to compute a measure. Even after reaching a conclusion as to the appropriateness of pooling a set of items for a respondent measure, researchers should not evaluate data immediately. This is because "raw numbers" cannot be immediately used for an analysis. For tests (e.g., true/false, multiple choice) in which students can earn 1 for a correct response or 0 for an incorrect response, the data should not be immediately evaluated. This is also the case for surveys in which a number is used to indicate the position on a rating scale that a respondent selected (e.g., "4" for *Strongly Agree*, "3" for *Agree*, "2" for *Disagree*, and "1" for *Strongly Disagree*). A simple third example, which is a mix of issues with right/wrong tests and attitude rating scales, is a partial credit test. In this case, a respondent is presented with a set of items and can earn partial scores for each item. Data collected with partial credit instruments also should not be immediately used for statistical analyses.

Formative Assessment Checkpoint #1

Question (True/False): It is okay to enter test data (right as a "1," wrong as a "0") and immediately compute the means of subgroups for a 30-item test.

Answer: No. The data entered as "1" and "0" only serve as labels to indicate which items were correctly answered by each respondent. All one knows is that a "1" for an item denotes more knowledge of a respondent along a trait relative to a "0." Linear measures must be computed using the Rasch model.

The analysis of partial credit test data and some rating scale data typically follows most of the steps and thought processes for our analysis of STEBI data. There are, however, very important "twists" to the use of some rating scale instruments and most partial credit test data instruments that are not often considered in fields of research. One such "twist" is that it may be beneficial to ask respondents questions using a number of different rating scales, but of course, the numbers used for coding should not be viewed as having the same meaning (e.g., if *Very Often* is coded as a "4" and *Strongly Agree* is coded as a "4"), but in some cases, the data can still be pooled for the computation of an overall person measure. Our discussion of this difficult issue is illustrated with the OSU McFarland data and the German data of this chapter.

An Example of Combining Data

In the Donna McFarland (at the Ohio State University, OSU) study, researchers wanted to know how students in middle school viewed scientists. To evaluate students' views of scientists, a number of students were asked to produce three separate drawings of a scientist "in action." Each drawing was then rated by a judge with regard to what is termed student sophistication. More specifically, three ratings were made of each drawing. First, a rating of the drawing was made with regard to what is called the scientist's "location." A second rating was made with regard to the "activity" presented, and a third rating was made with regard to the "appearance" of the scientist. A scale of 0, 1, 2, 3 was used to rate drawings with respect to appearance, location, and activity; thus, three ratings were produced for each drawing using the same four-step scale. Figure 19.1 presents the data for the first three students of a data set. Also provided is a key to indicate which column pertains to which of the three drawings and which of the three types of ratings (appearance, location, or activity). In other words, the students were evaluated with 9 items of sophistication on their "drawing a scientist." Recall that each student provided three drawings.

The data presented in Fig. 19.1 conform to a requirement of Rasch analysis. All nine items are all viewed as defining a sophistication level of a single trait, "drawing a scientist," and, second, the number values of the ratings are ordinal and nonlinear. There exists, however, a nuance that we wish to emphasize and point out to readers who have read the earlier chapters. The type of thinking that we will hope you will successfully do for this chapter will be difficult and will require some long, hard thinking, much as having to think about numbers for rating scales being just labels may have been difficult.

Why and how is there an issue that influences the analysis and therefore requires a small but important change in the control file to facilitate the computation of person measures? To understand the why and the how, we will walk together through what we, as education practitioners of Rasch, did to better understand the data

```
Col 1   Drawing 1 app rating
Col 2   Drawing 1 loc rating
Col 3   Drawing 1 act rating
Col 4   Drawing 2 app rating
Col 5   Drawing 2 loc rating
Col 6   Drawing 2 act rating
Col 7   Drawing 3 app rating
Col 8   Drawing 3 loc rating
Col 9   Drawing 3 app rating

222222232
111322121
321321321
```

Fig. 19.1 Three students' data and a key for identifying the data

Fig. 19.2 A trait line for the trait of sophistication

Fig. 19.3 The sophistication trait line with three aspects of the trait – location, appearance, and activity

from OSU. First, the team that collected the data was asked to draw a horizontal line representing sophistication ("less sophistication" was used to label the left side of the line, and "more sophistication" was used to label the right side of the line) (Fig. 19.2).

The authors of the study were then asked to mark where they thought each typical "location" item would fall along the line of the trait. They were then asked to mark where they thought each typical "activity" item would be on the trait and finally where each typical "appearance" item would fall on the trait. Above we provide three potential locations of these three item types (Fig. 19.3).

A second question was then posed to the OSU researchers: "Okay, we understand the data collection, three ratings per picture, and three pictures per student. We now need to discuss your rating scale. It goes from 0 to 1 to 2 to 3. Higher numbers are supposed to indicate higher sophistication, right? Okay, does moving from 0 to 1 to 2 to 3 for increased sophistication of location necessarily mean the same increase in sophistication as a movement from 0 to 1 to 2 to 3 for increase in sophistication of appearance? For activity?"

The question posed to the OSU researchers may be perceived as mundane – a waste of time. If, however, readers' sensitivities toward the pitfalls of poor measurement and the nuances of high-quality measurement have increased, then readers should sense a problem. Let us restate the question as follows: Is it reasonable to treat a rating of 0, a rating of 1, and a jump from 0 to 1 as having the same meaning in terms of the change in the amount of the trait (sophistication) for each of the three types of the items (location, appearance, activity)?

The OSU researchers concluded that, indeed, the rating scale increments from 0 to 3 should *not* be viewed as the same for the three types of items. Two additional points concurrently influenced the discussion of the research team: First, different rubrics were used to define the meaning of 0, 1, 2, and 3 for each of the three item types. Second, moving toward higher sophistication (e.g., from 1 to 2 for location) may not mean the same advancement as moving from 1 to 2 for a different part of the trait (e.g., from 1 to 2 for appearance). Figure 19.4 presents a schematic of the distribution of rating scale steps as a function of the trait. For our discussion, the most

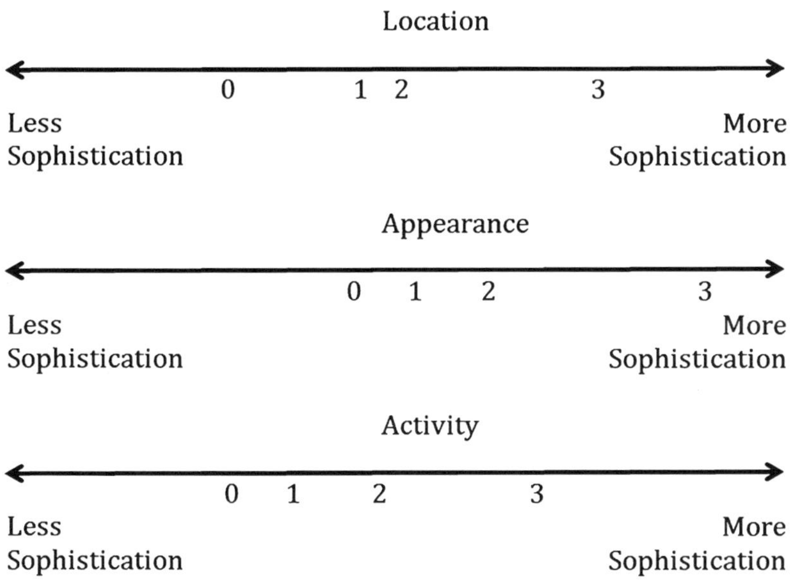

Fig. 19.4 A schematic showing the potentially different ways in which the rating scale steps of 0, 1, 2, and 3 mark the latent trait of sophistication

important observation to take note of is that the spacing from each rating category to the next is both uneven and unique for each of the three item types. Also important to note is a numerical rating (e.g., 2) of one item type is not necessarily at the same point on the continuum of the trait for another item type (the meaning of a "2" for location may not mark the same spot on the continuum at a "2" for appearance).

Figure 19.4 is critically important and typically difficult to understand when first encountered. We can personally testify to this! If, however, readers can pause and reflect, mastery of the concept should substantially increase your level of thinking about measurement. One will be able to sense and avert potential problems (e.g., mistakes, oversights) before these and other issues compromise the quality of your data collection and analysis.

The figure graphically presents the uneven "steps" of rating scales that we discussed to a degree in earlier chapters. Presenting such uneven steps is central to understanding the theory, usefulness, and necessity of Rasch measurement. It is so easy to be seduced by the numbers used to code responses. Coding a rating scale of SA, A, D, and SD with the numbers 4, 3, 2, 1 seems intuitively definitive and accurate. Given our everyday life experiences, we are immediately tempted to assume that we could subtract, add, divide, and multiply (maybe even find a square root!). However, as we have demonstrated in preceding chapters, the coding (SA, A, D, SD) portrays only an ordering with unknown spacing. The uneven spacing of the numbers of Fig. 19.4 shows pictorially what we mean.

There exists, however, a new part to this idea. In some instances, data are collected with regard to a single trait, but the rating scale used for different items may have different conceptual meanings with respect to the trait. Discussions with

Professor McFarland's research team about this sample data set revealed that, although all items were hypothesized to be along a single trait of "sophistication," using a scale with numbers from 0 to 1 to 2 to 3 to portray the meaning of moving from one level of sophistication to another could not be assumed to be the same for each of the three issues (location, appearance, activity) evaluated in student drawings. Rasch theory helps researchers think about the numbers we use to construct measures and in turn helps identify important nuances in the data, such as this one. In Fig. 19.4, all of this thinking comes together. Readers should notice not only an uneven spacing of rating category steps for each type of sophistication but also the misalignment of the numbers (e.g., 2) for each sophistication type. We have intentionally plotted the numbers not to line up. This is to show that the meaning of a "2" may be different for different parts of the trait, even though the same words were used to define the rating scale category. The different location of the number 2 for, say, "activity" and "location" can in essence be summarized as the following: The meaning of the number 2 (aka the meaning of the word Disagree) may be different for location relative to activity along the trait of sophistication.

Figure 19.5 presents a portion of the control file used to evaluate the OSU data set. Most lines of this control file will now be familiar to readers, but let's talk through each line in order to help readers master this new measurement challenge that can be noted by thinking about Rasch theory and then can be attacked by applying Rasch measurement through a program such as Winsteps.

- The line "&INST" is just the line that begins each control file.
- The line "TITLE" is, of course, the line we use to print a title on an output table. When performing several analyses of different data sets, a title helps researchers stay organized and saves time.
- The lines NAME1 = 1 and NAMELENGTH = 4 tell the program where the person identification information begins and how many columns contain person information, respectively. In this data set, the person "name" or ID begins in column 1 and has a total length of 4 columns. In other words, the first column of data for the person information is in column 1, and the last column of data for the person information is in column 4.
- The next four lines (ITEM1 = 5, NI = 9, XWIDE = 1, CODES = 0123) of the control file tell the program the location and meaning of data pertaining to items. The first column of data is located in column 5; there is a total of 9 items; each datum for each item is only one column or space wide; and the numbers used to indicate the rating a respondent received could be a 0, 1, 2, or 3. If any other numbers are read in the columns pertaining to items, then those data are ignored.
- We now reach a new, very important, line. This line tells Winsteps that specific subsets of items will be viewed as having the same rating scale, whereas other subsets of items will be viewed as having a different rating scale. The phrase "ISGROUPS" tells Winsteps that such a situation exists in the Draw a Scientist data (note one can use the phrase "GROUPS," and the result is identical with the use of "ISGROUPS"). In our data set, we choose to enter data for drawing 1 first, then the data for drawing 2, and finally the data for drawing 3. The pattern of data entry for a drawing was to first enter the appearance rating, then the location

Fig. 19.5 Part of the control file for evaluating the OSU data set

```
&INST
TITLE = 'Draw a Sci'
NAME1 = 1
NAMELENGTH = 4
ITEM1 = 5
NI = 9
XWIDE=1
CODES = 0123
ISGROUPS=ABCABCABC
CLFILE = *
1+0 lowest sophistication of app 0
1+1 low but not lowest sophistication of app 1
1+2 moderate sophistication of app 2
1+3 highest sophistication of app 3
2+0 lowest sophistication of loc 0
2+1 low but not lowest sophistication of loc 1
2+2 moderate sophistication of loc 2
2+3 highest sophistication of loc 3
3+0 lowest sophistication of act 0
3+1 low but not lowest sophistication of act 1
3+2 moderate sophistication of act 2
3+3 highest sophistication of act 3
*
&END
Q1 D1 app rating
Q2 D1 loc rating
Q3 D1 act rating
Q4 D2 app rating
Q5 D2 loc rating
Q6 D2 act rating
Q7 D3 app rating
Q8 D3 loc rating
Q9 D3 act rating
END NAMES
1011222222232
1022111322121
1032321321321
1042113232323
```

rating, and finally the activity rating. As a result, the coding "ABCABCABC" communicates this sequence of data entry. Thus, the first letter "A" represents the appearance rating for drawing #1 of a student and that data is the first piece of rating data presented for a student. The second presentation of the letter "A" for the line ISGROUPS tells the program that the 4th entry of rating data also involves an "appearance" rating, and that the rating scale will have the same "jumps" as that which will be used to evaluate the drawing #1 appearance rating scale data. The final presentation of the letter "A" simply helps indicate that the seventh column of rating scale data also concerns the "appearance" rating scale.

- The ISGROUPS is the key line for readers, for here we say in essence that rating data for "appearance" have a potentially unique structure. That is, the spacing of jumps from 0 to 1 to 2 to 3 for all appearance ratings is the same, but the spacing of the jumps for the "location" rating scale may not be the same as for "appearance." Both rating scales involve an assessment of "sophistication," but the way in which the rating scale works for "location" may not necessarily be the same for "appearance."

- The 14 lines following ISGROUPS are not required for an analysis of data. But, inclusion of these lines, in particular for the type of analysis required for this data set, helps keep one organized. For example, the act of typing each line forces researchers to double-check their coding of the ISGROUPS line, which helps save time when results are reviewed. The command "CLFILE" can be looked up in the "Help" option of Winsteps. The command CLFILE provides a way to identify the numbers for each type of rating for each item. The "*" at the beginning and end of the command tells Winsteps that there are multiple pieces or lines of information within the single command. For example, the first four lines under "CLFILE=*" are a synopsis of the meaning of each rating category for the appearance items, which are coded item type "1." Thus, "1+0 lowest sophistication of app 0" indicates that for item type 1, a "0" rating refers to the lowest level of sophistication for the appearance items. The line "1+1 low but not lowest sophistication of app 1" refers to the rating category of "1" for item type 1 as a low level of sophistication for appearance items.

Once a researcher has coded their data and written their control file, then an analysis is conducted as one would conduct earlier analyses described herein. Moreover, person measures are used just as one would use a person measure computed from earlier examples provided herein. Person measures are computed, placed in a data set, and evaluated using statistics. The key is the ISGROUPS command, details of which are presented below, now that we have presented the logic behind what we have done. The important thing for readers to note is that when one believes that data have been collected for one trait, but the data have been collected (in this case) using different rating scales, then it is possible to make use of all the data. Below we provide details for ISGROUPS from the Winsteps manual (Linacre, 2012):

> Items in the same "grouping" share the same dichotomous, rating scale or partial credit response structure. For tests comprising only dichotomous items, or for tests in which all items share the same rating (or partial credit) scale definition, all items belong to one grouping, i.e., they accord with the simple dichotomous Rasch model or the Andrich "Rating Scale" model. For tests using the "Masters' Partial Credit" model, each item comprises its own grouping (dichotomous or polytomous). For tests in which some items share one polytomous response-structure definition, and other items another response-structure definition, there can be two or more item groupings. (p. 151)

A Second Example of Combining Scales

Improved ability to detect the type of data nuance described in this chapter and to correct for such issues is an important step for researchers; therefore, we provide an additional example of a situation in which ISGROUPS could be used. Using knowledge gained by the analysis of the OSU data set, readers should find this example easier to understand.

Andrea Moeller is a professor of biology education at the University of Trier (Germany). She and collaborators collected data to investigate the issue of inquiry teaching in the life sciences. To collect these data, the research team authored 24 separate test items. Each item had an open response format and was scored with a rubric evaluating the trait of "student competence." An important aspect of the test is that 6 of the 24 items focused on "formulating questions" with respect to competence, another 6 items involved "generating hypotheses" with respect to competence, yet another 6 items targeted "planning an investigation" with respect to competence, and the fourth group of 6 items centered on "interpreting data" with respect to competence. A rubric from 0 to 5 (with rating scale steps of 0, 1, 2, 3, 4, and 5) was developed for each item type. The critical aspect is that moving up the scale numerically means increased competence regardless of item type, but a difference in receiving a "1" rating and a "2" for the 6 "formulating questions" items does not necessarily mean the same quantitative difference in competence as rating of a "1" and a "2" for the 6 "generating hypotheses" items. Of course, this situation (unequal quantitative differences between, for example, a "1" and a "2") holds across all item groups. Figure 19.6 presents a schematic that summarizes this issue, which influenced the analysis of these data. Specifically, the schematic illustrates the potentially different ways in which the rating scale steps of 0, 1, 2, 3, 4, and 5 mark the latent trait.

At first glance, these data may seem different than the OSU data; however, the measurement issue present in the two data sets is identical. Since the increase in competence when moving from 2 to 3 for "Formulating Questions" items and "Planning an Investigation" items is not viewed as necessarily the same increment of increase or the same position along the spectrum of competence, then a Rasch

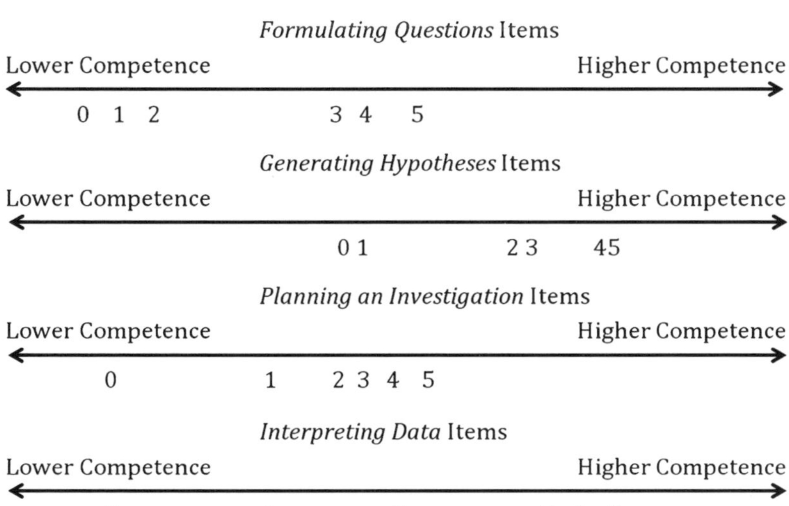

Fig. 19.6 A schematic showing the potentially different ways in which the rating scale steps of 0, 1, 2, 3, 4, 5 mark the latent trait "Competence in Scientific Inquiry"

analysis of these data must also use the control line "ISGROUPS=". In Fig. 19.7, we present part of the control file used for the analysis of this data set. We have included a few additional lines that were important for the analysis but do not directly relate to ISGROUPS. These additional lines will be useful to researchers as they evaluate real data. These new lines are discussed at the end of this chapter.

```
&INST
Title= "ISGROUPS T1-T2_ALL.sav"
ITEM1 = 1
NI = 24
NAME1 = 26
NAMLEN = 37
XWIDE = 1
CODES = 012345
ISGROUPS=AAAAAABBBBBBCCCCCCDDDDDD
STKEEP=Y
@FEMALE = $C49W1
&END ; Item labels follow: columns in label
A1A ITEM TYPE A
A3A ITEM TYPE A
A8A ITEM TYPE A
A11A ITEM TYPE A
A13A ITEM TYPE A
A16A ITEM TYPE A
B6A ITEM TYPE B
B7A ITEM TYPE B
B12A ITEM TYPE B
B13A ITEM TYPE B
B14A ITEM TYPE B
B17A ITEM TYPE B
C3A ITEM TYPE C
C7A ITEM TYPE C
C8A ITEM TYPE C
C9A ITEM TYPE C
C11A ITEM TYPE C
C16A ITEM TYPE C
D1A ITEM TYPE D
D3A ITEM TYPE D
D6A ITEM TYPE D
D12A ITEM TYPE D
D13A ITEM TYPE D
D17A ITEM TYPE D
END NAMES
....20.1...1..2......2..    1 1 ADJO05102 15 10 0 1 0 0 0 0 0
..1.1.1.2.......0.2.....    2 1 ADMI03 71 12  7 1 0 0 1 0 0 0
•

•
```

Fig. 19.7 A segment of the control file for evaluating the Trier data set

The form of the `ISGROUPS=AAAAAABBBBBBCCCCCCDDDDDD` line initially looks different than the ISGROUPS command line for the OSU data. However, closer inspection reveals that this apparent difference is due only to the organization of the Trier data set. A review of the item names reveals that the 6 items of type "A" are presented first, then the 6 items of item type "B," and so on. Therefore, the ISGROUPS line is organized by six-type A item, then six-type B items, and so on. We could have had an ISGROUPS line that had the following form: `ISGROUPS=WWWWWWKKKKKKPPPPPPQQQQQ`. The single important purpose is to tell the program which items correspond to which rating scale.

The two examples – the OSU data set with ratings of sophistication and the Trier data set with partial credit scores of competence – represent one situation in which researchers can and should not only make use of the Rasch model's capacity to correct for nonlinearity but also acknowledge that some rating scales in an analysis might not function in the same manner (jumps from say a "2" to a "3" might not be the same as in the location and appearance scale), but the data can still be used to compute a person measure.

A Third Example of Combining Scales

Risking a bit of "overkill" but hypothesizing that potential exists for reaching yet a deeper level of understanding, we present a third example of how ISGROUPS can be used. Of course, these education examples apply to research in other areas of education, medicine, and beyond.

In our third example, a number of clearly different rating scales are used with a set of items that involve the same trait. Again, we provide a sample control file as well as sample survey items. Readers will observe a mix of items that are often presented as part of an evaluation instrument. The technique that we discuss is one in which a mix of items can potentially be combined for an analysis of greater statistical power. In Fig. 19.8, we present a possible 17-item survey of attitudes toward science instructional techniques. Of course, the likely respondents would be teachers of science. For our purposes in this example as well as in this entire chapter, all items are viewed as being part of a single trait. However, in this example, five different rating scales (a two-step scale, a four-step scale, a five-step scale, and two different six-step scales) are used for the set of items.

Figure 19.8 presents a potential set of items that could be collected in an education research project that examines science teachers' attitudes toward various techniques for teaching science. All items involve a single trait, teaching science; however, a mix of rating scales is used for collecting the data. Our experience is that such surveys are commonly used in projects in many fields of research. Generally, presentation of a mix of items occurs when measurement specialists are not involved in planning the data collection at the onset of a project. As a result, a wide range of items is presented to respondents. Often, the project attempts to address multiple goals, and a number of items may be "pulled" from a published survey and inserted

Attitudes toward Selected Science Teaching Techniques

What is your level of agreement to the following three statements?

1-Most of a science lesson should consist of NOT lecturing.

Strongly-1 Agree	Agree-2	Neutral-3	Disagree-4	Strongly-5 Disagree

2-It is important for students to work in groups.

Strongly-1 Agree	Agree-2	Neutral-3	Disagree-4	Strongly-5 Disagree

3-Data analysis should be a central part of a science lesson.

Strongly-1 Agree	Agree-2	Neutral-3	Disagree-4	Strongly-5 Disagree

Should the following teaching techniques be used when teaching science?

4-Data Collection

Yes-1	No-2

5-Spread Sheets to Graph Data

Yes-1	No-2

How often should the following teaching techniques when teaching science?

6-Make use of computer probes

Always-1	Very -2 Often	Often-3	Sometimes-4	Rarely-5	Never-6

7-Make use of the National Science Standards

Always-1	Very -2 Often	Often-3	Sometimes-4	Rarely-5	Never-6

8-Make use of a colleague observing you and writing a summary of your science teaching

Always-1	Very -2 Often	Often-3	Sometimes-4	Rarely-5	Never-6

9-Make use of community resources

Always-1	Very -2 Often	Often-3	Sometimes-4	Rarely-5	Never-6

Fig. 19.8 A potential 17-item survey of attitudes toward techniques for teaching science that could be administered to teachers

into a "homemade" instrument. This type of procedure results in some data being collected to "document" how a project did or did not address its goals. Often the result is a mix of items that may be evaluated only at the item level (one by one), and items are viewed as not being able to be pooled together for a measure. However, when a set of items does involve the same trait, careful use of ISGROUPS opens the door to pool the items together for the computation of a single measure.

How important do you think the following are for teaching science?

10-Use 3 seconds of wait time when asking questions of students.

Very-1 Important	Important-2	Unimportant-3	Very-4 Unimportant

11-Use guided, scaffolded inquiry strategies

Very-1 Important	Important-2	Unimportant-3	Very-4 Unimportant

12-Use cooperative learning strategies

Very-1 Important	Important-2	Unimportant-3	Very-4 Unimportant

13-Perform discrepant event demonstrations

Very-1 Important	Important-2	Unimportant-3	Very-4 Unimportant

14-Lecture

Very-1 Important	Important-2	Unimportant-3	Very-4 Unimportant

15-Use frequent formative assessments

Very-1 Important	Important-2	Unimportant-3	Very-4 Unimportant

How often do you think science teachers should make use of the following resources?

16-Field Trips

Always-1	Very -2 Often	Often-3	Sometimes-4	Rarely-5	Never-6

17-Guest Speakers

Always-1	Very -2 Often	Often-3	Sometimes-4	Rarely-5	Never-6

Fig. 19.8 (continued)

Understanding the idea of a single trait can help researchers understand that *if* the same trait is being evaluated and *if* there is a way to mathematically address the issue of different rating scales, then it *is* possible to combine sets of items that involve different rating scales.

Following is Fig. 19.9, which contains a segment of the control file for this example. This control file could be used to evaluate the 17-item survey that includes 4 different rating scales. Item names are detailed with an item number (e.g., Q9), the number of rating scale steps for the item (e.g., 4S), the symbol used to identify the specific rating scale (just because an item has a rating scale of 4 steps does not mean it will have the same rating scale as all items with a 4-step rating scale), and the text of the item.

Figure 19.9 presents a control file for the analysis of the 17 items presented in the survey to respondents. The ISGROUPS control line indicates – as it should – that

```
&INST
Title= "Science Teaching Technique Evaluation"
ITEM1 = 1
NI = 17
NAME1 = 16
NAMLEN = 10
XWIDE = 1
ISGROUPS=AAABBCCCCDDDDDEE
CODES = 123456
&END
Q1-5S-A-Most of a science lesson should consist of NOT lecturing
Q2-5S-A-It is important for students to work in groups.
Q3-5S-A-Data analysis should be a central part of a science lesson.
Q4-2S-B-Data Collection
Q5-2S-B-Spread Sheets to Graph Data
Q6-6S-C-Make use of computer probes
Q7-6S-C-Make use of the National Science Standards
Q8-6S-C-Make use of a coll. obsers u & writing summary of your sci. teaching
Q9-6S-C-Make use of community resources
Q10-4S-D-Use 3 seconds of wait time when asking questions of students.
Q11-4S-D-Use guided, scaffolded inquiry strategies
Q12-4S-D-Use cooperative learning strategies
Q13-4S-D-Perform discrepant event demonstrations
Q14-4S-D-Lecture
Q15-4S-D-Use frequent formative assessments
Q16-6S-E-Field Trips
Q17-6S-E-Guest Speakers
END NAMES
54312651312142462
45522563643441415
55521651643442466
.
.
.
21522623631244256
```

Fig. 19.9 Part of the control file for evaluating science teaching

items 1, 2, and 3 use one particular rating scale, items 4 and 5 use a second rating scale, items 6–9 use yet another rating scale, items 10–15 use a unique rating scale, and items 16–17 also use a 6-step scale, but one that is different than that which was used for items 6–9. When performing a Rasch analysis, the two key assumptions that researchers must make – germane to the points we make in this chapter – are the following: (1) The rating scale steps for each type of item rating scale are potentially unique in spacing and (2) the set of items define a single trait.

Figure 19.10 displays a schematic of a potential spacing of the rating categories as a function of item type. Please note that the ISGROUPS line helps show that the two different 6-step rating scales (Scale C and Scale E as shown in the ISGROUPS line) should be viewed as different, unique scales. One 6-step scale does not equate to another 6-step scale.

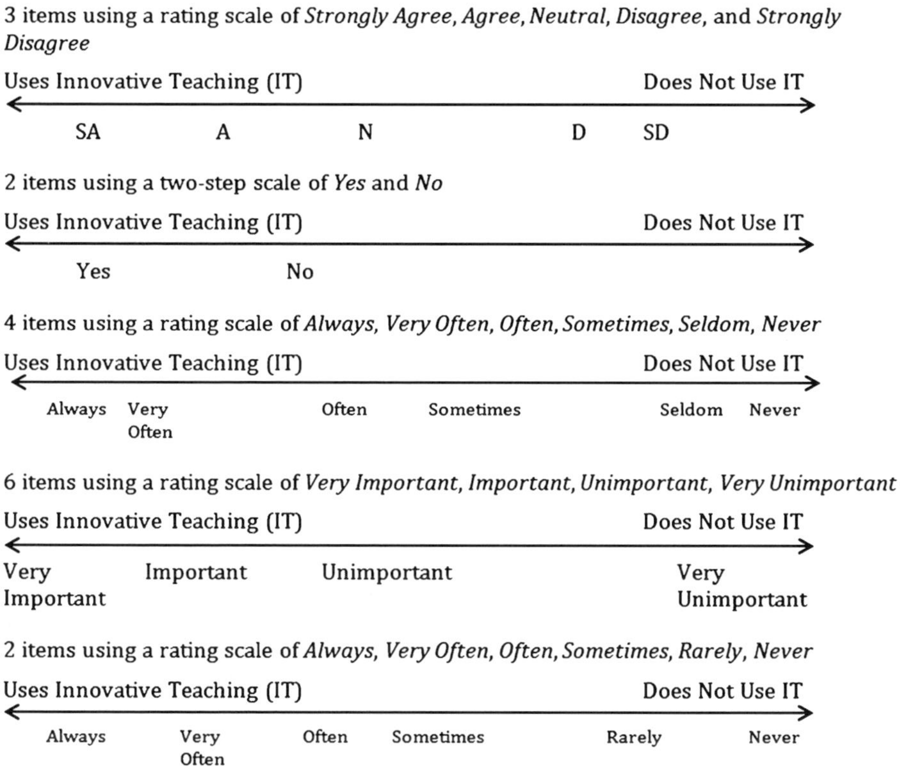

Fig. 19.10 A schematic display of potential spacing of the rating categories as a function of item type in the definition of a single trait

---

Formative Assessment Checkpoint #2

Questions: If survey data are collected using a number of different rating scales, then must the items only be evaluated one at a time? Must only items with the same rating scale be combined to compute measures?

Answers: No. If items involve the same trait, it is possible to use the command ISGROUPS (and the theory behind this command line) to potentially combine survey items in which more than one rating scale was utilized.

A Final Example

To close this chapter, we present one final use of ISGROUPS to combine items that may have rating scales that might define a single trait in a different manner. This example makes use of our old friend the STEBI (which has two subscales – the self-efficacy subscale and the outcome-expectancy subscale). As the developers of the STEBI detailed, the STEBI provides two measures. Let us pretend that specialists in self-efficacy (SE) and outcome-expectancy (OE) have detailed philosophical underpinnings that support the argument that the 13 SE items and the 10 OE items of the STEBI define a single trait called science teaching efficacy. The experts assert that the 23 items can be used together to provide a measure, but the experts also stress that the self-efficacy aspect of the 23-item scale may be different in some ways than the outcome-expectancy component of the scale. If this is the case, then Rasch techniques would facilitate an analysis of those data.

In Fig. 19.11, we provide a control file that could be used for the 23-item science teaching efficacy measure, if the two scales could be used together. Of particular importance is that the 13 SE items are designated as one type of rating scale structure and the 10 OE items are designated as potentially having a different rating scale structure.

If all items are viewed as defining a single trait (science teaching efficacy), then this control file would be used for an analysis of the 23 STEBI items. One caveat is that the rating scale might not be assumed to function in the same manner for OE and SE items, even though the same words are used to define the categories of the 6-step scale. This is what ISGROUPS allows us to do.

A Final Point

In this chapter, we have tackled a nuance of what is possible with Rasch measurement, more specifically, that when data are collected for a single trait, it is possible to make use of different rating scales for specific items. A number of different scenarios were presented. Perhaps the most common scenario involves presenting a number of items to a respondent using a number of items that differ in the rating scale used (our 17-item survey). In most cases, we have seen, even though many researchers are unaware of the nonlinearity of rating scales, the researchers are aware that they should not treat, for instance, the "1" of a yes/no scale, as the same as the "1" of a *Strongly Agree* to *Strongly Disagree* scale. However, most researchers do not know that under certain circumstances, it *is* possible to make use of items with differing rating scales. Just as the Wright Map opens up new worlds to researchers, we feel that the use and understanding of what can be accomplished through ISGROUPS will be greatly beneficial to researchers.

```
&INST
TITLE = 'IF IT WAS OKAY TO USE ALL 23 STEBI ITEMS TOGETHER'
NAME1 = 1
; The first column of data is the start of the person ID
; The name is 10 letters long
NAMELENGTH = 10
;
; The 11th column of data is the answer to the 1st item of the STEBI
ITEM1 = 11
; There are 23 items in total to the STEBI
;
NI = 23
CODES = "123456"
;
FORMAT=(10A1,1X,23(A1))
; This is an old fashioned way of reading data
; The "10A1" denotes the name of each data record.
; Since this is the first bit of information to appear
; in the format statement, it is information about the respondent
; name. Then one column of data is skipped. Then 23 items are read in
; succession.
;
ISGROUPS=BAABAABAABBABBBBAAAAAAA
; NAMING THE SE ITEMS AS ITEM TYPE A
; NAMING THE OE ITEMS AS ITEM TYPE B
;
&END
Q1oe
Q2se
Q3se-rc
Q4oe
Q5se
Q6se-rc
Q7oe
Q8se-rc
Q9oe
Q10oe-rc
Q11oe
Q12se
Q13oe-rc
Q14oe
Q15oe
Q16oe
Q17se-rc
Q18se
Q19se-rc
Q20se-rc
Q21se-rc
Q22se
Q23se-rc
END LABELS
21141    PR 46552655554254455545555
91052    PR 5645252533455566xxxxxxx
95793    PR 4665554654556554xxxxxxx
```

Fig. 19.11 Part of the control file for evaluating science teaching efficacy

Isabelle and Ted: Two Colleagues Conversing

Isabelle: *Okay Ted, help me, for once. I have this survey that was given to 10,000 teachers. But, in the rush to get it out, it has a really odd mix of items. By odd, I do not mean poorly written. The items involve one issue, but all sorts of different rating scales are used. I have 12 items that use a Likert scale (level of agreement) and another 7 items that use a frequency rating scale. I am not sure that the 7-item scale will result in useful person measures, for in looking over the data, I think there may be quite a bit of person measure error. What do I do?*

Ted: *I think you will be able to combine the items together, but the key issue is whether or not you think all the items all involve one trait. Do they?*

Isabelle: *I have thought this through, and I think I can make a strong argument that the items all involve one general issue.*

Ted: *Okay, if you give me the data, I can have them evaluated for you. I just need to know which columns of the data include the data for the 12-item scale and which columns include data for the 7-item scale. Then I will use the ISGROUPS command line in a Winsteps control file.*

Isabelle: *I have not used that line before; what does it do?*

Ted: *Basically, it tells the program which items to use as an identical rating scale. If we did not do this, then the program would think that all the numbers have the same meaning for the two rating scales.*

Isabelle: *I think I understand. So, I can take some very complex rating scale data in which, say, two different rating scales are used, and if the items all involve the same trait, then there might be a way to look at the data in the same control file and compute a person measure using items that do not have the same rating scale.*

Ted: *Right! There are some more complicated things that we can do, but that is right.*

Isabelle: *Just one more question … tell me how this might be used for tests?*

Ted: *Well, for example, my colleague in Trier was involved in collecting this huge data set. Kids took a range of test items, and kids could get scores ranging from 0 to 5 on test items. So, they could get partial credit on items. But the real kicker was this: There were 4 types of items across the 24 items. This meant that the meaning of improving from, say, a partial credit grade of 3 of possible points (pre) to 4 of 5 possible points (post) did not have the same exact meaning for all items.*

Isabelle: *Uh, say that again.*

Ted: *Here goes. Remember when you first helped me start to use Rasch in my research? You really stressed that I must not be tricked by numbers. You helped me appreciate that to move from 1 to 2 to 3 on a rating scale did not mean the jump from 1 to 2 is necessarily the same size as the jump from 2 to 3. This was because the numbers were only labels to indicate a rating scale category that was selected. The same is true for a partial credit test. Moving from earning 2 points to 3 points on an item is an improvement, but we cannot assume that moving from 2 to 3 is the same amount of improvement as moving from a 3 to a 4 on the partial credit point earning for that same item.*

Isabelle: *Okay, I get that. But, why wouldn't we just use an edited version of the STEBI control file to evaluate that data?*

Ted: *This is the cool thing. We need to remember that the meaning of moving from, say, a 2 to a 3 for one item type in the German data set may not mean the same amount of movement from a 2 to a 3 for another item type. We know that an increase in partial credit score is indeed better for each of the 24 items, but we do not want to assume that the jump from one point value to another is the same meaning for all item types.*

Isabelle (*interrupting*):*...so that is what we tell Winsteps in the control line with ISGROUPS?*

Ted: *Exactly.*

Isabelle: *I guess you are making me do what I always have you do!*

Ted: *What's that?*

Isabelle: *Think about what the numbers we collect mean and do not mean.*

Keywords and Phrases

ISGROUPS
Trait
Rating scale

Potential Article Text

As a component of a large national effort to improve science learning and teaching, a total of 2,345 students completed a set of competence assessments in a random sample of German Länder. Students completed both a pre- and post-assessment developed to measure the general construct of competence. The assessment consisted of 24 partial credit test items classified as tapping one of four central components of competency. Each of the four components involved the single construct of competence.

Masters' partial credit model (Wright & Masters, 1982) was employed to calibrate and prepare the data for further statistical analysis. The Masters' partial credit model is a specific example of a Rasch model. Rasch analysis facilitates the computation of linear person measures using ordinal data (e.g., partial credit data). Furthermore, due to the careful use of the Rasch model, respondents could complete different combinations of test items and still be measured on the same scale. The Winsteps program (Linacre, 2011) was utilized for analysis. One key analysis step taken for the evaluation of this data set was acknowledgment that the partial credit scale of each of the 4 item types (although all used to denote improvement in the same direction along the same unidimensional construct) did not necessarily represent the same amount of movement along the trait as a function of item type. A change from 1 point to 2 points along one of the competence scales did not necessarily mean the same amount of movement along the competence scale as for each of the other three competence scales. Rasch analysis allowed for corrections to be made in such a manner as to facilitate confident computation of person measures for parametric statistical analyses.

Quick Tips

Use the command ISGROUPS to specify the different types of rating scales that you wish to combine. For example, the command line ISGROUPS=ABCABCABC means that a survey (in the case of data in this chapter) has (1) items 1, 4, and 7 that have a rating scale in common; (2) items 2, 5, and 8 that have a rating scale in common; and (3) items 3, 6, and 9 that have a rating scale in common.

Data Sets: (go to http://extras.springer.com)

cf 23 items
cf competency data

Activities

Activity #1

Task: A survey has been developed which consists of 15 survey items. Items 1–5 use a 3-step scale, items 6–10 use a 4-step scale, and items 11–15 use a 3-step scale. However, the two 3-step scales do not use the same words to label each rating scale step. Author a control file and supply fake data for two respondents. Make sure to add comments to your control file to explain, as best you can, what each line does.

Answer: Below we present one possibility.

```
&INST
; Here is a line that puts a title
Title= "Activity 1 Control File"
; This next line tells the program that the first column
of data is for
; the first survey item
;
ITEM1 = 1
; This line tells the program that one has 15 items
NI = 15
; This line tells the program that the 16th column of data
; is the start of the person ID, this means the 16th column
; is not response data!
NAME1 = 16
; This line tells the program that the person ID infor-
mation, which
```

```
; starts in column 16 is a total of 4 columns wide. That
would mean that
; the last column of person ID data is in the 19th column.
;
NAMLEN = 4
; This command tells the program that each piece of data
is 1 column wide.
; Each answer to an item takes up only one column.
XWIDE = 1
; This is the line that tells the program that items 1-5
are one rating ;scale, items 6-10 are another rating
scale, and items 11-15 are another
; rating scale.
;
ISGROUPS=GGGGGSSSSSNNNNN
; This line is telling the program that the only valid
entries for ratings
; are the numbers 0, 1 and 2.
CODES = 012
&END
Q1-Rating type G
Q2-Rating type G
Q3-Rating type G
Q4-Rating type G
Q5-Rating type G
Q6-Rating type S
Q7-Rating type S
Q8-Rating type S
Q9-Rating type S
Q10-Rating type S
Q11-Rating type N
Q12-Rating type N
Q13-Rating type N
Q14-Rating type N
Q15-Rating type N
END NAMES
012210012120012
122210101010002
```

Activity #2

Task: We provide a control file entitled "cf 23 items." This control file contains STEBI data. We have provided the ISGROUPS line in the control file as if OE and SE items can be viewed and marking parts of the same trait. First, verify that the ISGROUPS line is correctly authored. Second, run Ministeps with this control file to confirm that the program runs and that person measures can be computed.

Answer: The program does indeed run. This is an analysis of the STEBI data as if the OE item rating scales and the SE item rating scales should be viewed as not necessarily the same, even though both scales contain the same number of steps and use the same words to label rating scale categories.

Activity #3

Task: We have provided a control file with a small portion of the competency data collected by our colleague Andrea Moeller (cf competency data) and colleagues. First, review the file and make sure that you understand the data layout as well as each command line. Pay particular attention to the ISGROUP line. Then run the control file to verify that you are able to compute person measures.

Answer: The same translation of each command line is present for this control file. Readers will see that not all items were answered by all respondents. This is because a so-called multimatrix design was used to collect data. As long as common items link respondents, and one is collecting data with regard to one trait, then it is possible for respondents to only be presented with a subset of items. The presentation of an item subset to respondents can be seen in the missing data noted with the dots (.) in the data portion of the control file.

References

Wright, B. D., & Masters, G. (1982). *Rating scale analysis*. Chicago: Mesa Press.

Linacre, J. M. (2012) Winsteps (Version 3.74) [Software]. Available from http://www.winsteps.com/index.html

Additional Readings

Read the section of the *Winsteps Manual* devoted to the command line ISGROUPS.

Linacre, J. M. (2012) Winsteps (Version 3.74) [Software]. Available from http://www.winsteps.com/index.html

Chapter 20
Multifaceted Rasch Measurement

Isabelle and Ted: Two Colleagues Conversing

Ted: Isabelle, I have heard that Rasch is being used by some companies and for a number of high-stakes medical exams, and I wondered if some of those applications might be germane to analyses we could conduct in education?

Isabelle: Absolutely. Medical candidates respond to questions in many high-stakes medical exams just like students who take tests or surveys. An added element in medical credentialing involves judges who evaluate the performance of the candidates on specific test items. These medical groups realize that, in order to protect the public and the candidates, it is important to correct for differences in the severity of judges.

Ted: That is quite different compared to our approach in education research. We try to train all judges to act in the same way. Why don't they just try that?

Isabelle: These groups discovered that it is better for each judge to be consistent. Experts in a field should not be viewed as individuals whose long held views can be trained to be homogeneous with other experts. So, a tough judge should be consistently tough on all candidates, and an easy judge should be consistently easy for all candidates. Using Rasch techniques, medical officials can correct for the mix of judges each candidate receives. So a candidate is not rewarded if she or he gets an easy judge and is not penalized for getting a tough judge. In summary, when Rasch techniques are applied in which candidates are evaluated by a pool of judges, the quality of their analysis is greatly improved. An entire book could be written on this type of analysis, but if one has a basic understanding of Rasch, one can begin to use these techniques and greatly improve the analysis of specific types of data.

Immediately below is a common scenario for some tests in the USA:

Biology test ($n = 25{,}000$)
25 multiple-choice items (graded by computer)
1 long essay (graded by one of 100 trained judges)
1 short essay (graded by one of 100 trained judges)

The judges who are randomly assigned to grade the essays exert a great impact on student scores. For any individual student who may pass or fail the test, the impact of judges' scoring can be profound. Decisions that come to mind are admission to a desired college or university, to medical school, or to law school. Judges therefore

W.J. Boone et al., *Rasch Analysis in the Human Sciences*,
DOI 10.1007/978-94-007-6857-4_20, © Springer Science+Business Media Dordrecht 2014

typically receive extensive training on scoring essays in the same manner, and essays may be graded a second time by a second judge.

Prior to Rasch measurement, scores and ratings on the types of tests described above were flawed not only in terms of the nonlinearity but also in terms of the steps taken (and not taken) to correct for the variance in the scoring of different judges. In this chapter, we will demonstrate how specific Rasch software developed by Mike Linacre, called *Facets*, can facilitate a Rasch analysis of the type of data described above. Think of Facets as enabling one to conduct a Rasch analysis when one has items, respondents, and a third type of data – that being data from judges. Many credentialing groups have used Rasch *Facets* software to address the types of measurement issues already been discussed herein and also to correct for differences in judges' severity in scoring. Numerous research problems exist in education and other research fields that could benefit from Rasch measurement and *Facets* software. We introduce this analysis technique in the context of a type of data in education research. Such data are produced when preservice science teachers become "judges" as they evaluate a sample of teachers who have been videotaped. In this scenario, the teacher–judges use a rating scale and a number of survey items.

The Problem

To begin, we find it helpful to consider the use of judges in the real-world setting of the Winter Olympics (this type of example has been used by our colleagues Mary Looney (1996, 2004), Mike Linacre, and Ben Wright). Most of us are familiar with judging men's and women's figure skating, although we might not know much about the sport. The final event in Olympic figure skating is the long program. Until 2006, when this scoring system was altered, skaters were evaluated by a number of judges with regard to two traits (technical merit and presentation) on a 6.0-point scale. The judges for each skater's long program produced a technical merit rating and a presentation rating. These ratings ranged from a low of 0.0 to a high of 6.0. When a total score was computed for a contestant, Olympic officials attempted to correct for easy judges and tough judges by dropping the highest and lowest score from the panel of judges. There was an appreciation (at some level) that "odd" (extreme) judges could impact the overall composite (from all judges) rating a skater would receive, and there was an attempt to "correct" for these odd or extreme judges.

Formative Assessment Checkpoint #1

Question 1 (True/False): It is possible to train all judges to be the same.

Question 2 (True/False): It is advantageous to train all judges to be the same.

Answer 1: No.

Answer 2: No. It is neither possible nor advantageous. Training all judges to be the same is like marking or cutting a meterstick in the same spot over and over. As Mike Linacre (person communication with the authors, 2013) has indicated, "we cannot train the judges to have the same leniency, but we can train the judges in the technical meaning of the rating categories, the mechanics of the judging operation, etc."

Each judge's technical rating and artistic ratings are considered as two rough total measures of a skater's technical skill and artistic skill in performing elements (e.g., jumps, spins, step sequences). In essence, each judge views the contestant's long program and rates the performer on numerous skills, where each skill rating can be viewed as a single survey item for a single trait. Each judge then marks his or her ratings for all parts of the "technical skill construct" and then produces a total score. There is a similar procedure followed for artistic merit, but to explain Facets (and the supporting multifaceted Rasch theory), we will focus only on what is taking place with the judging of technical merit! To better match the education example we presented above (preservice teachers viewing videos of 5 teachers and the preservice teachers using numerous criteria to evaluate each tape's teacher), we pretend that each preservice teacher–judge uses a scale of 1, 2, 3, 4, 5, where "5" is the top performance rating. Note that although Olympic judges may keep a tally of deductions, any judge's total score is not easily predicted. Humans always disagree in appraisals, which is seen when judges' scores are presented for a skater. Rarely do all judges provide the same rating. In Fig. 20.1 are the judges and hypothetical scores for a skater.

Judge	Performer	Tech Skill 1	Tech Skill 2	Tech Skill 3	Tech Skill 4	Total
Tweed	Cool	4	3	5	1	13
Adams	Cool	3	2	4	4	13
Ludlow	Cool	2	2	2	1	7
McAlpin	Cool	1	4	2	2	9
Stern	Cool	5	5	5	4	19
Maisel	Cool	1	1	1	1	4
Tweed	Airy	4	4	5	2	15
Adams	Airy	4	3	5	5	17
Ludlow	Airy	3	3	2	2	10
McAlpin	Airy	2	5	3	3	13
Stern	Airy	5	5	5	5	20
Maisel	Airy	2	2	3	2	9

Fig. 20.1 Skill ratings for a skater (The ratings for 6 judges who evaluate the performance of a skater (Cool) with respect to 4 different skills. The judges use a 5-step scale. One can observe the impact of differing judge severity, different skater skill, and differing skill difficulty)

A MFRM Analysis

Our colleague and friend Professor Scott Townsend of Eastern Kentucky University (USA) supplied the data set for this chapter. As part of Dr. Townsend's interest in improving his science methods class, he had his students evaluate DVDs of a number of teachers teaching different science lessons. For the evaluation of lessons, students utilized the Elementary Science Teaching Analysis Matrix (ESTAM) (Gallagher & Lindsey, 1997). This scale consists of 24 items that students answer using a 5-step scale: (1) didactic/teacher-centered, (2) hands-on/student-centered, (3) conceptual, (4) constructivist, and (5) constructivist/inquiry. Therefore, a higher number means more constructivist teaching and a lower number means less constructivist teaching in a teacher's lesson. A total of 150 preservice teachers completed the ESTAM for all 5 DVD science lessons. A total of 17,250 ratings were possible if all preservice teachers marked a rating for all items for all 5 teachers (150 preservice teachers × 5 teachers rated × 23 ESTAM items = 17,250 ratings). Figure 20.2 presents a comparison of the figure skating example, our education example, and a market research example.

Still more scenarios lend themselves to MFRM. For example, classroom teachers are often evaluated by colleagues or supervisors with a rating scale or checklist. In this example, the judge is the colleague/supervisor, the contestant is the classroom teacher, and the items are the criteria used to rate the teacher. Another example could of course be what takes place in market research. For example, 20 individuals (the judges) are hired to compare 5 different brands of a product, perhaps peanut butter (the contestants), using 12 different criteria (the items).

We mentioned earlier that MFRM is often used by Medical Certification Boards in the USA. The way in which Medical Boards use MFRM parallels how MFRM can be used in education. Furthermore, the rationale for using MFRM in medicine should, in our opinion, be identical for using MFRM in education. In many cases, medical board certification requires expert judges to evaluate a candidate. In one scenario, a candidate might be asked to submit a diagnosis based on interpretation of pathology slides. Each slide can be viewed as a test item for which the candidate might receive a score of totally correct (2), partially correct (1), or not at all correct (0). The slides would normally be evaluated by a number of independent judges, each alone in a room, evaluating each slide with a grading scale. Previously, great efforts would have been taken to train all judges to act in

	Ice Skating	Education	Market Research
Facet 1:	Judges	Preservice Teachers	Customers
Facet 2:	Skaters	Veteran Teachers on DVD	Products
Facet 3:	Technical Skills	ESTAM Items	Rated Product Characteristics

Fig. 20.2 Three scenarios in which RFRM should be used to take into account, among many issues, differences in judge severity

an identical manner, almost as if they were robots. A Cohen's kappa would have been computed to measure the inter-rater reliability. Depending upon the number of slides graded, all judges might have graded all slides. A subset of judges might have been randomly assigned to evaluate the slides of specific candidates. Still other scenarios are possible.

Formative Assessment Checkpoint #2

Question: When using judges as raters, is it sufficient for a researcher to compute a Cohen's kappa in order to establish the quality of the judges' data?

Answer: No. Among many things, Cohen's kappa uses raw data, which are nonlinear. Mike Linacre (personal communication, March, 2013) also mentions that "Cohen's Kappa treats the ratings as nominal, and compares the frequency of observed agreement between the judges with the frequency of agreement expected by chance."

Also, in rare settings, the judges might act in unison, but it is far better to correct for differences in judge severity.

Why Use MFRM?

Why did medical boards adopt MFRM, and what are the important parallels for rigorous research? Medical boards evaluate candidates for two primary goals. Goal 1 is to protect the public. Most importantly, inferior candidates should not receive board certification. Goal 2 is to ensure that the board is fair to all candidates. For instance, a competent candidate should not fail because he or she had the statistical misfortune to be randomly assigned the toughest judges.

Medical boards also adopted MFRM for a number of additional reasons that we presented and discussed at length earlier herein. First, any rating is ordinal and therefore nonlinear. Only through Rasch measurement (when the data fit the model) can ordinal data can be expressed on a linear, equal-interval scale. Second, different forms of a test can be linked. For example, suppose that a new high-stakes test is designed for medical board certification in Fall 2012. As data are collected and examined, the test may be altered for Fall 2013 as well as for each Fall beyond 2013 in light of new data. Despite Herculean efforts to make the test of similar difficulty, attaining an exact match is not realistic. If the Fall 2013 test is harder than the Fall 2012 test, then, to be fair to and to protect the candidates, the lowest passing raw score should be lower than for the Fall 2012 test. Similarly, if the Fall 2013 test is easier than the Fall 2012 test, then, to protect the public, the lowest passing raw score should be higher for the easier test. Third, logistics and cost must be considered. Usually, a small number of judges evaluate a large number of candidates. Judges are not only rare because of the limited time they can dedicate to a project, but also they are expensive. Readers should recall that, given the measurement properties of the Rasch model, students who are evaluated with regard to a construct need not take the same set of items. As long as some identical items are presented, groups of test takers can be measured on the same scale. By using a multimatrix design, we can administer a number of different short tests. In Germany, researchers have collected large data sets using tests of reasonable length, but not all students complete the same test items. This, in part, allows for a range of items to be presented to the cadre being tested. Developing the plan for the creation of the test booklets takes some time, but in the end a shorter amount of time is required for administering the tests, and valid/reliable measures can be computed. This, of course, pleases teachers and administrators and also limits the fatigue of students. Taken together, these factors can engender an increase in the quality of the data. What does this have to do with our examples concerning judges? First, because judges can be expensive and have limited time, using MFRM facilitates an advantage, in that all judges need not evaluate all candidates. Second, a multimatrix design of sorts needs to be developed to successfully link all candidates on the same scale, but when this design is used, not as many judges are needed. Third, Rasch measurement can be used when not all students take all test items. Due to the measurement properties of the Rasch model, test takers are not penalized for taking a harder set of items, nor are they rewarded for completing an easier set of test items. Also if a judge cannot longer "judge" due to getting sick, the candidate will not be rewarded or penalized.

The Data Set

Recall that the data contain the ratings to the 24-item ESTAM from the 150 preservice teachers who each evaluated 5 teachers. The preservice teachers evaluated five science teachers by watching digital recordings of their science lessons. Therefore, the preservice teachers were the "judges," the "candidates" were the five teachers, and the items were the "ESTAM items."

To run an MFRM analysis, one must use Rasch *Facets* software. A version of this software (Minifac) is available for readers as provided by Mike Linacre. This free version of the program does not permit readers to conduct all the analyses that are possible with the full version of *Facets*, but a sufficiently thorough analysis can be conducted to complete a very good beginners MFRM analysis.

Below readers can see that the *Facets* file for our data looks similar in structure to the Winsteps files. We used a *Facet*s file provided in Bond and Fox (2007) as a skeleton for our file rather than to write a file from scratch (Fig. 20.3).

Certain parts of the control file tell the program the location and type of data, just as is the case for the control files presented earlier herein. Such lines extend from the first line to the line that begins with the word "Model." Following this line is a line that begins with the word "Labels." The line that follows begins with "1" and a

```
; Group Project Data File for Facets
Title =
Facets = 3; three facets: judges(methods student judges), examinee(sampleteachers)items24ESTAM
traits)
Inter-rater = 1        ; facet 1 (methods student judges) is the rater facet
Positive = 2   ; examinees (teachers on trape being evaluated) have greater creativity with
greater score
Non-centered = 1       ; examinees and items are centered on 0 logits, judges are allowed to
float
Model = ?B,?B,?,R9     ; judges, examinees and items produce ratings with maximum rating of 5.
    ; A bias/interaction analysis, ?B,?B will report interactions between facets 1 (methods
students) and 2 (teachers on tape)
Labels =
1, Methods students who are judging   ; name of first facet: judges
LINES REMOVED FOR THIS EXAMPLE

98459=
98459=
98459=
98459=
98459=
72564=
72564=
72564=
72564=
72564=
18099=
18099=
18099=
18099=
18099=
83879=
83879=
83879=
83879=
83879=
*
2, Teachers on Tape   ; name of second facet: the teachers on the tape
 1 =All Sorts of Leaves                ;
 2 =Water Purification          ;
 3 =Completing the Circuit        ;
 4 =Force and Motion            ;
 5 =Water Cycle                 ;
*
3, Traits              ;
 1 = C1                ;
 2 = C2                ;
 3 = P1                ;
 4 = P2                ;
 5 = TVA1
 6 = TVA2
 7 = TVA3
 8 = TVA4
 9 = TVA5
 10 = SA1
 11 = SA2
 12 = SA3
 13 = SA4
 14 = A1
 15 = A2
 16 = A3
 17 = A4
 18 = A5
 19 = ER1
 20 = ER2
 21 = ER3
 22 = ER4
 23 = ER5
 24 = ER6
*
Data=
98459,1,1-24,4,4,3,5,5,4,4,5,4,5,5,5,4,4,5,5,5,4,4,4,3,4,2,4
98459,2,1-24,4,3,3,4,2,3,3,2,3,3,5,3,3,4,3,3,4,3,4,3,3,3,1,3
98459,3,1-24,2,2,2,1,1,2,2,2,2,2,2,2,1,1,2,2,2,1,1,1,2,2,1,1
98459,4,1-24,5,5,4,5,5,5,5,4,4,5,5,4,4,5,5,4,5,5,4,4,5,5,4,5
98459,5,1-24,1,2,1,1,2,1,1,2,2,1,2,1,1,1,2,1,2,1,1,1,1,2,1,1
72564,1,1-24,5,4,5,4,5,4,5,4,4,5,4,4,4,4,4,4,5,4,4,4,5,5,4,4
72564,2,1-24,4,4,4,3,5,4,5,4,4,4,4,4,3,4,5,4,5,3,4,4,4,5,3,5
72564,3,1-24,4,4,3,4,4,4,5,4,5,5,5,4,4,5,4,3,5,4,4,3,4,5,3,4
72564,4,1-24,5,5,4,4,5,4,5,4,5,5,4,4,4,4,3,4,4,4,4,4,4,4,3,4
72564,5,1-24,4,4,4,3,4,5,4,3,4,3,4,4,4,4,3,4,5,3,5,3,4,4,4,4
18099,1,1-24,4,4,4,4,4,3,4,3,4,4,4,3,4,4,4,3,4,4,4,3,4,4,2,4
18099,2,1-24,4,3,4,4,3,3,3,4,4,3,4,4,4,3,4,3,4,4,4,4,4,3,2,4
18099,3,1-24,4,4,4,3,3,4,4,4,3,5,5,3,4,5,5,4,4,4,4,4,3,5,3,4
18099,4,1-24,3,3,4,3,3,4,4,4,5,5,4,3,4,5,5,4,5,5,1,5,4,5,2,5
18099,5,1-24,3,3,3,2,3,3,3,3,3,2,3,2,3,3,2,2,3,3,4,4,3,3,3,2
83879,1,1-24,5,4,4,4,5,5,4,4,4,3,4,4,4,3,4,4,4,3,1,4,5,4,3,4
83879,2,1-24,4,3,3,3,2,2,3,2,3,3,5,1,1,3,2,3,3,2,3,4,4,3,2,4
83879,3,1-24,5,5,5,5,5,5,5,5,5,5,4,5,5,5,5,5,5,5,4,5,5,5,4,5
83879,4,1-24,2,3,3,3,4,4,3,3,3,5,5,2,4,3,3,4,5,3,4,3,5,5,2,2
83879,5,1-24,4,3,3,5,4,5,5,4,4,3,4,4,4,5,5,4,4,4,5,3,4,4,5,5
```

Fig. 20.3 Group Project Data File for Facets

comma. This tells the program that the following information is about the first "facet" of the data set (the Facets in our example will be judges, items, and those being judged). In our data set, the first facet will be the judges. Below this line (the line starting with the word "Labels") in the FACET control file, readers will indeed see this judge information, namely, the ID codes we created for each of the judges (the preservice teachers who evaluated the teachers using the ESTAM). Readers should note that each judge appears five times because the data are presented one line at a time for each judge's evaluation of each teacher. Thus, if one of our preservice teachers is named Judy, then we will see a line of data for Judge Judy with regard to the lesson of science teacher A, then a line of data from Judge Judy's evaluation of teacher B, and so on. In this data set, the judges are indeed similar to judges at an Olympic figure skating competition. Judges are evaluating performers using a range of criteria. Moreover, the structure of the data and the question that our colleague Scott Townsend wanted to answer resulted in an MFRM structure in which the preservice teachers are viewed as judges.

Following the presentation of the IDs (5 times per judge) is a section of the control file in which the second facet is described. In this data, the second facet is composed of the five science teachers who were evaluated via DVD. We named each teacher by using the topic each teacher taught. This facet is analogous to the figure skaters of our figure skating analogy. We have added a comment in our Facets control file to highlight the start of the information for each facet.

The third part of the control file presents the third facet, which are the items of the ESTAM. These items are analogous to the different aspects of skating that the Olympic judges consider when evaluating a performance. In our data, we can see the rating of each preservice science teacher (judge) for each survey item (trait) for each teacher (skater).

The fourth and final part of the Facets control file is the raw data. In this data set, a line of data is provided for each preservice teacher's (judge's) evaluation of each teacher (skater) with regard to all ESTAM items (skating criteria).

The form of these data lines is simple; individual pieces of information and data are separated by commas. Figure 20.4 provides five data lines that are presented in the control file. This is the data for one judge (judge 16599).

The first piece of information is a unique judge ID (in Fig. 20.4 that is judge "16599"). Next is the assigned number of the teacher who was evaluated (in

```
16599,1,1-24,5,5,5,5,5,4,5,5,4,5,5,3,4,5,5,5,5,5,4,5,5,5,2,5
16599,2,1-24,4,3,3,3,3,4,3,4,3,3,4,3,4,3,2,3,3,2,4,3,3,4,2,4
16599,3,1-24,5,4,4,5,5,4,5,5,4,5,4,3,4,5,5,5,5,5,4,5,4,5,4,5
16599,4,1-24,5,4,5,5,5,5,5,5,3,5,5,3,5,5,5,5,5,5,4,3,4,5,2,3
16599,5,1-24,1,2,1,1,1,1,1,1,1,2,2,1,1,3,2,1,3,3,1,1,1,2,1,1
```

Fig. 20.4 Five lines of data in the control file (A judge ID is first presented (16599) and then a number to indicate the ID of the teacher being evaluated (teachers 1, 2, 3, 4, or 5). Following this information is a code to indicate that 24 items were answered in each rating of each teacher by the judge (1–24). Finally a long list of the 24 items is provided)

Fig. 20.4, this is one of five evaluated teachers). The "1–24" shows that 24 items were presented to the preservice teachers (judges). Last are the item data, the responses provided by the preservice teachers to items 1–24. Each of the 24 survey items could be evaluated using a 5-step scale; thus, each of the ratings is a 1, 2, 3, 4, or 5. Once a control file is completed, the *Facets* program runs in a fashion similar to Winsteps. Although some differences exist, the tables and the output of Facets are also similar to those of Winsteps. Below are directions for running Facets, a brief discussion of tables and Wright Maps produced by *Facets*, and a discussion of the implications of these results for researchers within and beyond science education.

To Run Facets

1. To begin, double click on the Facets icon.
2. From the menu, select the option "Files."
3. Then select from the option "Specification File Name." This tells the program the name of the control file.
4. Then find your control file and select it for the analysis. To select the control file, first click on your file when it appears in the window provided by Facets.
5. Then click on the button named "Open." A box containing a number of colored squares will appear on the screen. The first square is green and contains the word "OK." When we run a basic Rasch analysis, we simply click the "OK" box. Since this chapter is only an introduction to an MFRM analysis, we have decided to keep our analysis very simple.

   The program takes your complete control file name and adds "out" to the end of the file to remind you that the file is an output file. Although this works, we also insert "out" at the start of the file name. We name our files in such a way because naming the file with "out" at the end of the name can be confusing, as sometimes you may think you have two control files with the same name.
6. After naming your output file, click the "Open." Then, the program will begin to run.

Interpreting and Using MFRM Output for a Research Project

Before we examine the output and then explain how these data helped Dr. Townsend in his science methods class, we want to review some important points. First, this Facets analysis provides measures for the 150 preservice teachers (judges), measures for the five teachers (skaters) presented in the five DVDs, and measures for each ESTAM item. All measures are expressed on the same linear logit scale, which means that the results are not biased by the use of raw data. As a result, parametric statistical tests can be carried out with confidence using any of the measures.

Also, the same equal-interval measurement scale in logit units is used to express judges, the 5 teachers being evaluated, as well as the 24 survey items. Almost all of the techniques that we have presented for Winsteps can be applied to this data set. Because one has items, judges, and those being evaluated, there may be some applications that might not be immediately apparent to researchers that one can now use. For example, questions such as the following can be answered: (1) Do female and male preservice teachers evaluate the 5 teachers in different ways (excluding severity)? Is more misfit observed among male judges? (2) Are male preservice teachers easier or more severe judges than female preservice teachers?

Before moving on to some nuts and bolts of specific analyses of output from Facets, let's consider some global measurement issues that link immediately to what we have seen in Winsteps. The Facets analysis provides Wright Maps for preservice teachers (judges), teachers on DVD (skaters), and ESTAM survey items (skating criteria). Most Wright Map techniques discussed in earlier chapters can be used. For example, the functioning of the ESTAM can be evaluated so that new items can be added to enhance measurement precision or confidently removed to reduce measurement redundancy and lessen time for the completion of the instrument.

The analysis also provides summary tables similar to those presented in Winsteps (e.g., item entry table, person entry table). In this analysis, the key tables are the equivalent of the entry measure tables in Winsteps, preservice teachers (judges), teachers (skaters), and ESTAM items (skating traits). The key issue to recall is that values reported for the measures of preservice, teachers, and ESTAM items are not only corrected for the issue of ordinal raw data rating scales but also take into consideration judge severity, item difficulty, as well as an understanding that all teachers presented on the DVD represented teaching that was not identical with regard to the overall trait being measure by the ESTAM (the trait of constructivist teaching).

Preservice Teacher Facet (aka Olympic Skating Judge)

Let's look at one of the output tables. Figure 20.5 presents the measures of the preservice teachers in terms of how tough or lenient they were as judges.

The title of the table was entered into the control file and informs the analyst that this is the table for the judge facet. The key column for a basic analysis of the data involves finding the preservice science teachers (judges) in the table. To locate the first judge, look at the far right hand side of the top row of the table and find the number "2669." This is the ID of a judge. Next, look to the left side of the table and find the column labeled "Total Count." This column reports the number of survey items answered by the preservice science teachers (judges). The maximum number is 120 because each preservice teacher could evaluate up to five (5) teachers using the 24-item ESTAM ($24 \times 5 = 120$). Lower numbers in this column represent data in which not all items were answered. Fortunately, part of the beauty of a thoughtful Rasch analysis is that missing data do not present serious problems, whereas missing data do present serious problems in a traditional analysis.

Table 7.1.1 Methods students who are judging Measurement Report (arranged by mN).

Total Score	Total Count	Obsvd Average	Fair-M Avrage	Measure	Model S.E.	Infit MnSq	Infit ZStd	Outfit MnSq	Outfit ZStd	Estim. Discrm	Correlation PtMea	Correlation PtExp	Exact Agree. Obs %	Exact Agree. Exp %	Num Methods students who are judging
27	24	1.1	1.73	1.46	.55	.73	-.1	.59	-.4	1.04	.38	.07	40.1	36.8	2669 2669
27	24	1.1	1.73	1.46	.55	.76	-.1	.67	-.2	1.02	.24	.07	39.7	36.8	9444 9444
269	118	2.3	2.13	1.00	.10	.79	-1.8	.71	-2.0	1.27	.71	.61	24.5	22.0	84761 84761
302	118	2.6	2.47	.67	.10	1.61	4.1	1.48	2.9	.45	.65	.65	28.4	25.4	43660 43660
78	24	3.3	2.52	.63	.21	.39	-2.7	.38	-2.7	1.67	.45	.26	28.7	24.8	943 943
79	24	3.3	2.57	.58	.22	.28	-3.5	.28	-3.5	1.88	.06	.26	25.9	25.4	9440 9440
317	120	2.6	2.57	.57	.10	.55	-4.2	.52	-4.1	1.60	.81	.65	26.7	26.2	15627 15627
324	120	2.7	2.65	.50	.10	.61	-3.6	.60	-3.3	1.26	.81	.66	29.2	26.9	84830 84830
327	119	2.7	2.70	.45	.10	1.01	.0	1.00	.0	1.01	.72	.67	30.4	27.4	77311 77311
339	120	2.8	2.81	.35	.10	1.18	1.4	1.10	.7	1.09	.82	.67	33.7	28.2	56661 56661
343	120	2.9	2.85	.31	.10	.36	-6.7	.38	-6.0	1.50	.80	.67	32.0	28.6	60081 60081
344	120	2.9	2.86	.30	.10	.91	-.6	.92	-.6	.91	.74	.67	28.1	28.6	39861 39861
348	119	2.9	2.93	.23	.10	.57	-3.9	.57	-3.7	1.53	.86	.68	36.7	29.2	98487 98487
358	120	3.0	3.00	.15	.10	.95	-.3	.92	-.5	1.08	.84	.68	34.1	29.7	98459 98459
361	120	3.0	3.03	.12	.10	.93	-.4	.88	-.9	1.24	.81	.68	37.3	30.0	26902 26902
362	120	3.0	3.04	.11	.10	1.45	3.1	1.33	2.3	.89	.82	.68	33.7	30.0	17129 17129
361	119	3.0	3.07	.08	.10	.42	-5.7	.45	-5.3	1.62	.85	.68	35.3	30.2	79854 79854
368	120	3.1	3.10	.05	.10	1.14	1.0	1.23	1.6	.79	.47	.69	30.6	30.4	26498 26498
371	120	3.1	3.13	.02	.10	1.03	.2	.99	.0	1.01	.75	.69	34.2	30.6	64468 64468

Fig. 20.5 Measures of the preservice teachers (aka judges) (An output table from Facets which provides data which describes the preservice teachers who judged 5 teachers. The first 19 judges are presented)

Our quick overview of this table then turns to the first column of reported data under the header "Total Score." These data are the total raw scores for the preservice science teachers. For example, functioning as a judge, student 2669 answered the ESTAM 24 times, and the raw score total of this student's answers is 27. This probably means that this "judge" only supplied very limited data, probably providing a rating of 1 to most ESTAM items for one teacher (skater) on the DVD. The maximum expected number in the total score column would be 600. If a judge (preservice teacher) provided 23 ratings for each skater (teacher on DVD) and if the judge used a "5" for each rating, the total is 600 ($120\times5 = 600$). The numbers that we see in the total score column seem quite reasonable. Numbers in the range of 300–399 seem quite reasonable in that one might predict that there would be a mix of constructivist teaching being exhibited on the DVDs. A total score of, say, 300 would mean that a possible set of ratings from a judge would have been half the ratings with a 3 and half the ratings with a 2 [(60 ratings × 2 pts) + (60 ratings × 3 pts)] = 300. But please remember that there are many combinations of ratings that would result in a raw score of 300.

The column headed "Observed Average" reports the raw mean of each preservice teacher (judge) (27/24 = 1.1). Since the scale for the ESTAM ranged from 1 to 5, and a 1 represents very traditional teaching, then one can see that this preservice teacher may have been a tough judge, indeed (we would have to look at which of the five teachers they judged to know for sure, but for now let's go on to some more data in the table). The third judge listed in the table "84761" is a judge who provided almost the maximum number of potential ratings (total count is 118). This judge's average rating is 2.3, perhaps a judge who does not give (on average) many high ratings. When we do our work, sometimes we look at the raw data, but only to gain a feel for some of the data, always remembering all the problems with raw data.

The measure value of the judge is found in the "Measure" column. In this example, the measure value of the most severe preservice science teacher–judge is 1.46 logits. By scanning the table, one can see that the easier a judge is, the lower the measure value.

Finally readers should note some other familiar terms (e.g., Outfit ZSTD, Outfit MNSQ) that can be used to evaluate the response patterns of the judges. The techniques outlined in previous chapters are just as valid in the MFRM case as in an analysis of data using Winsteps for surveys and tests. For now we will emphasize interpretation that is more specific to the new topics in this chapter.

What is the meaning of the first preservice science teacher measure of 1.46 logits, the second preservice science teacher measure of 1.46 logits, and the third preservice science teacher measure of 1.00 logits? To understand the meaning of a higher or lower measure as a judge, one should just look at the Fair-M Average, the measures in logits, and also think about the structure of the ESTAM instrument. First of all one can see that a lower Fair-M Average results in a higher measure value for a judge. Since the ESTAM uses a scale of 1, 2, 3, 4, 5 (1 means a low level of constructivist teaching, and higher values mean higher levels of constructivist teaching), a higher measure in logits for the preservice teachers (judges) represents tougher judges. In layperson's terms, the first judge in the table was a tougher judge who less often awarded higher ESTAM item ratings (e.g., 4 or 5) that reflected higher or highest levels of constructivist teaching. The third judge (ID # 84761) exhibits a lower measure, which means that he or she used higher values in the ESTAM rating scale more frequently compared to the first judge. Again, for any calculations, use the measures, not raw scores. It is fine to use the raw data to understand what "going up the measurement scale means," but never use the raw data for calculations.

Teacher (aka Skater) Facet

A measure table is presented in Fig. 20.6 for one of the two remaining Facets of this data set. The entire table is for the teacher facet (these are the teachers on the DVD who were evaluated, much as Olympic skaters are evaluated). Each teacher is identified by a phrase that summarized the science concept he or she taught. This table is examined in the same manner as the judges' table. The far right column lists the names of the five teachers. Teacher #4 is named "Force and Motion," and teacher #1 is named "All Sorts of Leaves." One could have of course named the teachers Brigitta, Hans, Paul, and Cornelia. The column labeled "Total Count" lists the sum of all ratings of each teacher (skaters) supplied by all preservice science teachers (judges). Like the table of judges, the key numbers to use for any calculations are the logit measures for the 5 teachers who taught a lesson (e.g., teacher #4 = .93 logits, teacher #1 = .74 logits). To understand the meaning of a higher or lower measure with respect to the teachers, one should conduct a review similar to our review for understanding the judges. Such a review helps a researcher understand that a higher logit measure for each of the five teacher measures means a higher rating of constructivist teaching. For example, teacher #4 has a raw average of 4.2 (e.g., teacher #4 13267/3155 = 4.2) and a measure of .93; teacher #1 has a raw mean of 4.1 and a measure of .74.

The total score column reports the total number of raw points that each teacher received. Remember this will be a very large number because each ESTAM item is rated using a scale that ranges from 1 to 5. One can compute the total possible maximum score by multiplying the highest possible value (5) on an ESTAM item by the number of ESTAM items and then multiplying that produce by the number of total judges. For instance, teacher #4's total possible score is 18,000 (150 preservice teachers as judges × 24 items × 5 points maximum score = 18,000).

As was the case for our discussion of the preservice science teachers who served as judges, any mathematical computations or graphical presentations must use the Rasch measures of the teachers. These values are not only expressed on a linear scale, but they take into consideration judge severity. Recalling our presentation of sample judges, we can see that our most severe (toughest) judge did not provide the maximum number of evaluations. Each teacher's measure takes into account, among other things, that the tough judge may have rated some teachers and may not have rated other teachers. The truly earth-shattering aspect of this table is that it presents not only linear measures that can be used for statistical analyses but also measures that take into consideration judges' severity.

Table 7.2.1 Teachers on Tape Measurement Report (arranged by mN).

Total Score	Total Count	Obsvd Average	Fair-M Avrage	Measure	Model S.E.	Infit MnSq	Infit ZStd	Outfit MnSq	Outfit ZStd	Estim. Discrm	Correlation PtMea	Correlation PtExp	N Teachers on Tape
13267	3155	4.2	4.24	.93	.02	1.26	9.0	1.25	8.7	.72	.29	.40	4 Force and Motion
13330	3253	4.1	4.13	.74	.02	.91	-3.7	.90	-3.9	1.07	.43	.40	1 All Sorts of Leaves
11678	3158	3.7	3.73	.15	.02	.89	-4.3	.90	-4.3	1.07	.43	.45	2 Water Purification
10681	3126	3.4	3.44	-.20	.02	.98	-.6	.98	-.8	1.05	.53	.47	3 Completing the Circuit
6627	3167	2.1	2.04	-1.62	.02	.98	-.7	1.03	1.1	1.09	.47	.46	5 Water Cycle

Fig. 20.6 The table for the teacher facet (A table from the MFRM analysis. A table which presents the 5 teachers who were evaluated by the preservice teachers. In this scenario, the 5 teachers can be viewed as “ice skaters” who are performing, and the preservice teachers are viewed as “judges”)

ESTAM Items (aka Skating Traits) Facet

Figure 20.7 presents the items facet table. This table is similar to the table of survey items that we used for our discussion of surveys. Our first comment is that the table is organized in an identical fashion as the two previous tables for the preservice science (judges) teachers and the teachers (skaters). By now some may be tempted to go immediately to the "Measure" column for the parts of the trait measured by each of the 24 items, but we encourage readers to carefully review the table as before. Doing so provides an effective technique to check the data (e.g., missing data). Making the best use of any data set requires intimate familiarity with data and a clear understanding of the meaning of numbers that will be used for subsequent analysis. So an investment in what might be viewed as a mundane task can really speed up later research work. You can more easily spot problems in data, and if you really know your data, you will find it is easier to write reports and papers with confidence.

The "Total Count" column lists the number of ratings produced for each survey item. Looking at the far right column, the 23rd ESTAM item was answered 660 times. This number is in the range of numbers of responses one would expect to see when 150 preservice science teachers answer this item for each of the 5 teachers (150 judges × 5 skaters = 750). A number below 750 in the Total Count column simply means that not all judges evaluated all skaters with this item.

The "Total Score" can be understood by recalling that the lowest possible rating for each ESTAM item is "1" and the highest possible rating is "5." This means that the rough range of total score column should extend from a low of 660 (660 × 1 = 660) to a high of 3,300 (660 × 3,300). We use the word "rough" because, if not all students answer an item, that value will vary. The next step is to find the measure for each survey item. In this case, the first measure listed in this table is the item measure of 1.01. As was the case in the other tables, the two columns preceding the measure column are raw means. One column contains non-corrected averages (non-corrected for differences in judge severity), and the other column contains corrected averages. As before, all parametric statistical procedures to be performed and all graphical displays must use the measure value logits, but these raw means can be used to make sure one understands the meaning of a more positive measure and a less positive measure. Using these techniques for decoding the table, readers can easily observe that a higher logit measure for a survey item describes teacher behavior that is increasingly difficult to exhibit in a constructivist classroom. Thus, item ER5 (measure = 1.01 logits; observed average = 2.7) received mean ratings that are lower than the item listed immediately below it. Item SA3 has a measure of .18 logits and a mean rating of 3.4. This means that as one moves to items of lower logit measures, one is moving toward items (teacher behaviors) that are easier to observe in a constructivist teacher in a classroom. In terms of our tie-in to skating, items with a low measure are those skating skills that are easier to exhibit. So if ESTAM item TVA4 and TVA2 were skating skills, one can see that skating skill TVA2 is an easier skill than TVA4. Remember, not only are the measures for

Table 7.3.1 Traits Measurement Report (arranged by mN).

Total Score	Total Count	Obsvd Average	Fair-M Avrage	Measure	Model S.E.	Infit MnSq	Infit ZStd	Outfit MnSq	Outfit ZStd	Estim. Discrm	Correlation PtMea	Correlation PtExp	Nu	Traits
1801	660	2.7	2.67	1.01	.04	1.69	9.0	1.99	9.0	-.09	.42	.69	23	ER5
2234	662	3.4	3.46	.18	.05	1.03	.5	1.06	.9	.93	.71	.70	12	SA3
2230	659	3.4	3.47	.18	.05	.74	-5.1	.80	-3.8	1.20	.68	.70	9	TVA5
2247	665	3.4	3.47	.18	.05	.89	-2.1	.93	-1.3	1.06	.69	.70	3	P1
2225	655	3.4	3.48	.16	.05	1.10	1.8	1.10	1.8	.94	.71	.70	20	ER2
2255	663	3.4	3.49	.15	.05	1.03	.5	.98	-.3	1.12	.71	.70	13	SA4
2287	661	3.5	3.56	.06	.05	.86	-2.6	.87	-2.5	1.11	.72	.70	8	TVA4
2293	662	3.5	3.56	.06	.05	.98	-.4	.98	-.4	1.04	.72	.70	5	TVA1
2285	657	3.5	3.58	.04	.05	.88	-2.2	.87	-2.3	1.16	.75	.70	21	ER3
2321	662	3.5	3.61	.00	.05	.81	-3.6	.81	-3.6	1.21	.67	.70	19	ER1
2332	658	3.5	3.65	-.05	.05	.96	-.8	.97	-.5	1.01	.69	.70	18	A5
2336	659	3.5	3.65	-.05	.05	.90	-1.8	.89	-2.0	1.21	.76	.70	16	A3
2351	660	3.6	3.67	-.07	.05	.95	-.9	.99	.0	1.03	.67	.70	24	ER6
2361	663	3.6	3.67	-.07	.05	1.15	2.6	1.15	2.5	.82	.68	.70	4	P2
2362	663	3.6	3.67	-.08	.05	1.05	.8	1.05	.9	.97	.72	.70	15	A2
2366	662	3.6	3.68	-.09	.05	1.10	1.7	1.05	.9	1.02	.71	.70	6	TVA2
2377	663	3.6	3.69	-.10	.05	.94	-1.0	.99	-.2	.98	.69	.70	2	C2
2378	661	3.6	3.71	-.12	.05	1.01	.1	1.03	.5	1.01	.71	.69	10	SA1
2395	664	3.6	3.72	-.14	.05	.94	-1.0	.93	-1.2	1.18	.73	.69	1	C1
2407	659	3.7	3.77	-.20	.05	1.01	.2	1.04	.7	.97	.70	.69	17	A4
2421	662	3.7	3.77	-.21	.05	1.09	1.6	1.08	1.4	.94	.72	.69	7	TVA3
2449	664	3.7	3.81	-.26	.05	.85	-2.9	.85	-2.7	1.15	.72	.69	11	SA2
2426	656	3.7	3.82	-.27	.05	.97	-.6	.96	-.7	1.09	.70	.69	14	A1
2444	659	3.7	3.83	-.29	.05	.86	-2.7	.88	-2.2	1.04	.73	.69	22	ER4

Fig. 20.7 The table for the items facet (The measure (in logits) of the 24 ESTAM items computed through use of Facets)

the ESTAM items presented on a logit scale but also the locations of items in essence take into consideration what types of judge evaluated a teacher (aka a skater). Just as it is important to remember that judges can skip items, it is sometimes useful to remember that you can think of items being skipped by judges. Again many of the techniques we have presented for better understanding the function of survey and test items transfer to our use of this table. For example, a high value of Outfit MnSq suggests that ESTAM item ER5 might need to be investigated in more detail. At the least, this item should perhaps be monitored over a data collection.

Wright Maps

Below is Fig. 20.8, a Wright Map, which summarizes the information from the preservice science teachers (judges), the teachers (skaters), and the items (skating criteria). The Wright Map shows how well each trait is defined. Having read and digested earlier material herein, readers should realize that the quality of measurements made with tests and surveys depends on a wide range of issues, many of them long ignored in research. First, the use of raw, nonlinear data often results in incorrect conclusions. This harsh fact is based on parametric tests' assumption of linearity of measurement scales. If a scale is nonlinear (e.g., ordinal), then a fundamental assumption of parametric statistical tests is violated. Second, how a scale is defined by its items greatly affects how well a scale measures. We now consider an analogy used by Boone, Townsend, and Staver (2011), and then we will extend our analogy to the use and implications of the Wright Map from Facets. Think of a blank piece of wood (1 m in length) being fed into a machine that can make only 5 cuts. Clearly, if multiple cuts are made in the same spot, the meterstick will not measure as well as if the cuts do not bunch up and overlap. With regard to items, researchers will usually want to have a range of items defining a trait to better measure survey/test respondents when one has no idea where the respondents will fall on the trait. So, think of one goal as having a range of item cuts along the length of blank wood. Now let's think of persons who are acting as judges. We hope it makes sense to readers that, just as having a range of items helps us define a trait, it should make sense that having a range of judges will also be advantageous. So think of a blank piece of wood as not only being cut by items but also being cut by judges, too! Finally, although our goals in the use of the ESTAM might have been to rate the 5 teachers, it should also make sense that if we think of the 5 teachers on DVD as helping to make cuts, then it is also to our advantage (as best we can) to have a range of teachers who will make "cuts" on the blank stick. In Townsend's research, since he is interested in investigating the ratings of the preservice teachers (who are serving as judges), it *is* important to have a range of teachers on DVD (skaters) who are being evaluated!

Let's now look at our Wright Map for the MFRM. The first column of data presents the judges (preservice science teachers). Even without knowing which end

```
+---------------------------------------------------------------------------------+
|Measr|-Methods students who are judging|+Teachers on Tape       |-Traits   |Scale|
|-----+---------------------------------+------------------------+----------+-----|
|   2 +                                 +                        +          + (5) |
|     |                                 |                        |          |     |
|     |                                 |                        |          |     |
|     |                                 |                        |          |     |
|     |                                 |                        |          |     |
|     | **                              |                        |          |     |
|     |                                 |                        |          |     |
|     |                                 |                        |          |     |
|     |                                 |                        |          |     |
|     |                                 |                        |          |  4  |
|   1 + *                               +                        + *        +     |
|     |                                 | Force and Motion       |          |     |
|     |                                 |                        |          |     |
|     | *                               | All Sorts of Leaves    |          |     |
|     | ***                             |                        |          |     |
|     | *                               |                        |          |     |
|     | *                               |                        |          | --- |
|     | ***                             |                        |          |     |
|     | **                              |                        | ****     |     |
|     | ****                            | Water Purification     | ***      |     |
*   0 * *******                         *                        * ***      *     *
|     | *****                           |                        | ******** |     |
|     | ********                        | Completing the Circuit | **       |  3  |
|     | ***************                 |                        | ***      |     |
|     | ***********                     |                        |          |     |
|     | *************                   |                        |          |     |
|     | ***********                     |                        |          | --- |
|     | ****************                |                        |          |     |
|     | **********                      |                        |          |     |
|     | *******                         |                        |          |     |
|  -1 + *****                           +                        +          +     |
|     | ****                            |                        |          |  2  |
|     | **                              |                        |          |     |
|     | *                               |                        |          |     |
|     | *****                           |                        |          |     |
|     | ******                          |                        |          |     |
|     |                                 | Water Cycle            |          |     |
|     | ***                             |                        |          |     |
|     | *                               |                        |          | --- |
|     |                                 |                        |          |     |
|  -2 +                                 +                        +          +     |
|     |                                 |                        |          |     |
|     | *                               |                        |          |     |
|     |                                 |                        |          |     |
|     |                                 |                        |          |     |
|     |                                 |                        |          |     |
|     |                                 |                        |          |     |
|     |                                 |                        |          |     |
|     |                                 |                        |          |     |
|     |                                 |                        |          |     |
|  -3 + *                               +                        +          + (1) |
|-----+---------------------------------+------------------------+----------+-----|
|Measr| * = 1                           |+Teachers on Tape       | * = 1    |Scale|
+---------------------------------------------------------------------------------+
```

Fig. 20.8 A Wright Map of the information from the preservice science teachers (judges), the teachers (skaters), and the items (skating criteria) (A Facets Wright Map in which preservice teachers (judges), teachers on DVD (skaters), and items (skills) are presented using computed logit measures)

	Ice Skating	Education	Market Research
Trait 1->	Judges	Preservice Teachers	Customers
Trait 2->	Skaters	Veteran Teachers On Video Tape	Products
Trait 3->	Technical Skills	ESTAM Items	Product Characteristics Which are Rated

Fig. 20.9 Three scenarios in which RFRM should be used to take into account, among many issues, differences in judge severity

is a tough judge or a lenient judge, this part of the Wright Map shows clearly that all judges *do not* act in a similar fashion in terms of their behavior. Recall the ruler comment; it is important to mark different portions of a ruler and not to mark a ruler more than once in the same spot, in that multiple marks at or very near the same spot are wasted and do not help in the computation of measures. One should think of these judges as marks on a ruler and understand that it is better to have all judges operate in different manners and worse to have them operate in the same manner. As Dr. Townsend reviewed these data, he found the judge distribution to be very interesting in terms of the bell shape of judge severity and the slight skew toward more severe judges. In our own work and reading, we are no longer amazed (but we were at first) to see how varied even "trained" judges are in how they use a rating scale. The moral for us is: (1) it is very important to use MFRM when judges supply data, (2) training judges to act in the same way rarely works, and (3) if one thinks of cutting a ruler, then there are advantages to having judges with a range of severity. Recall that the judges with the higher measures are those who were tough judges, less likely to rate one of the five teachers as exhibiting constructivist teaching. Those judges with lower measures were more lenient, thus more likely to rate one of the 5 teachers as constructivist. If readers look back at the table of judges (ordered by measure), one will be able to see the presence of two judges who were quite tough, those are the two judges who appear as two stars (*), at the top of the Facets Wright Map on the first column of data.

Now let's look at the middle of our Wright Map. That column displays the measures of the 5 teachers who were rated by the 150 preservice science teachers (aka the ordering of the 5 skaters determined by an analysis of the data from 150 judges). If readers again look at Fig. 20.8, one will see that the 5 teachers/skaters are ordered from most constructivist at the top to least constructivist at the base of the Wright Map. This overall ordering is, in itself, important for a methods class. For example, why might specific teachers (skaters) have been highly rated (or not as highly rated)? Does the overall ordering of the 5 teachers teaching a sample lesson match what would be expected by their instructor? If the ordering did not match what the instructor expected, then what are the implications of such mismatches?

There is another important implication of the ordering and spacing of the five teachers. Recall that one goal of the study was to gauge and document the overall assessment of each preservice science teacher (judge), in essence each preservice science teacher's location along the trait. The locations of the preservice science teachers are, of course, determined by the ratings each preservice science teacher gave for each item for each of the five teachers; however, the quality (the certainty) of the locations of the 5 teachers depends upon how well the trait is marked by the survey items and also by how varied the judges are in their overall severity. This multiple layering of the issues that impact the overall quality of measurement in studies has been really rarely considered at the level possible with Rasch measurement. And certainly, the consideration of measurement issues at play in any similar data collection not only is now possible, but it is something that must be done to provide reliable data for statistical analyses.

The far right side of the Wright Map presents the measure of each of the 24 ESTAM items used to define the trait. Just as readers could look at the preservice teachers as "judge" table (Fig. 20.5) and the teachers as skaters table (Fig. 20.6) and identify judges and skaters on the Wright Map, the same can be done for the traits. For example, review of the Wright Map reveals one item that exhibits a substantially higher measure than the other 23 ESTAM items. The Wright Map shows us that the item is about 1.0 logit. Reviewing Fig. 20.8 reveals that item is ER5 (an item that is the hardest for the 5 teachers [skaters] to receive a high constructivist rating on). From a measurement perspective, we can see in this Facets Wright Map that many ESTAM items measure the same portion of the trait of constructivism. This means that a subset of ESTAM items could be selected that would provide measures of similar certainty. Of course, one significant advantage of using a subset of these 24 items is decreasing time and work. We will make a final observation for the moment that ties into techniques we have introduced and used earlier herein. We observe a very large gap between the 23 closely grouped ESTAM items and the single ESTAM item that exhibits a high measure. If the ESTAM were revised, researchers should attempt to author items that would fill in this gap. This is analogous to thoughtfully cutting the blank piece of wood at our factory. Not only can redundant items be removed from a measurement scale, but also items can be authored to "fill the gap." Just as the announcements on the London tube (just before doors of trains closing) are "mind the gap," researchers using Wright Maps (in Facets and in Winsteps) should also take care to "mind the gap" (perhaps gaps in items, gaps in judges, gaps in those being evaluated). Gaps have measurement implications and policy implications among many things! So when reviewing a Wright Map, "mind the gap," a gap could indicate you have made a mistake in how you have conceptualized the variable.

Since it takes quite a bit of time to view a teacher's entire lesson and then provide a rating, it might be possible to add additional teachers to be rated (add skaters).

Then, by decreasing the number of items to be evaluated by the judges (preservice teachers), it might be possible to add teachers/skaters without adding to the workload of already stressed judges. Using a multimatrix design, it might also be possible to increase the total number of skaters, but each judge would not have to evaluate all teachers (the skaters).

In this study, the ESTAM was used to understand differences in the severity of the judges (preservice science teachers) (e.g., how much variance preservice science teachers exhibit in terms of their assessment of constructivist teaching). There was, of course, interest in how the 5 teachers who taught the 5 science lessons compared in terms of logit measures, but the primary goal of this MFRM analysis was to investigate differences between preservice science teachers and to consider implications for science teacher education. Clearly, there is another way in which these data can be considered in light of even more common issues in science education. That is, namely, the assessment of science teachers. For instance, one could easily use the ESTAM to evaluate preservice science teachers' presentations of science lessons. In that case, one would likely have a smaller number of judges (presumably veteran teachers) judging the preservice teachers who are teaching the lessons. A very similar setup of a control file would be used in such a case.

In different sections of this book, we have discussed the incredible amount and quality of diagnostic information that is available when ordinal (nonlinear) data can be expressed on an equal-interval scale. Equal-interval scales are linear, which means that parametric statistical procedures can be carried out with assurance that the chances of making a type 1 or type 2 statistical error are not elevated. There is, however, another extremely important aspect of linear scales. With regard to linear scales (as Bill Nye would say) "consider the following": Take your finger and mark the gap between the measure of Mr. Smith, the teacher who taught the Water Purification Cycle, and Mr. Jones, the teacher who taught the Completing the Circuit lesson. Then see how many times that gap fits between the gap between Mr. Smith and Mr. Jones. No matter how large (or wobbly) your fingers are, you will fit the small gap about 3X within the large gap. This comparison brings incredible meaning and guidance to researchers. This means that the difference in exhibited constructivist teaching, as assessed by the judges using the 24-item ESTAM, between Mr. Smith and Mr. Jones is 1/3 the difference in levels of constructivist teaching as measured for a comparison of Mr. Smith and Mr. Jones.

Fitting the Concept of FIT into an MFRM Analysis

Earlier chapters discussed the issue of fit (e.g., infit, outfit, ZSTD, MNSQ) for persons and items. All of the same techniques can be used to evaluate the measurement function of items, judges, and those being judged in an MFRM analysis. The only difference between what is considered in an MFRM analysis, in contrast to an analysis of a multiple-choice test or a survey, is the presence of three Facets (persons, items, and judges) to review and consider instead of two Facets (persons and items).

Isabelle and Ted: Two Colleagues Conversing

Isabelle: *Well Ted, what do you think of the multifaceted Rasch model?*

Ted: *To begin, I think it would have been hard for me to understand it if I had not read the earlier chapters. But, by using the earlier chapters, it was not that difficult.*

Isabelle: *How so?*

Ted: *I think the main thing that really helped me was thinking about Olympic skating or diving. I think that most people have watched the finals of Olympic figure skating competitions, and they remember the excitement when the scores are revealed. When the scores are revealed, a top score and a bottom score are dropped. Then there is often talk about what judge gave what sort of score, and whether or not a tough judge might have been tough on earlier skaters. So all of that really helped me remember that, with judges, things can be a little more complicated.*

Isabelle: *I agree. The main thing that really resonated with me was that it is better to have a judge be consistent (consistently tough or consistently easy). I had always thought that all judges should be trained to act in the same way, but now it makes sense to me that consistency is more important. Furthermore, just as a test should not have items all of the same difficulty, judges should represent a range of severity. If all judges were the same, then why have more than one judge?*

Ted: *Yes, yes, yes. And you know what? The code for a Facets analysis is not that hard. What the authors and their colleague did was just take some code in Bond and Fox and adapt it to their own data set. I can see in the Facets analysis that there are new pieces of data we did not have with Winsteps, but many of the terms that we used for understanding Winsteps are used in Facets.*

Isabelle: *Right. And it is sort of interesting. Right now I am doing a Facets analysis of some data I collected. I collected data from 1066 German students who were completing a gymnasium physics class. Each student was asked to solve five pretty tough physics problems. After those data were collected, I was able to hire 50 judges. Because it takes about 10 minutes to grade each set of 5 problems, it would not be financially feasible and there would not be enough time for all judges to grade all tests. So, by using a multimatrix design, I was able to have two judges grade each physics item. By using MFRM, I was able to fairly quickly get more than one judge to provide an opinion on each student's answer, and I was able to correct for differences in judge severity. And of course, I was able to compute linear scale scores for all the students!*

Ted: *You know Isabelle, everyone talks about learning progressions now. Just as we can think better about learning progressions by using Wright Maps and of course using linear measures, don't you think that with Facets we also have learning progressions?*

Isabelle: *Absolutely!*

Ted: *I bet one can even do an MFRM with more than 3 Facets!*

Isabelle: *Yes you can.*

Keywords and Phrases

Judges (judges), skaters (persons), skill items (items)
Consumers (judges), products (persons), product trait (Items)
MFRM (multifaceted Rasch analysis)

Potential Article Text

In an effort to inform instruction in an undergraduate secondary science methods class, researchers conducted a Multifaceted Rasch Measurement (MFRM) analysis of data collected from a sample of 150 preservice science teachers. Each preservice science teacher evaluated the constructivist teaching of five (5) full-time teachers using the 24-item ESTAM (Gallagher & Lindsey, 1997). The MFRM analysis was conducted using Rasch Facets software (Linacre, 2012). For researchers in the field of science education, one important reason for using MFRM is that the technique takes into consideration differences in judges' severity (or leniency) that are always present no matter what the extent of training judges to act as robots.

A Wright Map was constructed following the analysis of FACET data. Results of the Wright Map, as well as review of data quality indices, suggested reliable and valid measurement of preservice teacher–judges' severity, teacher performance with respect to level of constructivist science teaching, and the definition of the constructivist trait by the 24-item ESTAM. One gap in the items was observed.

The distribution of preservice science teachers as judges suggests a wide range in judge severity. Not all preservice science teachers are judged in the same manner. The measures of the five (5) teachers who were rated were quite varied. Finally, the distribution of ESTAM items reveals a potential measurement gap that could be filled in subsequent versions of the instrument. Extensive overlap of items suggests that the time required to administer the ESTAM in similar scenarios could be lessened through removal of redundant items.

Quick Tips

When considering MFRM, think of Olympic judges evaluating skaters with regard to a number of skills.

You can use what you have learned for tests and rating scale to help you understand the Facets output.

Use an existing Facets control file to start your work.

Data Sets: (go to http://extras.springer.com)

cf Facets Judges Skaters Traits
cf fussball is the best for Facets

Activities

Activity #1

Task: We have provided the full file for the analysis of the data discussed at length in this chapter (cf Facets Judges Skaters Traits). Examine the file and determine what most of the lines show. Identify the ID of the first judge. Identify how many items were answered by the first judge for his or her rating of the teacher who taught the circuit lesson. Find the rating that the first judge gave to the teacher of the circuit lesson for the 6th ESTAM item. Identify the abbreviation the research team made for this 6th ESTAM item.

Answer: The first judge in the data set is judge 16599. The data for this judge pertaining to the teacher of the circuit lesson are contained in the third line of data for this judge. One can see that this is the line for the circuit lesson's teacher in that in the top part of the control file, one can see that this teacher has a number of "3" (3=Completing the Circuit ;). In this line of data, there are 24 items answered, so no items were skipped! The rating of this judge for this teacher to the 6th ESTAM item was a "4." TVA2 is the abbreviation for the 6th item.

Activity #2

Task: We provide a control file (cf fussball is the best for Facets) that we made by altering the control file above. This file provides fictitious ratings from 5 newspapers' sports departments with regard to the soccer (football) skills of some famous teams and some not so famous teams. The teams are rated with respect to 24 different soccer (football) skills. Ratings vary from 1 to 5 (using the numbers 1, 2, 3, 4, 5) and a 5 is the best rating one can earn. Identify the missing data in the control file.

Answer: One can see that there are missing data in that there are numerous "X" in the data portion of the control file.

Activity #3

Task: In earlier sections of this book, we have written about the use of "multimatrix design" for data collection. This is a technique that is used quite often in Germany, and elsewhere, when data are collected. Explain briefly what the multimatrix design is and then explain how you are able to see that a multimatrix design was used for the football data collection.

Answer: Rasch analysis does not require that all respondents must answer all items on a test or survey. When judges are rating individuals (judges evaluating skaters using criteria), all judges do not have to evaluate all individuals. Moreover, judges need not use all criteria for the individuals they do evaluate. This aspect of Rasch analysis can be used to help projects in many ways. For example, by limiting the number of individuals judges evaluate, judges can focus their concentration on fewer tasks (they are only human). Also, researchers can save money in terms of payments to judges for their time. Since not all items need to be used by any one judge, it is possible to present a large number of items.

Multimatrix design can be seen in the data in that there is at least one "link" between each judge. This means at least two judges must evaluate each item. However, the common judging of two items is more than that, in that one can see that there is a link from one judge to another as one goes through the data set, line by line.

Activity #4

Task: Using the data file with the missing data, run a Facets analysis. Figure out which judge, after correction, was the toughest. Which judge was the most lenient? Then identify which team received the highest rating. Which team received the lowest rating? Finally, identify the easiest skill and hardest skill for teams to earn a high mark on.

Answer: Newspaper B is the toughest rater with an average rating of 2.6 on the 1–5 scale. The easiest sports department is Newspaper D with a rating of 4.4. Remember, if you were going to conduct any statistical analysis of the raters, you would use the measures of each newspaper (e.g., Newspaper D measure is −2.53 logits). The Digital Demons received the highest team rating (a raw average of 4.2 and a measure of 1.35). The team with the lowest rating is the Galactic Pride (a raw average rating of 1.6 and a measure of −2.25). In terms of skills, the highest rated skill (the skill that appears to be the easiest) is shooting. The skill that appears to be the hardest is Goalkeeper.

Activity #5

Question: How does the difference in the overall rating of the Digital Demons and the Big Maulers compare to the overall rating of the Sharks and the Mundane Unicorns (as well as the Big Maulers)?

Answer: The exact ratio can be computed by using the tables from Facets, which provide the exact measures of the 5 teams. However, using one's fingers on the Wright Map, it looks as if the gap between the Digital Demons and the Big Maulers is about 1/3 the gap as that seen between the Sharks and the Big Maulers (and the Mundane Unicorns). This means that it will take really 3X as much movement for Sharks to catch up to the Big Maulers and the Mundane Unicorns compared to what it will take for the Big Maulers and the Mundane Unicorns to catch up to the Digital Demons.

References

Bond, T. G., & Fox, C. M. (2007). *Applying the Rasch model: Fundamental measurement in the human sciences* (2nd ed.). Mahwah, NJ: Lawrence Earlbaum Associates.

Boone, W., Townsend, S., & Staver, J. (2011). Using Rasch Theory to Guide the Practice of Survey Development and Survey Data Analysis in Science Education and to Inform Science Reform Efforts: An Exemplar Utilizing STEBI Self-efficacy Data. *Science Education, 95*(2), 258–280.

Gallagher, J., & Lindsey, S. (1997). *Elementary science teaching analysis matrix*. Unpublished document.

Linacre, J. M. (2012). *Facets computer program for many-facet Rasch measurement*. Beaverton, OR: Winsteps.com.

Looney, M. A. (1996). Figure skating fairness. *Rasch Measurement Transactions, 10*(2), 500.

Looney, M. A. (2004). Evaluating judge performance in sport. *Journal of Applied Measurement, 5*(1), 31–47.

Additional Readings

Myford, C. M., & Wolfe, E. W. (2003). Detecting and measuring rater effects using many-facet Rasch measurement: Part I. *Journal of Applied Measurement, 4*(4), 386–422.

Chapter 21
The Rasch Model and Item Response Theory Models: Identical, Similar, or Unique?

Isabelle and Ted: Two Colleagues Conversing

Ted: *Isabelle, I need your help here. I am looking at a number of articles that have used Rasch to analyze data. Sometimes the authors use the term "Rasch analysis," and sometimes they use the term "IRT" or "Item Response Theory." Are those words interchangeable? Also, there is another thing; I noticed that sometimes people write about the Rasch model as being the 1-parameter model, and in the same breath, they write about the 2-parameter model and the 3-parameter model. What is going on?*

Isabelle: *You know Ted, I wrestled with the same issue when I first started my work. It took me a while to sort things out, and now I understand the differences, but it would have been a lot easier if someone had taken me aside and explained the issues.*

Introduction

In this chapter, we present some critical information concerning the uniqueness of the Rasch model. Readers will note a few formulas and some philosophy, but no data sets. Our goal in this chapter is to help readers understand that Rasch models, in our minds, are substantially different in many ways from Item Response Theory (IRT) models. They are so different that we encourage researchers to not describe or portray the Rasch model as an IRT model. Also, we explain the thinking that led us to choose this model as the one to use for all our data analysis.

W.J. Boone et al., *Rasch Analysis in the Human Sciences*,
DOI 10.1007/978-94-007-6857-4_21,

Formative Assessment Checkpoint #1

Question: Is the Rasch model merely one of the three IRT models?

Answer: No. The mathematics of the Rasch model may look like the 1-parameter model, but there is a fundamental philosophical difference. This difference is so substantial that we encourage readers to not refer to the Rasch model as an IRT model.

Rasch: A Family of Models

Let's begin by noting that different Rasch models exist (e.g., dichotomous, rating scale, partial credit, multifaceted), but the family of Rasch models is quite different from other models that some researchers employ to evaluate test and survey data. Within the hallways of psychometric departments, we realize that experts have engaged and may continue to engage in heated discussions regarding what model to use to evaluate data sets such as multiple-choice data. In this chapter, we present our perspective as science educators, former science teachers, and former scientists and explain why the Rasch model is the one we have selected for our work.

Two central overarching characteristics of the Rasch model most influenced our decision to use Rasch. First, the philosophy of Rasch is very much aligned with what scientists do as they develop measurement instruments and collect data. Second, Rasch theory is in fact aligned with many of the standards that detail what K-12 science teachers should do to help students learn science. Many of the "habits of mind" that are discussed in the field of science education are indeed the habits of mind (habits of measurement if you will) that are fundamental to thinking and applying Rasch measurement.

Importance of Measurement

To detail our discussion, we first ask readers to recall basic measurement devices that scientists use and have used to advance knowledge and scientific understandings. Examples that come to mind include the meter stick, the thermometer, and the double pan balance. These three devices produce measures in linear, equal-interval units (e.g., meters, decimeters, centimeters, and millimeters; degrees Celsius; and grams, respectively), thereby permitting scientists to employ mathematics with confidence as they compare measurements of length, temperature, and mass at a single point in time or over a specific span of time. A second, equally important aspect of these scientific measurement instruments is that they were designed and built to be as robust and demanding as possible, in that objects of varying length, temperature, and mass can be measured. If objects varying from .001 to 100,000 g are to be weighed, it is possible without too much difficulty to build a scale that could precisely measure such a range of mass. The Rasch model attempts to do the same for measurements of persons. It transforms ordinal, non-equal-interval scores into linear, equal-interval units (e.g., the mass of an apple might be expressed as 35.7 g) and holds to principles of objective measurement, which in our minds mean that not only can one weigh an apple but one can also weigh different apples and if the apple is light (e.g., think of a student who incorrectly answers most items) or heavy (e.g., think of a student who correctly answers most items), the measurement is not influenced. If researchers take time to review the national standards of their country (or state or territory or Land) of choice, they will see that the carefully crafted standards pertaining to measurement and data collection that students are to master are exactly the skills that science educators must apply in their research. And, Rasch measurement theory allows science educators to apply the skills that we advocate students should master in their own research. The attention to "linearity" and the so-called sample independence in Rasch measurement are two core concepts that have been addressed via Rasch measurement for many years, and the philosophical underpinnings of these discussions in the context of Rasch measurement are well aligned with goals detailed in science standards. For example, the *National Science Education Standards* (National Research Council, 1996) describe expectations for learners in grades K-4 as follows:

> As children develop facility with language, their descriptions become richer and include more detail. Initially no tools need to be used, but children eventually learn that they can add to their descriptions by measuring objects—first with measuring devices they create and then by using conventional measuring instruments, such as rulers, balances, and thermometers. By recording data and making graphs and charts, older children can search for patterns and order in their work and that of their peers. For example they can determine the speed of an object as fast, faster, or fastest in the earliest grades. As students get older, they can represent motion on simple grids and graphs and describe speed as the distance traveled in a given unit of time. (pp. 126–127)
>
> In grades 5–8, students observe and measure characteristic properties, such as boiling and melting points, solubility, and simple chemical changes of pure substances, and use those properties to distinguish and separate one substance from another. (p. 149)

As the next generation of science standards are being developed and vetted, readers can see the continued importance of measurement in *A Framework for K-12 Science Education: Practices, Crosscutting Concepts, and Core Ideas (National Research Council*, 2012):

> Typically, units of measurement are first introduced in the context of length, in which students can recognize the need for a common unity of measurement – even develop their own before being introduced to standard units—through appropriately constructed experiences. Engineering design activities involving scale diagrams and models can support students in developing facility with this important concept.

> Once students become familiar with measurements of length, they can expand their understanding of scale and the need for units that express quantities of weight, time, temperature and other variables. They can also develop an understanding of estimation across scales and contexts, which is important for making sense of data. As students become more sophisticated, the use of estimation can help them not only to develop a sense of the size and time scales relevant to various objects, systems, and processes but also to consider whether a numerical result sounds reasonable. Students acquire the ability as well to move back and forth between models at various scales, depending on the question being considered. They should develop a sense of the powers-of-10 scales and what phenomena correspond to what scale, from the size of the nucleus of an atom to the size of the galaxy and beyond.
>
> Well-designed instruction is needed if students are to assign meaning to the types of ratios and proportional relationships they encounter in science. Thus the ability to recognize mathematical relationships between quantities should begin developing in the early grades with students' representations of counting (e.g., leaves on a branch), comparisons of amounts (e.g., of flowers on different plants), measurements (e.g., the height of a plant), and the ordering of quantities such as number, length, and weight. Students and then explore more sophisticated mathematical representations, such as the use of graphs to represent data collected. The interpretation of these graphs may be, for example, that a plant gets bigger as time passes or that the hours of daylight decrease and increase across the months. (pp. 90–91)

Measurement Defined

Most science educators will recognize the name "Thurstone" as an early pioneer in psychometrics. In 1928, Louis Leon Thurstone described the requirements for a measurement device, namely, that a measurement device must be *independent* of the group measured with regard to the trait. According to L.L. Thurstone (1928):

> The scale must transcend the group measured... One crucial experimental test must be applied to our method of measuring attitudes before it can be accepted as valid. A measuring instrument must not be seriously affected in its measuring function by the object of measurement. To the extent that its measuring function is so affected, the validity of the instrument is impaired or limited. If a yardstick measured differently because of the fact that it was a rug, a picture, or a piece of paper that was being measured, then to that extent the trustworthiness of that yardstick as a measuring device would be impaired. Within the range of objects for which the measuring instrument is intended, its function must be independent of the object of measurement. (p. 547)

This powerfully simple assertion – a measurement device must be independent of what it measures – is certainly aligned with what we as science educators believe must be true when students collect data in an experiment. The importance of the Rasch model is its documented status as the single model that meets the requirement set forth by Thurstone for scale validity (Wright, 1989).

Formative Assessment Checkpoint #2

Question (True/False): The measurement that takes place through the use of surveys and tests will always be quite different than the research that takes place in a scientist's laboratory.

Answer: False. Rasch measurement provides a path that allows the same rigorous measurement that takes place in laboratories to be applied to the measurement of humans.

Rasch and IRT: Philosophical Difference

Rasch measurement is often classified under the umbrella of Item Response Theory (IRT) models. However, a core philosophical difference exists between the Rasch model and the IRT models (often referred to as the 1-parameter, 2-parameter, or 3-parameter models). Whereas the IRT models are altered (more parameters added) to fit the data, the Rasch measurement model is not altered to fit the data and is thus viewed as a definition of measurement.

Examination of the 1-parameter IRT model reveals that it looks identical to the Rasch model. Consequently, some researchers refer to the Rasch model as the 1-P model or as the 1-P IRT Rasch model. We view such references as mistakes because of the immense philosophical difference, in that one model, IRT, is altered to fit data and one model, Rasch, is not altered to fit data. Therefore, Rasch is the model that is consistent with the definition of measurement as set forth by Thurstone over 80 years ago.

Two Linkages: Science Education Research and Measurement and K-12 Science Teacher Education and Measurement

We now return to linkages between science education research and measurement and linkages between measurement and K-12 science teacher education. Because science and mathematics are tightly linked (mathematics is science's most valuable tool), it may prove beneficial to look briefly at measurement standards for K-12 mathematics education. Such standards can be found in *Principals and Standards for School Mathematics* (National Council of Teachers of Mathematics, 2000):

> Measurement Standard: Instructional programs from prekindergarten through grade 12 should enable all students to —
>
> - Understand measurable attributes of objects and the units, systems, and processes of measurement;

- Apply appropriate techniques, tools, and formulas to determine measurements.
  Measurement is the assignment of a numerical value to an attribute of an object, such as the length of a pencil. At more-sophisticated levels, measurement involves assigning a number to a characteristic of a situation, as is done by the consumer price index. Understand what is a measurable attribute as and becoming familiar with the units and processes that are used in measuring attributes is a major emphasis in this Standard. Through their school experience, primarily in prekindergarten through grade 8, students should become proficient in using measurement tools, techniques, and formula in a range of situations.
  The study of measurement is important in so many aspects of everyday life. The study of measurement also offers an opportunity for learning and applying other mathematics, including number operations, geometric ideas, statistical concepts, and notions of function. It highlights connections within mathematics and between mathematics and areas outside of mathematics, such as social studies, science, art, and physical education. (p. 44)

In science and mathematics classrooms, we help students and teachers understand that if a measurement instrument has been carefully developed, then students and teachers can collect data with that instrument, and we do not just change the instrument to fit the data.

We have emphasized extensively and repeatedly herein the necessity that scores be on a linear scale. Indeed, the Rasch model provides us with such scales. Moreover, as science education researchers, science teacher educators, teachers of science to university students, and teachers of science to P-12 students, we view the Rasch model as the best representative of measurement as a core process of science as a way of knowing. To fully understand science as a way of knowing, students must construct and apply the concept that when they collect scientific data via measurement, the function of the instrument should remain independent of the objects being measured. Readers of this book have by now worked their way through many chapters of Rasch and have developed a facility with the thinking needed to conduct measurement. As a finale to this chapter, we ask that readers find a specific article and read the words of Ben Wright as he clearly and logically (step-by-step) explained why the Rasch model should be used and why the 2P and 3P model should not be used (Wright, 1992). This article contains Benjamin Wright's opening remarks in his invited debate with Ron Hambleton, Session 11.05, AERA Annual Meeting 1992.

Formative Assessment Checkpoint #3

Question: Is the Rasch model just an Item Response Theory (IRT) model?

Answer: The mathematics of the Rasch model looks identical to the so-called 1P (short for one parameter) IRT model. However, there is a crucial, fundamental difference between not only the 2P and 3P IRT models but also the seemingly identical 1P model. The Rasch model is viewed as a definition of measurement and is not altered to fit a data set. The IRT models (2P and 3P) are altered to fit a data set. The Rasch model is, among many things, a demanding model.

Formative Assessment Checkpoint #4

Question (True/False): The 2P and 3P models look more complicated than the Rasch model; therefore, those models must be better.

Answer: Greater complexity does not automatically mean better. No one would assert that Newton's law of $F=ma$ looks too simple and thus must be incorrect.

Isabelle and Ted: Two Colleagues Conversing

Isabelle: *Ok, it's quiz time Ted... there are these other models called 2P and 3P, and these models do not look all that different from the Rasch model. What is the big deal? I mean the model is a little different, and it looks more complex, so perhaps those two models are more sophisticated and better?*

Ted: *Well, from my reading, the biggest difference is that when one uses models other than the Rasch model, the models might look similar to the Rasch model, but there is a fundamental difference. When the 2P and 3P models are used, the models are altered to fit the data. When the Rasch model is used, the model is not altered to fit the data. The Rasch model is very demanding, and it is the only model that addresses the requirements of scientific measurement. Not altering the model to fit the data is the thing that makes the most sense to me. I would not ask Newton to alter F=ma to fit a data set that someone collected.*

Isabelle: *Ok that makes sense to me, but why are there some people who might be very vocal about using the 2P or 3P model? Is there something that they might not get?*

Ted: *Honestly Isabelle, I came to psychometrics as a physicist. If you throw out the name of any area in physics, I can name a measurement instrument that is used for data collection. It makes total sense to me that you do not alter a model to fit a data set (and as a result, one would use the Rasch model). My guess is that those who have used the 2P and 3P model sort of miss the importance of conducting measurement that does not depend on a sample of data. I think some of the problem may also just be human nature. If one has been using the 3P model for 20 years, then it is really hard to be reflective and just step back and think. I really do think if those who used the 3P model stepped back and reflected deeply, they would see that the Rasch model might be a little harder to use (since it is not altered to fit a data set), but it really is the model to use.*

Keywords and Phrases

Rasch
The Rasch model should not be called an Item Response Theory (IRT) model.
Thurstone
The Rasch model is not altered to fit a data set.

Potential Article Text

The Rasch model was utilized to evaluate a data set of 10,000 students who completed a 25-item multiple-choice test. Developed by George Rasch and applied by the University of Chicago's Ben Wright, Rasch is the only model that meets the requirements of objective measurement set forth by Thurstone (1928). Researchers in the social sciences (PISA), medical research, and medical credentialing have employed the Rasch model in their work.

Quick Tips

Always remember that it is misleading to classify the Rasch model as an IRT model (often people will refer to the Rasch model as the "1P IRT Model"). There is huge philosophical difference in the IRT approach and the Rasch approach. The Rasch model is viewed (and has been shown) to be a definition of measurement. The model is not altered to fit a data set.

Activities

Activity #1

Task: Read the excerpt from Nunnally, J.C. (1967). *Psychometric Theory*.

Question: In light of Nunnally's comments, why is it important to be able to conduct measurement?

"The major advantage of measurement is that it takes the guesswork out of scientific observation. A key principle of science is that any statement of fact made by one scientist should be independently verifiable by other scientists. The principle is violated if there is room for disagreement among scientists about the observation of empirical events. For example, since we have no standardized measure of "ego strength," two psychologists could disagree widely about the ego strength of a particular person. Obviously, then, it is not possible to make scientific tests of theories concerning ego strength. Thus theories concerning atomic particles, temperature of stars, intelligence of children, drive level in rats, and so on are testable to the extent to which there are unambiguous procedures for documenting empirical events.

A case could be made that the major problem in psychology is that of measurement. There is no end of theories, but the theories are populated with terms (hypothesized attributes) which presently cannot be adequately measured; consequently the theories go untested. This is the problem with Freudian theory. There are no agreed-on procedures for observing and quantifying such attributes as ego strength, libidinal energy, narcissism, and others. In fact it seems that major advances in psychology, and probably in all sciences, are preceded by breakthroughs in measurement methods. This is attested to by the flood of research following the development of intelligence tests. Recent advances in techniques for measuring the electrical activity of individual nerve cells provide another example of how the development of measurement methods spurs research. Scientific results inevitably are reported in terms of functional relations among measured variables, and the science of psychology will progress neither slower nor faster than it becomes possible to measure important variables." (p. 5)

Answer: Nunnally talks of taking the guesswork out of measurement. One of his points is that measurement takes the guesswork out of science, and he also points out that a statement by one scientist should be verifiable by other scientists. Having and using a measurement model (the Rasch model) that is not data dependent provides an opportunity to conduct measurement in the manner it is carried out in science labs.

Activity #2

Task: Find examples of "standards" that involve the topic of "measurement" and "measurement devices" (sometimes you may only need to look for words such as meter stick, thermometer, and balance). So if you conduct research in field XYZ, look at the standards of your field.

Question: Once you have found such standards, are there parallels that you can identify between what is emphasized in the standards and what is emphasized and carried out by those who have selected the Rasch model?

Answer: Standards for both K-12 science classrooms as well as standards for science teacher education will vary, but you will find examples of standards that do stress the importance of using instruments that provide a common metric and instruments that can be used for measuring a range of items. In many other fields, you will find similar standards. The Rasch model, because it is not altered to fit a data set, provides a measurement device that does not change from measurement to measurement. And the measurement device provided by the application of the Rasch model facilitates measurement of the type that K-12 science students and science teachers are encouraged to conduct.

Activity #3

Task: Search the Online and find some examples of authors using the term IRT, 1P, 2P, and 3P. Also look for discussions of the Rasch model and, for instance, the 3P model.

Question: What arguments do authors of those discussions make?

Answer: Readers will find arguments presented for both sides of the coin. Moreover, readers may find some authors who assert that in the end there really is no difference between Rasch measurement and IRT models (in particular the 2P and 3P model). Such articles range from the rare introductory article to very advanced arguments. We believe the important point to remember is that high-quality measurement should not depend upon the sample being measured, and the Rasch model is sample independent.

Activity #4

Task: Write out a paragraph that explains why you have selected the Rasch model for your analysis. Support your argument by utilizing both requirements of measurement outlined by Thurstone. Also, make use of observations and articles by Ben Wright in which he explains the confusion some individuals have with respect to the IRT perspective and the Rasch perspective. One article that we encourage you to read to complete this activity are the remarks of Benjamin Wright's opening remarks in his invited debate with Ron Hambleton, Session 11.05, AERA Annual Meeting 1992. These remarks can be found at Wright (1992). This article, as well as all *Rasch Measurement Transactions*, is available online.

References

National Council of Teachers of Mathematics. (2000). *Principals and standards for school mathematics*. Reston, VA: National Council of Teachers of Mathematics.

National Research Council. (1996). *National science education standards*. Washington, DC: National Academy Press.

National Research Council. (2012). *A framework for K-12 science education: Practices, crosscutting concepts, and core ideas*. Washington, DC: National Academies Press.

Thurstone, L. L. (1928). Attitudes can be measured. *American Journal of Sociology, 33*, 529–554.

Wright, B. D. (1989). Rasch model for Thurstone's scaling requirements. *Rasch Measurement Transactions, 2*(1), 13–14.

Wright, B. D. (1992). IRT in the 1990s: Which models work best? *Rasch Measurement Transactions, 6*(1), 196–200.

Additional Readings

Wright, B. D. (1995). 3PL or Rasch? *Rasch Measurement Transactions, 9*(1), 408.

Chapter 22
What Tables to Use?

Isabelle and Ted: Two Colleagues Conversing

Ted: In all of these Rasch programs such as Winsteps, there are so many tables. It makes my head spin. Which ones should I use? Is there a "best order" for tables?

Isabelle: You know, Ted, all data sets are different. Whenever an individual does an analysis, there is probably an order that the person uses (more or less), but it could be quite different than the order someone else uses. Also, a big factor is the research question the person is trying to answer as well as questions that come up as the data are analyzed.

Introduction

We have tried to stress throughout this book that (1) Rasch is theory, (2) Rasch is philosophical, (3) Rasch is conceptual, (4) Rasch is qualitative as well as quantitative, (5) Rasch attempts to duplicate "best practices" of scientists in their labs, and (6) all data sets contain noise. If researchers carefully consider such issues, they can make some sense of a data set, use Rasch (if the data fit the model), and then conduct parametric statistical procedures to appropriately evaluate the data.

A plethora of tables are presented in all Rasch software, and differences exist across software packages. Sometimes one software package provides a unique table, and sometimes a specific kind of table can be found in similar forms across software packages. For example, a column header is provided in one package but not in another package. Our purpose in this chapter is to present an overview of how and why we have used certain tables more often than others. There are some tables that we do not use often, but we hypothesize that particular types of data sets might require intensive use of a particular table. As we begin, we raise one caveat to readers: Beginning Rasch users can make great strides by using only a few tables. An introductory Rasch analysis, such as the presentation of a Wright Map, can be sufficiently important to result in a publishable paper. Using Rasch to compute scale scores, which are then used for parametric statistical testing, can yield more rigorous results, higher-quality manuscripts, and better publishing opportunities.

W.J. Boone et al., *Rasch Analysis in the Human Sciences*,
DOI 10.1007/978-94-007-6857-4_22, © Springer Science+Business Media Dordrecht 2014

Selecting a Table Depends on the Researcher's Purpose

When a researcher creates a control file, his or her necessary next step is to figure out if the data were read correctly. In so doing, a researcher should select one of the "person" tables provided by Winsteps. These tables present data on all respondents in an analysis. Winsteps provides different versions of these tables. Let's look first at the "entry order" table, which presents "person measures" organized in the row order in the original data set – for instance, Excel or SPSS. Thus, if Bob is the second person in a data set and his ID is 007, then the second person to appear in the "person entry order" table should be Bob with an ID of 007.

Figure 22.1 is a segment of Winsteps Table 18.1, the "person entry table" for the STEBI data set. The column on the far right in Fig. 22.1 has the heading "Person." This column lists the name that was read into the Winsteps analysis in the control file.

A segment of the control file is displayed immediately below in Fig. 22.2. The parts of the control file that read the data line and identify it as "Person" are the lines NAME and NAMELENGTH, which are in bold font.

These two lines tell Winsteps that the "name" of each respondent begins in the first column of raw data and has a length of 10 columns. In Fig. 22.3, we provide the data for persons 1, 2, and 3 in the data set. Careful counting of columns should help readers understand that use of NAME1 and NAMELENGTH in the control file results in particular portions of a data line being used for an ID.

Looking at the "person entry table," and in particular looking first at what is presented in the Person column, is a quick technique that we use to ensure that our data were correctly read. When data are not correctly read, the correct person labels frequently do not appear in this column, and we may see nonsensical IDs that we can quickly identify as bogus. Referring to Fig. 22.1 (Winsteps Table 18.1), one sees a label of 91052 PR in the Person column for the second person in the data set. If we had seen a series of numbers (and no letters) such as this (123432321), we would be able to quickly identify a problem with one ID. Another way to double-check IDs is to look for symbols in the ID that are known not to be in the coding used for responses. In this example, one has the letters PR in the person ID. Seeing those letters appear in the ID and seeing those letters appear in the same location of many IDs elevates confidence that the data were read in correctly.

After reviewing the first few "Person" labels, we usually scroll through the entire table. This data set has 75 people, so reviewing all of the names should not take long. Even in cases where data sets contain thousands of respondents, we sometimes quickly scroll through this table. By focusing on this column, researchers can quickly identify a change in a pattern in characters and thus detect a possible error in the data. If most respondents were correctly entered and read, then a strange ID probably represents a small problem with a part of the data set, not what the Rasch program has done.

Our usual second and third steps are to look at the TOTAL SCORE and TOTAL COUNT columns. In our example, we see in Fig. 22.1 that the 1st and 4th persons

```
TABLE 18.1 SCIENCE TEACHER EFFICACY BELIEFS         ZOU077WS.TXT  Sep  5 12:46 2011

INPUT: 75 PERSON  23 ITEM  MEASURED: 75 PERSON  13 ITEM  6 CATS  WINSTEPS 3.70.6
--------------------------------------------------------------------------------
PERSON: REAL SEP.: 2.52  REL.: .86 ... ITEM: REAL SEP.: 7.00  REL.: .98

         PERSON STATISTICS:  ENTRY ORDER

------------------------------------------------------------------------------------------------
|ENTRY   TOTAL  TOTAL           MODEL|   INFIT  |  OUTFIT  |PT-MEASURE |EXACT MATCH|             |
|NUMBER  SCORE  COUNT  MEASURE  S.E. |MNSQ  ZSTD|MNSQ  ZSTD|CORR.  EXP.| OBS%  EXP%| PERSON      |
|-----------------------------------+----------+----------+-----------+-----------+-------------|
|     1     60     13  582.04  23.77|1.71   1.5|1.63   1.4|  .52   .57| 46.2  50.2| 21141    PR|
|     2     27      6  547.76  33.61| .71   -.3| .57   -.6|  .83   .60| 50.0  46.1| 91052    PR|
|     3     33      6  713.09  54.41| .62   -.5| .57   -.5|  .61   .46| 66.7  66.9| 95793    PR|
|     4     47     13  496.79  19.80| .31  -2.6| .28  -2.5|  .84   .67| 46.2  40.0| 08453    PR|
------------------------------------------------------------------------------------------------
```

Fig. 22.1 A part of the person entry table for the STEBI data set

```
&INST
TITLE = 'SCIENCE TEACHER EFFICACY BELIEFS
NAME1 = 1          ; First column of person label
NAMELENGTH = 10;  Length of person label
ITEM1 = 11         ; First column of responses in data file
```

Fig. 22.2 A segment of the control file for analysis of the STEBI data

Fig. 22.3 Data for persons 1, 2, and 3 in the STEBI data set

```
21141   PR 4655265555425445554555
91052   PR 5645252533455566xxxxxxx
95793   PR 4665554654556554xxxxxxx
```

Fig. 22.4 Data for persons 1, 2, 3, and 4 in the STEBI data set

```
21141   PR 4655265555425445554555

91052   PR 5645252533455566xxxxxxx

95793   PR 4665554654556554xxxxxxx

08453   PR 5535255555444444423352
```

answered 13 items and the second and third persons answered 6 items. We also observe that the 1st person's total score is 60, whereas the 2nd, 3rd, and 4th persons' total scores are 27, 33, and 47, respectively.

Returning to our control file, which contains the data for each person or returning to the original Excel or SPSS data set, we can see that, indeed, persons 1 and 4 answered 13 items (The data for persons 1–4 are presented in Fig. 22.4). Also, we note that persons 2 and 3 answered only 6 items. Remember, one of the marvelous benefits of Rasch is that missing data often do not cause problems. Even without looking at the original survey sheets, we hypothesize that the 6 answers by persons 1 and 4 are likely due to not answering items on the back page of STEBI.

```
21141    PR z65z26z5zzz2zzzz5545555
91052    PR z64z25z5zzz5zzzzxxxxxxx
95793    PR z66z55z6zzz5zzzzxxxxxxx
08453    PR z53z25z5zzz4zzzz4423352
```

Fig. 22.5 Data for persons 1, 2, 3, and 4 in the STEBI data set, with outcome-expectancy items marked "z"

Our final step is to do a quick "by hand" calculation of each person's raw score total. The sum of the entire first line of data is 106, but we need a total only for those 13 items in the self-efficacy scale of the STEBI, not the total of all 23 STEBI items. Remember, 10 items are for the outcome-expectancy scale which is also presented in the STEBI. To get the sum we want, we need to identify the position of each self-efficacy item in the row. In Fig. 22.5, we present the data for the same 4 respondents shown in Fig. 22.4, but we place a "Z" for the 10 outcome-expectancy scale items we will not use. The sum of the first line of data is 6+5+2+6+5+2+5+5+4+5+5+5+5=60, the total self-efficacy scale raw score for person 1.

Having organized and simplified the data for each line, we can compute the raw score total for each line. Doing so, we find that each line's total matches the "total" in the person entry table.

Another step that researchers can perform to verify data is to look at the header at the top of this table and in fact a header that is provided at the top of all Winsteps tables. This header from the table in Fig. 22.1 is provided in Fig. 22.6 for readers' convenience.

This header shows that 75 people were analyzed and 23 items were identified. To only look at the self-efficacy items, either one could create a control file using only the SE items or the non-SE items could be not looked at in the Rasch analysis. This second option is accomplished through the command IDFILE noted below.

```
IDFILE=*
1
4
7
9-11
13-16
18-23
*
```

Additional information in the header provides confirmation that our data were correctly read. Researchers should note the phrase "6 CATS" in the header. This phrase tells us that Winsteps used 6 rating scale categories for the analysis. Because we know that the rating categories were SD, D, BD, BA, A, and SA, this is further feedback that our data were correctly read.

What other tables can be used to evaluate if data were run correctly? Earlier we presented Winsteps Table 13.1 (Item STATISTICS: MEASURE ORDER). In Fig. 22.7, we present Winsteps Table 13.1 again and explain how one should use this table to check data. Review of the person entry data suggests that the data set

```
TABLE 18.1 SCIENCE TEACHER EFFICACY BELIEFS        ZOU077WS.TXT  Sep  5 12:46 2011
INPUT: 75 PERSON  23 ITEM  MEASURED: 75 PERSON  13 ITEM  6 CATS  WINSTEPS 3.70.6
--------------------------------------------------------------------------------
PERSON: REAL SEP.: 2.52  REL.: .86 ... ITEM: REAL SEP.: 7.00  REL.: .98
```

Fig. 22.6 Header from Winsteps Table 18.1 in Fig. 22.1

```
TABLE 13.1 SCIENCE TEACHER EFFICACY BELIEFS        ZOU077WS.TXT  Sep  5 12:46 2011
INPUT: 75 PERSON  23 ITEM  MEASURED: 75 PERSON  13 ITEM  6 CATS  WINSTEPS 3.70.6
--------------------------------------------------------------------------------
PERSON: REAL SEP.: 2.52  REL.: .86 ... ITEM: REAL SEP.: 7.00  REL.: .98

         ITEM STATISTICS:  MEASURE ORDER
```

ENTRY NUMBER	TOTAL SCORE	TOTAL COUNT	MEASURE	MODEL S.E.	INFIT MNSQ	INFIT ZSTD	OUTFIT MNSQ	OUTFIT ZSTD	PT-MEASURE CORR.	PT-MEASURE EXP.	EXACT MATCH OBS%	EXACT MATCH EXP%	ITEM
19	206	68	603.19	8.99	1.10	.7	1.21	1.2	.67	.69	44.8	40.8	Q19se-rc
5	258	75	569.60	8.38	1.28	1.7	1.34	1.9	.54	.67	37.8	41.2	Q5se
23	260	69	544.95	8.75	.80	-1.3	.86	-.8	.70	.63	44.1	41.1	Q23se-rc
20	262	69	542.68	8.76	1.07	.5	1.13	.8	.65	.63	39.7	41.3	Q20se-rc
17	277	69	525.27	8.97	.69	-2.1	.72	-1.7	.71	.61	39.7	42.8	Q17se-rc
12	310	75	514.25	8.78	.99	.0	.96	-.2	.56	.60	52.7	44.2	Q12se
3	317	75	506.14	8.93	1.69	3.5	1.64	3.1	.52	.59	33.8	44.5	Q3se-rc
21	295	69	502.85	9.41	.77	-1.4	.84	-.9	.65	.58	45.6	44.9	Q21se-rc
18	298	69	498.88	9.51	.66	-2.1	.62	-2.3	.62	.57	47.1	46.0	Q18se
6	352	75	459.26	10.27	.96	-.2	.89	-.5	.52	.53	51.4	54.5	Q6se-rc
8	369	75	429.77	11.43	1.03	.2	.97	-.1	.55	.50	59.5	57.9	Q8se-rc
22	364	69	367.55	14.34	.97	-.1	.93	-.3	.51	.44	64.7	63.7	Q22se
2	410	75	323.03	15.03	1.06	.4	1.08	.5	.31	.41	55.4	64.4	Q2se
MEAN	306.0	71.7	491.34	10.12	1.00	.0	1.01	.1			47.4	48.2	
S.D.	53.9	3.1	75.80	2.10	.26	1.5	.26	1.4			8.6	8.4	

Fig. 22.7 (Winsteps Table 13.1): Item STATISTICS: MEASURE ORDER table for the STEBI data

has been evaluated correctly. To make sure there are no problems, we recommend that researchers consult such a table.

In this type of table, we often look at items at (or toward) the top of the table (e.g., Q5se) and at (or toward) the base of the table (e.g., Q22se, Q2se). Of these three items, which should be easier and harder to agree with? This question can be answered by reviewing the text of the three items and considering the definition of the trait being investigated. Q2 (I will continually find better ways to teach science) and Q22 (When teaching science, I will usually welcome student questions) are items that would perhaps be easier for respondents to agree with than item Q5se (I know the steps necessary to teach science effectively). Now, we usually mark up those items that are flipped by inserting a word into the text that mimics the flip. Then we review all survey items and ask, does it make sense that Q2 and Q22 are at one end of the continuum? And, does it make sense that Q5se is at the other end of the continuum? When we are satisfied, we use this table to verify the reading of the data as we review the column headed "TOTAL COUNT." Numbers in this column reflect the total number of data for each item. The maximum number in this column is 75, which matches the 75 persons in the data set. Some items have a TOTAL COUNT smaller than 75. This is normally not a cause for concern, for the lower number is usually a reflection of items that were skipped or not clearly answered by respondents. A final technique is to review the TOTAL SCORE column. For instance, the raw score of Q5se is 258. To check the validity of this number, researchers may return to the original data set and compute the raw score total for all respondents to item Q5se. If data have been read in correctly, that value should be 258. We have just described what tables we often use to check the reading of data; in Chap. 24, we describe potential steps one might take for an initial analysis.

Isabelle and Ted: Two Colleagues Conversing

Isabelle: So, tell me how might I check quickly to make sure my data have been read correctly.

Ted: Easy, very easy. There are several tables in Winsteps. But, if you go to the person entry table first, you can see what the program thinks are the data for each person. Go to the last column and check if the person ID appears correct for each person. If you don't see what you wanted to identify as your "Name" in your control file, then, more than likely, you have made a mistake and you are telling Winsteps that your ID is in a different spot.

Isabelle: Helpful.

Ted: Also, if I find that my ID is screwed up, then it might be the case that most or all of the other data from each person were likely not read in correctly. For example, one time I ran some data and the person IDs were goofed up, and I only had information for every other item in the item measure table. That told me I had made a mistake.

Isabelle: What about the header? How does that help you?

Ted: The header at the top of the person entry table as well as the item measure table is presented in a lot of tables. It shows you how much data the program has read in. And, it also reports how much data were evaluated as well as how many rating categories were used.

Isabelle: How cool.

Ted: One last thing. Out of paranoia, I sometimes do "double-checking calculations" using SPSS and Excel. Of course, I am not doing Rasch with SPSS, but double-checking helps me verify that data were read carefully.

Keywords and Phrases

Many tables present the same information, but it is organized in differing ways.

Potential Article Text

Data from 143 respondents were entered into an SPSS spreadsheet. Each item was labeled. Missing data were entered as a "z." Following data entry, an initial Winsteps Rasch control file was created, and an initial Rasch analysis was conducted. A number of data quality steps and data analysis quality steps were then conducted. In SPSS, raw score totals for both items and respondents were computed. Following reverse coding of appropriate items, the sum of the raw scores of all of the first person's responses was computed. Item total raw scores were also computed for all items, and individual item totals (e.g., Q5se) were checked. Computing these numbers with the original data and then comparing those values with Winsteps tables provides a check as to whether or not data were read in correctly. The steps taken verified that data were correctly read in by Winsteps.

Quick Tips

There are a number of key tables that you usually will want to consult. These tables allow you to check your data to see if you have read the data correctly or if there has been some sort of error in the coding of the data. We have found that the tables that we look at most often at the start of an analysis are the item entry table, the person entry table, the summary statistic stable, the score measure table, and the Wright Map.

Data Sets: (go to http://extras.springer.com)

cf SE for chp what table to use scaled 1 to 1,000

Activities

Activity #1

Task 1: Take the supplied control file that was used to generate the Rasch analysis for this data set (cf SE for chp what table to use scaled 1 to 1,000). Run an analysis and find the TOTAL SCORE for the 6th and 7th person in the data set. Also find the ID for these two people. After you have done this, open the control file using either the edit option in the Winsteps tool bar or using a word processing program to open the file. Find the data for the 6th and 7th person in the data set.

Question: Do you have agreement with the IDs you have found in the table?

Task 2: Make a by-hand calculation of the TOTAL SCORE using the raw data that are present in the control file for persons 6 and 7 in the data set.

Answers: You should identify person 6 with the ID of 85453 PR with a TOTAL SCORE of 64. Person 7 has an ID of 46328 PR with a TOTAL SCORE of 21. This person has a low TOTAL SCORE because she or he only answered 6 of the 13 SE items. In part because Rasch involves a single latent trait, even though the 7th person answered only 6 items, she or he can be expressed on the same scale as if she or he had answered all the SE items.

Activity 2

Task: Write a dialog, similar to our friends Isabelle and Ted in which one person explains why we use the term "person measures" in Rasch measurement. Why don't we just use person total score?

Answer: In Rasch measurement, one of our concerns is to conduct rigorous measurement of the sort that is carried out by scientists. We take care to use the term person measures to remind ourselves (and others) that the raw values that have been used in past research are far removed from the measures that scientists take in labs. Rasch person measures are expressed on a linear metric, which can be used for parametric tests, while raw scores of respondents are not necessarily linear. A dialog should present this issue. You might even write a sample dialog you might use to explain "person measures" at a conference such as ESERA.

Additional Readings

Many of the tables that are provided in Winsteps are actually plots. Also in Winsteps it is possible to plot a range of data very quickly. The following article provides some discussion of the reasons why creating graphical data presentations can be of great use.

Linacre, J. M. (2001). Correspondence analysis and Rasch. *Rasch Measurement Transactions, 15*(3), 829–830.

Chapter 23
Key Resources for Continued Expansion of Your Understanding of Rasch Measurement

Books

Bond, T., & Fox, C. M. (2007). *Applying the Rasch model: Fundamental measurement in the human sciences*. Mahwah, NJ: Lawrence Erlbaum.

Smith E. V., Jr., & Smith, R. M. (Eds.). (2004). *Introduction to Rasch measurement: Theory, models, and applications*. Maple Grove, MN: JAM Press.

Wright, B. D., & Stone, M. H. (1979). *Best test design*. Chicago: MESA Press.

JAM Press has produced a number of books that concern Rasch measurement. We encourage readers of this book to visit the JAM Press WWW and review the Rasch books that are available.

Software Manuals

Linacre, J. M. (2012). *A user's guide to Winsteps* [User's manual and software]. http://www.winsteps.com/winsteps.htm

Journals

Rasch Measurement Transactions [*RMT*]. http://www.rasch.org/rmt/

Journal of Applied Measurement (*JAM*). http://www.jampress.org/

Peer-Reviewed Science Education Articles

Boone, W., Abell, S., Volkmann, M., Arbaugh, F., & Lannin, J. (2011). Evaluating selected perceptions of science and mathematics teachers in an alternative certification program. *International Journal of Science and Mathematics Education*, *9*(3), 551–569. doi:10.1007/s10763-010-9205-8.

Boone, W. J., & Scantlebury, K. (2006). The role of Rasch analysis when conducting science education research utilizing multiple-choice tests. *Science Education*, *90*(2), 253–269.

W.J. Boone et al., *Rasch Analysis in the Human Sciences*, DOI 10.1007/978-94-007-6857-4_23,

Boone, W. J., Townsend, J. S., & Staver, J. R. (2011). Using Rasch theory to guide the practice of survey development and survey data analysis in science education and to inform science reform efforts: An exemplar utilizing STEBI self-efficacy data. *Science Education, 95*(2), 258–280.

Donnelly, L. A., & Boone, W. J. (2007). Biology teachers' attitudes toward and use of Indiana's evolution standards. *Journal of Research in Science Teaching, 44*(2), 236–257.

Eggert, S., & Bögeholz, S. (2010). Students' use of decision making strategies with regard to socioscientific issues – An application of the Rasch partial credit model. *Science Education, 94*, 230–258.

Liu, X., & Boone, W. J. (2006). An introduction to Rasch measurement. In X. Liu & W. J. Boone (Eds.), *Rasch measurement in science education* (pp. 1–22). Maple Grove, MN: JAM Press.

Liu, X., & Lesniak, K. (2005). Students' progression of understanding the matter concept from elementary to high school. *Science Education, 89*(3), 433–450.

Neumann, K., Fischer, H. E., & Kauertz, A. (2010). From PISA to educational standards: The impact of large scale assessments on science education in Germany. In Fou-Lai Lin (Ed.), *International Journal of Science and Mathematics Education, 8*(3), 545–563.

Neumann, I., Neumann, K., & Nehm, R. (2011). Evaluating instrument quality in science education: Rasch-based analyses of a nature of science test. *International Journal of Science Education, 33*(10), 1373–1405.

Peer-Reviewed Health Field Articles

Baylor, C. R., Yorkston, K. M., Eadie, T. L., Miller, R. M., & Amtmann, D. (2009). Developing the communicative participation item bank: Rasch analysis results from a spasmodic dysphonia sample. *Journal of Speech, Language, and Hearing Research, 52*, 1302–1320. doi:1092-4388/09/5205-1302.

Cole, J. C., Rabin, A. S., Smith, T. L., & Kaufman, A. S. (2004). Development and validation of a Rasch-derived CES-D short form. *Psychological Assessment, 16*(4), 360–372. doi:10.1037/1040-3590.16.4.360.

Comins, J., Brodersen, J., Krogsgaard, M., & Beyer, N. (2008). Rasch analysis of the knee injury and osteoarthritis outcome score (KOOS): A statistical re-evaluation. *Scandinavian Journal of Medicine & Science in Sports, 18*, 336–345.

Covic, T., Pallant, J. F., Conaghan, P. G., & Tennant, A. (2007). A longitudinal evaluation of the Center for Epidemiologic Studies-Depression scale (CES-D) in a rheumatoid arthritis population using Rasch analysis. *Health and Quality of Life Research, 5*, 41. doi:10.1186/1477-7525-5-41.

De Morton, N. A., Keating, J. L., & Davidson, M. (2008). Rasch analysis of the Barthel index in the assessment of hospitalized older patients after admission for an acute medical condition. *Archives of Physical Medicine and Rehabilitation, 89*, 641–647. doi:10.1016/j.apmr.2007.10.021.

Duncan, P. W., Bode, R. K., Lai, S. M., & Perera, S. (2003). Rasch analysis of a new stroke-specific outcome scale: The stroke impact scale. *Archives of Physical Medicine and Rehabilitation, 84*, 950–963. doi:10.1016/S0003-9993(03)00035-2.

Hagquist, C. (2001). Evaluating composite health measures using Rasch modeling: An illustrative example. *Social and Preventive Medicine, 46*(6), 369–378. doi:10.1007BF01321663.

Itzkovich, M., Tripolski, M., Zeilig, G., Ring, H., Rosentul, N., Ronen, J., et al. (2002). Rasch analysis of the Catz-Itzkovich spinal cord independence measure. *Spinal Cord, 40*, 396–407. doi: 10.1038/sj.sc.3101315.

Latimer, S., Covic, T., Cumming, S. R., & Tennant, A. (2009). Psychometric analysis of the Self-Harm Inventory using Rasch modeling. *BMC Psychiatry, 9*, 53.

Stelmack, J., Szlyk, J. P., Stelmack, T., Babcock-Parziale, J., Bemers-Turco, P., Williams, R. T., et al. (2004). Use of Rasch person-item map in exploratory data analysis: A clinical perspective. *Journal of Rehabilitation Research & Development, 41*(2), 233–242.

Teslo, L. (2003). Measuring behaviors and perceptions: Rasch analysis as a tool for rehabilitation research. *Journal of Rehabilitation Medicine, 35*, 105–115.

Technical Reports

A second very helpful resource are technical reports. Some technical reports are available as the result of large-scale, high-stakes testing. Rasch measurement is employed in many countries to guide the development of high-stakes education tests. Furthermore, those Rasch techniques are used to compute scale scores, which are then used for parametric statistical tests. The examples we provide are by no means an inclusive list. The technical reports cited below are each organized in a different manner. Each report includes details of a Rasch analysis that is more sophisticated than those presented herein. However, many portions of these reports are written in a clear, nontechnical manner. We hypothesize that such documents are likely a result of authors' interests in producing reports that can, at least partially, be understood by nontechnical audiences, such as policy makers. These technical reports sometimes contain an outline of Rasch analysis steps taken to develop an instrument and evaluate a data set. Quite useful are tables that present high-stakes testing results using both raw scores and scale scores. We have found that such reports often help authors improve the manner in which they explain Rasch measurement in a science education article, be it text or tables/figures.

Below, we provide a number of website addresses to access technical reports for a number of US states that use Rasch to analyze high-stakes No Child Left Behind (NCLB) data. Readers should note that State Departments of Education in the USA frequently reorganize their websites; as a result, the location of archived documents can change. We also provide a very useful citation for the use of Rasch in PISA.

Organization for Co-operation and Economic Development. (2009). The Rasch Model. In OECD (Ed.), *PISA data analysis manual*: *SPSS* (2nd edn.). Paris: OECD Publishing. doi:10.1787/9789264056275-6-en

State of Ohio K-12 Testing

http://www.ode.state.oh.us/GD/Templates/Pages/ODE/ODEDetail.aspx?page=3&TopicRelationID=285&ContentID=9479&Content=60228 Then select a technical report such as "March 2008 OGT Statistical Summary".

State of Texas K-12 Testing

http://www.tea.state.tx.us/index3.aspx?id=4326&menu_id3=793. Then select a date, for instance, "Technical Digest 2007-2008". Then select a chapter such as 15. http://ritter.tea.state.tx.us/student.assessment/resources/techdigest/2008/chapter_15.pdf

State of Pennsylvania K-12 Testing

First go here for Pennsylvania http://www.pde.state.pa.us/a_and_t/site/default.asp Then for a sample State of Pennsylvania technical report select "Technical Analysis". The URL for the "Technical Analysis" is http://www.pde.state.pa.us/a_and_t/cwp/view.asp?a=108&Q=108328&a_and_tNav=|6395|&a_and_tNav=|

Then select a technical report such as "2008 Reading and Mathematics PSSA Technical Report." This is the URL for this particular technical report. http://www.pde.state.pa.us/a_and_t/lib/a_and_t/2008_Math_and_Reading_Technical_Report.pdf

State of California K-12 Testing

California Standards Tests CSTs Technical Report. Spring 2008 Administration. http://www.cde.ca.gov/ta/tg/sr/documents/csttechrpt08.pdf

State of Illinois K-12 Testing

Draft 2006 PSAE Technical Manual. http://www.isbe.state.il.us/assessment/pdfs/2006_PSAE_tech_manual.pdf

Workshops

The third and final suggested "next step" is to consider additional workshops that present an overview of Rasch measurement techniques. We bashfully suggest our workshops as well as in-person and online workshops offered by Rasch practitioners listed on the RMT-website.

Chapter 24
Where Have We Been and What's Next?

Isabelle and Ted: Two Colleagues Sharing Thoughts

Ted: *Well Isabelle now that I have finished this book, I have so many new ideas of things I can do with my data. I don't know if I will ever have to collect any more data.*

Isabelle: *Are you serious?*

Ted: *Sort of.... Given the immense amount of data that I have, here is a quick list of things I could present and write articles about: How to design tests/surveys, interpreting Wright Maps to provide guidance to practitioners, explaining what it means to measure, why counting is not measuring, etc.*

Isabelle: *(Interrupting) Ok, you have made your point. If you had to think Ted, what might have been the hardest concept for you to master?*

Ted: *Honestly it was that counting and just entering a number into a spreadsheet is not measuring.*

Isabelle: *What, then, might have been your favorite concept that you have mastered?*

Ted: *At least for me, it is that in so many fields, one would not only compute a raw score total, but one would have no idea what the raw score total would mean. With Rasch measurement, we can compute person measures and item measures and explain what the meaning of a particular person measure is. With Wright Maps we can explain qualitatively and conceptually what a person can do and cannot do.*

Introduction

In this book we have presented an introduction to the application of Rasch techniques, step-by-step. Our hope was, and remains, to help anyone who is interested in collecting, evaluating, and interpreting a wide range of data. The aspects of Rasch measurement that we chose to present are only a sample of the full array of Rasch techniques, and the issues that we chose to discuss are only a sample of the range of issues one can confront with Rasch techniques. However, we think that all researchers can use these chapters to understand what it means to measure, why it is important to measure, and how, with Rasch, one can measure.

W.J. Boone et al., *Rasch Analysis in the Human Sciences*,
DOI 10.1007/978-94-007-6857-4_24, © Springer Science+Business Media Dordrecht 2014

Using Rasch Is Performing High-Quality Measurement

There are now thousands of published articles that consider Rasch measurement. Some are highly technical, while others are less technical. Our book's mark on the trait line is in the "less technical" region. By no means is our book meant to be the final word on any topic (e.g., fit). We do, however, hope it is clear to readers why this book may be a final word for those who might consider using only raw data from a test or raw data from a survey for a parametric statistical analysis.

We attempted herein to review the techniques we used to explain Rasch over many many years to a wide variety of individuals (e.g., doctors, scientists, psychologists, market researchers) and present the techniques in a step-by-step fashion. We also searched for wonderful insights and tips provided in articles and in the detailed and extensive Winsteps user manual. Our goal was to provide guidance to those who are not necessarily in the field of psychometrics, but realize why Rasch measurement is crucial to their work.

The authors' prior experiences have involved many different fields. One group of individuals we commonly work with is the broad scientific community (e.g., biologists, chemists, physicists, geologists). Through this work, we have developed a great appreciation for the quality of measurement that is done in the indoor and outdoor laboratories of our science colleagues. However, in many fields of human research, sloppy and haphazard measurement has been prominent for decades. We strongly believe that high-quality measurement can thrive when test and survey data are collected and examined with Rasch measurement techniques of the type we have presented in this book.

In many chapters, we presented and analyzed rating scale survey data, but also we presented and analyzed test data. We did from time to time switch to tests in which items could be answered and graded as right or wrong. We did so due to wishing to present a number of book topics and to limit the length of some chapters. We already have a long long list of the topics to present in a second book. In that book we will extend some of our chapters and also present additional topics that have been important to our workshop participants and students. That same book would also be an applied book.

Of course, this book would not have been possible without the work of George Rasch and countless individuals who worked in the field for many years. Two individuals, however, have greatly influenced our book, Ben Wright and Mike Linacre. Both Ben and Mike have enthusiastically supported those of us who want to "measure," be it through technical articles and books, user-friendly advice, or software. Our deepest, most sincere thanks go to Ben and Mike.

There are many aspects to Rasch measurement. We think some of the key ones that our colleagues wish to master are how to design an instrument, how to evaluate the function of the instrument, how to compute person measures and item measures, how to create and interpret a Wright Map, and how to link instruments. Of course your own "measuring" goals will have unique components, but we feel that if readers just read and practice the thinking and activities we have outlined, you will be on your way to making a significant and positive mark in your field. You will be able to develop instruments of the upmost quality. You will be able to compute "measures," both for persons and items. You will be able to interpret your data with Wright Maps and bring qualitative and conceptual as well as quantitative meaning to your papers and explanations at conferences. When you review the work of others, you will see so much more than when you first read Chap. 1 herein. You will now know that some of the parametric statistical tests presented at a conference may not be of much use if raw scores were used.

Lessons Learned from Past Students

Before we listen in on Isabelle and Ted, we wished to provide segments of a set of thoughts that were provided to us by our past students. In doing so, we hope you might see in their own words how they sifted through and constructed an understanding of the information we provide in this book.

> Student #1: Measurement is necessary. Without a tape measure, how would an interior designer know how many square feet of tile to buy? How would an architect know how tall to construct a building? How would a cook know how much of an ingredient to add? How would a doctor know how much medicine to give a person? How would we shop for clothes if we didn't know what size to get (although women's clothing in particular is subject to interpretation on sizing)? We measure the physical world around us every day. I measure out half a cup of food twice a day for each of my two cats. If I am sick, I use a thermometer to make sure I don't have a fever. In setting up jumps for a course, I have to measure the distances between them to ensure the horse takes a certain number of strides. We can't get away from using measurement in our daily lives for tangible objects, so it makes sense that we measure things that are less concrete as well.
>
> Student #2: When I entered your classroom in May, I expected to learn specific ideas about measurement. I felt certain that we would be debating whether to assess students through multiple-choice questions or essays, the validity of standardized tests, or the benefits of a Montessori school. Perhaps the next day we would go over how to construct a fool proof rubric, one that is clear and effective for students while still usable and specific for us. We might even talk about what an A really means or what a B means and how they compare to each other. Instead we learned something far more valuable: what measurement is and how to measure well.
>
> Had we spent a week discussing when to give which letter grade, we would have been doing nothing but spouting hot air. Before this semester, I would never have considered what an A communicated to us about Bob and how that compared to what Betty's B communicated. Grade letters mean very little, just like raw scores.
>
> Our tendency in education, and in other areas of analysis, is to rely on raw scores as a concrete method of measurement. Raw scores on standardized tests are sent out to school systems and to parents. Teachers go over the latest exam score in parent teacher conferences. "Billy Sue only got an 80 on this test, but if you look here, Bobbie Lee got a 95." Although those numbers tell us that one person missed more questions than another person, they explain very little. Perhaps Billy Sue missed an easy question because she was not paying close enough attention. She might have gotten more difficult questions correct. Bobbie Lee, on the other hand, might have had a couple of lucky guesses. We cannot know unless we analyze the data through Rasch.

Rasch creates good measurement because it creates a "meter stick" on which to measure people. Rather than looking at raw scores as if they are the answers to all, Rasch looks at individual answers and at the difficulty of the test item in order to organize items and people in such a way as to be marks on a meter stick. By aligning those marks, we can tell parents that Billy Sue has a few problems with the more difficult concepts: She can't add multiple digit numbers together when the problem involves carry over. Or perhaps the analysis will reveal that Billy Sue appears to have some difficulty in concentrating on the test, because although she got the most difficult items right, she missed problems involving single-digit addition with no carry over. This is clearly more information than simply stating, "Billy Sue got an 80."

However, having more information does not necessarily mean good measurement; good measurement is dependent on the type of information. In order to consider what good measurement is, we have to first think of what the purpose of measurement is. I do not believe that measurement is about creating a competition among students by ranking them from best to worst. Measurement is meant to communicate the total range of student knowledge, how far spread that range is, what individual students are struggling with, and where they are succeeding. Rasch is able to communicate this information by converting raw scores into meaningful measures. We are able to use a ruler to tell us that the table has one leg two inches shorter than the other three. This raw data serves as a measure because of the simple nature of measuring length. We are used to that ease, but we cannot expect to treat human beings and their learning in that way. Two inches shorter tells us exactly what must be done to fix the table: Stick a two inch thick book under the leg. We need to know what the student knows, where her knowledge is likely to break down (i.e., her position on the Wright Map. Where Rasch would predict her ability level is equal to the item difficulty), and what the student does not know. A 55 on a test does not tell us what the student needs to learn. And if the test is faulty, a 55 compared to a 70 might not reflect that much difference in knowledge.

That is another beauty of the Rasch model. Rasch is good measurement, and we have discussed why, but Rasch also actively improves the tools teachers use to measure. Through analysis with Rasch, educators can find when there are flaws in the test. Perhaps the math teacher has asked too many single-digit addition problems on her unit test or maybe she is testing her students on a level too far below or above their knowledge to really understand anything about their performance. Rasch, particularly the Wright Maps, will help reveal these problems. Because of the layout of the map, educators can see where they need more questions and where they need fewer questions over various difficulty levels. The map should look similar to a ruler, with items ticking off fairly equal and small spaces over the length of the map, just as the centimeters in a meter stick appear. Rasch will show when there are large gaps that indicate imprecision in measurement and when there are too many questions at the same level that are a waste of both the students' and the teachers' time.

Rasch has taught me a lot about what the purpose behind assessment is and how I should allow that to affect my teaching. I know now that when I sit to write a test, I should be cognizant of each question's likely difficulty and how much I am spreading out the difficulty of the questions. When I look at numbers now, I won't be thinking in my head that Joe's 55 and Sam's 50 show the same kind of difference that Sarah's 90 and Stacy's 95 show. I also won't assume that a 55 shows that Joe knows nothing or that Sarah's 90 shows that she knows everything. I have learned that raw scores don't mean much, and that knowledge will affect who I am as a teacher. In looking beyond the surface of the numbers, I will be able to find ways to push my students further and to improve my tools for assessment.

Isabelle and Ted: Two Colleagues Sharing Thoughts

This will not be the last we hear from Ted and Isabelle, but for the time being let's listen in one last time....

Isabelle: Well Ted what is that on your desk, a paper you are going to submit?

Ted: Actually yes, and I received an acceptance yesterday! The paper uses one of those data sets which I collected after I finished the Boone, Staver, and Yale book. In the paper I detail the steps I took to think about the variable, and then I explain how I authored items for the 30-item multiple-choice test and the 20-item rating scale survey I had the students complete. I also outlined how I used fit statistics to evaluate data quality (both students and items). Here on page 15, you can see I present two Wright Maps, one for the test and one for the rating scale. The lines you see added are the mean measures of the treatment group and the control group in my study. What worked out very well was that I had 5 multiple-choice items that fell between the two lines, and I had 4 survey items that fell between the treatment group mean and the control group mean. That really helped me not only talk about the statistical difference that I found between the two groups, but I was also able to explain the meaning of the difference. I also explained in the article that I chose the Rasch model because it is a definition of measurement and that the Rasch model is not altered to fit the data.

Quick Tips

Remember, you are conducting measurement for a number of reasons: to develop an instrument, to evaluate the quality of your instrument, and to compute linear measures that you can use for parametric statistical tests.

Remember, the Rasch model is a definition of what it means to measure; the model is not altered to fit the data. We do not view the Rasch model as the simplest form of IRT models, and we do not view the Rasch model as the 1-P IRT model. The Rasch model is the model that needs to be used for measurement.

Remember, because items and persons are expressed on the same scale, you can bring immense meaning to your work. By using Wright Maps, you can explore the ordering and spacing of items. Also, you can of course also look at what it means for a person to have a specific measure.

Remember, in Rasch measurement we start with theory. What does it mean to measure a trait? Using Rasch measurement, we conduct many investigations to see if our data fit the model.

Rasch measurement allows you to measure over time by linking forms of a survey or test.

A potential series of steps you might use to go from "A" to "B" are the following:

Detail the variable you will be measuring. Draw a line and be able to explain with words what it means to improve from less to more along the line of the trait.

Provide support for your assertions as to why you think one task might be "easy" and another task might be "harder." Can you find literature that supports your assertions? Have colleagues review your theories. Revise as needed.

For the case of a rating scale, review potential rating scales that have been used in instruments. Often you can change wording in all your items so that a scale you find might work for your items. Remember, too many categories will overwhelm, and too few categories will change your item into a right/wrong item.

Write your items and be able to predict where the items fall on the line of the variable (trait). Make sure to avoid asking about more than one issue in an item.

We feel strongly that little is gained with items that need to be flipped. Yes, there might be respondents who just circle everything, but you will be able to detect these people in your data set as individuals who are too predictable.

Collect some pilot data from a group of 30–50 respondents if possible. Even before you enter your data and conduct a Rasch analysis, do you see rating scale categories that are not used? If some categories are not used, it might be that you will want to change the wording in some items.

Enter your data in a spreadsheet, for instance SPSS or Excel. At this point you might want to develop very short sets of phrases for each item. This is because those phrases at the top of an Excel column can be read into Winsteps as an item name.

As you enter your data, make sure to keep track if you use any codes beyond those that "label" each rating scale. For instance, are you using a code of, say, 77 for missing data? Or, are you using a code of 9 for an answer that you could not read? You might consider putting in a Mr. Perfect and a Mr. Not at all Perfect in your data set!

To help you double-check your analysis, you might, at this time, compute a raw score total for a respondent and also a raw score total for an item.

Create your control file. Remember the two red lines are key.

Once you have created your control file, look it over. Do you see all your items? Does it look like you have all your data? Now make sure to look at the CODES line in the control file. For your rating scale, you will want to use only the labels that indicate a rating. That means if you used a "9" for unreadable data, remove the number "9" from the codes line, remove a "77" you might have used for missing data! Now save your control file if you have edited it.

Run your analysis.

Bring up the item entry table. At the top of the table, what is the number of people and items that you see listed as "input"? What do you see as "output"? Does it match what you would predict? Now look at the columns "total score," "total count," and "measure." Do the number of people answering your item match about what you know from your data? You know the coding you used in your rating scale, so now you should be able to see what a greater measure means in terms of your rating scale. Now look at the item Outfit for ZSTD and MNSQ. For very large samples, just look at MNSQ. What do you see? Are there some items with ZSTD above 2.0 or below −2.0? What do you see in terms of MNSQ? Do you see items above 1.3 or 1.5? If any items are flagged by these values, you will want to think about these items and do some sleuthing. Are there a few respondents who were unexpected (or too expected) on these flagged items? Does the wording of the flagged items suggest that a structural improvement in the item is needed?

Carry out the same type of analysis for the respondents.

Now look at the summary statistics table. Write down and think about what you see in terms of person reliability, person separation, item reliability, and item separation. Is your data set allowing you to separate items with confidence and separate respondents with confidence? With what level of confidence?

Use Table 7.1 responses (this table allows you to see how expected an answer was for each respondent to each item) to identify the odd responses for the worst fitting item. Experiment carefully with making an unexpected response as missing. Then run your analysis and review Table 3.1 (the summary statistics table). Write down the person and item reliability and separation values. Do you see a change?

Now go back to your item entry table and look at the second part of the table, the section entitled "ITEM CATEGORY/OPTION/DISTRACTOR FREQUENCIES: ENTRY ORDER." Look at the mean measures for each rating scale category for each item. Do the mean measures increase as they should?

Now, in preparation for your review of your Wright Map, review the Score Table (Table 20). Do you understand what you are seeing? Remember, this table provides you with the "measure" for any possible raw score on your instrument. You may not have all these raw scores in your sample, but this table provides all possibilities. Can you find the UMEAN and USCALE values that you would use to rescale from linear logits to a linear scale of your choice (e.g., 0 to 1,000)? Can you find the ogive that shows you the relationship between the nonlinear raw scores and the linear logit measures? Using this table, you will be able to make sure you understand what a higher person measure means. Write that down!

Go back to the item entry table and figure out the meaning of a higher item measure. In a test in which items are coded as "1" for correct and as "0" for incorrect, a higher person measure will be a more capable student, and a higher item measure will be a harder item. For a rating scale of 1, 2, 3, and 4, where 4 is *Strongly Agree*, 3 is *Agree*, 2 is *Disagree*, and 1 is *Strongly Disagree*, a high person measure will be someone who is more agreeable, and a high item measure will be an item that is harder to agree with.

Now look at your Wright Map! Mark what the meaning of going up or down the scale is for items. Do the same for persons. Where is the mean item measure? Where is the mean person measure? If the two Ms are not near each other, what can you conclude about your instrument in terms of the targeting of items to respondents? Do you see gaps in how your trait is defined? Do you have lots of marks made by items in some locations? Is the item ordering and spacing what you would predict (construct validity)? Is it not what you would predict? Why? Is the ordering of respondents what you might predict? Can you identify a specific person in the Wright Map? To do so, pick a person, look up her/his approximate person measure on the Wright Map, then try to find that measure in the person entry table.

Pretend that you have computed the means of two subgroups of respondents (e.g., males and females) and mark those two means on your map. Then draw two horizontal lines at those marks. Pretend we have evaluated a test in which items are scored right and wrong. The items that fall between the lines are those items that show you the difference in what the higher performing group can accomplish compared to the lower performing group.

You then will conduct many of the other steps that we have detailed in this book, for example, DIF.

Ultimately, you are confident you have an instrument that functions well. So, use output files and create a spreadsheet with your person and item measures. Then place those measures in your statistical package of choice. Then, move on to your statistical analyses.

Index

W.J. Boone et al., *Rasch Analysis in the Human Sciences*,
DOI 10.1007/978-94-007-6857-4, © Springer Science+Business Media Dordrecht 2014

Printed by Printforce, the Netherlands